NEW FRONTIERS IN BLACK HOLE ASTROPHYSICS

IAU SYMPOSIUM 324

COVER ILLUSTRATION:

Logo of the IAU Symposium 324 on the background of NGC 6240 (Logo Credit: Branko Žalar, Photo Credits: X-ray: NASA/CXC/MIT/C. Canizares, M. Nowak; Optical: NASA/STScI)

IAU SYMPOSIUM PROCEEDINGS SERIES

INTERNATIONAL ASTRONOMICAL UNION

UNION ASTRONOMIQUE INTERNATIONALE

International Astronomical Union

NEW FRONTIERS IN BLACK HOLE ASTROPHYSICS

PROCEEDINGS OF THE 324th SYMPOSIUM OF THE INTERNATIONAL ASTRONOMICAL UNION HELD IN LJUBLJANA, SLOVENIA SEPTEMBER 12–16, 2016

Edited by

ANDREJA GOMBOC

University of Nova Gorica, Slovenia

CAMBRIDGE
UNIVERSITY PRESS

CAMBRIDGE UNIVERSITY PRESS
University Printing House, Cambridge CB2 8BS, United Kingdom
1 Liberty Plaza, Floor 20, New York, NY 10006, USA
10 Stamford Road, Oakleigh, Melbourne 3166, Australia

First published 2017

Printed in the UK by Bell & Bain, Glasgow, UK

Typeset in System LaTeX 2_ε

A catalogue record for this book is available from the British Library Library of Congress Cataloguing in Publication data

This journal issue has been printed on FSC™-certified paper and cover board. FSC is an independent, non-governmental, not-for-profit organization established to promote the responsible management of the world's forests. Please see www.fsc.org for information.

ISBN 9781107169944 hardback
ISSN 1743-9213

Table of Contents

Topic 3: Tidal Disruption Events

Topic 4: Active Galactic Nuclei

Topic 5: Tests of fundamental theories of physics using black hole systems

Topic 6: Technology Drivers and Future Capabilities

Preface

Black holes lie at the heart of some of the most fascinating astrophysical phenomena. The International Astronomical Union Symposium 324: "New Frontiers in Black Hole Astrophysics" marked the 100th anniversary of Schwarzschild's solution of Einstein's field equations, which set the scene for the theoretical prediction of black holes. Although they were at first considered as a purely mathematical curiosity, it is now well established that they are ubiquitously present in the Universe. In the last few decades our understanding of black holes has come an impressively long way, with the last major discovery in this field being coalescing black holes producing gravitational waves. Gravitational waves, also predicted in 1916, remained elusive for a long time - therefore, it was our outmost pleasure to mark during the symposium the first anniversary of the first direct gravitational wave detection, which started a new era in astronomy.

In addition to historical perspective, the symposium was timely also in terms of scientific and technological points of view. Technological advances across the electromagnetic spectrum and beyond to multi-messenger signature are coming of age. The new generation of gravitational wave detectors is operational, improved sensitivity in neutrino detectors is available, the multi-wavelength community has an impressive suite of ground- and space-based facilities covering a wide range of energy bands and timescales. And on the theoretical side, astrophysics communities are providing new testable predictions from advances in numerical simulations.

IAU symposium 324 was held from September 12 to 16, 2016 in Slovenia's capital city Ljubljana. It was organized by the University of Nova Gorica and opened by the Slovenian Minister of Education, Science and Sport, Dr. Maja Makovec Brenčič. The symposium brought together 130 observational and theoretical experts from 30 countries across the globe to discuss the current state-of-the-art in the astrophysics of black-hole driven systems and their exploitation in testing fundamental theories of physics. The scientific program began with a historical introduction given by Virginia Trimble and concluded with the symposium summary by Carole Mundell. The program consisted of 24 invited talks, 57 contributed talks and 30 posters. Topics of presentations and discussions spanned a wide range and included similarity and diversity of black hole systems, gamma ray bursts, tidal disruption events, active galactic nuclei, gravitational waves, black hole systems as multi-messenger sources and tools for testing theories of gravity and elementary particles, current and future large experimental facilities, and opening of new observational horizons.

Scientific highlights of the symposium included presentations and discussions on:

· *Similarities and diversities of black hole systems*, how a large-scale disk dynamo creates large-scale, long-lived jets; influence of time-dependent collapse history, growth of supermassive black holes and seed black holes.

· *Magnetic field in black hole systems*, in particular spatially resolved images of jets, polarisation, multi-messenger signals, and advances in numerical simulations of the magnetic field effects.

· *Tidal Disruption Events* and origin of UV/optical emission in them, rareness of relativistic events, possible unification scheme, host galaxies properties, estimation of type and mass of disrupted star, and many other puzzles unresolved due to a small number of these events discovered so far.

· *The birth of gravitational wave astronomy*, the information contained in gravitational wave signals (inspiral, merger and ringdown phase), potential electromagnetic signals, and possibilities of constraining the neutron star equation of state.

· *Tests of fundamental physics and black holes*, including quantum gravity 'zoo' and advances in theoretical modelling of gravitational waves.

Poster's presenters were given an opportunity to give a short highlight talk of their work and to take part in the best poster competition with prizes drawn from traditional Slovenian natural produce.

Social program of the symposium included a walking tour of Ljubljana, symposium dinner and trip to Postojna Caves and Lanthieri mansion in Vipava.

Symposium program was accompanied by several events for general public. In a public lecture on Sep 14, 2016, exactly on the 1st anniversary of the first direct detection of gravitational waves, Sheila Rowan presented a captivating story about detection of gravitational waves, in front of the packed full symposium hall. Exhibitions about black holes were on view during the symposium week in the National Assembly of the Republic of Slovenia and in the Cankar Centre, the symposium venue. After the symposium, one exhibition was put on display in Slovenian Museum of Natural History, while the other one is travelling and visiting Slovenian schools. To maximize educational outreach we invited all interested teachers to the symposium, to view the posters and listen to review talks. We organized a special teachers workshop on black holes in collaboration with the Society of Mathematicians, Physicists and Astronomers of Slovenia. General public and media exhibited large interest in the symposium, with all major TV stations and journals reporting news about the symposium and/or publishing interviews with symposium participants.

We were indeed very happy and honoured to host this symposium! It was not only the first International Astronomical Union Symposium held in Slovenia, but also the largest professional astronomy meeting in Slovenia so far.

I hope that these proceedings will serve as a summary of the highly interesting contributions and fruitful discussions held during the symposium, and that they will be especially useful to researchers and graduate students engaged in many different, yet all very exciting, research fields related to black holes.

Andreja Gomboc
Chair of the SOC and LOC
Ajdovščina, March 8, 2017

Acknowledgements

The symposium was sponsored by travel grants from the International Astronomical Union and supported by the International Astronomical Union Divisions A (Fundamental Astronomy), B (Facilities, Technologies and Data Science), C (Education, Outreach and Heritage), D (High Energy Phenomena and Fundamental Physics), J (Galaxies and Cosmology), and Commission 44 (Space & High Energy Astrophysics).

We are grateful to members of the Scientific Organizing Committee and the Local Organizing Committee for their highly valuable contributions.

Many special thanks to Carole Mundell for co-chairing the symposium!

Financial support by Aresis Ltd. and IPZ d.o.o. is gratefully acknowledged.

Sponsors:
Aresis Ltd.
IPZ d.o.o.

Cover image text:
Logo of the IAU Symposium 324 on the background of NGC 6240 (Logo Credit: B. Žalar, Photo Credits: X-ray: NASA/CXC/MIT/C. Canizares, M. Nowak; Optical: NASA/STScI)

CONFERENCE PHOTOGRAPH

Photo 1. Introductory review of the black hole concept and development of black hole research field given by Prof. Virginia Trimble (12 Sep 2016, photo: D. Novakovič/STA).

Photo 2. Enrico Ramirez-Ruiz reviewing numerical simulations of tidal disruption events (13 Sep 2016, photo: G. Kukec Mezek).

Photo 3. Discussions during a coffee break (photo: G. Kukec Mezek).

Photo 4. Symposium trip to Postojna Cave (14 Sep 2016, photo: G. Kukec Mezek).

Photo 5. Announcement of the poster competition winners (15 Sep 2016, photo: G. Kukec Mezek).

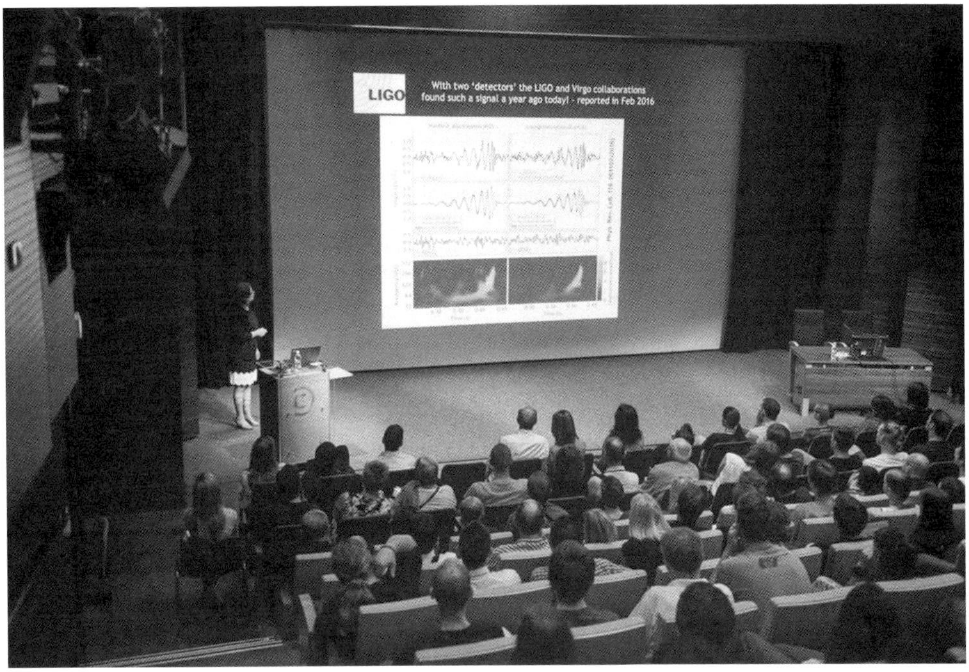

Photo 6. Public lecture on gravitational waves given by Sheila Rowan on the first anniversary of the first direct detection by the LIGO observatory (14 Sep 2016, photo: N. Tejić/STA).

Photo 7. Exhibition about black holes in the National Assembly of the Republic of Slovenia (12-16 Sep 2016, photo: B. Peršolja).

PARTICIPANTS

Tal **Alexander**	Weizmann Inst. of Science, Israel	tal.alexander@weizmann.ac.il
Cornelia **Arcaro**	INFN Padova & Padova Univ., Italy	cornelia.arcaro@pd.infn.it
Farruh **Atamurotov**	Ulugh Beg Astronomical Institute, Uzbekistan	farruh@astrin.uz
Patrycja **Bagińska**	Adam Mickiewicz Univ., Poland	patibag@gmail.com
Charles **Bailyn**	Yale Univ. / Yale-NUS College, USA	charles.bailyn@yale.edu
Maxim **Barkov**	University of Potsdam, Germany	barmv05@gmail.com
Josefa **Becerra Gonzalez**	NASA Goddard SFC, USA	jbecerragonzalez@gmail.com
Alessio **Berti**	INFN, University of Trieste, Italy	alessioberti90@gmail.com
Geoffrey **Bicknell**	RSAA, Australian National Univ., Australia	geoff.bicknell@anu.edu.au
Akos **Bogdan**	Harvard-Smithsonian CfA, USA	abogdan@cfa.harvard.edu
Edi **Bon**	Astronomical Observatory, Serbia	ebon@aob.rs
Nataša **Bon**	Astronomical Observatory, Serbia	nbon@aob.rs
Iztok **Bončina**	Astronomsko društvo Javornik, Slovenia	iztok.boncina@guest.arnes.si
Željka Marija **Bošnjak**	University of Rijeka, Croatia	bosnjak@uniri.hr
Martin **Bourne**	Kavli Institute for Cosmology, IoA, UK	mabourne@ast.cam.ac.uk
Katja **Bricman**	University of Ljubljana, Slovenia	katjabricman@gmail.com
Alberto **Carraminana**	INAOE, Mexico	alberto@inaoep.mx
Brad **Cenko**	NASA GSFC, USA	brad.cenko@nasa.gov
Paula **Chadwick**	University of Durham, UK	p.m.chadwick@durham.ac.uk
Sunil **Chandra**	TIFR, India	sunil.chandra355@gmail.com
Marisa **Charisi**	Columbia University, USA	mc3561@columbia.edu
Surajit **Chattopadhyay**	PCMT, India	surajitchatto@outlook.com
Aurora **Clerici**	University of Nova Gorica, Slovenia	aurora.clerici@ung.si
Carlos **Coimbra Araujo**	Univ. Federal do Paraná, Brazil	carlos.coimbra@ufpr.br
Stefano **Covino**	INAF / Brera, Italy	stefano.covino@brera.inaf.it
Andrej **Čadež**	University of Ljubljana, Slovenia	andrej.cadez@fmf.uni-lj.si
Miha **Černetič**	University of Ljubljana, Slovenia	miha@filetki.si
Colin **Degraf**	IoA, University of Cambridge, UK	cdegraf@ast.cam.ac.uk
Barbara **De Lotto**	University of Udine & INFN, Italy	barbara.delotto@uniud.it
Anja **Dobravec**	University of Ljubljana, Slovenia	anja.dobravec@gmail.com
Dijana **Dominis Prester**	University of Rijeka, Croatia	dijana@phy.uniri.hr
Ioana **Duţan**	Institute of Space Science, Romania	idutan@spacescience.ro
Viacheslav **Emelyanov**	KIoT, IoTP, Germany	viacheslav.emelyanov@kit.edu
Dimitrios **Emmanoulopoulos**	University of Southampton, UK	d.emmanoulopoulos@soton.ac.uk
Dunja **Fabjan**	University of Ljubljana, Slovenia	dunja.fabjan@fmf.uni-lj.si

Vandad **Fallah Ramazani**	University of Turku, Finland	vafara@utu.fi
Juan Antonio **Fernández Ontiveros**	Inst. de Astrof. de Canarias, Spain	jafo@iac.es
Sandor **Frey**	Satellite Geod. Observatory, Hungary	frey.sandor@fomi.hu
Krisztina **Gabányi**	Satellite Geod. Observatory, Hungary	gabanyik@sgo.fomi.hu
Stefano **Gabici**	Laboratoire APC, France	gabici@apc.in2p3.fr
Mikhail **Garasev**	Institute of Applied Physics, Russian Federation	garasev@appl.sci-nnov.ru
Sarah **Gibson**	University of Leicester, UK	slg44@le.ac.uk
Andreja **Gomboc**	University of Nova Gorica, Slovenia	andreja.gomboc@ung.si
Rene **Goosmann**	Obs. astron. de Strasbourg, France	rene.goosmann@astro.unistra.fr
Diego **Götz**	CEA Saclay, France	diego.gotz@cea.fr
Minfeng **Gu**	Shanghai Astronomical Obs., CAS, China	gumf@shao.ac.cn
Mark **Hannam**	Cardiff University, UK	mark.hannam@astro.cf.ac.uk
Carlos **Herdeiro**	Aveiro University, Portugal	herdeiro@ua.pt
Talvikki **Hovatta**	Aalto Univ. Metsahovi Radio Obs., Finland	talvikki.hovatta@aalto.fi
Xiaoli **Huang**	Guangxi University, China	1536756072@qq.com
Samo **Ilc**	University of Ljubljana, Slovenia	samo.ilc37@gmail.com
Susumu **Inoue**	RIKEN, Japan	susumu.inoue@riken.jp
Kunihito **Ioka**	YITP, Kyoto University, Japan	kunihito.ioka@yukawa.kyoto-u.ac.jp
Željko **Ivezić**	University of Washington, USA	ivezic@astro.washington.edu
Agnieszka **Janiuk**	Center for Theor. Physics, PAS, Poland	agnes@cft.edu.pl
Taj **Jankovič**	University of Ljubljana, Slovenia	taj.jankovic@gmail.com
Pavel **Jefremov**	ZARM, Univ. Bremen, Germany	paulefremov@gmail.com
Helen **Jermak**	Lancaster University, UK	h.jermak@lancaster.ac.uk
Aleksej **Jurca**	Gimnazija Bežigrad, Slovenia	
Anton **Jurca Vassilis Karamanavis**	Max-Planck-Inst. für Radioastr., Slovenia Germany	vkaraman@mpifr-bonn.mpg.de
Sergey **Karpov**	SAO RAS, Russian Federation	karpov.sv@gmail.com
Shai **Kaspi**	Tel Aviv University, Israel	shai@wise.tau.ac.il
Nobuyuki **Kawai**	Tokyo Inst. of Technology, Japan	nkawai@phys.titech.ac.jp
Norita **Kawanaka**	University of Tokyo, Japan	norita@astron.s.u-tokyo.ac.jp
Minjin **Kim**	Korea Astr. and Space Sc. Inst., Republic of Korea	mkim@kasi.re.kr
Shiho **Kobayashi**	ARI, Liverpool JMU, UK	S.Kobayashi@ljmu.ac.uk
Serguei **Komissarov**	University of Leeds, UK	s.s.komissarov@leeds.ac.uk
Stefanie **Komossa**	MPIfR, Germany	stefanie.komossa@gmx.de
Uroš **Kostić**	Aalta Lab d.o.o., Slovenia	uros.kostic@aalta-lab.com
Gašper **Kukec Mezek**	University of Nova Gorica, Slovenia	gasper.kukec@ung.si
Pawan **Kumar**	Univ. of Texas at Austin, USA	pk@astro.as.utexas.edu
Jutta **Kunz**	University of Oldenburg, Germany	jutta.kunz@uni-oldenburg.de
Samo **Kupper**	Fizikalno društvo Kočuha, Austria	samo.kupper@expi.at
Gavin **Lamb**	ARI, Livrepool JMU, UK	g.p.lamb@2010.ljmu.ac.uk
Robert **Lasenby**	Perimeter Institute, Canada	robertlasenby@gmail.com

Shuangliang **Li**	Shanghai Astronomical Obs., China	lisl@shao.ac.cn
Enwei **Liang**	Guangxi University, China	LEW@GXU.EDU.CN
Elina **Lindfors**	Tuorla Obs., Univ. of Turku, Finland	elilin@utu.fi
Tamilan **Mageshwaran**	Indian Institute of Astrophysics, India	mageshwaran@iiap.res.in
Pablo **Marchant**	Universität Bonn, Germany	pamarca@gmail.com
Paola **Marziani**	INAF - Astron. Obs. Padova, Italy	paola.marziani@oapd.inaf.it
Jin **Matsumoto**	RIKEN, Japan	jin.matsumoto@riken.jp
Tatsuya **Matsumoto**	Kyoto University, Japan	matsumoto@tap.scphys.kyoto-u.ac.jp
Kevin **Meagher**	Univ. libre de Bruxelles, Belgium	kmeagher@ulb.ac.be
Attila **Meszaros**	Charles University, Czech Republic	meszaros@cesnet.cz
Felix **Mirabel**	CEA-Saclay-France, France	felix.mirabel@cea.fr
Carole **Mundell**	University of Bath, UK	c.g.mundell@bath.ac.uk
Saeede **Nafooshe**	Trieste University, Italy	saeede.nafooshe@ung.si
Priyamvada **Natarajan**	Yale University, USA	priyamvada.natarajan@yale.edu
Lubos **Neslusan**	AISAS, Slovakia	ne@ta3.sk
Zsolt **Paragi**	Joint Inst. for VLBI ERIC, Netherlands	zparagi@jive.eu
Qiuhe **Peng**	Nanjing University, China	qhpeng@nju.edu.cn
Bradley **Peterson**	Ohio State Univ./STSI, USA	peterson.12@osu.edu
Pauli **Pihajoki**	University of Helsinki, Finland	pauli.pihajoki@helsinki.fi
Peter **Raffai**	Lorand Eotvos University, Hungary	praffai@bolyai.elte.hu
Enrico **Ramirez-Ruiz**	UCSC, USA	enrico@ucolick.org
Antti **Rantala**	University of Helsinki, Finland	antti.rantala@helsinki.fi
Sheila **Rowan**	University of Glasgow, UK	sheila.rowan@glasgow.ac.uk
Richard **Saxton**	XMM SOC / ESAC, Spain	rsaxton@sciops.esa.int
Luka **Seliškar**	University of Ljubljana, Slovenia	lmdogg@gmail.com
Roberto **Serafinelli**	Univ. of Rome 'Tor Vergata', Italy	roberto.serafinelli@roma2.infn.it
Samo **Stanič**	University of Nova Gorica, Slovenia	samo.stanic@ung.si
Iain **Steele**	ARI, Liverpool JMU, UK	i.a.steele@ljmu.ac.uk
Lisa K. **Steinborn**	Univ.-Sternwarte München, Germany	steinborn@usm.lmu.de
Petra **Sukova**	CTP PAS, Poland	psukova@cft.edu.pl
Tachibana **Yutaro**	Tokyo Inst. of Technology, Japan	tachibana@hp.phys.titech.ac.jp
Hiromichi **Tagawa**	University of Tokyo, Japan	tagawahr@nao.ac.jp
Ataru **Tanikawa**	University of Tokyo, Japan	tanikawa@ea.c.u-tokyo.ac.jp
Nial **Tanvir**	University of Leicester, UK	nrt3@le.ac.uk
Alexander **Tchekhovskoy**	UC Berkeley, USA	atchekho@berkeley.edu
Tomislav **Terzić**	University of Rijeka, Croatia	tomislav.terzic@gmail.com
Kenji **Toma**	Tohoku University, Japan	toma@astr.tohoku.ac.jp
Virginia **Trimble**	UC Irvine & Queen Jadwiga Obs., USA	vtrimble@astro.umd.edu
Marta **Trini**	University of Nova Gorica, Slovenia	marta.trini@ung.si
Audrey **Trova**	AIASCR, Czech Republic	trova@asu.cas.cz
Rok **Vogrinčič**	University of Ljubljana, Slovenia	rokvogrincic@gmail.com
Serguei **Vorobiov**	University of Nova Gorica, Slovenia	sergey.vorobyev@ung.si

Amri **Wandel**	Hebrew Univ. of Jerusalem, Israel	amri@huji.ac.il
Vojtěch **Witzany**	ZARM, Univ. Bremen, Germany	witzany@zarm.uni-bremen.de
Lukasz **Wyrzykowski**	Warsaw Universit, Poland	lw@astrouw.edu.pl
Lili **Yang**	University of Nova Gorica, Slovenia	lili.yang@ung.si
Farhad **Yusef-Zadeh**	Northwestern University, USA	zadeh@northwestern.edu
Gabrijela **Zaharijas**	University of Nova Gorica, Slovenia	gabrijela.zaharijas@ung.si
Luca **Zampieri**	INAF-Ast. Obs. Padova, Italy	luca.zampieri@oapd.inaf.it
Sebastjan **Zamuda**	Gimnazija Bežigrad, Slovenia	
Danilo **Zavrtanik**	University of Nova Gorica, Slovenia	danilo.zavrtanik@ung.si
Shuqing **Zhong**	Guangxi University, China	1548771042@qq.com
Tomaž **Zwitter**	University of Ljubljana, Slovenia	tomaz.zwitter@fmf.uni-lj.si
Maruša **Žerjal**	University of Ljubljana, Slovenia	marusa.zerjal@fmf.uni-lj.si

New Frontiers in Black Hole Astrophysics
Proceedings IAU Symposium No. 324, 2016
Andreja Gomboc, ed.

doi:10.1017/S1743921317001983

Old/Past/Ancient/Historic Frontiers in Black Hole Astrophysics

Virginia Trimble

Department of Physics and Astronomy, University of California Irvine CA 92697 and Queen Jadwiga Observatory, Rzepiennik, Poland
email: vtrimble@astro.umd.edu

Abstract. Basic questions about black holes, some of which are fairly old, include (1) What is a black hole? (2) Do black holes exist? And the answer to this depends a good deal on the answer to (1), (3) Where, when, why, and how have they formed? and (4) What are they good for? Here I attempt some elaboration of the questions and partial answers, noting that general relativity is required to described some of the phenomena, while dear old Isaac Newton is OK for others.

Keywords. black hole, general relativity, horizon, gravitational redshift, binary stars, quasars, gamma-ray bursts, gravitational radiation, advection, singularity

---

1. Introduction

The phrase "black hole" in its modern meaning first appeared in print more than 60 years ago (Ewing 1964). Bartusiak (2015) has chased the words back to their Princeton pairing and popularization by Robert Dicke and John Wheeler from 1963 onward. Yes, the Black Hole of Calcutta 1756 event is probably part of the story, and Priya Natarajan (Natarajan 2016) tells us about it (as well as providing an Indian point of view on the zodiac and creation myths). These myths are probably not part of our story, unless you want to consider our universe coming out of a white hole before the epoch of inflation.

The best early description I heard of the phenomenon came in spring 1963 (when there were no black hole theorists by that name and even neutron stars were hardly part of the universe) from the late Thornton Leigh Page (whose father, Leigh Page, was (also) a distinguished physicist) in a UCLA course on stellar evolution. "It folds" he said, "a blanket of space-time around itself, and quietly goes to sleep." He thought that these sleepers and neutron stars were more likely end points for massive star evolution than white dwarfs, which he described as elephants being required to turn into white mice. Elephant color not specified.

It was this folding and sleeping to which Eddington (1926) (p. 6) objected , writing, "the mass would produce so much curvature of the space-time metric that space would close up around the star, leaving us outside (i.e. nowhere)." Other sorts of objections will turn up as we proceed. On creation "ex nihilo" and the alternatives, see Leeming & Leeming (1984).

2. What is a black hole?

The arithmetic here is particularly easy, because the Newtonian expression for escape velocity = speed of light gives you the same answer as the relativistic expression with $(1-2GM/Rc^2)^{\frac{1}{2}}$ downstairs, $R = 2GM/c^2$. This brings us to the oldest, simplest definition.

Something with $R \lesssim$ gravitational radius. The recent custom of beginning discussions of black holes with a brief historical introduction (including this one) has made the names of John Michell and Pierre-Simon de Laplace more familiar in this context today than they were a few decades ago. Each wrote (Michell 1784, Laplace 1795) roughly the same thing, that a body with the density of Earth or Sun and a size comparable with the diameter of Earth's orbit would have an escape velocity greater than or equal to the speed of light, and so "all light ... would be made to return towards it by its own proper gravity" (Michell), or "il est donc possible que les plus grands corps lumineux de l'univers, soient par cela même, invisibles." (Laplace).

Laplace is the sort of person who has whole books written about him (Gillispie 1997, the same chap who edited the entire first edition of the *Dictionary of Scientific Biography*), though the deep question of why he left his grands corps out of later editions of his book remains unanswered. Michell remains less celebrated, and it seems (McCormmach 2012) that no portrait of him survives. He wrote for a journal that did not include abstracts, which presumably accounts for his title "On the Means of Discovering the Distance, Magnitude, &c. of the Fixed Stars, in consequence of the Diminution of the Velocity of their Light, in case such a Diminution should be found to take place in any of them, and such other Data should be procured from Observations, as would be farther necessary for that Purpose. By the Rev. John Michell, B.D.F.R.S. In a Letter to Henry Cavendish, Esq. F.R.S. and A.S." The B.D. was bachelor of divinity and he was elected FRS for his work on propagation of earthquake waves. The lower case letter s's (except at the ends of words) are stretched out to look like f's with the cross piece missing.

Before moving on to general relativity, let's check that Newton plus a particle theory of light is enough to estimate small gravitational redshifts. A test mass m leaves a large body (mass M, size R) with initial kinetic energy $\frac{1}{2}mv^2 = \frac{1}{2}mc^2$. It arrives at infinity having gained potential energy from $-GMm/R$ to 0, so

$$\frac{1}{2}mv^2 = \frac{1}{2}mc^2 - \frac{GMm}{R},\ v^2 = c^2 - \frac{2GM}{R},\ \frac{v}{c} = (1 - \frac{2GM}{Rc^2})^{\frac{1}{2}} \tag{2.1}$$

which is approximately $1 - GM/Rc^2$. Put in properties of a white dwarf $M = 1.5 \cdot 10^{33}$ g, $R = 10^9$ cm, G and c in cgs units to find:

$$\frac{GM}{Rc^2} \approx 1.5 \cdot 10^{-4} \text{ or } \Delta V \approx (1.5 \cdot 10^{-4})(3 \cdot 10^5 \text{ km/s}) \tag{2.2}$$

or $\Delta V \approx 45$ km/s, which is what we see.

Something with a Horizon. This brings us to Einstein and GR, with his four classic papers all submitted on Thursdays (Gutfreund & Renn 2015), and the shortest textbook ever, that by Weber (1961). The first exact solution to the field equations came from Karl Schwarzschild (1916), submitted by Einstein (also on Thursdays), because Karl himself had volunteered for active duty with the German forces in WWI . He died that same year; perhaps less well known is that a number of other classic GR papers date from 1916-19, some written from the trenches. Astrophysically, the important things about horizons are that any material or radiation that crosses them going inward cannot get out (which happens quickly for an observer riding with the infall), but that looking from outside, the infall goes slower and slower, the information about it arrives more and more redshifted, and the external observer thinks the process takes forever. Kerr black holes (Kerr 1965, though the talk was given in 1963) are characterized by angular momentum as well as mass, also have horizons, but also have ergospheres, from which one can extract the portion of the apparent mass-energy that is associated with the rotation (Penrose 1969, Blandford & Znajek 1977, discovery and application respectively).

Stuff disappears at the horizon. Stuff means either matter or information in this context, and the arguments have been more complex and aggressive about the information. Whether entropy can disappear is even more contentious. The case where energy is carried down the tubes is called advection-dominated-accretion.

Information/matter/entropy hangs up at/around the horizon. This is the membrane paradigm, usually associated with Kip S. Thorne, though he credits others (Thorne 1994) for pioneering the idea. The important associated idea is that, if stuff hangs up just outside $R = 2GM/c^2$ (or the Kerr equivalent), then it probably comes back out in due course, with information (who was that hadron I saw you with last night?) and entropy intact. Modern black holes (well, perhaps modern theories) can also act like conductors, permitting the support of magnetic fields in their neighbourhood, with which, in turn, jets can be collimated and other wondrous things occur.

Strange things happen inside. First, if you look at the line element, space and time exchange their properties, and you can traverse a space-like interval but not a time-like interval. Perhaps also your wrist watch turns into a ruler and your meter stick into a chronometer, but I haven't checked. Second, physical objects get torn, shredded, chaotically mixed, and destroyed. Third, at the center there is a singularity (a point for the non-rotating Schwarzschild case, a small circle for the more realistic Kerr case). A good, modern GR textbook like Hartle (2003) indexes such matters under "Kerr geometry." Thorne (1994) devoted, appropriately, chapter 13 to "Inside Black Holes." And you will find some discussion of the issues in volumes at many levels of technicality (Begelman & Rees 2010, Scharf 2012, Bartusiak op. cit., Thorne op. cit., Hawking & Israel 1987, Hartle op. cit. and so forth). A singularity sounds bad enough when you say just that density and space-time curvature will be infinite there. Even worse, world lines end there. Physics (at least general relativity) loses all predictive power, and many physicists have said that quantum gravity must prevent anything of the sort from happening. The key "singularities happen" paper is Hawking and Penrose (1970), with credit also due to Werner Israel and George Ellis on related topics. In any case, a general relativistic expanding universe must have a singularity in its past, and a body that collapses inside its gravitational (Schwarzschild, Kerr) radius has a singularity in its future, no matter how asymmetric, chaotic, or messy the collapse is.

3. Do Black Holes exist?

This is clearly the key question from an astrophysical point of view! The answer will depend somewhat on the definition we choose from Section 1, but in any case one can imagine three pieces: can they form? or if not, could they have existed from before or during the Big Bang? and do they persist? Let's give the theorists the first word, before going on to what astronomical observations might tell us.

J. Robert Oppenheimer, about whom many things can be said that are not relevant to our story, having dealt with neutron stars in a collaboration with George Volkoff (Oppenheimer & Volkoff 1939) sat down (well one's image has him pacing around continuously) with Hartland Snyder to calculate "On continued gravitational contraction" (Oppenheimer & Snyder 1939). Published the day World War II broke out, this so annoyed Einstein that he rushed into print (Einstein 1939) in an attempt to show that, at least for a cluster of mass points in circular orbits, angular momentum would forever prevent the collapse. This was sufficiently wrong that it appears only as part of an exercise for the student in Misner, Thorne & Wheeler (1971).

Physicists who had been young when WWII broke out contributed to bomb development, radar, and many other things, and some came back to establish relativity groups.

A younger generation (now not so young) settled down to establish the reality of gravitational collapse and singularities. Penrose (1965), less than 3 pages long was crucial. Lifshitz & Khalatnikov (1961) thought they had proven that a sufficiently disordered system could not yield a singularity. Thorne (1994) (p. 465 ff) tells the story of how they changed their minds.

So, collapse will make singularities with black holes and horizons around them. What about primordial black holes? The sub-stellar mass black holes suggested by Hawking (1971) must be of that type, since there is no pressure strong enough to form them now. We currently have no evidence for these, nor of other much more massive ones that could go back before the beginning (but need not), and could be relevant to LIGO events (but need not).

As for persistence, a speaker at GR21 in New York in July said firmly that a Kerr black hole will eventually spin down to a Schwarzschild black hole, but that the Schwarzschild one was forever, because the time scale for evaporation by Hawking radiation would always exceed the time scale for growth by capture of CMB photons, for some BH mass above some calculable limit. Unfortunately, I have been unable to identify the speaker from the conference program and pages of notes I took at the time.

Which brings us to the observations, sometimes filtered through theoretical minds.

Sizes close to gravitational radii. Michell (1784) thought one might detect his dark stars by their gravitational pull on luminous bodies, and Zwicky (1937) supposed that gravitational lensing would reveal all forms of mass (both correct). And then, as previously noted, there was a war, and GR crept gradually back into prominance after a long latency period (Blum *et al.* 2015).

Observational radio and X-ray astronomy leaped ahead of theory for a while, so that we pick up in 1964 with Salpeter (1964) and Zeldovich & Novikov(1964) proposing that the energy source for the recently-discovered quasi-stellar radio sources (now quasars, even the radio-silent ones) might be accretion onto a central very compact mass. An early worry was that spherically-symmetric accretion would take all the carefully-accumulated potential energy down the tubes with the gas, and so accretion disks were invented. Over the next 20 years, something of the sort became the standard model for active galactic nuclei (Rees 1984).

Close binary stars slopped gas back and forth starting at about the same time (Hoyle 1964, Paczyński 1971), and Zeldovich & Guseinov (1965) soon set out to look for cases where the recipient was a collapsed or frozen star; neutron stars were added to the search space after the discovery of pulsars. No, they didn't find any, nor did Trimble & Thorne (1969) with a larger catalogue to examine. But the method was right – check for X-ray emission from systems where the optically-invisible component was too massive to be a white dwarf. That HDE 226868 was not in the Batten (1967) catalogue of spectroscopic binary orbits is a sad story for another time. But it was the optical identification for Cyg X-1, on which I wrote (Trimble *et al.* 1973) the last fundamentally wrong paper. And since then, the best mass estimate for the accretor has gradually crept up from about $6M_{\odot}$ to $10M_{\odot}$ or more. Not a white dwarf; not a neutron star.

Horizons. On the assumption that accretors of more than about $4M_{\odot}$ must be BHs, one can then say that, in X-ray binaries, they never display the sorts of X-ray bursts (called Type l) associated with stuff hitting a neutron star surface. Spectral signatures are not reliable, and BHXRBs sometimes show quasi-periodic oscillations not so very different from the NSXRBs, but there is never a signature of a constant rotation period. Analysis of eclipses in BHXRBs reveal distortions connected with space-time being dragged around, by angular momenta from about 0.3 to 0.95 of the maximum possible. The assortment of radio installations being called the Event Horizon Telescope should soon record radio waves being bent almost all the way around by the black holes in our own Sgr A* and M87.

Energy disappearing. Having invented accretion disks to make sure that the energy released by infall to black holes came out as radiation (Shakura & Sunyaev 1973), the community then had to back off and deal with BH emitters of X-rays at levels very considerably below their Eddington luminosities. At least three excuses are possible. First, you may see material coming in at a distance, but it all gets turned around into outflow or collapsed into stars. Second, there truly is very little around to accrete (until an AGN-BH tears up another star or an XRB BH ruffles its disk). Or, third, secreting gas is carrying with it a great deal of the thermal energy. This is called advection-dominated accretion (ADA). Narayan *et al.* (1998) is a very nice status report. The clearest cases they discuss are, for XRBs, A00620-00, V404 Cyg, and Nova Muscae 1991 and, for AGNs Sgr A*, M87, and NGC 4258 (the one with the maser disk providing a good measurement of the BH mass). The three BHXRBs have flared up near to the Eddington luminosities and fallen back to 10^{-4} or less, with quiescent accretion rates proportionally depleted. The AGN BHs are even fainter, though at least for SgrA*, the ADA explanation (Fabian & Rees (1995), Rees (1982)) may not be the likeliest explanation for faintness now and much higher luminosity within the past 100 years (Ponti *et al.* 2014). The main point made by Narayan *et al.*, is that the ADA hypothesis provides much better fits to observed spectra than does a standard accretion disk. The same claim is made for the X-ray light curve of Nova Muscae 1991 (data from Ebisawa *et al.* 1994), but look for yourself, before you endorse or confute my skepticism.

Exciting happenings just outside or inside the horizon. Black hole entropy, thermodynamics, and radiation are old enough to count as history (Bekenstein 1972, Hawking 1974), but there do not seem be any very relevant observations, except that one can say either the universe is not bound by low-mass primordial black hole, Hawking radiation doesn't happen, or both. Thorne (1994) has a whole chapter (Ch. 13) on "inside", and Amos Ori (Ori 1991, Ori 1992) has attempted to look inside relatively realistic geometries. But I learned my lesson from Penrose (1962), "Don't go there!". And indeed, although quite a number of our graduate students seem to have descended past horizons, none of them has ever submitted a thesis.

4. What are they good for?

This is just a list with comments, because all suggestions to date remain viable for some set of parameters.

• Dark Stars, as suggested by Michell and Leverrier, although theirs would have had masses near $10^8 M_\odot$.

• Galactic nuclei, for a very large fraction of galaxies, though most are not active at present, because of starvation, advection dominated accretion, or masses, hence horizon radii, too large for tidal disruption of captured stars.

• Gamma ray bursts, if not when PBHs boil away, then more common sorts of GRBs associated with collapses of rapidly-rotating cores of massive stars and/or mergers of neutron star pairs.

• Dark matter, in moderate amounts formed from material that was previously baryonic, but must have taken matter out of the baryon inventory before BB nucleosynthesis if BHs are to be the dominant dark stuff.

• Bursts of gravitational waves, both at formation and when merging in binary star systems.

The last item led directly to a bunch of photographs (from about 1965 to 1985) showing the first generations of detectors for gravitational radiation and Joseph Weber who built them, in company with Howard Laster and Gart Westerhout (former directors of physics and astronomy at the University of Maryland); Darrell Gretz and another of the technician/programmers who helped operate those early detectors; Nathan Rosen (of Einstein *et al.* 1935); who later in his career (played out in Israel) came to doubt not only gravitational waves but GR in general; Ron Drever (a major contributor to the design and construction of LIGO); Fang LiZhi; distinguished theorist David Finklestein; and yours truly.

5. Conclusion

There is, of course, no end to "history", it merely hands over to "current events" (as at least one designer of time machines discovered to his distress), but a nice end-point for my generation is the flowchart, drawn by Roger F. Griffin, to accompany the text of Martin J. Rees's (1978) Halley Lecture. It also appears in the more accessible Rees (1984), and in full, glorious color in Begelman & Rees (2010). The key point is that it is really quite difficult to avoid forming a black hole sooner or later, when gravity is the dominant force.

Acknowledgements

I am indebted to organizers Andreja Gomboc and Carole Mundell for finding time on a crowded program for what I had to say and for agreeing to cope with a typed (on paper!!!) contribution toward the proceedings. My views on black holes have been shaped by very many colleagues, most more knowledgeable and distinguished than I. The most important, in chronological order, have been Kip Stephen Thorne, Martin John Rees, and Joseph Weber.

References

Bartusiak, M. 2015, *Black Hole*, Yale University Press

Batten, A. H. 1967, *Publ. Dom. Astrophys. Obs. Victoria*, 13, 119

Begelman, M. & Rees, M. 2010, *Gravity's Fatal Attraction*, 2nd Ed. Cambridge University Press

Bekenstein, J. D. 1972, *Lettere al Nuovo Cimento*, 4, 757

Blum, A. S., Lalli, R., & Renn, J. 2015, *ISIS*, 106, 598

Blandford, R. D. & Znajek, R. L. 1977, *MNRAS*, 179, 433

Ebisawa, K., Ogawa, M., Aoki, T., Dotani, T., Takizawa, M., Tanaka, Y., Yoshida, K., Miyamoto, S., Iga, S., Hayashida, K., Kitamoto, S., & Terada, K. 1994, *PASJ*, 46, 375

Eddington, R. S. 1926, *The Internal Constitution of the Stars* (Dover reprint), p. 5

Einstein, A. 1939, *Annals of Mathematics*, 40, 922

Einstein, A., Podolsky, B., & Rosen, N. 1935, *Physical Review*, 47, 777

Ewing, A. 1964, *Science News Letters*, 18 Jan. 1964, p. 39 ("Black Holes in Space")

Fabian, A. C. & Rees, M. J. 1995, *MNRAS*, 277, L5

Gillispie, C. C. 1997, *Pierre-Simon Laplace (1749-1827)*, Princeton Univ. Press

Gutfreund, H. & Renn, J. 2015, *The Road to Relativity*, Princeton Univ. Press

Hartle, J. B. 2003, *Gravitation: An Introduction to Einstein' s General Relativity*, Addison Wesley

Hawking, S. W. 1971, *MNRAS*, 152, 75

Hawking, S. W. 1974, *Nature*, 248, 30

Hawking, S. W. & Israel, W. Eds. 1987, *300 Years of Gravitation*, Cambridge Univ. Press

Hawking, S. W. & Penrose, R. 1970, *Proceedings of the Royal Society London* A314, 529

Hoyle, F. 1964, *Royal Observatory Bulletin*, 82, 90

Kerr, R. 1965, in: I. Robinson, A. Schild, & E. Schucking (eds.), *Quasi-Stellar Sources and Gravitational Collapse* (Univ. of Chicago Press), Chapter 9, 99 and *Physical Review Letters*, 11, 237

Laplace, P. S. 1795, *Exposition du Systeme du Monde*, Paris, Imprimerie du Cercle-Social, Tome II, p. 305

Leeming, D. & Leeming, M. 1984, *A Dictionary of Creation Myths*, Oxford Univ. Press

Lifshitz, E. M. & Khalatnikov, I. M. 1961, *JETP* 12, 108 & 558 (Russian original 1960, Journal of Experimental and Theoretical Physics)

Loinger, A. & Marsico, T. 2016, Preprint dated 13 February 2016

Marolf, D. & Polchinski, J. 2013, *Physical Review Letters*, 111, 171301

McCormmach, R. 2012, *Weighing the World*, Springer

Michell, J. 1784, *Philosophical Transactions of the Royal Society*, 74, 35-57

McCormmach, R. 2012, *Weighing the World*, Springer

Misner, C. W., Thorne, K. S., & Wheeler, J. A. 1971, *Gravitation*, W. H. Freeman

Narayan, R., Mahaevan, R., & Quartaert, E. 1998, in: M. A. Abramowicz, G. Bjornsson & J. E. Pringle (eds.), *Theory of Black Hole Accretion Disks* (Cambridge Univ. Press), p. 148-182

Natarajan, P. 2016, *Mapping the Heavens*, Yale University Press

Oppenheimer, J. R. & Snyder, H. 1939, *Physical Review*, 56, 455

Oppenheimer, J. R. & Volkoff, G. 1939, *Physical Review*, 54, 540

Ori, A. 1991, *Physical Review Letters*, 67, 789

Ori, A. 1992, *Physical Review Letters*, 68, 2117

Paczyński, B. 1971, *ARAA*, 9, 183

Penrose, R. 1962, in: A. Lichnerowicz &. H. Tonnelat (eds.), *Theories Relativistes de la Gravitation* (Proceedings of the 1959 Royaumont conference, CNRS, Paris)

Penrose, R. 1965, *Physical Review Letters*, 14, 57

Penrose, R. 1969, *Revista Nuovo Cimento*, 1, 252

Ponti, G., *et al.* 2014, in: L.O. Sjouwerman *et al.* (eds.) IAUS 303, *The Galactic Center* (Cambridge Univ. Press), p. 333

Rees, M. J. 1978, *Observatory Magazine*, 98, 210

Rees, M. J. 1982, in: G. R. Riegler & R. D. Blandford (eds.), *The Galactic Center* (American Inst. of Physics, New York), p. 18

Rees, M. J. 1984, *ARAA*, 22, 471

Salpeter, E. E. 1964, *ApJ*, 140, 796

Scharf, C. A. 2012, *Gravity's Engine*, Allen Lane

Schwarzschild, K. 1916, *Sitz. König. Preuss. Akad. Wiss. Berlin*, 424-39 & 951-59

Shakura, N. I. & Sunyaev, R. A. 1973, *A&A*, 24, 337

Thorne, K. S. 1994, *Black Holes and Time Warps*, W. W. Norton & Co.

Trimble, V. 2016, *EPJH* in press

Trimble, V. L. & Thorne, K. S. 1969, *ApJ*, 156, 1013

Trimble, V., Rose, W. K., & Weber, J. 1973, *MNRAS*, 162, p. 1P - 3P

Weber, J. 1961, *General Relativity and Gravitational Waves*, Interscience Publishers Inc. New York

Zeldovich, Ya. B. 1964, *Sov. Phys. Doklady*, 9, 195

Zeldovich, Ya. B. & Guseinov, O. H. 1965, *ApJ*, 144, 841

Zeldovich, Ya. B. & Novikov, I. D. 1964, *Doklady Akad. Nauk SSSR*, 158, 811

Zwicky, F. 1937, *Physical Review*, 51, 290 & 679

Topic 1. Similarity and Diversity of Black Hole Systems

New Frontiers in Black Hole Astrophysics
Proceedings IAU Symposium No. 324, 2016
Andreja Gomboc, ed.

doi:10.1017/S1743921317001338

Similarity and diversity of black holes - view from the Very High Energies

Elina Lindfors

Tuorla Observatory, Department of Physics and Astronomy, University of Turku, Finland
email: elilin@utu.fi

Abstract. Active galactic nuclei, hosting supermassive black holes and launching relativistic jets, are the most numerous objects on the gamma-ray sky. At the other end of the mass scale, phenomena related to stellar mass black holes, in particular gamma-ray bursts and microquasars, are also seen on the gamma-ray sky. While all of them are thought to launch relativistic jets, the diversity even within each of these classes is enormous. In this review, I will discuss recent very high energy gamma-ray results that underline both the similarity of the black hole systems, as well as their diversity.

Keywords. very high energy gamma-rays, active galactic nuclei, tidal disruption events, microquasars, gamma-ray bursts

1. Introduction

The known black hole systems cover both stellar mass and supermassive black holes. In Active Galactic Nuclei, the activity is driven by the accretion of matter to supermassive black hole in the centre of the galaxy. In inactive galaxies the black hole may occassionally shine up due to tidal disruption events, which occur when a star gets too close to the supermassive black hole and is bulled apart by the black hole's tidal forces. In extreme and rare cases (only two have been observed Cenko *et al.* 2012, Komossa 2015) this launches a collimated jet of particles. On stellar black hole mass scales, the microquasar phenomenon consists of a black hole feeding from companion star and launching a jet. Finally, we assume that long gamma-ray bursts are death crowls of massive stars, creating a black hole and collimated jets of particles.

The common astrophysical ingredients of these systems, the spinning black hole, the accretion disk, and the collimated jets of particles have led several scientists to look for similarities among the different systems. What properties simply scale with the black hole mass and could this be signature of something fundamental? For example, in the early work by Merloni *et al.* (2003) a fundamental plane of black hole activity was established, when the authors found that active galaxies and galactic black holes lie on a plane in three dimensional space (radio luminosity, X-ray luminosity and mass of the black hole). More recently, Nemmen *et al.* (2012) found that jets created by black holes maintain the same coupling between the total power carried by the jet and power radiated away. In our work, we have investigated if the jets of blazars and microquasars are similar in terms of outbursting mechanism and jet parameters that can be derived from the decomposition of the radio to optical light curves (Türler *et al.* 1999). We found that indeed the outburst of the microquasars and quasars are well described by shock-in-jet model, and that the jet parameters derived were rather similar for both types of systems (Türler & Lindfors 2007).

In recent years, astroparticle physics has opened new observational window to black hole systems and during our symposium we heard several interesting presentations

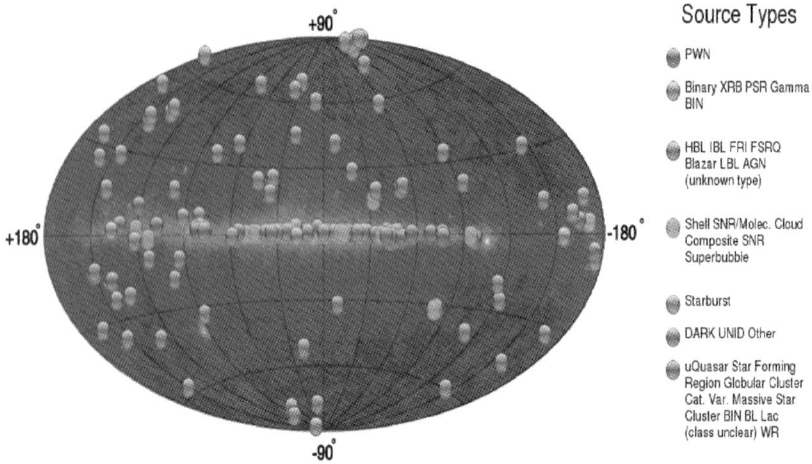

Figure 1. The Very High Energy gamma-ray sky. The map from http://tevcat.uchicago.edu.

coming from this community. Very High Energy (VHE, $E > 100\,GeV$) γ-ray experiments have revealed > 170 astrophysical sources, and black hole systems are well represented among these sources. IceCube has started the era of neutrino astronomy with the discovery of neutrinos of astrophysical origin (Aartsen *et al.* 2013) and while their origin is still unknown, black hole systems are certainly among the candidates. Finally, the LIGO experiment measured gravitational waves from merging black holes (Abbott *et al.* 2016). In this paper, I concentrate on the new observations of black hole system in the very high energies, covering the observations of active galactic nuclei, tidal disruption events, microquasars and gamma-ray bursts.

2. Supermassive black holes at very high energies

2.1. *Active Galactic Nuclei*

Active galactic nuclei (AGN) are the most numerous sources on the extragalactic VHE γ-ray sky. However, there is quite some diversity among this population. The most numerous ones are blazars, a type of active galactic nuclei, where the relativistic jet points very close to our line of sight. The blazars divide into several subcategories, all of which are seen at the VHE γ-ray sky. Furthermore, also several radio galaxies are detected.

Blazar spectral energy distribution (SED) shows two bumps, the low energy bump extends from radio to ultraviolet–X-rays while the high energy bump extends from X-rays to VHE γ-rays. The low energy emission is synchrotron emission by the electrons spiralling in the magnetic field of the jet, while the high energy emission is in most cases inverse Compton emission. The location of the synchrotron peak is used to divide the blazars in sub-categories. The sources having the peak at UV-X-ray energies are the most numerous sources in the extragalactic VHE γ-ray sky, but nowdays also low synchrotron peaking objects have been detected.

Blazars are variable in all wavelengths from radio to VHE γ-rays in timescales ranging from minutes to years. While short timescale variability is expected, as the jet is pointing so close to our line of sight, it becomes challenging when the timescales are as short as minutes. However, current instruments have observed minute-scale variability in VHE γ-rays from several sources, from all blazar classes and also from a radio galaxy (Aharonian *et al.* 2007, Albert *et al.* 2007a, Aleksic *et al.* 2011, Arlen *et al.* 2013, Aleksic *et al.* 2014a).

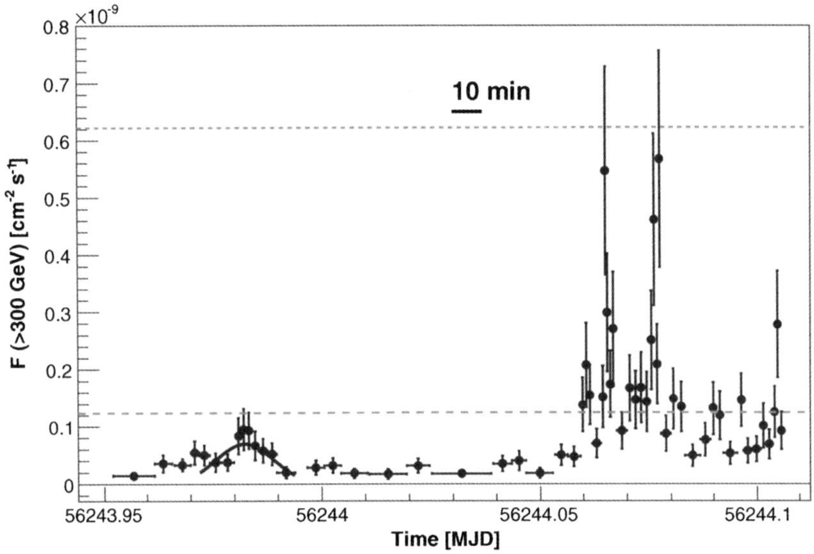

Figure 2. The light curve of IC 310 on the night of 12/13 November 2012 as observed with the MAGIC telescopes. Figure from Aleksic *et al.* 2014a.

The minute-scale variability is challenging for models, as a huge energy has to be radiated within an extremely compact region. The time scale is actually shorter than the scale expected from the central black hole's horizon (which for $10^9 M_{sun}$ black hole is an order of one hour). Furthermore, it is a challenge for particle acceleration and emission models. Many solutions to this dilemma has been suggested, such as strong recollimation of the jet or very compact region embedded within large scale jets (spine-sheath, jets-in-jet, ring-of-fire) (e.g. Ghisellini *et al.* 2005, Giannios *et al.* 2009, MacDonald *et al.* 2015).

In many cases the fast VHE γ-ray variability is detected during periods when the source has been showing enhanced flux levels in all wavebands already prior to the detection of the fast variability. It is therefore evident that these events somehow connect to the overall activity within the relativistic jet. In the particular case of flat spectrum radio quasars, it is also evident that the VHE γ-ray emission cannot originate very close to the central black hole, as it is surrounded by broad line emission clouds (at the distance of $\sim$1 parsec) that are very efficient in absorbing VHE γ-rays via pair production (e.g. Böttcher & Els 2016 and references therein).

One particularly interesting case of fast variability is radio galaxy IC 310 in Perseus cluster. MAGIC Telescopes detected a huge flare in the VHE γ-ray band in November 2013. The flux reached several Crab Units, and the light curve revealed variability with doubling timescales faster than 4.8 minutes (Aleksic *et al.* 2014a). As the viewing angle of the jet is rather well constrained to $10-20$ degrees, the general solution of introducing $\Gamma > 50$ to explain the fast variability does not work. In general it was concluded that models, where such a bright flare with such fast variability would be produced within the jet, were not viable. Therefore it was suggested that the flare actually originated in the magnetosphere of the black hole and it was dubbed a black hole lightning. The magnetospeheric model for IC 310 was discussed in detail in Hirotani & Pu (2016), who concluded that it is feasible to produce the observed flare if the black hole at the time of the flare was accreting at very low rate. The feasibility of this model was discussed also during the symposium (see Barkov, this volume).

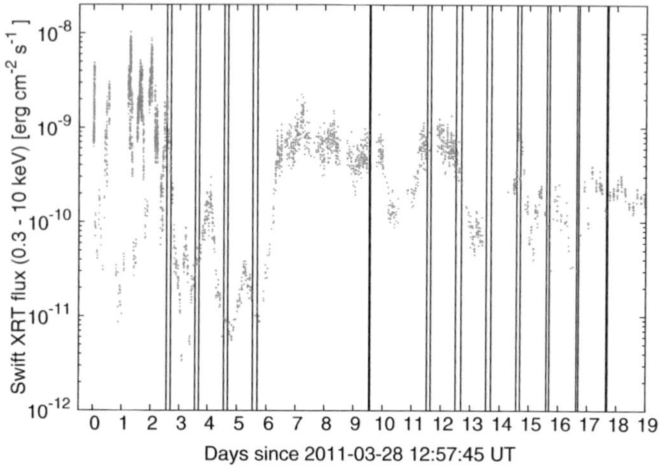

Figure 3. The light curve of Sw 1644+57 from *Swift*-XRT. The MAGIC observation windows of the event are shown as vertical lines. MAGIC detected no VHE γ-rays from this TDE. Figure from Aleksic *et al.* 2013.

Beyond the fast variability of VHE γ-rays, also the slower timescale variability of the VHE emission, the shape of the VHE γ-ray spectrum and in more general terms the population of active galaxies that we see at the highest energies tell us important stories about supermassive black holes. Looking at the variability of the light curves in different bands as well as the snap-shot spectral energy distributions have shown that the emission takes place in multiple emission regions within the relativistic jets (e.g. Aleksic *et al.* 2014b, Aleksic *et al.* 2015a, Lindfors *et al.* 2016). It has also been shown that, at least occasionally, the main energy dissipation region must move further out in the jet, at least to distances beyond the broad line region clouds (Ghisellini *et al.* 2013, Aleksic *et al.* 2014c, Ahnen *et al.* 2015).

2.2. *Tidal Disruption Events*

In addition to AGN, the VHE γ-ray telescopes have been pointed to a normally inactive supermassive black hole, to the famous tidal disruption event Sw 1644+57. Bloom *et al.* (2011) and Burrows *et al.* (2011) argued that this event arose from the activation of a beamed jet that was hypothesized to be the result of a tidal disruption of a star by a $\sim 10^6 - 10^7 M_\odot$ black hole.

The VERITAS Telescopes started observing Sw J1644+57 approximately 22.5 hours after the first BAT trigger and followed it for 18 days (Aliu *et al.* 2011). The MAGIC telescopes observed the Sw J1644+57 during the flaring phase, starting observations nearly 2.5 days after the trigger time (Aleksic *et al.* 2013). Neither of the telescopes found evidence for emission above the energy threshold of 100 GeV and the upper limits were in agreement both with the synchrotron and inverse Compton scattering scenarios for the X-ray emission (Aliu *et al.* 2011, Aleksic *et al.* 2013). However, in principle the VHE γ-ray observations can be very constraining for the conditions of the jet in such tidal disruption events, in particular for the synchrotron origin of the X-ray emission and for the lorentz factor of the newly formed jet.

In addition to Sw 1644+57, only one other possible case of jetted tidal disruption event is known (Cenko *et al.* 2012, Komossa 2015), but to my knowledge it was not followed by the VHE γ-ray telescopes.

2.3. *Supermassive black hole in the centre of our Galaxy*

Galactic center is a strong and well established source of VHE γ-rays (Aharonian *et al.* 2004, Albert *et al.* 2006a, Archer *et al.* 2014). It was already early suggested that the point source in the galactic centre would correspond to a central black hole (Aharonian *et al.* 2004), but as there are also other candidates, such as supernova remnant Sgr A East, the question of direct emission of the black hole in the center of our galaxy is not yet resolved.

Very recently H.E.S.S. Collaboration showed, using 10 years of VHE γ-ray observations of the galactic centre, that the black hole in the centre of our galaxy is the first established PeVatron, i.e particle accelerator that can accelerate particles up to PeV energies (Abramowski *et al.* 2016).

3. Stellar mass black holes at very high energies

3.1. *Microquasars*

X-ray binaries are binary systems with a compact object (a black hole or a neutron star) feeding from a companion. In microquasars this feeding launches a relativistic jet and as an analogy to more massive quasars, these jets could accelarate particles to energies high enough to emit VHE γ-rays.

There were two detections of VHE γ-ray emission from binary systems that were first considered as possible microquasars, LS5039 and LSI +61 303 (Aharonian *et al.* 2005, Albert *et al.* 2006b), but later observations have supported binary pulsar model for these sources (e.g. Albert *et al.* 2008). Up to date, the hint of VHE γ-ray emission from a microquasar Cyg X-1 (Albert *et al.* 2007b) is the only indication of VHE emission, while other observations have resulted only in upper limits (Aleksic *et al.* 2010a, Abdalla *et al.* 2016). The reason can be that the conditions are not favorable to emission of VHE γ-rays, given that it always requires acceleration of particles to extreme energies and typically also presence of photons to be up-scattered. Other possible explanation is an observational bias, as it is evident that the high energy emission from the microquasar jets is a transient phenomena.

Cyg X-1 and Cyg X-3 are both detected in the lower gamma-ray energies by the Fermi satellite and the origin of this emission seems to be the jet (Abdo *et al.* 2009, Zanin *et al.* 2016, Zdziarski *et al.* 2016). Therefore, it is to be expected that during flares the emission would extend also to energies > 100 GeV, unless the gamma-rays are always emitted very close to jet base, where strong photon field absorbing VHE γ-rays is present. If the microqusars were to be analogies to quasars, one would expect that occassionally the energy dissipation region would move further out and VHE γ-rays could escape. To catch such a epoch, which is assumably an order of an hour in duration or shorter, with ground-based telescopes (limited duty cycle and limited field of view) is of course challenging.

3.2. *Gamma-ray bursts*

Gamma-ray bursts (GRBs) launch short lived, but extremely fast and luminous jets. As the duration of the prompt emission from GRBs is typically an order of a minute or less, fast pointing to the location of the GRBs is in a key role. Up to now there is no detection of VHE γ-ray emission from GRBs by the Imaging Air Cherenkov Telescopes (Albert *et al.* 2006c, Tam *et al.* 2006, Albert *et al.* 2007c, Aharonian *et al.* 2009, Aleksic *et al.* 2010b, Acciari *et al.* 2011, Aleksic *et al.* 2014d), but the detection of high energy photons by the *Fermi*-LAT from some GRBs (e.g. Ackermann *et al.* 2014) certainly gives

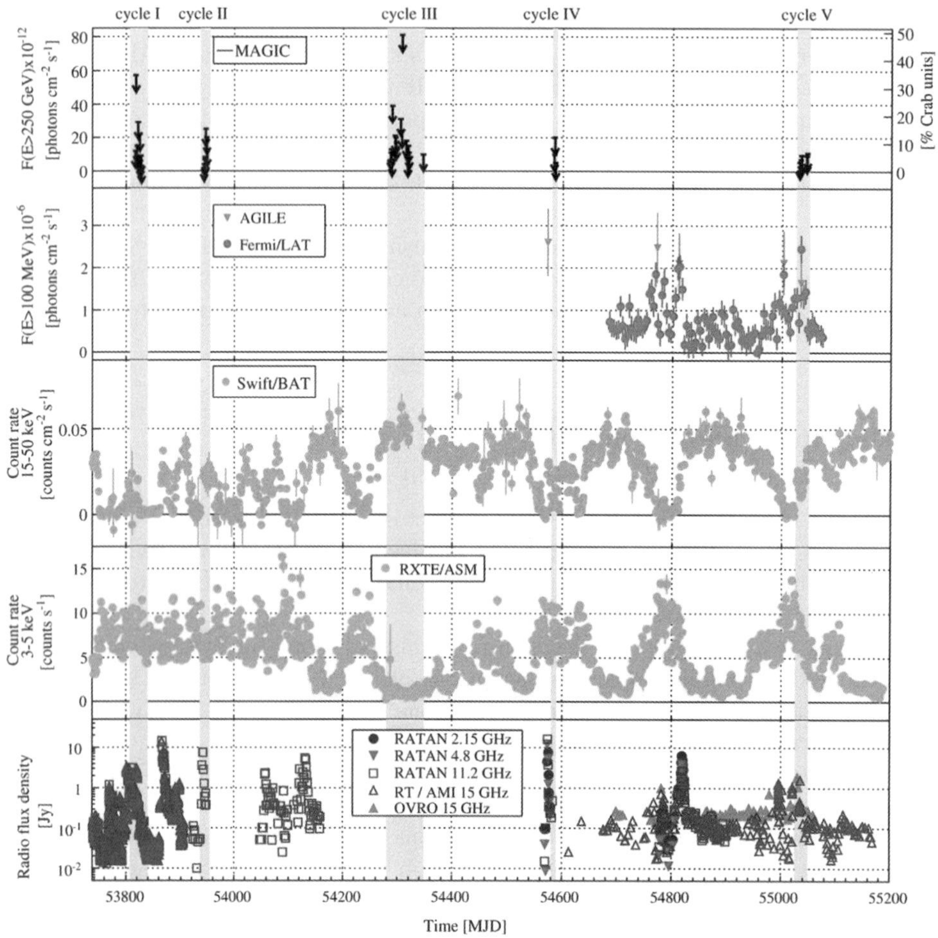

Figure 4. The light curve of Cyg X-3 as observed with the MAGIC telescopes (top), textitFermi-LAT and AGILE satellites, *Swift*-BAT, RXTE-ASM and OVRO and AMI telescopes (bottom) from 2006 to 2009. The gray band show the periods corresponding to MAGIC observations. The MAGIC observations covered different X-ray/radio spectral states. No VHE γ-ray emission from the source was detected. Figure from Aleksic *et al.* 2010a.

hope also to ground-based experiments. VHE γ-ray observations are a powerful tool for emission processes and physical conditions in GRBs and could also give important insight on the similarity of the emission processes in GRB and blazar jets.

4. Summary

Phenomena related to black holes at VHE γ-rays are mostly transient or extremely variable and therefore a huge observational challenge. As this window to black hole systems is still rather new, there are many open questions and also significant discovery potential in this energy range. The most prominent black holes at VHE γ-rays are the active galactic nuclei, where we have already started to see diversity as the number of known sources is steadily increasing (currently > 70). For the rest; tidal disruption events, microquasars and gamma-ray bursts, we will have to wait for further observations at VHE γ-rays, with the current generation of telescopes and with the future Cherenkov

Telescope Array, to be able to say something about the similarity of the all black hole systems at very high energies.

References

M. G. Aartsen, R. Abbasi, Y. Abdou, *et al.* (IceCube Coll.) 2013, *Science*, 342, 6161

B. P. Abbott, R. Abbott, T. D. Abbott *et al.* (LIGO Coll.) 2016, *PhysRevLett*, 116, id.061102

A. A. Abdo, M. Ackermann, M. Ajello *et al.* (Fermi-LAT Coll.) 2009, *Science*, 326, 1512

H. Abdalla, A. Abramowski, F. Aharonian *et al.* (H.E.S.S. Coll.) 2016, *A&A*, accepted, arxiv:1607.04613

A. Abramowski, F. Aharonian, F. A. Benkhali *et al.* (H.E.S.S. Coll.) 2016, *Nature*, 531, 476

V. A. Acciari, E. Aliu, T. Arlen *et al.* (VERITAS Coll.) 2011, *ApJ*, 743, 62

M. Ackermann, M. Ajello, K. Asano *et al.* (Fermi-LAT Coll.) 2014, *Science*, 343, 42

F. Aharonian, A. G. Akhperjanian, K.-M. Aye *et al.* (H.E.S.S. Coll.) 2004, *A&A*, 425, 13

F. Aharonian, A. G. Akhperjanian, K.-M. Aye *et al.* (H.E.S.S. Coll.) 2005, *Science*, 309, 746

F. Aharonian, A. G. Akhperjanian, A. R. Bazer-Bachi *et al.* (H.E.S.S. Coll.) 2007, *ApJ*, 664, 71

F. Aharonian, A. G. Akhperjanian, U. Barres *et al.* (H.E.S.S. Coll.) 2009, *ApJ*, 690, 1068

M. Ahnen, S. Ansoldi, L. A. Antonelli *et al.* (MAGIC Coll.) 2015, *ApJ*, 815, 23

J. Albert, E. Aliu, H. Anderhub *et al.* (MAGIC Coll.) 2006a, *ApJ*, 638, 101

J. Albert, E. Aliu, H. Anderhub *et al.* (MAGIC Coll.) 2006b, *Science*, 312, 1771

J. Albert, E. Aliu, H. Anderhub *et al.* (MAGIC Coll.) 2006c, *ApJ*, 641, 9

J. Albert, E. Aliu, H. Anderhub *et al.* (MAGIC Coll.) 2007a, *ApJ*, 669, 862

J. Albert, E. Aliu, H. Anderhub *et al.* (MAGIC Coll.) 2007b, *ApJ*, 665, 51

J. Albert, E. Aliu, H. Anderhub *et al.* (MAGIC Coll.) 2007c, *ApJ*, 667, 358

J. Albert, E. Aliu, H. Anderhub *et al.* (MAGIC Coll.) 2008, *ApJ*, 684, 1351

J. Aleksic, L. A. Antonelli, P. Antoranz *et al.* (MAGIC Coll.) 2010a, *ApJ*, 721, 843

J. Aleksic, L. A. Antonelli, P. Antoranz *et al.* (MAGIC Coll.) 2010b, *A&A*, 517, 5

J. Aleksic, L. A. Antonelli, P. Antoranz *et al.* (MAGIC Coll.) 2011, *ApJ*, 730, 8

J. Aleksic, L. A. Antonelli, P. Antoranz *et al.* (MAGIC Coll.) 2013, *A&A*, 552, 112

J. Aleksic, S. Ansoldi, L. A. Antonelli *et al.* (MAGIC Coll.) 2014, *Science*, 346, 1080

J. Aleksic, S. Ansoldi, L. A. Antonelli *et al.* (MAGIC Coll.) 2014, *A&A*, 567, 135

J. Aleksic, S. Ansoldi, L. A. Antonelli *et al.* (MAGIC Coll.) 2014, *A&A*, 569, 46

J. Aleksic, S. Ansoldi, L. A. Antonelli *et al.* (MAGIC Coll.) 2014d, *MNRAS*, 437, 3103

J. Aleksic, S. Ansoldi, L. A. Antonelli *et al.* (MAGIC Coll.) 2015, *A&A*, 578, 22

E. Aliu, T. Arlen, T. Aune *et al.* (VERITAS Coll.) 2011, *ApJ*, 738, 30

T. Arlen, T. Aune, M. Beilicke *et al.* (VERITAS Coll.) 2013, *ApJ*, 762, 92

A. Archer, A. Barnacka, M. Beilicke *et al.* (VERITAS Coll.) 2014, *ApJ*, 790, 149

J. Bloom, D. Giannios, B. Metzger *et al.* 2011, *Science*, 333, 203

D. N. Burrows, J. A. Kennea, G. Ghisellini *et al.* 2011,

M. Böttcher & P. Els 2016, *ApJ*, 821, 102

S. B. Cenko, H. A. Krimm, & A. Horesh 2012, *ApJ*, 753, 77

G. Ghisellini, F. Tavecchio, & M. Chiaberge 2005, *A&A*, 432, 401

G. Ghisellini, F. Tavecchio, L. Foschini *et al.* 2013, *MNRAS*, 432, 66

D. Giannios, D. Uzdensky, & M. Begelman 2009, *MNRAS*, 395, 29

K. Hirotani & H. Pu 2016, *ApJ*, 818, 50

S. Komossa 2015, *Journal of High Energy Astrophysics*, 7, 148

E. Lindfors, T. Hovatta, K. Nilsson *et al.*, 2016 *A&A*, 593, 98

N. MacDonald, A. Marscher, S. Jorstad *et al.*, 2015 *ApJ*, 804, 111

A. Merloni, S. Heinz, & T. di Matteo, 2003 *MNRAS*, 345, 1057

R. S., Nemmen, M. Georganopoulos, S. Guiriec *et al.* 2012, *Science*, 338, 1445

P. H. Tamm, S. J. Wagner, & G. Pühlhofer (on behalf of H. E. S. S. Coll.) 2006, *Il Nuovo Cimento B*, 121, 1595

M. Türler, T. J.-L. Courvoisier & S. Paltani 1999, *A&A*, 349, 45

M. Türler & E. Lindfors 2007, *Proceedings of IAU Symposium #238, Cambridge, UK: Cambridge University Press*, 238, 305

R. Zanin, A. Fernandez-Barral, E. de Ona-Wilhelmi *et al.* 2016, *A&A*, accepted, arxiv:1605.05914

A. A. Zdziarski, D. Malyshev, M. Chernyakova *et al.* 2016, *MNRAS*, submitted, arxiv:1607.05059

New Frontiers in Black Hole Astrophysics
Proceedings IAU Symposium No. 324, 2016
Andreja Gomboc, ed.

doi:10.1017/S1743921316012849

Where is the electric current driven in the Blandford-Znajek process?

Kenji Toma[1] and Fumio Takahara[2]

[1]Frontier Research Institute for Interdisciplinary Sciences, Tohoku University, Sendai 980-8578, Japan; email:toma@astr.tohoku.ac.jp

[2]Department of Earth and Space Science, Graduate School of Science, Osaka University, Toyonaka 560-0043, Japan

Abstract. The Blandford-Znajek process, the steady electromagnetic energy extraction from a rotating black hole, is widely believed to work for driving relativistic jets, although it is still under debate where the electric current is driven. We address this issue analytically by investigating the time-dependent state in the Boyer-Lindquist and Kerr-Schild coordinate systems. This analysis suggests that a non-ideal magnetohydrodynamic region is required in the time-dependent state, while not in the steady state.

Keywords. black hole physics, MHD, plasmas, galaxies: jets, gamma rays: bursts

1. Introduction

The driving mechanism of relativistic jets associated with active galactic nuclei, gamma-ray bursts, and Galactic microquasars is one of the major problems in astrophysics. A widely discussed model is based on the Blandford-Znajek (BZ) process, the energy extraction from a rotating black hole (BH) along magnetic field lines threading it (Blandford & Znajek 1977). This process produces Poynting-dominated jets, in which the plasma particles can be accelerated to relativistic speeds (e.g., Komissarov *et al.* 2009, Lyubarsky 2009, Toma & Takahara 2013). This theoretical picture is now being quantitatively compared with the recent observational results from the M87 jet (e.g., Asada & Nakamura 2012, Kino *et al.* 2015, Porth & Komissarov 2015, Mertens *et al.* 2016).

The BZ process was proposed in the pioneering paper of Blandford & Znajek (1977), in which were found steady, axisymmetric, force-free solutions of the Kerr BH magnetosphere in the slow rotation limit where the Poynting flux is nonzero along the field lines threading the event horizon. This was followed by demonstrations of numerical and approximate analytical solutions with finite speed of BH rotation (e.g., Komissarov 2001, Koide *et al.* 2002, McKinney & Gammie 2004, Barkov & Komissarov 2008, Tchekhovskoy *et al.* 2011, Ruiz *et al.* 2012, Contopoulos *et al.* 2013, Beskin & Zheltoukhov 2013). It is also generically proved that the steady, axisymmetric situation of no Poynting flux cannot be maintained for open field lines threading the ergosphere (Toma & Takahara 2014). However, the physical mechanism for the creation of the flux in the electromagnetically dominated plasma has not been clearly explained.

The issue is where the steady electric current (which produces the toroidal magnetic field) is driven in the BZ process. All the previous studies on this issue focused on the steady state condition, as far as we are aware, while we argue that this issue cannot be solved without investigating the time-dependent state towards the steady state. Then we suggest that the electric current is generally driven in the non-ideal magnetohydrodynamic (MHD) region in the time-dependent state, and such regions are hidden near (or within) the event horizon in the steady state. The steady current circuit is maintained

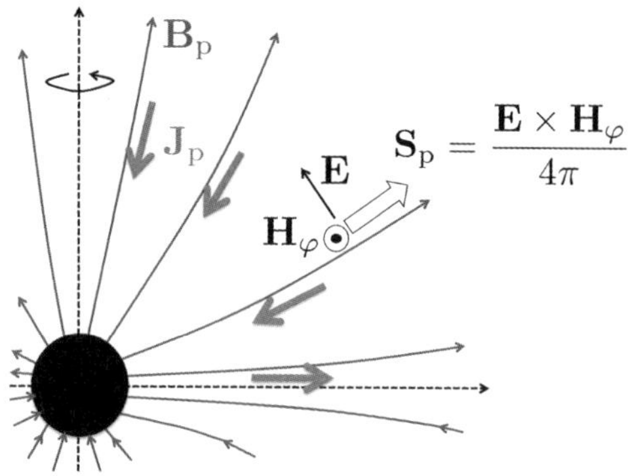

Figure 1. Schematic picture of the electromagnetic structure of the BZ process.

without any current driving source, because the force-free plasma has no resistivity. In this article we explain the issue in more details and briefly summarize our argument.

2. The issue

The BZ process assumes that the poloidal magnetic field $\mathbf{B}_{\mathrm{p}}$ produced by the plasma filling the magnetosphere (not by the BH itself) threads the event horizon, as shown in Figure 1. In the case of non-zero BH rotation, the steady axisymmetric state has the electric field $\mathbf{E}$ perpendicular to $\mathbf{B}_{\mathrm{p}}$ (with $E_{\varphi} = 0$) and the toroidal magnetic field $\mathbf{H}_{\varphi}$, which form the outward Poynting flux $\mathbf{S}_{\mathrm{p}} = \mathbf{E} \times \mathbf{H}_{\varphi}/4\pi$, where $\mathbf{E}$ and $\mathbf{H}$ are fields as measured in the coordinate basis. One should first note that $\mathbf{H}_{\varphi}$ is produced by the poloidal current $\mathbf{J}_{\mathrm{p}}$, and not produced in vacuum just by the rotation of space-time (Wald 1974). Second, in contrast to the BZ process, the origin of outward Poynting flux of pulsar winds has been definitely identified: The rotation of the matter-dominated central star (pulsar) generates $\mathbf{E}$ ($= -\mathbf{V}_{\varphi} \times \mathbf{B}_{\mathrm{p}}$) and drives $\mathbf{J}_{\mathrm{p}}$ via the unipolar induction process (Goldreich & Julian 1969). We can write $\nabla \cdot \mathbf{S}_{\mathrm{p}} = -\mathbf{E} \cdot \mathbf{J}_{\mathrm{p}}$, and the source term is positive in the star, where $\mathbf{J}_{\mathrm{p}}$ is driven to flow in the direction of $-\mathbf{E}$. The BZ process, working in the electromagnetically dominated plasma, does not include any matter-dominated region in which $\mathbf{B}_{\mathrm{p}}$ is anchored. The current cannot cross the field lines, which means $\nabla \cdot \mathbf{S}_{\mathrm{p}} = 0$. Then it is unclear where $\mathbf{J}_{\mathrm{p}}$ is driven.

The membrane paradigm (Thorne *et al.* 1986) interprets the event horizon as a rotating conductor that generates $\mathbf{E}$ and drives $\mathbf{J}_{\mathrm{p}}$ in an analogy with the unipolar induction of pulsar winds. However, the horizon does not actively affects its exterior, but just passively absorbs particles and waves (Punsly & Coroniti 1989). The flux production has to work outside the horizon.

For such a causal flux production, some theorists proposed a scenario in which certain types of negative energies (as measured in the coordinate basis) created outside the horizon (and within the ergosphere) flow towards the horizon, resulting in the positive outward $\mathbf{S}_{\mathrm{p}}$. This is an analogous to the mechanical Penrose process. However, MHD simulations demonstrate that no regions of negative particle energy are seen in the steady state (Komissarov 2005). The concept of negative electromagnetic energy inflow (Komissarov 2009, Lasota *et al.* 2014, Koide & Baba 2014) is not physically essential, because the sign of electromagnetic energy density ($e = -\alpha T^{t}_{\mathrm{EM},t}$, where α is the lapse function of Kerr metric) depends on the coordinates (Toma & Takahara 2016). The total energy flux per unit mass flux treated in the analytical ideal MHD model can be negative even outside the ergosphere (Takahashi *et al.* 1990). The inertial drift current in the ideal

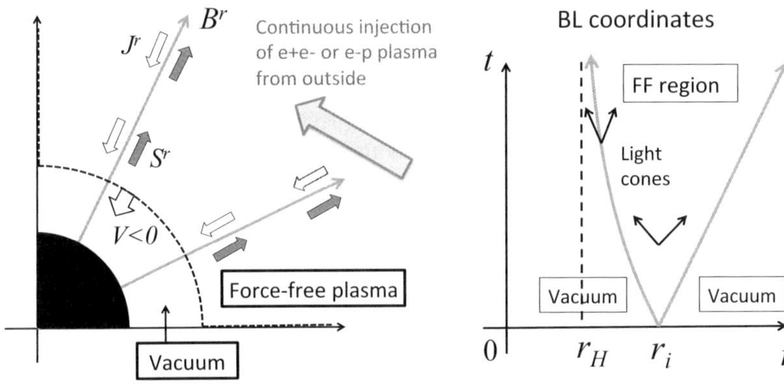

Figure 2. Our toy model of the time-dependent state (focusing on the inflow region) (*left*) and its schematic space-time diagram in the Boyer-Lindquist coordinates (*right*).

MHD model can cross the field lines, but it cannot produce all of $\mathbf{S}_\mathrm{p}$ (Hirotani *et al.* 1992).

The current $\mathbf{J}_\mathrm{p}$ might be driven in the narrow non-ideal MHD region which steadily exists outside the horizon (Okamoto 2006, Blandford & Znajek 1977). This possibility may not be excluded. If this is the case, such a non-ideal MHD region may have electric field component parallel to the magnetic field, which accelerates the particles and produces characteristic electromagnetic radiation. Thus the present issue appears to be not only theoretical but also related to observations.

3. Our argument

To help understanding the present issue, let us consider how the Schwarzschild gravitational field is formed. Its origin is definitely the mass inside the horizon. However, the information that the mass exists inside the horizon never comes out to the exterior. Then how is the Schwarzschild gravitational field formed? This question cannot be properly answered only by considering the static state. The answer is that this gravitational field has been produced by the mass before it falls across the horizon.

This consideration suggests that the origin of steady electromagnetic field surrounding the horizon cannot either be understood without investigating the time-dependent state towards the steady state. All the previous studies on the current driving mechanism (as far as we are aware) investigated only the steady state.

Although such a time-dependent state should be analyzed numerically, we use a toy model to qualitatively illustrate the process of building the poloidal current circuit (Toma & Takahara 2016). In order to find the essential physics, our analysis is performed in the Boyer-Lindquist (singular at the horizon) and Kerr-Schild (regular at the horizon) coordinate systems. As shown in Figure 2 (left), we initially consider a vacuum in Kerr space-time, and then begin the continuous injection of force-free plasma particles between the two light surfaces as a Gedankenexperiment. The inflow (outflow) fills the vacuum near the horizon (at infinity). Then we can see a process building the poloidal current circuit. Our toy model assumes that $\mathbf{B}_\mathrm{p}$ is fixed to be split-monopole in the whole region, that the two boundaries between the force-free region and the vacuum are geometrically thin and move radially, and that the force-free region and the vacuum have their steady-state structures, but the values of the physical quantities (E, H_φ, etc.) keep being updated as determined by the varying conditions of the inner and outer boundaries. These assumptions enable us to integrate the 3 + 1 Maxwell equations. Then we found that $\mathbf{J}_\mathrm{p}$ has to flow on the boundaries (in the directions for making the circuit). More importantly, we found that the displacement current at the inner boundary also regulates $\mathbf{J}_\mathrm{p}$ in the force-free region, i.e., $H^{\mathrm{ff}}_\varphi = \sqrt{\gamma}V(D^\theta_{\mathrm{ff}} - D^\theta_{\mathrm{vac}}) - 4\pi\sqrt{\gamma}\eta^\theta$, where the subscript and superscript "ff" and "vac" denote the force-free and vacuum regions, respectively, and $\mathbf{D}$, V, η^θ, and γ are the electric field as measured by the zero angular momentum observers, the velocity of inner boundary, the surface current on the inner boundary, and the determinant of spatial metric, respectively.

The outward Poynting flux in the force-free region is produced at the inner boundary as $\nabla \cdot \mathbf{S}_\mathrm{p} = -\partial_t e - \mathbf{E} \cdot \mathbf{J}_\mathrm{p}$. This process takes place outside the horizon, which does not violate causality. Figure 2 (right) is the schematic space-time diagram of the two boundaries in the Boyer-Lindquist coordinates. When the inner boundary approaches the horizon, the outward signal from it propagates slower and slower and it can hardly affect the force-free region. This will lead to the steady state, and one has $\nabla \cdot \mathbf{S}_\mathrm{p} = 0$, i.e., there is no driving source of $\mathbf{J}_\mathrm{p}$ in the steady state. Since the force-free plasma is assumed to have no resistivity, the current can keep flowing steadily.

In conclusion, the driving source of $\mathbf{J}_\mathrm{p}$ or the non-ideal MHD region exists outside the horizon in the time-dependent state, while it is not required in the steady state. See Toma & Takahara (2016) for more details.

References

Asada, K. & Nakamura, M. 2012, *ApJ*, 745, L28
Barkov, M. V. & Komissarov, S. S. 2008, *MNRAS*, 385, L28
Beskin, V. S. & Zheltoukhov, A. A. 2013, *Astron. Lett.*, 39, 215
Blandford, R. D. & Znajek, R. L. 1977, *MNRAS*, 176, 465
Contopoulos, I., Kazanas, D., & Papadopoulos, D. B. 2013, *ApJ*, 765, 113
Goldreich, P. & Julian, W. H. 1969, *ApJ*, 157, 869
Hirotani, K., Takahashi, M., Nitta, S., & Tomimatsu, A. 1992, *ApJ*, 386, 455
Kino, M., Takahara, F., Hada, K., *et al.* 2015, *ApJ*, 803, 30
Koide, S. & Baba, T. 2014, *ApJ*, 792, 88
Koide, S., Shibata, K., Kudoh, T., & Meier, D. L. 2002, *Science*, 295, 1688
Komissarov, S. S. 2001, *MNRAS*, 326, L41
Komissarov, S. S. 2004, *MNRAS*, 350, 427
Komissarov, S. S. 2005, *MNRAS*, 359, 801
Komissarov, S. S. 2009, *J. Korean Phys. Soc.*, 54, 2503
Komissarov, S. S., Vlahakis, N., Königl, A., & Barkov, M. V. 2009, *MNRAS*, 394, 1182
Lasota, J.-P., Gourgoulhon, E., Abramowicz, M., *et al.* 2014, *PRD*, 89, 024041
Lyubarsky, Y. 2009, *ApJ*, 698, 1570
McKinney, J. C. & Gammie, C. F. 2004, *ApJ*, 611, 977
Mertens, F., Lobanov, A. P., Walker, R. C., & Hardee, P. E. 2016, *A & A*, 595, 54
Okamoto, I. 2006, *PASJ*, 58, 1047
Porth, O. & Komissarov, S. S. 2015, *MNRAS*, 452, 1089
Punsly, B., & Coroniti, F. V., 1989 *PRD*, 40, 3834
Ruiz, M., Palenzuela, C., Galeazzi, F., & Bona, C. 2012, *MNRAS*, 423, 1300
Takahashi, M., Nitta, S., Tatematsu, Y., & Tomimatsu, A. 1990, *ApJ*, 363, 206
Tchekhovskoy, R., Narayan, R., & McKinney, J. C. 2011, *MNRAS*, 418, L79
Thorne, K., *et al.* 1986, *The Membrane Paradigm* (Yale University Press, New Haven)
Toma, K. & Takahara, F. 2013, *PTEP*, 2013, 3E02
Toma, K. & Takahara, F. 2014, *MNRAS*, 442, 2855
Toma, K. & Takahara, F. 2016, *PTEP*, 2016, 3E01
Wald, R. M. 1974, *PRD*, 10, 1680

New Frontiers in Black Hole Astrophysics
Proceedings IAU Symposium No. 324, 2016
Andreja Gomboc, ed.

doi:10.1017/S1743921316012953

Transonic structure of slowly rotating accretion flows with shocks around black holes

Petra Suková[1,2], Szymon Charzyński[1,3] and Agnieszka Janiuk[1]

[1]Center for Theoretical Physics, Polish Academy of Sciences,
Al. Lotników 32/46, 02-668 Warsaw, Poland
[2]Astronomical Institute, Czech Academy of Sciences
Boční II 1401, 141 00 Prague, Czech Republic
email: petra.sukova@asu.cas.cz
[3]Chair of Mathematical Methods in Physics, University of Warsaw
ul. Pasteura 5, 02-093 Warsaw, Poland

Abstract. We present recent results of the studies of low angular momentum accretion of matter onto Schwarzschild black hole using fully relativistic numerical simulations. We compare the resulting 2D structure of transonic flows with results of 1D pseudo-Newtonian computations of non-magnetized flow. The research has observable consequences on black holes on the whole mass scale, in particular it is related to the time-scale and shape of luminosity flares in Sgr A* or to the evolution of QPO frequency during outbursts of microquasars.

Keywords. accretion, accretion disks, hydrodynamics, shock waves, X-rays: binaries

1. Introduction

With the growing number of available X-ray data especially from the dedicated X-ray satellites, the complicated behaviour of the microquasars during their outbursts have been spotted. It has been realized, that the accretion through the simple Shakura-Sunyaev accretion disc is not able to explain different phenomena occurring in the those sources, e.g. quasiperiodic oscillations (QPOs) and flares or the spectral and timing properties of the lightcurves. Hence, different models were proposed to describe such behaviour.

Usually, in those models, two different components of the accretion flow are present, namely the cold Keplerian disc and a hot corona, which are responsible for different timing properties in soft and hard X-rays. However, the origin and geometry of the hot component are still under debate and different ideas have been set in the literature. Because of the short timescales in the hard X-ray component, the hot flow should exhibit high radial velocities and fast inflow. Therefore, the reasonable expectation is that it has a low value of angular momentum.

The flows with constant low angular momentum have been studied theoretically and the authors showed the possibility of standing shock appearance for certain subspace of the parameter space, for which there exist two different solutions going through the inner sonic point located very close to the black hole and the outer sonic point located far away from the central region (for discussion about low angular momentum flows see (Chakrabarti & Titarchuk, 1995) and references therein). The dependence of the shock location, which connects the two branches of solution, and shock strength on the parameters, especially on the angular momentum λ, energy ϵ and adiabatic index γ of the flow, leads to the possible movement and oscillations of the shock front with changing properties of the inflowing matter. Because at the shock front the dissipation processes

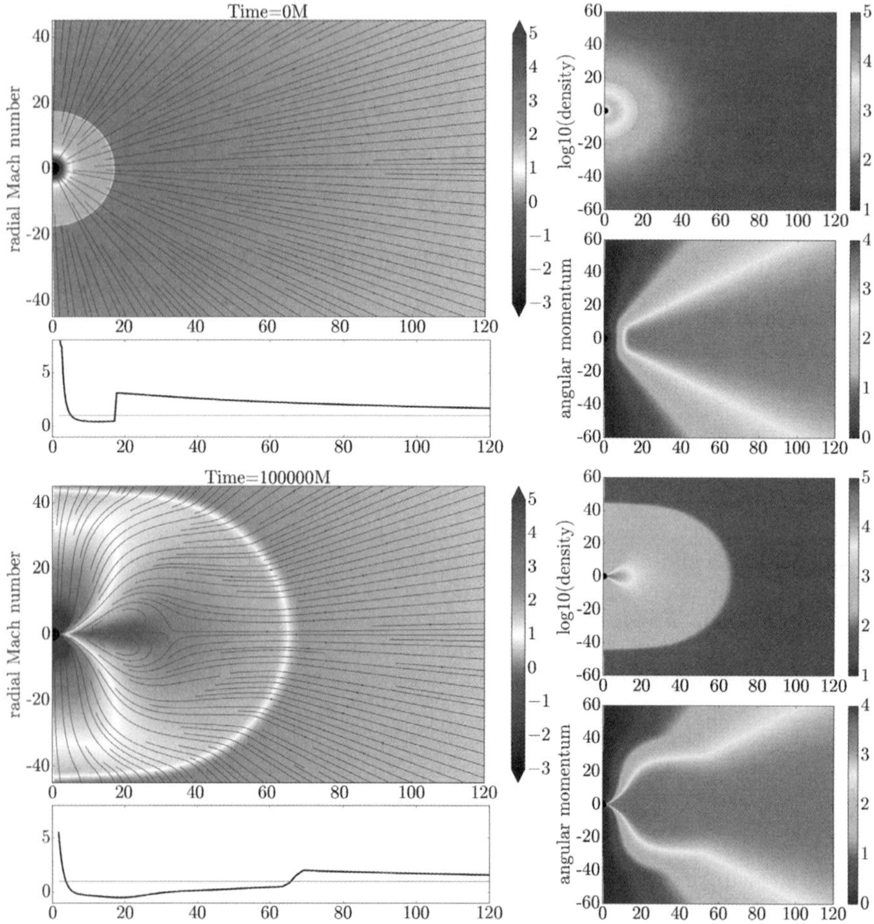

Figure 1. Top set of four panels: zoom of the initial conditions for simulation ($\lambda_{\max} = 3.54M, \epsilon = 0.0025, \gamma = 4/3$, $r_s^{\rm in} = 14M$) – radial Mach number $\mathcal{M}$ and its equatorial profile on the left, and density ρ and angular momentum λ on the right. The x and y axes are measured in M, as well as λ, ρ is given in arbitrary units and $\mathcal{M}$ is dimensionless. Bottom set of four panels: the same for $t = 100\,000M$.

release energy in form of radiation, the movement of the shock front can have observational effects, such as QPOs with changing frequency. The propagating oscillatory shock (POS) has been used to explain the rise and decline of the low frequency QPO in several microquasars during their outburst (Nandi *et al.* 2012), however, there the behaviour of the shock (the shock strength and the shock front velocity) was inferred as a fit to the observational data. As such, this phenomenological model lacks the explanation of the shock appearance and motion.

We are interested in the appearance of the shock front and its movement in low angular momentum flows. In (Suková & Janiuk, 2015) we showed that the steady 1D solution of the slowly rotating flow is achieved in the corresponding 1D pseudo-newtonian hydrodynamical simulations with the code ZEUS. We found the shock front stable for lower angular momentum and oscillating around the theoretical position of the shock for higher values of angular momentum. We also observed the recurring formation and disappearance (very fast accretion) of the shock front when the angular momentum of the incoming matter was changed periodically.

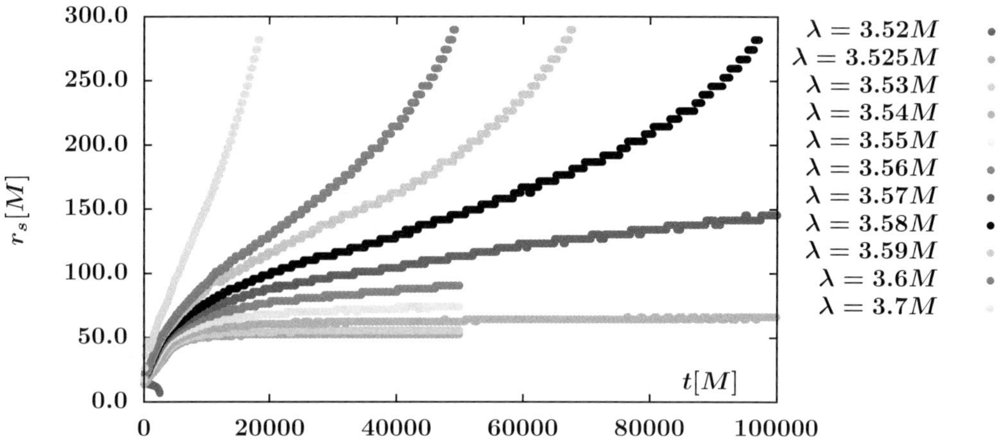

Figure 2. Time dependence of the location of the shock front in the equatorial plane for different λ. Other parameters are kept constant ($\epsilon = 0.0025, \gamma = 4/3$).

2. Numerical setup and initial conditions

Here, we study the behaviour of the shock in the framework of general relativity. For that purpose we use the code HARM, which has been equipped by 3D capabilities and MPI interface (Tchekhovskoy, Narayan & McKinney, 2011; and references therein). Our goal is to find out, if the 1D results can be generalised to realistic flows on the black holes, in particular, if there exist a stationary solution with standing shock or a solution with oscillating shock in higher dimension. Because we will use the cylindrically symmetric initial conditions, it is sufficient to compute 2D slice with constant ϕ. We use logarithmic grid in r with $r_{\rm out} = 20\,000M$ and the resolution is 256 x 128.

In (Suková, Charzyński & Janiuk, 2016) we showed, that if we start with the Bondi initial conditions for density ρ, pressure p and inward radial velocity u^r, which is modified by additional slow rotation given by a nonzero ϕ component to the four-velocity of the gas in the outer region according to $u^\phi = \frac{\lambda}{r^2}\sin^2\theta$, then the solution converges to the outer branch of solution, the so called "bondi-like" accretion. There, the sonic point lies far away from the center and the properties of the flow are similar to the Bondi solution, modulated by the centrifugal barrier close to the horizon.

Therefore, another type of initial conditions has to be used, if the shock front† properties should be studied. We start with the simplest generalization of our 1D pseudo-newtonian model and we took the profile of density, pressure, and inward velocity given by the solution with $\lambda_{\rm max}, \epsilon, \gamma$ for each streamline with constant angle θ. However, we use the initial location of the shock $r_{\rm s}^{\rm in}$ (i.e. the location where we jump from the inner branch to the outer branch of solution) as a free parameter, in other words we initially glue the branches of the solution at a chosen radius, not necessarily the one given by the 1D solution. Moreover, we prescribe the angular momentum $\lambda = -u_\phi/u_t$ according to the relation $\lambda = \lambda_{\rm max}\sin^2\theta$, where $\theta = \pi/2$ in the equatorial plane, hence we decrease the rotation towards the axis to avoid diverging velocities close to the axis. It is obvious, that our initial state cannot correspond to the stationary state. We expect a short transient period at the beginning of the computation, during which the flow will retain the corresponding shape.

† The shock front is the surface, where the radial Mach number $\mathcal{M} = u^r/c_s$ jumps from supersonic ($\mathcal{M} > 1$) to subsonic ($\mathcal{M} < 1$) regime). Here u^r is the radial component of the four velocity in Boyer-Lindquist coordinates and $c_s = \gamma p/(\rho + \gamma p/(\gamma - 1))$ is the local sound speed.

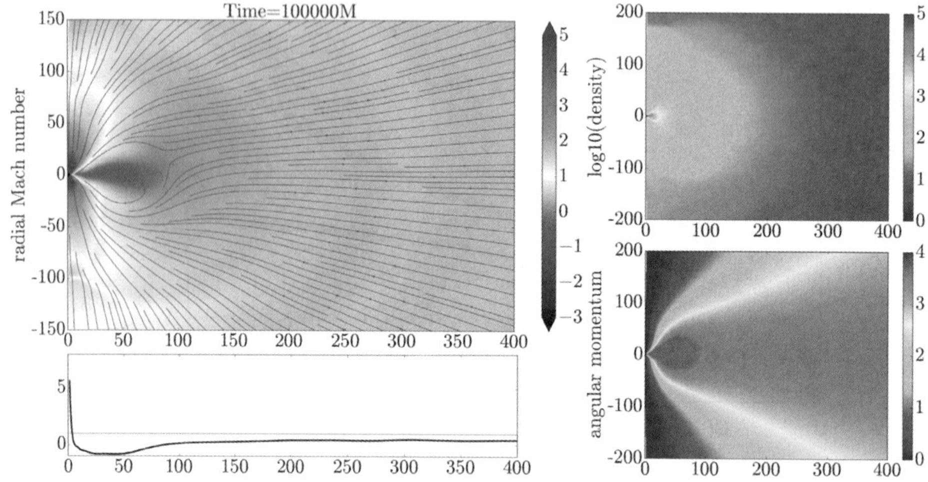

Figure 3. The same as in Fig. 1 for $\lambda = 3.6M$, $r_s^{in} = 18M$ and $t = 100\,000M$.

3. Results

For our study we have chosen the energy $\epsilon = 0.0025$ and the polytropic index $\gamma = 4/3$, while we are varying λ. Our 1D model predicts for this case the existence of shock solution for $\lambda \in (3.52 - 3.7)M$ with the shock locations $r_s \in (14 - 140)M$ and the outer sonic point location around 284M. The recent 2D simulations show, that for $\lambda_{max} = 3.52M, r_s^{in} = 14M$ the shock front is accreted onto the black hole. However, already for $\lambda_{max} = 3.525M, r_s^{in} = 18M$ the shock bubble grows and converges to stationary state with $r_s = 53M$, hence the position of the shock is much farther. The example of initial state and final converged state with $\lambda_{max} = 3.54M, r_s^{in} = 18M$ is given in Fig. 1.

With increasing angular momentum, the final shock location moves up and for $\lambda_{max} = 3.56 - 3.57M, r_s^{in} = 18M$ we do not obtain the converged state during our simulation time ($t_f = 50\,000M$ or $t_f = 100\,000M$ – see Fig. 2). For $\lambda_{max} \geqslant 3.58M$, the shock bubble grows so much, that it meets the outer sonic point and the shock front disappears. The example of such evolution is given in Fig. 3 for $\lambda_{max} = 3.6M, r_s^{in} = 18M$.

The presented set of simulations suggests, that the general relativistic 2D simulations put the shock front farther away from the black hole and so the interval of λ with stationary shock existence is smaller. We will continue our research with more detailed study of the shock behaviour depending on different initial geometries and on other important physical effects (spin of the black hole, magnetic field).

4. Acknowledgements

This work was supported in part by the grant DEC-2012/05/E/ST9/03914 from the Polish National Science Center and from the Czech Science Foundation Center of Excellence titled "Albert Einstein Center for Gravitation and Astrophysics" (No. 14-37086G).

References

Chakrabarti, S. & Titarchuk, L. G., 1995, *ApJ*, 455, 623
Nandi, A., Debnath, D., Mandal, S., & Chakrabarti, S. K., 2012, *A&A*, 542, A56
Suková, P., Charzyński, S., & Janiuk, A. 2016, *preprint arXiv:1602.07215*
Suková, P. & Janiuk, A. 2015, *Mon. Not. R. Astron. Soc.*, 447, 1565
Tchekhovskoy, A., Narayan, R., & McKinney, J. C. 2011, *Mon. Not. R. Astron. Soc.*, 418, L79

New Frontiers in Black Hole Astrophysics
Proceedings IAU Symposium No. 324, 2016
Andreja Gomboc, ed.

doi:10.1017/S1743921316012576

Stellar progenitors of black holes: insights from optical and infrared observations

I. F. Mirabel[1,2]

[1]Institute of Astronomy and Space Physics. CONICET - Universidad de Buenos Aires, Ciudad Universitaria, Av. Cantilo S/N, 1428 Buenos Aires - Argentina
email: mirabel@iafe.uba.ar
[2]Laboratoire AIM-Paris-Saclay, CEA/DSM/Irfu CNRS, CEA-Saclay, pt courrier 131, 91191 Gif-sur-Yvette, France
email: felix.mirabel@cea.fr

Abstract. Here are reviewed the insights from observations at optical and infrared wavelengths for low mass limits above which stars do not seem to end as luminous supernovae. These insights are: (1) the absence in archived images of nearby galaxies of stellar progenitors of core-collapse supernovae above 16-18 $M_{\odot}$, (2) the identification of luminous-massive stars that quietly disappear without optically bright supernovae, (3) the absence in the nebular spectra of supernovae of type II-P of the nucleosynthetic products expected from progenitors above 20 $M_{\odot}$, (4) the absence in color magnitude diagrams of stars in the environment of historic core-collapse supernovae of stars with $\geqslant$20 $M_{\odot}$. From the results in these different areas of observational astrophysics, and the recently confirmed dependence of black hole formation on metallicity and redshift of progenitors, it is concluded that a large fraction of massive stellar binaries in the universe end as binary black holes.

Keywords. black hole physics, gravitational waves, X-rays: binaries, supernovae: general

1. Introduction

The question on the range of masses of stars that may collapse to form black holes (BHs) without exploding as energetic supernovae (SNe) is of topical interest for the incipient Gravitational-Wave Astrophysics. Strong natal SN kicks disrupt massive binary systems and eject black holes from their site of birth, reducing the numbers of binary black holes (BBHs), which may be prolific sources of gravitational waves.

Most models on the evolution of massive stars predict that above ~16-18 $M_{\odot}$ a large fraction of stars end as BHs by direct (Sukhbold *et al.* 2016 and reference therein), or failed supernova collapse (Lovergrove & Woosley 2013). In the following are reviewed the current results from observations at optical and infrared wavelengths that are used to contrast these theoretical predictions.

2. Missing progenitors of $\geqslant$18 $M_{\odot}$ among core-collapse supernovae

Direct identifications of the progenitor stars of core-collapsed SNe discovered in nearby galaxies are being carried out by means of high resolution archival images from space and ground based telescopes. So far, among the 45 supernovae with either detected progenitors or upper limits, there has been a remarkable deficit of stars above an apparent limit of log $L/L_{\odot}$~5.1 dex (Smartt 2015). Type II-P SNe are the explosions of red supergiant stars (RSGs) which are believed to have masses in the range of 10 $M_{\odot}$ to 40 $M_{\odot}$. Figure 1 illustrates the absence of luminous RSGs progenitors with estimated masses $\geqslant$ 16-18 $M_{\odot}$, from which it has been suggested (Smartt 2015) that the bulk of stars of $\geqslant$ 16-18 $M_{\odot}$ end as BHs with no visible supernova. Furthermore, with the lack of detected

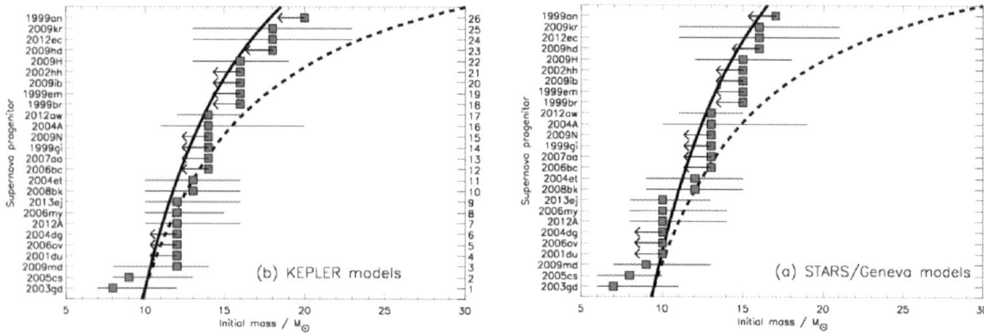

Figure 1. Mass of stellar progenitors of core collapse SNe in the context of the STARS, Geneva, and KEPLER models of stellar evolution. The detections are marked with error bars, the limits with arrows, and the lines extend from the minimum to the maximum masses from cumulative Salpeter IMFs. Allowing the mass function to vary up to 30 $M_\odot$, the mass distribution need to be truncated at masses of ~16.5 $M_\odot$ and ~18$M_\odot$, respectively. From Smartt (2015).

high mass progenitors in type IIb SNe, it has been proposed that "the missing high mass problem" has become relevant to all type II SNe (Smartt, 2015).

3. Disappearance of massive stars without bright supernovae

An alternative approach is the search for massive stars that quietly disappear without bright supernovae by means of repeated observations of nearby galaxies (Kochanek *et al.* 2008). Following this strategy, from a survey of 27 galaxies with the Large Binocular Telescope (Gerke *et al.* 2015), and a systematic analysis of archival Hubble Space Telescope images of 15 galaxies (Reynolds *et al.* 2015), were found respectively, one candidate now named N6946-BH1 with an estimated mass of ~25 $M_\odot$, likely associated with a failed SN (Figure 2), and one candidate of 25-30 $M_\odot$ named NGC3021-1 that underwent an optically dark core-collapse (Figure 3). The optical disappearance of the ~25 $M_\odot$ RSG first identified in the survey with the Large Binocular Telescope, now named N6946-BH1, has recently been confirmed using new and archival Hubble Space Telescope imaging (Adams *et al.* 2016).

The light curves of both candidates showed long time lasting faint transient displays before disappearing, as predicted in the model of very faint SNe from neutrino mass loss ([Lovergrove & Woosley (2013)). From N6946-BH1 remains faint near infrared emission, and the mid-infrared emission measured with Spitzer slowly decreased to the lowest levels since the first measurements. The authors propose that this event is due to the ejection of the loose envelope of the red RSG that collapsed as a failed SN with a late-time emission due to fall-back accretion onto a newly formed black hole. Given the detection of 6 successful SNe in the sample of 27 monitored galaxies and one likely failed SN, the implied fraction of core-collapses that result in failed SNe is f~0.14±0.20 at 90% confidence, and if the current candidate is ultimately rejected, there is a 90% confidence upper limit on failed SN fraction of f⩽0.35.

4. Indirect strategies for the mass limits of ccSN stellar progenitors

An indirect strategy was applied (Williams *et al.* 2014) using resolved stellar photometry from archival Hubble Space Telescope imaging to generate color magnitude diagrams of stars within 50 pc of the location of 17 historic core-collapse supernovae that took place in galaxies within a distance of 8 Mpc. Fitting the color magnitude distributions

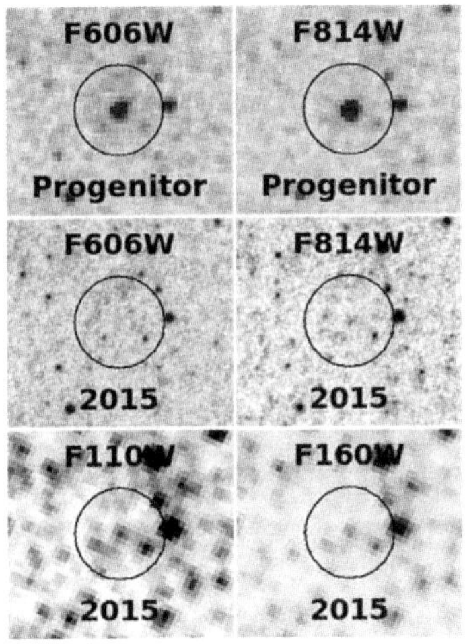

Figure 2. HST images of the region surrounding N6946-BH1. The top row shows the progenitor images with the filters WFPC2 F606W (588.7 nm) on the left, and F814 (802.4 nm) on the right, respectively. The middle row shows the corresponding 2015 WFC3 images. The bottom row shows the images with the WFC3/IR F110W (1153.4 nm) filter on the left, and F160W (1536.9 nm) on the right. The circles have a radius of 1 arcsec. The progenitor has dramatically faded in the optical but in 2015 there is still faint near-IR emission. From Adams *et al.* 2016.

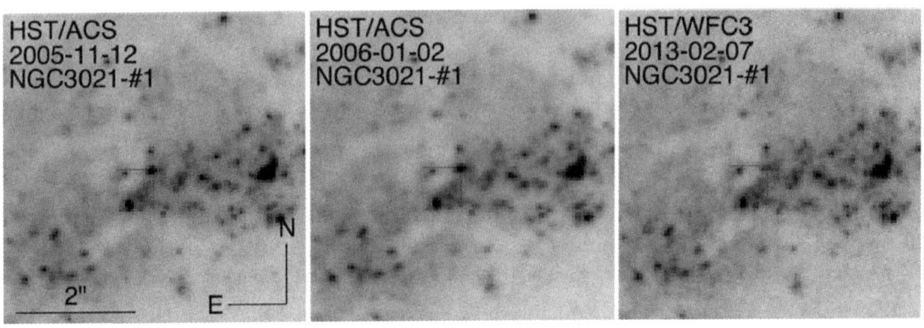

Figure 3. Selected HST F814W image cuts centered on the position of a candidate in the galaxy NGC3021 with location indicated with tick marks. From Reynolds *et al.* (2015)

with stellar evolution models to determine the best-fit age distribution of the young population, the authors conclude that so far there is not a single high-precision measurement of a SN progenitor of $\geqslant$20 $M_\odot$.

Furthermore, the nucleosynthetic products in the nebular spectra of supernovae can also provide constraints on the mass of the exploding star. In particular, it has been realized (Jerkstrand *et al.* 2014) that the observed evolution of the cooling lines of oxygen are difficult to reconcile with the expected nucleosynthesis products from progenitors of Type II-P SN with $\geqslant$20 $M_\odot$.

Caveats: It has been argued that the optical and infrared observational approaches described above, and the conclusions inferred from them, may have several biases, such as those possibly due to the influence of circumstellar dust, magnitude variations of RSGs, luminosity-mass analysis, sample selection, and limited numbers statistics. Although those caveats could apply to particular cases, it is unlikely that such biases could account for the remarkable general stellar mass limit of 18-20 $M_\odot$ for the progenitors of Type II SNe found by the different observational methods at optical and infrared wavelengths presented here, and at radio wavelengths presented by Mirabel (2017) at another contribution of these proceedings. However, there may be mass islands of progenitor explodability as proposed by current models (e.g. Sukhbold *et al.* 2016) that can produce successful supernovae for different ranges of progenitor masses.

5. Metallicity and redshift dependence of black hole formation

The metallicity (Z) dependence of BH formation has been observationally confirmed using a large set of extended off center x-ray luminosities (mostly coming from accreting BHs in high mass x-ray binaries) in nearby galaxies with known Zs and star formation rates (SFRs). For a given SFR the off-nuclear x-ray luminosities are typically ten times in low-Z galaxies ($\leqslant$20% solar) than in $Z_\odot$ galaxies (Douna *et al.* 2015). Besides, the X-ray luminosity due to accreting stellar BHs in normal galaxies show out to z=2.5 a redshift (z) an evolution given by $L_{2-10keV}$/(HMXB)/SFR α (1 + z) due to declining Z with increasing z (Lehmer *et al.* 2016).

6. Conclusion

A large fraction of stars above 18-20 $M_\odot$ may collapse by implosion and most likely end as BHs. This is inferred from searchers for SN progenitors in archived images at optical and infrared wavelengths, from massive stars that quietly disappear in the dark, from the largest stellar masses in the young stellar populations that host historic SN remnants, and from the absence of nucleosynthetic products of very massive stars in the nebular spectra of core-collapse SNe.

The detection of an x-ray source at the position of a disappearing luminous star as proposed by Adams *et al.* (2016) may solve the problem of the missing high-mass SN progenitors, and open the possibility for the observation of BH formation in real time. Because most massive stars belong to multiple stellar systems, black holes recently formed in situ may accrete from nearby stars and produce x-rays and synchrotron radio waves, as BHs in high mass x-ray binaries. To witness the birth and early infancy of BHs, the sensitivities and angular resolutions of the James Webb Space Telescope for observations at 10-20 microns, Athena for observations at x-rays, and SKA for the observations at radio wavelengths may be required.

From the metallicity dependence of BH formation and dynamical evolution of massive stellar systems, it is inferred to first order approximation that ~30% of stellar binaries with Z$\leqslant$0.1 $Z_\odot$ and primaries above 40 $M_\odot$ end as binary black holes.

References

Adams, S. M., Kochanek, C. S., Gerke, J. R., *et al.* 2016, *arXiv:1609.01283v1*
Douna, V. M., Pellizza, L. J., Mirabel, I. F., *et al.* 2015, *A&A*, 579, 44
Gerke, J. R., Kochanek, C. S., & Stanek, K. Z. 2015, *MNRAS*, 450, 3289
Jerkstrand, A., Smartt, S. J., Fraser, M., *et al.* 2014, *MNRAS*, 439, 3694
Kochanek, C. S., Beacom, J. F., Kistler, M. D., *et al.* 2008, *ApJ*, 684, 1336
Lehmer, B. D. Basu-Zych, R., Mineo, S., *et al.* 2016, *ApJ*, in press
Lovergrove, E. & Woosley, S. E. 2013, *ApJ*, 769, 109
Mirabel, I. F. 2017, *New Frontiers in Black Hole Astrophysics. Proceedings IAU Symposium No 324, Andreja Gomboc & Carole Mundell, eds.*, in press
Reynolds, T. M., Fraser, M., & Gilmore, M. 2015, *MNRAS*, 453, 2885
Smartt, S. J. 2015, *PASA*, 32, 16
Sukhbold, T., Ertl, T. Woosley, S. E., *et al.* 2016, *ApJ*, 821, 38
Williams, B. F., Peterson, S., Murphy, J., *et al.* 2014, *ApJ*, 791, 105

New Frontiers in Black Hole Astrophysics
Proceedings IAU Symposium No. 324, 2016
Andreja Gomboc, ed.

doi:10.1017/S1743921316012850

Investigating ultraluminous X-ray sources through their multiwavelength variability and broadband spectra

Luca Zampieri[1], Elena Ambrosi[2] and Amid Nayerhoda[2,3]

[1]INAF-Astronomical Observatory of Padova
Vicolo dell'Osservatorio 5, 35122 Padova, Italy
email: luca.zampieri@oapd.inaf.it
[2]Department of Physics and Astronomy, University of Padova,
Vicolo Osservatorio 3, 35122 Padova, Italy
[3]Institute of Nuclear Physics PAN,
ul. Radzikowskiego 152, 31-342 Kraków, Poland

Abstract. The population of ultraluminous X-ray sources encompasses accreting compact objects with significantly different masses, from black holes of intermediate mass, to black holes of stellar origin, to neutron stars. Investigating these sources will help us answering crucial questions on the distribution of black hole masses and on mass accretion above the Eddington limit in the local Universe, that have potential implications in other astrophysical areas. In order to perform a detailed investigation of ultraluminous X-ray sources, an accurate modeling of their evolution and multiwavelength emission properties is needed. We report some preliminary results of the activities that we are carrying out at present in this area.

Keywords. accretion, accretion disks, methods: numerical, X-rays: binaries

---

1. Introduction

Ultra-Luminous X-ray sources (ULXs) are point-like off-nuclear X-ray sources in nearby galaxies, intrinsically powerful but observationally rather faint. Their luminosity exceeds (although not necessarily all the time) the Eddington limit for spherical accretion onto a $\sim 10\,M_\odot$ black hole ($L > 10^{39}$ erg s^{-1}). Hundreds of sources have been discovered and reported in various surveys/catalogues to date. The observational evidence points towards the identification of the majority of ULXs ($\sim 60-70\%$) with accreting binaries (see e.g. Feng & Soria 2011 for a review).

Two dynamical mass measurements of the compact objects in ULXs have been recently performed through dedicated spectral and/or photometric monitorings of their optical counterparts. M 101 ULX-1 has a black hole (BH) with mass $M_{BH} > 5\,M_\odot$ (Liu *et al.* 2013), while the mass of the compact object in NGC 7793 P13 is $< 15\,M_\odot$ (Motch *et al.* 2014). The latter system was very recently discovered to show X-ray pulsations and then hosts a neutron star (Fuerst *et al.* 2016, Israel *et al.* 2016a). Other two pulsating neutron stars (NSs) have been identified among ULXs (M 82 X-1, Bachetti *et al.* 2014; NGC 5907 ULX-1, Israel *et al.* 2016b), suggesting that pulsar ULXs may not be uncommon.

On the other hand, a few extremely luminous or hyperluminous sources, most notably HLX-1 in ESO 243-49 (with a peak luminosity of 10^{42} erg s^{-1}; Farrell *et al.* 2009), have been identified as very good candidates intermediate mass BHs. For HLX-1, BH mass estimates based on X-ray spectroscopy and radio jet emission are in the ballpark of $10^4-10^5\,M_\odot$ (Servillat *et al.* 2011, Webb *et al.* 2012).

These findings indicate that ULXs may contain compact objects of different types and that they may have a key role in exploring the origin and unknown distribution of BH masses in the local Universe. They could help us understanding what are the heaviest BHs that can form through direct collapse of massive stars, whether the BHs in ULXs may have something to do with the massive BHs recently discovered with the Laser Interferometer Gravitational-Wave Observatory, whether we can find further evidences of the existence of intermediate mass BHs, or what is the maximum luminosity and spin-up rate of magnetized NSs accreting above the Eddington limit.

2. ULX binaries and accretion above the Eddington limit

After the discovery of the gravitational wave event GW150914 with the Laser Interferometer Gravitational-Wave Observatory (Abbott *et al.* 2016), we know that massive ($\sim 30\,M_\odot$) BHs exist. As ULXs appear to host compact objects with significantly different masses (and origin), a natural question is whether BHs in a similar mass range could reside also in ULXs. A possible indirect evidence in favour of the existence of massive ($30-80\,M_\odot$) BHs in ULXs comes from their association with low metallicity environments (e.g. Mapelli *et al.* 2010), where they may form through the direct collapse of massive progenitor stars that have retained a significant fraction of their initial mass (Mapelli *et al.* 2009, Zampieri & Roberts 2009, Belczynski *et al.* 2010).

To produce an active ULX phase, massive BHs and more conventional stellar-mass BHs have to cross a Roche-lobe-fed evolutionary phase. We investigated the evolution of BH binaries with massive companions in clusters including metallicity-dependent stellar evolution and found that dynamical interactions play a significant role (Mapelli & Zampieri 2014, and references therein). As a consequence, both stellar-mass and massive BHs of stellar origin can power a non-negligible number of Roche-lobe-fed ULXs.

Besides the nature of their compact objects, the other challenge in understanding ULXs is the character of their accretion flow, which is able to produce luminosities 10-100 times larger than the Eddington limit. Even accounting for a reasonable beaming factor ($\gtrsim 10\%$), the accretion rate needed to sustain these luminosities is larger than 10^{19} g s^{-1}. The question is then what is the accretion physics for BHs or NSs emitting/accreting at this pace. In this respect, ULXs represent laboratories for extreme accretion environments (where photon advection/trapping becomes important), that may be relevant for the first generation of Quasars at very high redshift.

Supercritical accretion onto BHs has been studied by means of 2D magneto-hydro simulations. They show the formation of an advection-dominated disc and an outflow region, with powerful clumpy winds driven by radiation pressure (e.g. Takeuchi *et al.* 2013). This structure can explain some basic facts concerning the X-ray spectral components and the short-term variability at high energies observed in some ULXs (Middleton *et al.* 2015). This interpretation is supported also by the detection of emission lines and blueshifted (0.2c) absorption lines from highly ionized Fe, O and Ne in high-resolution X-ray spectra of NGC 1313 X-1 and NGC 5408 X-1, although the cumulative significance is only slightly above 5σ (Pinto *et al.* 2016).

Supercritical accretion onto NSs is less studied. In order to account for the observed luminosity and long-term spin-up of the pulsar, the accretion rate and the magnetic field B at the base of the accretion column have to be sufficiently high. However, the latter should not be too high to trigger the propeller stage and prevent accretion at the magnetospheric radius. These constraints give values of B in the range $\sim 10^{12}-10^{14}$ G and suggest that the magnetic field may have multipolar components (Bachetti *et al.* 2014, Israel *et al.* 2016a,b). If the donor is more massive than the NS, a Roche-lobe-fed

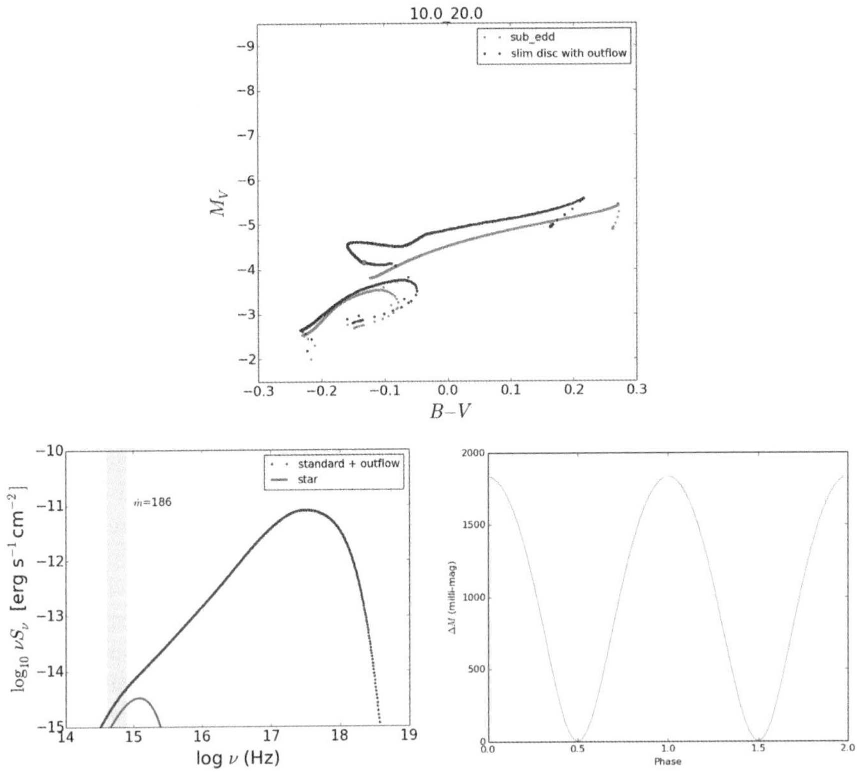

Figure 1. *Top*: Evolution on the colour-magnitude diagram of a ULX binary system with a BH mass $M_{BH} = 20\,M_\odot$ and a donor mass $M_* = 10\,M_\odot$ (at zero age main sequence). The lower track (*gray* dots) represents the evolution calculated assuming standard sub-Eddington accretion (Patruno & Zampieri 2008). The upper track (*blue* dots) represents the new evolution computed for super-Eddington accretion. *Bottom-left*: Optical-through-X-ray spectrum of the system at the phase marked with a bigger (*red*) dot on the evolutionary track for super-Eddington accretion ($\dot{m} = 186$ and the actual mass of the star is $4.4\,M_\odot$). The upper (*red*) points represent the spectrum of the self-irradiated outer disc plus the outflow and the innermost slim disc, while the lower (*green*) line is the spectrum of the X-ray heated donor. The vertical (*light blue*) strip marks the optical band. *Bottom-right*: Orbital modulation of the optical light curve caused by the irradiated donor for an inclination angle of 30^0.

system is unstable. However, it may still be relatively long lived, if (thermal timescale) mass transfer is stabilized by stellar and disc winds/outflows (Fragos *et al.* 2015).

3. A comprehensive multi-wavelength variability emission model

Our goal is to investigate the emission properties at super-Eddington rates and use them to constrain the physical properties of ULXs. Thus, we developed a model to calculate the optical-through-X-ray emission of ULX binaries during their evolution, accounting for super-Eddington accretion (Ambrosi & Zampieri, in preparation). The starting point is the model of Patruno & Zampieri (2008), that includes X-ray irradiation. For accretion rates below the Eddington limit, the treatment of the accretion flow is that of Patruno & Zampieri (2008). However, for accretion rates above Eddington, a bimodal structure is assumed: an inner flow with non-standard (slim) disc geometry and temperature profile (vertical scale height comparable to the radius and $T \propto r^{-1/2}$), and an outer flow with standard structure. The transition to the standard disc occurs where the inward advected heat is equal to the viscously dissipated heat. Following Poutanen *et al.* (2007), in our last implementation an optically thick outflow is also included in the treatment when $\dot{m}$ (the accretion rate in units of the Eddington rate) exceeds unity. The portion of the inner disc that illuminates the outer regions, where the UV-optical-IR radiation is produced, changes significantly from that considered in previous works.

Our model incorporates also a calculation of the expected orbital modulation of the optical light curve caused by the emission of the geometrically distorted and X-ray irradiated companion (Nayerhoda & Zampieri, in preparation). The donor is assumed to fill or slightly underfill its Roche lobe. This calculation is being generalized at present to include the contribution of the optical emission produced by the irradiated outer disc.

The implementation of the model is in progress. In Figure 1 we show some preliminary results of the calculation of the expected emission properties for a ULX binary system with a BH mass $M_{BH} = 20\,M_{\odot}$ and a donor mass $M_{*} = 10\,M_{\odot}$ (at zero age main sequence). A snapshot of the optical-through-X-ray spectrum and of the optical light curve during a post-main sequence super-Eddington phase is also shown. Owing to the fact that the accretion rate is above the Eddington limit, in the post-main sequence phase the evolutionary track is brigther and bluer than that computed assuming Eddington-limited accretion.

The next step will be to test our model against ULXs with measured or constrained compact objects masses, as NGC 7793 P13, that contains a neutron star, is accreting above Eddington, and has detailed optical light curve and broadband spectral measurements available. To this end, we plan to inlude the emission from the accretion column of a magnetized neutron star to determine the X-ray continuum spectrum. The model will then be used for calculating the emission of self-consistent ULX binary system configurations computed following their evolution in clusters through N-body simulations (extending our previous work in this area; see Mapelli & Zampieri 2014).

We acknowledge financial contribution from the grant ASI-INAF I/037/12/0.

References

Abbott, B. P., Abbott, R., Abbott, T. D., *et al.* 2014, *PhRvL*, 116, article id. 061102
Bachetti, M., Harrison, F. A., Walton, D. J., *et al.* 2014, *Nature*, 514, 202
Belczynski, K., Bulik, T., Fryer, *et al.* 2010, *ApJ*, 714, 1217
Farrell, S. A., Webb, N. A., Barret, D., *et al.* 2009, *Nature*, 460, 73
Feng, H. & Soria, R. 2011, *New Astronomy Reviews*, 55, 166
Fragos, T., Linden, T., Kalogera, V., & Sklias, P. 2015, *ApJ*, 802, L5
Fuerst, F., Walton, D. J., Stern, D., *et al.* 2016, *ApJL*, in press (arXiv:1610.00258)
Israel, G. L., Papitto, A., Esposito, P., *et al.* 2016, *MNRAS Letters*, in press (arXiv:1609.06538)
Israel, G. L., Belfiore, A., Stella, L., *et al.* 2016, *ArXiv e-prints*, (arXiv:1609.07375)
Liu J.-F., Bregman, J. N., Bai, Y., *et al.* 2013, *Nature*, 503, 500
Mapelli, M., Colpi, M., & Zampieri, L. 2009, *MNRAS*, 395, L71
Mapelli, M., Ripamonti, E., Zampieri, L., Colpi, M., & Bressan, A. 2010, *MNRAS*, 408, 234
Mapelli, M. & Zampieri, L. 2014, *ApJ*, 794, article id. 7
Middleton, M. J., Heil, L., Pintore, F., *et al.* 2015, *MNRAS*, 447, 3243
Motch, C., Pakull, M. W., Soria, R., *et al.* 2014, *Nature*, 514, 198
Patruno, A. & Zampieri, L. 2008, *MNRAS*, 386, 543
Pinto, C., Middleton, M. J., & Fabian, A. C. 2016, *Nature*, 533, 64
Poutanen, J., Lipunova, G., Fabrika, S., *et al.* 2007, *MNRAS*, 377, 1187
Servillat, M., Farrell, S. A., Lin, D., *et al.* 1995, *ApJ*, 743, article id. 6
Takeuchi, S., Ohsuga, K., & Mineshige, S. 1995, *PASJ*, 65, article id. 88
Webb, N., Cseh, D., Lenc, E., *et al.* 2012, *Science*, 337, 554
Zampieri, L. & Roberts, T. P. 2009, *MNRAS*, 400, 677

New Frontiers in Black Hole Astrophysics
Proceedings IAU Symposium No. 324, 2016
Andreja Gomboc, ed.

doi:10.1017/S1743921317002083

Black Hole Binaries in Quiescence

Charles D. Bailyn

Department of Astronomy, Yale University,
PO Box 208101, New Haven CT, 06520-8101 USA
email: charles.bailyn@yale.edu
Yale-NUS College
16 College Avenue West, Singapore, 138527

Abstract. I discuss some of what is known and unknown about the behavior of black hole binary systems in the quiescent accretion state. Quiescence is important for several reasons: 1) the dominance of the companion star in optical and IR wavelengths allows the binary parameters to be robustly determined — as an example, we argue that the longer proposed distance to the X-ray source GRO J1655-40 is correct; 2) quiescence represents the limiting case of an extremely low accretion rate, in which both accretion and jets can be observed; 3) understanding the evolution and duration of the quiescent state is a key factor in determining the overall demographics of X-ray binaries, which has taken on a new importance in the era of gravitational wave astronomy.

Keywords. binaries: close, black hole physics, gravitational waves, X-rays: binaries

---

1. Introduction

Most black hole X-ray binaries are transients. They spend years to decades in a very low state, and then increase in luminosity over a few days by as much as eight orders of magnitude. The outbursts last for weeks to months before the source returns to quiescence. Quiescent black hole binaries (hereafter QBHBs), have optical/IR emission dominated by the companion star, which can be used to precisely determine the orbital parameters, including the mass of the black hole (see Kreidberg *et al.* 2012 for details and a discussion of potential errors). But there is also emission from the residual accretion flow and from jets, which plays an important role in understanding the changes in accretion state during the outburst. Finally, the inferred demographics of X-ray binaries, which have assumed a heightened importance in the dawning era of gravitational wave astronomy (Abbott *et al.* 2016), depend critically on the duration of quiescence.

Observational astrophysicists commonly employ a variety of metaphors in explaining the importance of their work. Accordingly I will organize the rest of this paper by means of three of the more common metaphors, whose literal manifestations are shown in Figure 1. I will take as an example the prototypical black hole X-ray transient A0620-00 (hereafter A0620), which underwent a huge outburst in 1975 (Elvis *et al.* 1975) reaching a peak flux of over 50 Crabs. and has been in quiescence ever since.

2. The Smoking Gun: Evidence for Black Holes in QBHBs

A "Smoking Gun" is an observation that confirms something that was already suspected to be true. In the case of QBHBs, the confirmed truth is that the accreting object in many X-ray binaries is a black hole.

The companion stars which dominate the optical/IR light in QBHBs have optical absorption spectra similar to those of ordinary main sequence or giant stars. By applying traditional binary star techniques to these stars, the orbital parameters of the

Figure 1. Three well-known astrophysical metaphors

system can be robustly obtained. In many cases, the mass of the compact accreting star can be shown to be greater than $3M_\odot$, the upper limit on the mass of a neutron star. Such objects are sometimes referred to as "dynamically confirmed black hole candidates".

The radial velocity curve of the companion can be used to determine the mass function $M_f = PK^3/(2\pi G)$, which is a strict lower limit on the mass of the compact object. The first mass function greater than $3M_\odot$ was obtained by McClintock & Remillard (1986) for A0620; since then better data have been obtained for A0620 (Nielsen *et al.* 2008) and over a dozen other systems have been observed (Corral-Santana *et al.* 2016). The radial velocity curves are beautifully sinusoidal, as expected from tidally circularized systems, and there is little doubt that they represent the true motion of the companion star.

To obtain a mass, rather than a mass function, for the compact object, information about the orbital inclination i must be obtained. This is generally done by studying the ellipsoidal variations of the companion star, caused by the tidally distorted shape of the companion, which results in changes in the cross-sectional area of the star as a function of orbital phase. The interpretation of such observations requires accounting for the contribution of the accretion flow to the observed flux (e.g Cantrell *et al.* 2010 for A0620). When this is done, the derived binary parameters fix the geometric size of the companion quite precisely, and the observed colors and absorption spectrum fix the temperature. Therefore the intrinsic luminosity of the companion is well-determined, leading to robust determinations of the distance.

One relevant case is that of GRO J1655-40, which shows very clearly defined ellipsoidal variations (Greene, Bailyn & Orosz 2001, Beer & Podsiadlowski 2002). The well-determined physical size and temperature of the companion require a distance of $\gtrsim 3$ kpc. An alternative distance of $\lesssim 1.7$ kpc (Foellmi *et al.* 2006) would require the companion star to underfill its Roche lobe by a factor of 2, which would result in a nearly spherical star that cannot produce the observed ellipsoidal variations. The lower distance estimate is based on the assumption that the observed spectral type implies a star with mass and radius similar to those of isolated field stars. But the evolution and surface gravity of a star losing mass through Roche lobe overflow differ dramatically from those of isolated stars, and the observed masses and radii of QBHB companions differ from field stars of similar spectral type in many well-studied systems, including A0620 (Cantrell *et al.* 2010) and V4641 Sgr (MacDonald *et al.* 2014), as well as GRO J1655-40.

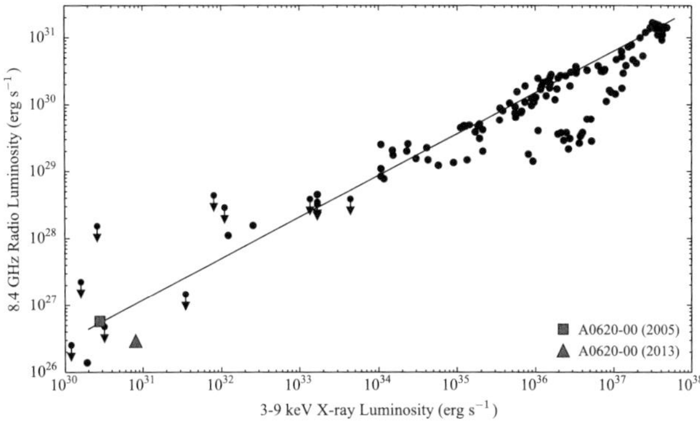

Figure 2. Radio/X-ray correlation for X-ray binaries — Dinçer *et al.* in preparation. The line represents the X-ray/radio correlation defined by Corbel *et al.* 2013.

3. The Rosetta Stone: The Disk-Jet Connection

A "Rosetta Stone" is an observation that connects two hitherto disparate phenomena. QBHBs may provide such a connection between accretion physics and jet outflows.

During the transition from quiescence to outburst and back, X-ray binaries change not only the amount of accretion, but the nature and state of the accretion flow. These so-called "state changes" have been the subject of considerable recent work. The general consensus (see Remillard & McClintock 2006 for a review) is that there is a low-hard state, with a power-law X-ray spectrum and radio emission from a jet, and a high-soft state, in which the X-ray luminosity is dominated by a thermal accretion disk, which quenches the jet. There are also a variety of intermediate states.

The presence of a jet in the low-hard state has led to a scenario in which the accretion in that state is in the form of a radiatively inefficient radial flow, which allows the formation of a jet (Fender *et al.* 2003). But it should be noted that the empirical support for such accretion-jet scenarios rests on the radio/X-ray correlation. The possibility that there might be a "universal" correlation in which $L_R \propto L_X^{0.6}$, and, with suitable correction for the mass of the accretor, extends to the AGN, generated considerable excitement (Corbel *et al.* 2013). Observations of A0620 (Gallo *et al.* 2006) held down the low luminosity end of this correlation nicely. However, further investigation revealed multiple correlations and branches, and a recent reobservation of A0620 has shown that the source has moved in the (L_X, L_R) plane in a direction perpendicular to the general correlation (see Figure 2). So the connection between accretion and jet physics is not yet as empirically well-founded as one might have hoped.

4. The Tip of the Iceberg: QBHB Demographics

The "Tip of the Iceberg" is an object or observation which implies the existence of a large unseen population. The current set of observed X-ray transients is certainly the tip of the iceberg of many currently unobserved galactic black hole binaries.

X-ray transients are generally discovered during their brief outbursts, which requires that the outburst occur at a time when it can be observed. Since X-ray astronomy is only ≈ 50 years old, objects that have not gone into outburst during that time cannot have been identified. Many objects have been observed in outburst only once, which implies the existence of objects that haven't gone into outburst at all. If such objects are common, the number of QBHBs in the galaxy will scale with the typical outburst recurrence time, a quantity for which there is currently no empirical constraint.

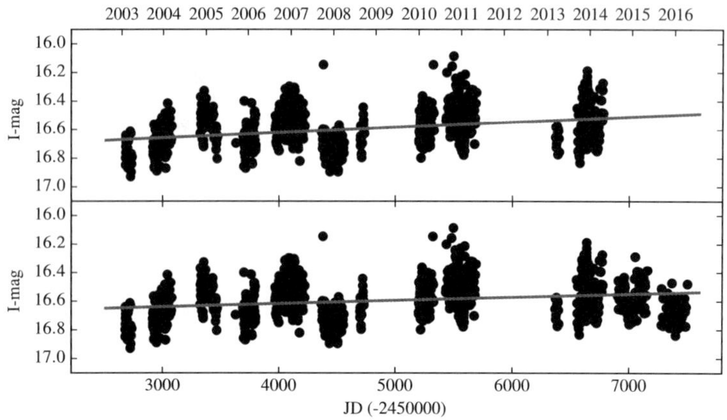

Figure 3. Photometry of A0620 from 2003 through 2014 and 2016 with an overall linear fit. Note the decrease in the slope when the data from 2015-16 are included.

Theoretical arguments can be made regarding the outburst recurrence time based on assumptions associated with the disk instabilities responsible for the outbursts (e.g. White & van Paradijs 1996). However the variable recurrence timescales displayed by objects that have multiple outbursts demonstrate that we still have much to learn about these instabilities. But we now have decades of monitoring of QBHBs, and we can begin to see how they evolve as the disk builds up toward its next outburst. Recently, a decade-long brightening trend has been found in Nova Muscae 1991 (Wu *et al.* 2016). It is also notable that the passive quiescent state in A0620 (Cantrell *et al.* 2010) has disappeared over the past ten years in favor of the brighter active quiescent state. We therefore examined all of the quiescent data from A0620 accumulated since 2003. As can be seen in Figure 3, the initial finding was very encouraging — however when data from 2015-16 were included, the inferred increase diminishes dramatically (although it is still significantly > 0). As in many situations when studying quasi-periodic phenomena, it is important to beware mistaking short term trends for periodic cycles or long-term evolution.

References

Abbot, B. P., *et al.* 2016, *ApJ*, 818, L22.
Beer, M. E. & Podsiadlowski, P. 2002, *MNRAS*, 331, 351.
Cantrell, A. G., *et al.* 2010, *ApJ*, 710, 1127.
Corbel, S., *et al.*] 2013, *MNRAS*, 428, 2500.
Corral-Santana, J. M., *et al.* 2016, *A&A*, 587, 61.
Elvis, M., Page, C. G., Pounds, K. A., Ricketts, M. J., & Turner, M. L. J. 1975, *Nature*, 257, 656.
Fender, R. P., Gallo, E., & Jonker, P. G. 2003, *MNRAS*, 343. L99.
Foellmi, C., Depagne, E., Dall, T. H., & Mirabel, F. 2006, *A&A*, 457, 249.
Gallo, E., *et al.* 2006, *MNRAS* 370, 1351.
Greene, J., Bailyn, C. D., & Orosz, J. A. 2001, *ApJ*, 554, 1290.
Kreidberg, L., Bailyn, C. D., Farr, W. M., & Kalogera, V. 2012, *ApJ*, 757, 36
MacDonald, R. K. D., *et al.* 2014, *ApJ*, 784, 2.
McClintock, J. E. & Remillard, R. A. 1986, *ApJ*, 308, 110.
Nielsen, J., Steeghs, D., & Vritilek, S. D. 2008, *MNRAS*, 384, 849.
Remillard, R. A. & McClintock, J. E. 2006, *ARA&A* 44. 49.
White, N. E. & van Paradijs, J. 1996, *ApJ*, 473, L25.
Wu, J., *et al.* 2016, *ApJ*, 825, 46.

New Frontiers in Black Hole Astrophysics
Proceedings IAU Symposium No. 324, 2016
Andreja Gomboc, ed.

doi:10.1017/S1743921316012710

The search for isolated BH candidates based on kinematics of pulsars - their former companions in disrupted binaries

E. Chmyreva, G. Beskin, V. Dyachenko and S. Karpov

Special Astrophysical Observatory, Nizhnij Arkhyz, Russia 369167
email: lisa.chmyreva@mail.ru

Abstract. We propose searching for isolated stellar-mass black hole (BH) candidates based on the fact that more than 50% of radio pulsars have originated in binary systems, where the other component could have evolved into a BH prior to the second supernova event of the system, which caused its disruption. We selected isolated, relatively young radio pulsars with known parallaxes and proper motions and traced their trajectories back to their presumed birth locations. These locations were then analyzed for possible BH candidates based on the available positional, photometric, and spectral data. We present the first results for 2 pulsars, J0139+5814 and J0922+0638. Seven BH candidates were selected for further analysis.

Keywords. (stars:) binaries (including multiple): close, (stars:) pulsars: general, (stars:) pulsars: individual (J0139+5814, J0922+0638)

1. Introduction

At least 50% of all stars are known to be members of binary and multiple systems (Batten, 1967; Duquennoy & Mayor, 1991; Halbwachs *et al.*, 2003) and, accordingly, many of the young pulsars and stellar-mass BH must have formed in high-mass binary systems. In BH+NS pairs, the BH, being more massive, would have formed first. About 40% of the pairs withstand this first supernova explosion (Bethe & Brown., 1998), whereas the second supernova event, in which a pulsar is born, generally disrupts the system. The kick that pulsars receive in the process would explain their high velocities (see, e.g., Hobbs *et al.*, 2005). Tracing the pulsars back to their birthplaces can therefore help localize the region where stellar-mass BH candidates may be located (see, e.g., Chmyreva *et al.*, 2010; Prokhorov & Popov, 2002).

2. Method

Using a sample of isolated radio pulsars with measured proper motions and parallaxes, relatively small spin-down ages, and no previous associations that fit our criteria, we simulated 100,000 trajectories for each, tracing them back to their probable birthplaces. We use the birthplaces of pulsars J0139+5814 and J0922+0638 (Fig. 1) obtained in our simulations to search for BH candidates and to illustrate the procedure. Theories predict that isolated BHs should have featureless flat spectra covering the entire electromagnetic range (see Beskin & Karpov, 2005, and references therein). We therefore used publicly available databases to select all non-optical sources that fall within the 3-sigma contours of the derived pulsar birth locations, cross-matching them within their positional error ellipses. We then searched for optical counterparts for these sources. Given the large position errors of x-ray and radio sources, more than one optical object would usually be found within the error ellipse of a given source. Photometric, spectral, and proper motion

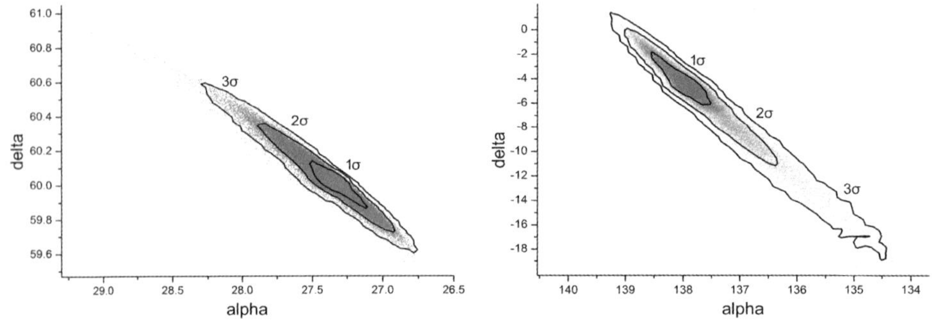

Figure 1. Birthplaces of pulsars J0139+5814 and J0922+0638. The grey dots are the endpoints of each simulated trajectory, traced back according to the spin-down age of the pulsar. The outlines show the 1-, 2-, and 3-sigma levels.

data were then used to classify the possible matches and reject the candidates that do not fit the criteria. The candidates that are promising were then selected for further investigation. A critical condition for these objects to be considered BH candidates would be their super fast variability - the subject of our future study.

3. Results and discussion

The birthplace of pulsar J0139+5814 did not yield any BH candidates. It contains 3 x-ray sources, each having several possible optical matches, but no spectra are available, and the area is contaminated by bright stars causing large photometric errors. We therefore can report no conclusive matches. For the second pulsar, J0922+0638, an automated cross-match yielded 19 matches, and a further 20 matches were found during visual inspection. We have individually inspected each of these sources and rejected the ones with stellar, galactic, or quasar spectra, visual galaxies, and extended sources. For objects with no available spectra, we plotted color-color diagrams. Since BH candidates are expected to have observational properties similar to BL Lacertae and DC-dwarfs, we kept the objects located in the region of the diagrams dominated by quasars and white dwarfs and well-described by a power law, discarding the regular stars and galaxies. Further data (spectra, variability) are needed to classify these objects with certainty. The 7 remaining sources are: 1RXS J091407.9-015949, 1RXS J091249.6-034034, 1RXS J090939.1-051030, 1RXS J091052.4-061129, 1RXS J090809.5-072652, SDSS J091605.73+000802.2, and WD 0913+005. Proper motions, confirming that the sources belong to our Galaxy, have been detected for all sources except 1RXS J091407.9-015949. These 7 sources will be studied further in our upcoming work.

References

Batten, A. H. 1967, *ARA&A*, 5, 25
Duquennoy, A. & Mayor, M. 1991, *A&A*, 248, 485
Halbwachs, J. L., Mayor, M., Udry, S., & Arenou, F. 2003, *A&A*, 397, 159
Bethe, H. A. & Brown, G. E. 1998, *ApJ*, 506, 780
Hobbs, G., Lorimer, D. R., Lyne, A. G., & Kramer, M. 2005, *MNRAS*, 360, 974
Prokhorov, M. E. & Popov, S. B. 2002, *AstL*, 28, 536
Chmyreva, E. G., Beskin, G. M., & Biryukov, A. V. 2010, *Astronomy Letters*, 36, 116
Beskin, G. M. & Karpov S. V. 2005, *A&A*, 440, 223

New Frontiers in Black Hole Astrophysics
Proceedings IAU Symposium No. 324, 2016
Andreja Gomboc, ed.

doi:10.1017/S1743921316012606

Prospects for the Discovery of Black Hole Binaries without Mass Accretion with Gaia

Norita Kawanaka[1,2], Masaki Yamaguchi[3], Tsvi Piran[4] and Tomasz Bulik[5]

[1]Department of Astronomy, Graduate School of Science, Kyoto University, Kitashirakawa-Oiwake cho, Sakyo-ku, Kyoto 606-8502, Japan
email: norita@kusastro.kyoto-u.ac.jp
[2]Department of Astronomy, Graduate School of Science, University of Tokyo, 7-3-1 Hongo, Bunkyo-ku, Tokyo 113-0033, Japan
[3]Institute of Astronomy, Graduate School of Science, University of Tokyo, 2-21-1 Osawa, Mitaka, Tokyo 181-0015, Japan
[4]Racah Institute for Physics, The Hebrew University, Jerusalem 91904, Israel
[5]Astronomical Observatory, Warsaw University, Ujazdowskie 4, 00-478 Warsaw, Poland

Abstract. We study the prospect for Gaia to detect black hole binary systems without the mass transfer from their companion stars. Gaia will be able to discover Galactic black holes without mass accretion by detecting the proper motion of their companion stars. We evaluate the number of such black hole binaries which have the orbital period short enough to be detected by Gaia during its operation, taking into account the binary evolution model.

Keywords. black hole physics, astrometry, binaries: general, stars: evolution

---

From the star formation rate ($\sim 3.5M_\odot$ yr^{-1}) and the initial mass function (IMF), one can estimate the number of stellar-mass black holes in our Galaxy as $\gtrsim 10^{8-9}$ (Brown & Bethe 1994; Timmes *et al.* 1996). Those whose existence is confirmed from the observations form X-ray binaries, the systems of a black hole and a companion star with mass accretion. However, only a few dozens of such black holes have been detected (Remillard & McClintock 2006), and so we can expect that in our Galaxy there are much more black holes without mass accretion from their companions. Then, how can we discover such black holes? One of the ways is to detect the motion of companion stars by astrometric observations, such as *Gaia*, *JASMINE*, etc. *Gaia* was launched at the end of 2013, and it can perform absolute astrometric measurements with a great precision on objects with $V \lesssim 20$ (Barstow *et al.* 2014). In this work we estimate the number of non-accreting black hole binaries using *Gaia* during its operation (~ 5 years).

The necessary conditions for a black hole binary without mass accretion to be detected by *Gaia* are the followings:

(i) The binary should contain a black hole and a main sequence star/giant star/white dwarf. In other words, the binary should be older than the lifetime of a primary star (i.e., the progenitor of a black hole) and younger than the lifetime of a secondary star.

(ii) The radius of a secondary star should be smaller than the Roche lobe radius of the binary so as to avoid the mass transfer.

(iii) The orbital period of the binary should be short enough to be detected by Gaia within its operation time (e.g. $\lesssim 2$ years).

We assume the initial mass function (IMF) as

$$\Psi(\bar{M}_1)d\bar{M}_1 \propto \begin{cases} \bar{M}_1^{-1.3}d\bar{M}_1, & 0.08M_\odot \leqslant \bar{M}_1 < 0.5M_\odot, \\ \bar{M}_1^{-2.2}d\bar{M}_1, & 0.5M_\odot \leqslant \bar{M}_1 < 1.0M_\odot, \\ \bar{M}_1^{-\alpha_{\rm IMF}}d\bar{M}_1, & 1.0M_\odot \leqslant \bar{M}_1 < 150M_\odot, \end{cases} \tag{0.1}$$

a flat mass ratio distribution ($\Phi(q) = 1$), and a log-flat orbital separation distribution ($\Gamma\left(\bar{A}\right) \propto 1/\bar{A}$). As for the stellar evolution processes, we take into account the common envelope phase that would be important when the mass ratio is less than ~ 0.5. During this phase, the orbital separation becomes larger because the orbital energy is used to expel the envelope of a primary star. We can estimate the number of black hole binaries without mass accretion that are detectable by Gaia, N, from the following equation:

$$\begin{aligned} N &= \frac{f_{\rm bin}}{1+f_{\rm bin}} \int_{M_{\rm min,BH}}^{M_{\rm max}} d\bar{M}_1 \Psi\left(\bar{M}_1\right) \int_{q_{\rm min}}^{q_{\rm max}} dq\, \Phi(q) \\ &\times \int_0^1 d\bar{e}\, \Xi\left(\bar{e}\right) \int_{t_{\rm L,1}}^{t_{\rm L,2}} dt \int_{\bar{A}_{\rm RL}}^{\bar{A}_{\rm 2yr}} d\bar{A}\, \Gamma\left(\bar{A}\right), \end{aligned} \quad (0.2)$$

where $f_{\rm bin} = 0.5$ is the binary fraction, $M_{\rm min,BH} = 20M_{\odot}$ is the minimum ZAMS mass for a star to become a black hole after its death, $t_{{\rm L},i}$ is the lifetime of a primary ($i = 1$) and a secondary ($i = 2$) star. Here, the lower limit of the orbital separation integral, $\bar{A}_{\rm RL}$, is determined from the condition (ii), and the upper limit, $\bar{A}_{\rm 2yr}$ is determined from the condition (iii).

According to our calculations, we found that the number of black hole binaries without mass accretion in our Galaxy is $\sim 1.5 \times 10^5$, i.e. Gaia may detect $\lesssim 0.1\%$ of Galactic stellar mass black holes ($\sim 10^{8-9}$) by astrometry. If we can discover a large number of Galactic black holes in this way, we will be able to construct their mass function with better statistics, which is important to understand the stellar evolution processes and the binary evolution processes.

References

Barstow, M. A., *et al.* 2014, arXiv:1407.6163
Brown, G, E. & Bethe, H. A. 1994, *ApJ*, 423, 659
Remillard, R. & McClintock, J. 2006, *ARA&A*, 44, 49
Timmes, F. X., Woosley, S. E., & Weaver, T. A. 1996, *ApJ*, 457, 834

New Frontiers in Black Hole Astrophysics
Proceedings IAU Symposium No. 324, 2016
Andreja Gomboc, ed.

doi:10.1017/S1743921317000011

Two optical emission components with different variability in V404 Cygni

Yutaro Tachibana, Taketoshi Yoshii and Nobuyuki Kawai

Department of Physics, Tokyo Institute of Technology, 2-12-1, Ohokayama, Tokyo, Japan
email: tachibana@hp.phys.titech.ac.jp

Abstract. V404 Cygni went into an outburst again on June 15, 2015 after 26 years of quietness. Soon after the notifications, we started intense optical observation campaign of this source. The spectral index between R_C and I_C-band was stable over the outburst, whereas that between g' and R_C-band varied violently. With the time domain analysis of the multi-color optical light curves, we successfully decomposed optical variations into three components: highly-variable component (HVC), little-variable component (LVC). The loci of the LVC in the color-color diagram is consistent with that of a single temperature blackbody radiation or a multi-color blackbody radiation from a standard accretion disk, while those of the HVC trace that of power-law spectra.

Keywords. X-rays: binaries – Stars: individual (V404 Cygni) – Techniques: photometric

1. Introduction

V404 Cygni, also known as GS 2023+338, is a black hole binary (BHB), originally discovered during the 1989 outburst by the *GINGA* satellite (Kitamoto *et al.* 1989, and Makino *et al.* 1989). In contrast to typical blackhole transients having a fast rise and exponential decay profile, V404 Cygni showed multiple flaring activity during one outburst (Tanaka and Shibazaki. 1996, Ferrigno *et al.* 2015). The nature of the variation is still under discussion (*e.g.*, Tanaka *et al.* 2016, Martí *et al.* 2016, Gandhi *et al.* 2016, and Kimura *et al.* 2016). Our aim is to derive information about the nature of the complex optical variation.

2. Analysis & Result

Light Curve and Flux-flux Plots

Fig.1 shows the optical (g', R_C, and I_C-band) light curves during MJD 57192–57194 (left side), and the flux-flux plot constructed by the g' and I_C-band photometric data (right side).

The amplitudes of optical flux variations reached ~3 mag on a timescale shorter than an hour, and the variations in the optical bands were perfectly correlated. The data in the flux-flux plot is well expressed by a broken- linear model (left side). Linear branches below and above a break, colored by red and blue respectively, imply that the existence of two variable components with distinct spectral indices. We named these components as little-variable component (LVC) and highly-variable component (HVC), respectively.

SED Modeling

The spectral indices of a variable components can be directly derived from slopes of straight lines in flux-flux plot. The color-color diagram for the decomposing components indices of LVC are consisted with that of black body or standard disk radiation, while those of the HVC trace out a power-law spectrum.

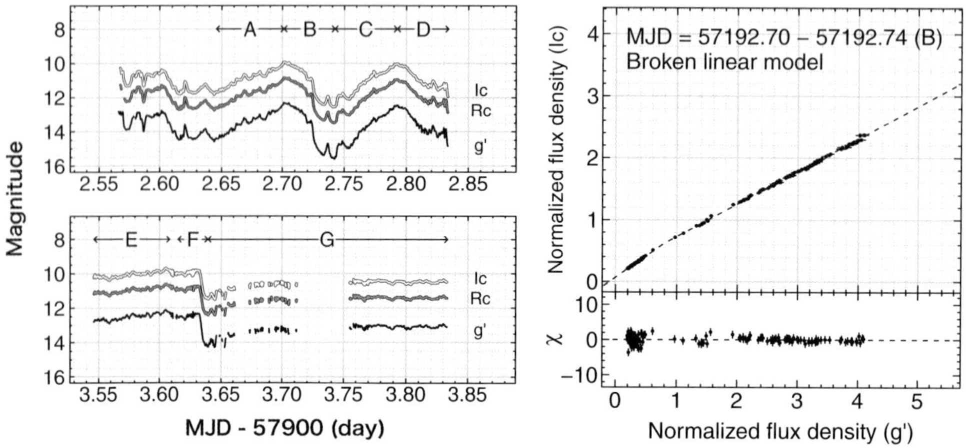

Figure 1. Left figure: g′, R_C, I_C and K_s light curves. **Right figure**: Flux-flux plots and fits for fluxes in g′ and I_C-band at "B" portion.

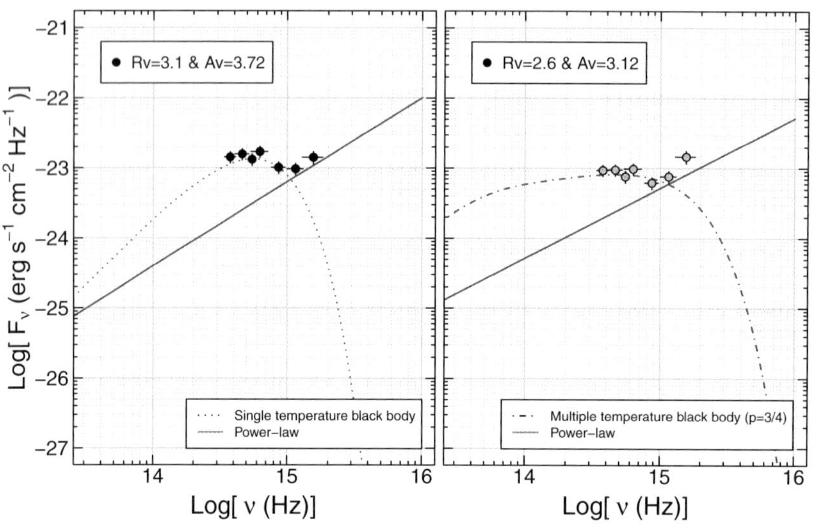

Figure 2. SED from optical to UV with two variable components at MJD = 57194.

Fig. 2 shows the SED from optical to UV. In the left panel, the dotted line and the solid line denote the black body spectrum and power-law spectrum respectively, and the dash-dotted line denote the standard disk spectrum in the right panel. The SED is well reproduced with our model.

References

Ferrigno, C., Bozzo, E., & Boissay, R. 2015, *ATel*, 7731, 1
Gandhi, P., Littlefair, S. P., Hardy, L. K., *et al.* 2016, *MNRAS*, 631, 976
Kimura, M., Isogai, K., Kato, T., *et al.* 2016, *Nature*, 529, 54
Croat, T. K., Stadermann, F. J., & Bernatowicz, T. J. 2005, *ApJ*, 631, 976
Makino, F., 1989, *IAU Circ.*, 4782, 1
Martí, J., Luque-Escamilla, P. L., & García-Hernández, M. T. 2005, *A&A*, 586, A58
Tanaka, Y. & Shibazaki, N. 1996, *ARA&A*, 34, 607
Tanaka, Y. T., Itoh, R., & Uemura, M., *et al.* 2016, *PASJ*, 823, 8

New Frontiers in Black Hole Astrophysics
Proceedings IAU Symposium No. 324, 2016
Andreja Gomboc, ed.

doi:10.1017/S1743921317000035

A pseudo-Newtonian description of any stationary space-time

Vojtěch Witzany[1] **and Claus Lämmerzahl**

ZARM, Universität Bremen,
Am Fallturm, 28359, Bremen, Germany
[1]email: witzany@zarm.uni-bremen.de

Abstract. Since the first investigations into accretion onto black holes, astrophysicists have proposed effective Newtonian-like potentials to mimic the strong-field behavior of matter near a Schwarzschild or Kerr black hole. On the other hand, the fields of neutron stars or black holes in many of the alternative gravity theories differ from the idealized Schwarzschild or Kerr field which would require a number of new potentials. To resolve this, we give a Newtonian-like Hamiltonian which almost perfectly mimics the behavior of test particles in any given stationary space-time. The properties of the Hamiltonian are excellent in static space-times such as the Schwarzschild black hole, but become worse for space-times with gravito-magnetic or dragging effects such as near the Kerr black hole.

Keywords. gravitation, relativity, black hole physics, accretion, gravitational lensing

---

1. Introduction

In the description of accretion onto compact bodies and violent relativistic events such as tidal disruptions either through analytical or numerical means one must employ a number of approximations to be able to give any observational predictions. One of the popular ways how to simplify the description of all the complicated physics of the matter, electromagnetic fields, and radiation in such processes is to use a completely Newtonian framework and only modify the gravitational field to somehow mimic the relativistic gravitational field of a compact object (e.g. Paczyńsky & Wiita (1980)). This, typically *ad hoc*, simplification is called the pseudo-Newtonian approach.

Here, we present an approach to "pseudo-Newtonization" which is systematic, taking only the space-time metric in a convenient set of coordinates and giving a pseudo-Newtonian prescription without prior knowledge of the structure of orbits which are supposed to be mimicked by the pseudo-Newtonian dynamics. This approach is inspired by previous works of Tejeda & Rosswog (2013, 2015), Sarkar *et al.* (2014), and Ghosh *et al.* (2014), but expands the stated works in generality and elucidates the pattern and theoretical foundation of these pseudo-Newtonian prescriptions. The following is only an overview of the results and a full description can be found in Witzany & Lämmerzahl (2016). In the following we use the $G = c = 1$ geometrized units

2. Overview

We can formally "Newtonize" a space-time by taking every orbit of a particle in it and rescaling its velocity by its individual γ-factor, $\gamma = 1/\sqrt{1 - v^2}$. In return, the trajectories approaching the speed of light in the original, relativistic picture approach infinite velocities in the new, "Newtonized" picture.

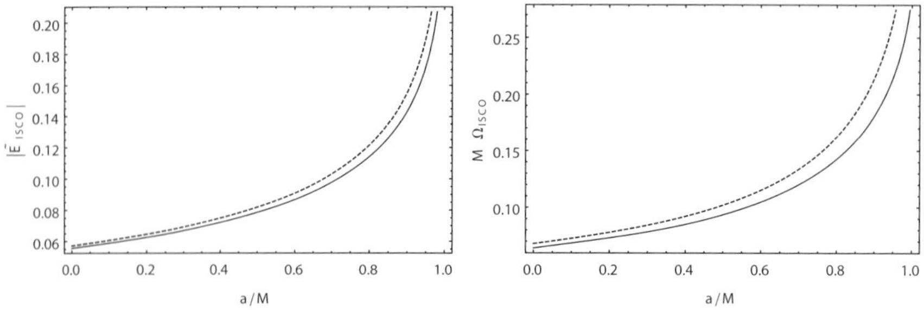

Figure 1. Specific binding energy $\tilde{E}_{\rm ISCO}$ and angular frequency $\Omega_{\rm ISCO}$ in Kerr space-time (dashed) and the pseudo-Newtonian dynamics (full line) as a function of black hole spin a/M.

In curved space-time, the quantity generalizing the notion of the γ-factor is the specific energy $-u_t$, where $u^\mu \equiv {\rm d}x^\mu/{\rm d}\tau$ is the particle four-velocity and $t = x^0$ is the Killing time (the time with respect to which the space-time is stationary). The "Newtonization" in curved space-time can then be done by using the following Hamiltonian as the new fundamental test-particle Hamiltonian

$$H_{\rm PN} = -\frac{1}{2}\frac{g^{ij}}{g^{00}}u_i u_j - \frac{1}{2}\left(\frac{1}{g^{00}} + 1\right) + \omega^i u_i(\omega^i u_i - \sqrt{\mathcal{D}})\,, \tag{2.1}$$

where $\mathcal{D} = (\omega^i u_i)^2 - (g^{ij}u_i u_j + 1)/g^{00}$ and $\omega^i \equiv g^{0i}/g^{00}$ is the dragging velocity. In static space-times ($g^{0i} = 0$) we obtain the Lagrangian

$$L_{\rm PN} = -\frac{1}{2}\frac{g_{ij}}{g_{00}}\dot{x}^i\dot{x}^j + \frac{1}{2}(g_{00} + 1), \tag{2.2}$$

which generalizes the previous results of Tejeda & Rosswog (2013, 2015) and Sarkar *et al.* (2014). Furthermore, the theoretical construction also gives direct estimates of various errors in frequency and energy with respect to the relativistic counter-part, allows to construct a description of light-rays, and leads to an improved set of pseudo-Newtonian fluid equations in the Schwarzschild space-time (see Witzany & Lämmerzahl (2016) for details).

An astrophysically important application of this formalism is in the Kerr space-time where the formula (2.1) yields the distribution of angular momenta over circular orbits identical to the Kerr space-time, and the energies and frequencies are given within a few-percent error. This is shown in Fig. 1 on the properties of the innermost stable circular orbit (ISCO) at different black-hole spins a.

3. Conclusion

The herein presented systematic pseudo-Newtonization of stationary space-times generally works very well for the description of test particles; especially in the case of static space-times, and with promising features in the Kerr space-time.

References

Ghosh, S., Sarkar, T., & Bhadra, A. 2014, *Mon. Not. R. Astron. Soc.*, 445, 4463
Paczyńsky, B. & Wiita, P. J. 2014, *Mon. Not. R. Astron. Soc.*, 445, 4463
Sarkar, T., Ghosh, S., & Bhadra, A. 2014, *Phys. Rev. D*, 90, 063008
Tejeda, E. & Rosswog, S. 2013, *Mon. Not. R. Astron. Soc.*, 433, 1930
Tejeda, E. & Rosswog, S. 2015, arXiv:1402.1171
Witzany, V. & Lämmerzahl, C. 2016, arXiv:1601.01034

Topic 2: Gamma Ray Bursts

New Frontiers in Black Hole Astrophysics
Proceedings IAU Symposium No. 324, 2016
Andreja Gomboc, ed.

doi:10.1017/S1743921317001788

Gamma-ray Bursts Progress and Problems

N. R. Tanvir
University of Leicester, Department of Physics and Astronomy,
University Road, Leicester, LE1 7RH, United Kingdom
email: nrt3@le.ac.uk

Abstract. Our understanding of gamma-ray bursts (GRBs) has come a long way in the past fifty years since their first detection. We now know that GRBs arise in distant galaxies and that there are at least two distinct sub-classes, the long-duration class being produced by some rare massive star core collapse and the short-duration class likely by compact binary mergers involved neutron stars. In both cases, the final remnant will be a stellar-mass black-hole or a massive neutron star. The bursts themselves are associated with ultra-relativistic jetted outflows created by these events, and their afterglows by the impact of these outflows on the surrounding circumburst material. Increasingly GRBs are also being used as probes of the universe, both for understanding galaxy evolution back to the era of reionization, and for the physics of gravitational wave sources. However, many aspects of GRBs remain poorly understood, some pointers to which are given here.

Keywords. gamma rays: bursts, galaxies: high-redshift, galaxies: ISM

1. Introduction

Gamma-ray bursts have provided an enduring source of intrigue and mystery since their discovery in the late 1960s (Klebesadel *et al.* 1973). Their rapid time variability suggested an origin in a small (less than $\sim 10^3$ km) emitting region, but otherwise the lack of a distance scale led to a wide range of plausible progenitors being considered (Nemiroff 1994).

Whilst the possibility of a cosmological origin for GRBs was discussed early (Paczynski 1986), it did not become widely favoured until their isotropic distribution on the sky was convincingly demonstrated by the *CGRO*/BATSE experiment (Meegan *et al.* 1992). Subsequently, the first X-ray afterglow detections by the BeppoSAX satellite (Costa *et al.* 1997) enabled optical afterglow location (van Paradijs *et al.* 1997), and ultimately direct redshift measurements (Metzger *et al.* 1997). Thus GRBs were shown to be extremely powerful events in distant galaxies.

The necessary $\sim 10^{52}$ erg energy requirement can in principle be met from the potential energy reservoir associated with a massive star or compact binary (involving neutron stars or a neutron star and black-hole), providing there is sufficiently high efficiency in converting this energy to radiation. From the inferred compactness of the source, to allow emission of gamma-rays without attenuation from photon-photon pair production and consequent high opacity, it is required that the radiation be produced by an ultrarelativistic outflow: in the standard picture this is a jet with bulk Lorentz factor ~ 300. For an extensive recent review of the physics of GRB jets, see Kumar & Zhang (2015).

It was realised early-on that the observed GRB population splits into two sub-classes, the short-duration harder-spectrum class and the long-duration softer-spectrum class, with a dividing line at roughly 2 s (Mazets *et al.* 1981, Kouveliotou *et al.* 1993). In fact, arguments have been made for further possible sub-classes, including intermediate duration (e.g. de Ugarte Postigo *et al.* 2011), and most recently ultra-long duration (e.g. Levan *et al.* 2014). Both of these rarer proposed sub-classes are more controversial, although the latter has received much attention.

In this short review I will summarise some recent important developments in the field, focussing particularly on observational breakthroughs, many of which have been relied on the large samples of GRBs localised by the *Swift* satellite, and highlight a selection of outstanding problems going forward.

2. Long-duration bursts

2.1. *The progenitor*

From the first few long-GRB host galaxies identified, it was clear that they were all actively star forming, thus suggesting an association with massive star core-collapse. This received remarkable support with the observation that GRB 980425 was accompanied by a bright, broad-lined supernova, SN1998bw (Galama *et al.* 1998).

However, GRB 980425 was an unusually faint GRB and occurred in a galaxy at a redshift of only $z \approx 0.008$. At more typical redshifts, it is unfeasible to search for the much fainter supernova light, and the question remained whether high-luminosity long-GRBs were also accompanied by such SNe. This was settled in the affirmative by the spectroscopic confirmation of similar broad-lined supernovae accompanying first GRB 030329 (e.g. Hjorth *et al.* 2003), and most recently by the very bright GRB 130427A (Xu *et al.* 2013, Levan *et al.* 2014b). Although reassuring as this is, in supporting the standard collapsar picture of long-GRB production, it does remain puzzling that the observed supernova events associated with GRBs seem generally quite similar to each other (e.g. Cano 2016), even for extremely different GRBs (e.g. Tanvir *et al.* 2010).

2.2. *Probes of galactic and chemical evolution*

Long-duration gamma-ray bursts provide an important means of selecting star-forming galaxies over a large range of redshifts. The benefits of LGRB selection is that it does not depend on the luminosity of the host in any waveband, thus sampling the whole luminosity function. This benefit is partially lost in many GRB redshift samples, since they themselves are biased against dusty sight lines where the afterglow proves too faint for spectroscopic follow-up. However, in recent years, various campaigns have attempted to overcome this problem by selecting samples based only on their high energy properties, and attempting to complete the redshift determinations through identification and spectroscopy of the host galaxies.This has only been possible in the *Swift* era, since it relies on the few arcsec X-ray positions it provides to allow deep host searches. Even then, there is potentially some ambiguity about the faintest hosts.

Samples remain modest in size: the TOUGH sample of 69 galaxies (Hjorth *et al.* 2012), BAT6 sample (Salvaterra *et al.* 2012) and most recently the SHOALS sample (Perley *et al.* 2016). In many cases spectroscopy of the GRB afterglow provides not just a redshift, but also information about gas phase abundances, dust and molecular content, even of galaxies too faint to be otherwise detected. Fig. 1 shows a summary of the metallicity measurements made for LGRB hosts based on both absorption line analysis where the afterglow was observed (e.g. Thöne *et al.* 2013). This is challenging below $z \sim 2$ as Ly-α moves out of the optical band. However, in many cases these hosts are bright enough for emission line diagnostics of metallicity, and these are also shown on the same figure (Kruehler *et al.* 2015).

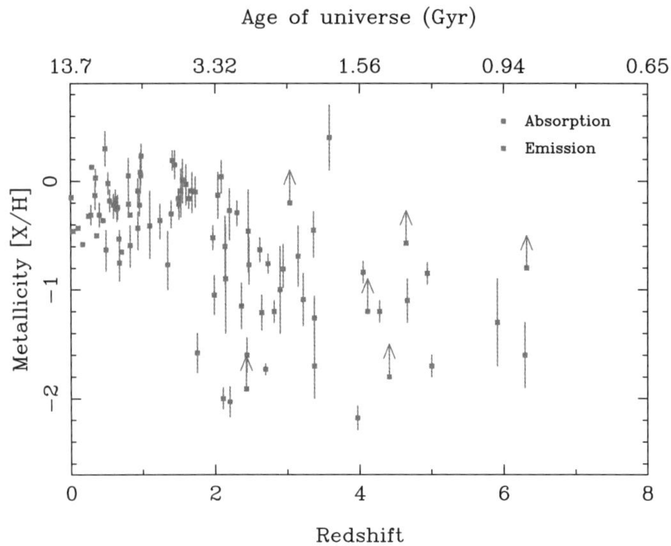

Figure 1. Measurements of metallicity via both galaxy emission lines and afterglow absorption lines for a sample of GRB hosts.

3. Short-duration bursts

3.1. *The progenitor*

Short-duration bursts are not only rarer, comprising only $\sim 10\%$ of detected *Swift* GRBs for example, but also have on average much fainter afterglows. When the first were localised, it became clear that they were associated with a much wider range of stellar populations, including some with only ancient stars (Gehrels *et al.* 2005). Furthermore, in some cases, the burst appeared at a large offset of 10s of kpc from their hosts (e.g. Fong *et al.* 2013), and lacked associated supernova counteparts (e.g. Hjorth *et al.* 2005). These properties are consistent with the bursts being produced during the mergers of compact objects, specifically binaries consisting of either two neutron stars or a neutron star and a black hole. Although these compact objects are formed during core-collapse events, the in-spiral times for such binaries can be very long (some systems would not merge in the lifetime of the universe), leading to mergers occurring long after a star formation episode.

A potential breakthrough in confirming this picture came about with the discovery of an apparent excess near-infrared flux in the days following short-GRB 130603B at $z = 0.36$ (Tanvir *et al.* 2013, Berger *et al.* 2013). This enhancement over the signal expected from the steeply declining afterglow is consistent with predictions of a so-called "kilonova" (or "macronova"), which is expected to follow a binary merger in which a small proportion of the neutron star material has been ejected. Heavy line blanketing should largely attenuate the optical flux.

Low redshift bursts for which such kilonova searches can be conducted are rare, and it remains of key importance to study further examples and explore their range of behaviour. A recent example of the difficulties faced is illustrated in Figure 2, which was a short-GRB originally associated with a $z = 0.3$ host galaxy, but for which deep *HST* imaging revealed an underlying, and likely more distant galaxy, which is more likely the real host.

Figure 2. The field of short-GRB 150424A. The leftt panel ground-based image shows the afterglow (circled) close by a redshift $z = 0.30$ spiral galaxy that was initially presumed to be the host. However, deep *HST* imaging, shown in right panel, revealed an underlying, and likely more distant, faint galaxy (circled) for which a redshift is not currently available.

3.2. *Gravitational wave counterparts*

The first, epoch making detections of gravitational waves were made by the Advanced-LIGO detectors in September 2015. This source, and the other early detections and possible detections, were found to be consistent with binary black-hole mergers (Abbott *et al.* 2016a), and thus unlikely to produce significant electromagnetic emission. Nonetheless, a world-wide campaign of follow-up observations (Abbott *et al.* 2016b) demonstrated the capabilities that are now ready to search for the first unambiguous electromagnetic counterpart to a gravitational wave detection, presumed to be cases where at least one component of the binary is a neutron star.

4. Other populations

4.1. *Low luminosity GRBs*

The first GRB-related supernova, GRB 980425/SN1998bw, gave strong support to massive star progenitor models, but also high-lighted the huge range of intrinsic luminosities of long-GRBs, of $\sim 10^6$. Furthermore it has been suggested that the low-luminosity events form a separate population with much higher space density (Liang *et al.* 2007, Chapman *et al.* 2007), despite the rareness in current samples due to the limited range at which they can be detected. Furthermore, it has been suggested they are produced by shock-breakout rather than directly from the relativistic jets (Bromberg *et al.* (2011)). Clarification of the relationship between the low and high luminosity populations of long-GRBs is a high priority.

4.2. *Ultra-long GRBs*

The longest duration prompt emission seen by *CGRO*/BATSE was less than ~ 1500 s (e.g. Tikhomirova & Stern 2005), although longer events would have been hard to detect due to Earth occultation.

The prototype of a new class of what has come to be known as ultra-long duration was GRB 101225A. The event triggered *Swift* as an "image trigger", which provides sensitivity to transients that are spread out in time such the peak flux does not necessarily reach the threshold for triggering a "rate trigger", but for which reconstruction of the sky image over a timescale of ~minutes shows the presence of a new source. In this case, the event was detectable by *Swift*/BAT for at least ∼7000 s, and exhibited a strongly variable X-ray light-curve for ∼ 10 hr before fading rapidly. The burst was ultimately shown to be associated with a $z = 0.85$ star-forming host galaxy, which was faint and compact even by the standards of typical long-GRB hosts (e.g. Levan *et al.* 2014).

Several further examples (and candidates) of ultra-long GRBs have subsequently been identified, although the sample remains small and the ∼ 90 min low-Earth orbit of *Swift* is not ideal for finding such long-lived events. In all cases to-date where a host galaxy has been detected it is actively star forming, suggesting again a massive star progenitor, and a possible accompanying superluminous supernova signature has been seen in one case, namely GRB 111209 (Greiner *et al.* 2015). The relationship of these events to the bulk of long-duration bursts, and hence how their progenitors are distinguished, is a fascinating open question.

References

Abbott, B. P., *et al.*, 2016a, *PhRvL*, 116, 061102
Abbott, B. P., *et al.*, 2016b, *ApJ*, 826, L13
Berger, E., Fong, W., & Chornock R. 2013, *ApJ*, 774, L23
Bromberg, O., Nakar, E., & Piran, T., 2011, *ApJ*, 739, L55
Cano, Z. 2016, *LPI Contributions*, 1962, 4116
Chapman, R., Tanvir, N. R., Priddey, R. S., & Levan, A. J., 2007, *MNRAS*, 382, L21
Costa, E., Frontera, F., Heise, J., *et al.* 1997, *Nature*, 387, 783
de Ugarte Postigo A., Horvath, I., Veres, P., *et al.* 2011, *A&A*, 525, A109
Eichler, D., Livio, M., Piran, T., & Schramm, D. N. 1989, *Nature*, 340, 126
Fong, W., Berger, E., Chornock, R., *et al.* 2013, *ApJ*, 769, 56
Fruchter, A. S., Levan, A. J., Strolger, L., *et al.* 2006, *Nature*, 441, 463
Galama, T. J., Vreeswijk, P. M., van Paradijs, J., *et al.* 1998, *Nature*, 395, 670
Gehrels, N., Sarazin, C., O'Brien, P., *et al.*, 2005, *Nature*, 437, 851
Greiner, J., Mazzali, P., Kann, D. A., *et al.* 2015, *Nature*, 523, 189
Hjorth, J., Sollerman, J., Møller, P., *et al.* 2003, *Nature*, 423, 847
Hjorth, J., Sollerman, J., Gorosabel, J., *et al.* 2005, *ApJ*, 630, L117
Hjorth, J., Malesani, D., Jakobsson, P., *et al.* 2012, *ApJ*, 756, 187
Krühler T., *et al.*, 2015, *A&A*, 581, A125
Klebesadel, R. W., Strong, I. B., & Olson, R. A. 1973, *ApJ*(Letters), 182, L85
Kouveliotou, C., Meegan, C. A., Fishman, G. J., *et al.* 1993, *ApJ*(Letters), 413, L101
Kumar, P. & Zhang, B., 2015, *PhR*, 561, 1
Levan, A. J., Tanvir, N. R., Starling, R. L. C., *et al.* 2014a, *ApJ*, 781, 13
Levan, A. J., Tanvir, N. R., Fruchter, A. S., *et al.* 2014b, *ApJ*, 792, 115
Liang, E., Zhang, B., Virgili, F., & Dai, Z. G., 2007, *ApJ*, 662, 1111
Mazets, E. P., Golenetskii, S. V., Ilinskii, V. N., *et al.* 1981, *ApSS*, 80, 3
Meegan, C. A., Fishman, G. J., Wilson, R. B., *et al.* 1992, *Nature*, 355, 143
Metzger, M. R., Djorgovski, S. G., Kulkarni, S. R., *et al.* 1997, *Nature*, 387, 878
Nemiroff, R. J. 1994, *ComAp*, 17, 189
Paczynski, B. 1986, *ApJ*(Letters), 308, L43
Perley, D. A., Krühler, T., Schulze, S., *et al.* 2016, *ApJ*, 817, 7
Salvaterra, R., Campana, S., Vergani, S. D., *et al.* 2012, *ApJ*, 749, 68
Tanvir, N. R., *et al.*, 2010, *ApJ*, 725, 625
Tanvir, N. R., Levan, A. J., Fruchter, A. S., *et al.* 2013, *Nature*, 500, 547
Thöne C. C., *et al.*, 2013, *MNRAS*, 428, 3590
Tikhomirova, Y. Y. & Stern, B. E. 2005, *AstL*, 31, 291
van Paradijs, J., Groot, P. J., Galama, T., *et al.* 1997, *Nature*, 386, 686
Xu, D., de Ugarte Postigo, A., Leloudas, G., *et al.* 2013, *ApJ*, 776, 98

New Frontiers in Black Hole Astrophysics
Proceedings IAU Symposium No. 324, 2016
Andreja Gomboc, ed.

doi:10.1017/S1743921317002010

Gamma-Ray Bursts Polarization

Diego Götz[1] and Stefano Covino[2]

[1]CEA Saclay - DRF/Irfu/Service d'Astrophysique,
Orme des Merisiers, Bât. 709, F-91191 Gif-sur-Yvette, France
email: `diego.gotz@cea.fr`

[2]INAF / Brera Astronomical Observatory,
Via Bianchi 46, 23907, Merate (LC), Italy
email: `stefano.covino@brera.inaf.it`

Abstract. We review the current observational and theoretical status of the polarization measurements of Gamma-ray Bursts at all wavelengths. Gamma-Ray Bursts are thought to be produced by an ultra-relativistic jet, possibly powered by a black hole. One of the most important open point is the composition of the jet: the energy may be carried out from the central source either as kinetic energy (of baryons and/or pairs), or in electromagnetic form (Poynting flux). The polarization properties are expected to help disentangling main energy carrier. The prompt emission and afterglow polarization are also a powerful diagnostic of the jet geometry.

Keywords. Polarization, gamma rays: bursts, gamma rays: observations, techniques: photometric

---

1. Introduction

Polarimetric measurements can provide useful complementary (wrt. timing and spectroscopy) information about the physical processes at work in Gamma-Ray Bursts (GRBs). Indeed several different possible scenarios have been invoked to interpret the large amount of observational data now available for GRBs. Most of the theoretical efforts have been applied to the so-called "standard model", which, although a fully satisfactory picture is still missing, offers the best interpretative scenario for the polarimetric observations.

In this paper we first summarize the observational status of the GRB prompt emission polarimetric measurements, and then compare them to theoretical predictions. Then we will do the same concerning the GRB afterglow phase, and we conclude with some perspectives on the domain. For more details on GRB polarimetric measurements and their implications also on fundamental physics, see the recent review by Covino & Götz (2016).

2. Polarization in the prompt phase

The measurement of polarization during the prompt phase of GRBs has always been challenging. This is mainly due to the fact that no wide field gamma-ray polarimeter with a large effective area has yet been flown, and that many of the measurements attempted to date have been performed with instruments which have some polarimetric capabilities, but have not an explicit polarimetric oriented design. In addition at odd to the afterglow emission, the GRB prompt emission is very limited in time, mostly less than a few hundreds of seconds, and hence, in spite of the high average flux of GRBs, the total number of collected photons is often too limited to derive statistically stringent limits for polarization.

2.1. *Early Results*

The first attempt to measure linear polarization in the prompt emission of GRBs was reported by Coburn & Boggs (2003). They used the Reuven Ramaty High Energy Solar Spectroscopic Imager (RHESSI) observations of GRB 021206. RHESSI has been designed to study solar flares in the 3 keV–17 MeV energy range. In the soft gamma-ray energy range (0.15–2.0 MeV) the dominant photon interaction in RHESSI is Compton scattering. Polarization at high energies can be measured, thanks to the polarization dependency of the differential cross section for Compton scattering

$$\frac{d\sigma}{d\Omega} = \frac{r_0^2}{2}\left(\frac{E'}{E_0}\right)^2\left(\frac{E'}{E_0} + \frac{E_0}{E'} - 2\sin^2\theta\cos^2\phi\right) \tag{2.1}$$

where r_0^2 is the classical electron radius, E_0 the energy of the incident photon, E' the energy of the scattered photon, θ the scattering angle, and ϕ the azimuthal angle relative to the polarization direction. Linearly polarized photons scatter preferentially perpendicularly to the incident polarization vector. Hence by examining the angles of scattering of the photons among the different RHESSI detectors, one can in principle derive the degree and angle of linear polarization of the incident photons.

Coburn & Boggs (2003) reported a high level of linear polarization of Π=80±20% (close to and beyond the theoretical value) at a high level of confidence ($> 5.7\sigma$) for GRB 021206. GRB 021206 was a quite bright GRB with a fluence of 1.6×10^{-4} erg cm^{-2} in the 25–100 keV energy band, and a peak flux of 2.9×10^{-5} erg cm^{-2} s^{-1}. Coburn & Boggs (2003) interpreted the angular modulation measured in the data as a high-level polarization signal. However subsequent re-analyses of the same data set could not confirm this result reporting a polarization level compatible with zero (Rutledge *et al.* 2004; Wigger *et al.* 2004). These authors show that the number of suitable events for polarization analysis has been over-estimated by a factor 10 (830±150 versus 9840±96), since spurious coincidences had been counted as Compton scattering events, implying in the end a too small statistics for being able to measure any polarization signal, even for a 100% polarized source.

2.2. *IBIS and SPI on board INTEGRAL*

At the time of its discovery GRB 041219A was among the top 1% in terms of GRB fluence. This prompted different attempts to measure its polarization with the instruments that observed it. The first attempt was performed by McGlynn *et al.* (2007) using the SPI spectrometer on board *INTEGRAL*. They were able to measure the degree of linear polarization over the brightest pulse of the GRB (lasting 66 s) to $\Pi = 63^{+31}_{-30}\%$ and the polarization angle to $PA = 70^{+14}_{11}$ degrees. GRB041219A was also observed by the Imager on Board the INTEGRAL Satellite (IBIS). Thanks to its two superposed pixellated detection planes, IBIS can be used as a Compton Polarimeter. By examining the scatter angle distribution of the detected photons in the two planes

$$N(\phi) = S[1 + a_0\cos 2(\phi - \phi_0)], \tag{2.2}$$

one can derive the polarization angle, $PA = \phi_0 - \pi/2 + n\pi$, and the polarization fraction $\Pi = a_0/a_{100}$, where a_{100} is the amplitude expected for a 100% polarized source derived by Monte Carlo simulations. The IBIS results are reported in Table 1.

Using the same method polarization could be measured for two other GRBs (061122† and 140206A) with IBIS by Götz *et al.* (2013), Götz *et al.* (2014), see Table 2, but no

† A consistent polarization measurement has been obtained with SPI.

Table 1. Polarization results for GRB 041219A for the different time intervals. From Götz *et al.* (2009)

Name	T_{start} U.T.	T_{stop} U.T.	Π %	PA degrees	Image SNR
First Peak	01:46:22	01:47:40	<4	–	32.0
Second Peak	01:48:12	01:48:52	43±25	38±16	20.0
P6	01:46:47	01:46:57	22± 13	121±17	21.5
P8	01:46:57	01:27:07	65±26	88±12	15.9
P9	01:47:02	01:47:12	61±25	105±18	18.2
P28	01:48:37	01:48:47	42±42	106±37	9.9
P30	01:48:47	01:48:57	90±36	54±11	11.8

Errors are given at 1 σ c.l. for one parameter of interest.

Table 2. Summary of recent GRB polarization measurement by IBIS/SPI and GAP.

GRB	Π (68% c.l.)	Peak energy (keV)	Fluence (erg cm^{-2})	Energy Range	Redshift z	Instrument
041291A	65±26%	201^{+80}_{-41}	2.5×10^{-4}	20–200 keV	$0.31^{+0.54}_{-0.26}$	IBIS, SPI
06122	>60%	188±17	2.0×10^{-5}	20–200 keV	$1.33^{+0.77}_{-0.76}$	IBIS, SPI
100826A	27±11%	606^{+134}_{-109}	3.0×10^{-4}	20 keV–10 MeV	0.71–6.84[1]	GAP
110301A	70±22%	107±2	3.6×10^{-5}	10 keV–1 MeV	0.21–1.09[1]	GAP
110721	84^{+16}_{-28}%	393^{+199}_{-104}	3.5×10^{-4}	10 keV–1 MeV	0.45–3.12[1]	GAP
140206A	>48%	98±17	2.0×10^{-5}	15–350 keV	2.739±0.001	IBIS

[1] redshift based on empirical prompt emission correlations, not on afterglow observations.

time-resolved analysis could be performed due to the limited statistics, making GRB041219A the only GRB for which a time variable polarization signal could be measured to date with IBIS.

2.3. *GAP*

High levels of linear polarization could be measured also for three GRBs (100826A, 110301A and 110721) by the Gamma-Ray Burst Polarimeter experiment on board the IKAROS spacecraft (Yonetoku *et al.* 2011 & 2012). In particular for GRB 100826A, which had a similarly high fluence as 041219A, Yonetoku *et al.* (2011) were able to measure a change in the polarization angle by dividing the GRB in two ~50 s long time intervals: the angle changed from 159±18° to 75±20° (1 σ c.l. for two parameters of interest) with a significance of 3.5 σ for the change.

2.4. *Interpretation*

The expected level of polarization of the prompt γ-ray emission in GRBs has been estimated by several authors for different models, or variations within them. In most cases, the observed γ-ray emission is due to the synchrotron radiation from relativistic electrons. To have a high radiative efficiency and to allow for the short time scale variability in the GRB light curves, these electrons have to be in the fast cooling regime. The intrinsic polarization level of the synchrotron radiation (assuming a time-averaged electron distribution $n(\gamma) = \gamma^{-p}$ with $p' = p+1$ above Γ_m and $p' = 2$ below) $\Pi_{\rm syn} = (p'+1)/(p'+7/3)$ is then of the order of $\Pi_{\rm syn} = (p+2)/(p+10/3) \simeq 75\%$ above $\nu_{\rm m}$ and $\Pi_{\rm syn} = 9/12 \simeq 70\%$ below, where $\nu_{\rm m}$, the peak of the spectrum in νF_ν, is the synchrotron frequency of electrons at $\Gamma_{\rm m}$ (the minimum Lorentz factor of the injected electrons). High polarization levels can also be reached if inverse Compton scatterings are the dominant radiative process.

Actually different scenarios in terms of radiation processes and observer's viewing angle can be envisaged to explain the presence of polarized emission during the prompt phase of GRB emission. They can be roughly divided in two families: intrinsic models and geometric models, for which peculiar observing conditions are required.

(*a*) *Synchrotron emission from shock-accelerated electrons in a relativistic jet with an ordered magnetic field contained in the plane perpendicular to the jet expansion.* This scenario leads to a maximum time-averaged polarization in long intervals of $\Pi/\Pi_{\rm max} \sim 0.6$, i.e. $\Pi \sim 45\%$ in this case (Granot 2003, Granot & Königl 2003, Nakar *et al.* 2003). The main requirement for this model to apply is to have a uniform magnetic field in space, i.e. with a coherence spatial scale $R\theta_B$ with $\theta_B \gtrsim 1/\Gamma$. If θ_B is smaller, so that a number $N \sim (\Gamma\theta_B)^{-2}$ of mutually incoherent patches are present in the visible region, the level of polarization will decrease, but the variability (both of the polarization level and angle) will increase (Granot 2003). Indeed if the radiating electrons are accelerated in internal shocks, the Lorentz factor associated with the individual shells is necessarily varying in the outflow, which can be an additional source of variability for the polarization. This scenario could hence produce time variable polarization, as long as the coherence scale $\theta_{\rm B}$ of the field is larger than $1/\Gamma$ in most of the emitting regions.

(*b*) *Synchrotron emission from a purely electromagnetic outflow.* In this scenario the GRB is powered by the rotational energy of a magnetar-like progenitor (e.g. Metzger *et al.* 2011). The estimated level of polarization in this case is comparable with the previous scenario (up to $\sim$50%) (Lyutikov *et al.* 2003). In addition, a magnetic field with a large coherence scale is naturally expected in such a purely electromagnetic outflow.

(*c*) *Synchrotron emission from shock-accelerated electrons in a relativistic jet with a random field generated at the shock and contained in the plane perpendicular to the jet velocity.* A high level of polarization can be obtained even with a random magnetic field if the jet is observed from just outside its edge (Ghisellini & Lazzati 1999, Waxman 2003). The polarization at the peak of a given pulse can reach $\Pi/\Pi_{\rm syn} \simeq 0.8$, i.e. $\Pi \simeq 60\%$ resulting in a time-integrated value of the order of $\Pi/\Pi_{\rm syn} \simeq 0.5 - 0.6$, i.e. $\Pi \simeq 40 - 45\%$ (Granot 2003, Granot & Königl 2003, Nakar *et al.* 2003). However these high values are obtained if the jet is seen with $\theta_{\rm obs} \simeq \theta_{\rm j} + 1/\Gamma$, where $\theta_{\rm j}$ is the opening angle of the jet and $\theta_{\rm obs}$ the angle between the line-of-sight and the jet axis. Such viewing conditions are rare, except if $\theta_{\rm j} \sim 1/\Gamma$.

(*d*) *Synchrotron emission from shock-accelerated electrons in a relativistic jet with an ordered magnetic field parallel to the jet velocity.* This case has been studied by Granot & Königl 2003 and gives very similar results to the previous model. The viewing conditions have to be the same and it suffers the same difficulties as listed above;

(*e*) *Inverse Compton emission from relativistic electrons in a jet propagating within a photon field ("Compton drag" model).* The level of polarization in this scenario can be even higher than for the synchrotron radiation and reach $60 - 100\%$, but only under the condition that the jet is narrow with $\Gamma\theta_j \lesssim 5$, see Lazzati *et al.* (2004). The maximum level of polarization is again obtained for $\theta_{obs} \simeq \theta_j + 1/\Gamma$. These viewing conditions are very similar to those of model (c). Again, the polarization is reduced if the edges of the jet are not sharp enough. Variability of the Lorentz factor will again result in a varying polarization, with the same difficulties regarding the final level of polarization than in model (c). However, variations of the Lorentz factor could possibly be less large in this scenario as part of the variability of the light curve can be related to the inhomogeneity of the ambient photon field.

(*f*) Independently from the emission process (synchrotron or inverse Compton), *fragmented fireballs* (shotguns, cannonballs, sub-jets) can produce highly polarized emission, with a variable polarization amplitude. The fragments are responsible for the single pulses and have different intrinsic properties (such as Lorentz factors), opening angles, orientations with respect to the observers and magnetic domains. (e.g. Lazzati & Begelman 2009). In this case the most polarized pulses are those which have about one tenth of the flux of the main pulse, i.e. an anti-correlation between the polarization level and the GRB pulse flux is expected.

3. Afterglow emission

Due to the large data set of polarimetric observation during the GRB afterglow phase we address the reader to the Covino & Götz (2016) paper for a complete picture, and here we restrict ourselves only to the most relevant measurements.

3.1. *GRB 990510*

Polarimetry at a few percent level can be demanding for rapidly fading sources as GRB afterglows, and therefore is not a surprise that the first positive detections in the optical band came a few months after the first unit of the VLT, with its collecting area and flexibility, become operational. GRB 990510 was observed two times by two independent teams (Covino *et al.* 1999, Wijers *et al.* 1999) at 18-21 hours after the burst with the ESO-VLT, providing a small but highly significant polarization level at $P = 1.7 \pm 0.2\%$. A later measurement one day after gave a result consistent with a non-variability of the observed polarization.

Polarization at this level is not common for extragalactic sources, however it is possible that it is due to polarization induced by dust grains interposed along the line of sight, which may be preferentially aligned due to the galactic magnetic fields. Large variations are anyway expected, and observed, for specific line of sights. On the other hand, the effect due to dust in the Galaxy can be in principle easily checked and removed if a sufficiently large number of stars are observed in the same field of view. Electron scattering can also lead to some polarization, as observed in supernove and usually attributed to asymmetries in their photospheres (Wang & Wheeler 2008). However, the degree of induced polarization is of the order of the electron optical depth that cannot be more than $\sim 10^{-6}$ a day after the event (Wijers *et al.* 1999).

GRB 990510 was also the first GRB with an achromatic (at least in the optical band) steepening, a jet-break, of the afterglow light-curve clearly observed (Israel *et al.* 1999, Harrison *et al.* 1999, Stanek *et al.* 1999). The observations with solid polarization detections were performed before the jet-break. Due to the large Lorentz factor of the outflow, Γ, only a fraction $\sim 1/\Gamma$ of the emitting region is accessible to the observer. Photons produced in regions at an angle $1/\Gamma$ with respect to the line of sight are emitted, in the comoving frame, at $\sim 90°$ from the velocity vector. A comoving observer at this angle can then see a compressed emitting region (Ghisellini & Lazzati 1999) and a projected magnetic field structure with a preferred orientation. If the gradual steepening of the light curve is a jet-break, we would observe only regions at a viewing angle $1/\Gamma$ at variance with an axis-symmetric situation, and this asymmetry can be the cause of the observed linear polarization that therefore becomes the "smoking gun" of synchrotron emission for GRB afterglows.

3.2. *Time evolution measurements*

The attempt to identify a time-evolution of the polarization degree and position angle generated a richer dataset for several events. GRB 020405 was observed with the VLA by Granot & Taylor 2005, the VLT by Masetti *et al.* (2003) and Covino *et al.* (2003) and the Multiple Mirror Telescope by Bersier *et al.* (2003) between one and three days from the burst. The polarization level was observed at about 1.2-2% for the VLT observations but at the MMT a much higher polarization at about 10% was detected. Only mild upper limits were obtained at the radio frequencies. The position angle possibly showed a slow change ($\sim 10^\circ$) from the first to the last observations. The observations derived with the VLT before and after the MMT observations are substantially consistent with a constant afterglow polarization, possibly also with an important contribution of dust in the host galaxy. The rapid variation required to move from the $\sim 1\%$ to $\sim 10\%$ in a timescale of about one hour is essentially inconsistent with basically all the geometric models and also with the patchy-shell idea (Ghisellini & Lazzati 1999, Sari 1999, Gruzinov & Waxman 1999). In principle a micro-lensing phenomenon Loeb & Perna (1998) could be responsible for the polarization "flare", although the rapid time scale, the almost constant position angle and the lack of an analogous brightening in the total flux curve make even this interpretation unlikely. The high polarization observed by Bersier *et al.* (2003) only about one hour after the observation carried out by Masetti *et al.* (2003) is therefore still of difficult interpretation.

3.3. *GRB 030329*

A fundamental breakthrough in the observational activities of GRB afterglows occurred with GRB 030329 (see Covino & Götz (2016) for a complete set of references). GRB 030329 was discovered by the HETE II satellite and was one of the few cases of low redshift GRBs ($z \sim 0.17$). Being in addition a regular cosmological GRB, i.e. not part of the category of low-luminosity low-redshift events, it showed an optical brightness sufficiently high to allow about one month of uninterrupted polarimetric observations in the optical with the ESO-VLT, the CAHA, the NOT, the IAG-USP and much longer in the radio with the VLBA. The afterglow polarization showed a strong variability in polarization degree and position angle. The polarization was typically in the 0.3-2.5% range. The light-curve of GRB 030329 was characterized by numerous bumps and wiggles, and after about 10 days a supernova component also affected the observations. The modeling of these polarization data is beyond the capabilities of any scenario discussed so far, lacking for instance any clear correlation between polarization and light-curve behaviour. Possibly, the observed emission and polarization is therefore due to the superposition of different phenomena that make a proper modeling difficult to achieve. The vanishing radio polarization at late-times might be due to much less ordered than expected magnetic fields and/or Faraday rotation depolarizing the emission at the radio bands.

3.4. *GRB 121024A*

GRB 121024A was intensively observed with the VLT (Wiersema *et al.* 2014) starting from a few hours after the GRB and obtaining a positive and highly unexpected detection of circular polarization, $P_{\rm circ} = 0.61 \pm 0.13\%$, together with an extensive linear polarimetric monitoring. The first observations showed a rather high polarization level, $P_{\rm lin} \sim 5\%$, with a global decreasing trend with time and a constant position angle. Observations carried out the night after, showed a lower polarization level with a clear 90° rotation of the position angle. Analysis of the light-curve allowed to identify a jet-break between the two sets of observations and this is a very clear identification of the polarization angle swing predicted to occur around the jet-break time of a homogeneous jet that is not spreading sideways (Rossi *et al.* 2004). During the circular polarimetry measurement the linear polarization was about 4%, and therefore the circular to linear polarimetry ratio turned out to be $P_{\rm circ}/P_{\rm lin} \sim 0.15$, a very high value, order of magnitudes greater than the theoretical expectations (Toma *et al.* 2008). If the emission process is synchrotron the expected polarization is indeed $P_{\rm circ} \sim \gamma_{\rm e}^{-1}$, where $\gamma_{\rm e}$ is the random Lorentz factor of the accelerated electrons emitting the observed radiation. This relation holds under the assumption of isotropic pitch-angle distribution and ordered magnetic fields (Toma *et al.* 2008), and the high value of measured circular polarization poses a challenge to this assumption. Furthermore, a detailed analysis carried out by Nava *et al.* (2016) suggests that under the hypothesis of optically thin synchrotron emission such a high value of circular to linear polarization ratio is not possible even with extremely isotropic pitch-angle distribution. A satisfactory interpretation of this striking result is still missing.

4. Conclusions and Perspectives

The large set of observations available for the afterglows, mainly but not only in the optical, allows us to derive a few important conclusions. First of all, the simple observation of variable polarization implies that the afterglow radiation is intrinsically polarized, thus offering one of the strongest observational evidence for the synchrotron origin of the afterglow emission. The observations of specific patterns (i.e. the position angle swing) during the evolution of the afterglows in polarimetry that have been predicted in advance give also confidence to the general interpretative scenario, although exceptions are present. And the detection of circular polarimetry at a level much higher than expected instead poses a formidable challenge to our present GRB afterglow emission interpretation.

The success of recent observational campaigns clearly shows that a massive approach, trying to follow the afterglow evolution from the early-time, with intermediate-size robotic telescope, to the late phases, with the biggest available facilities, is required. And the parameter space for discoveries is still huge. Radio observations are promising, in particular with future high-sensitivity facilities, and mm observations with ALMA can help to dramatically extend the energy range of the observations and the testing capabilities of the various interpretative scenarios.

For the prompt phase the situation is less clear, but also offering perspective for exciting discoveries in the near future. A final answer to distinguish between intrinsic and geometric models could be obtained by accumulating more observations. Indeed, models (a-b, f) – as defined in Section 2 – predict a polarized emission for all bursts, whereas models (c-e) would predict that only a small fraction of GRBs are highly polarized. This shows the importance of accumulating polarimetric measurements for the understanding of intrinsic properties of GRBs, but the current instrumentation is statistically limited and can provide measurements just for the brightest events.

Recently, a few polarization measurements of the very early optical afterglow have been reported in some cases while the prompt high-energy phase was still on going . The most intruguing case is GRB 120308A for which a high level, Π=28±4%, of linear optical polarization in the early afterglow has been reported by Mundell *et al.* (2013). The latter measure allowed to point out the presence of a magnetized reverse shock with an ordered magnetic field, confirming the presence of high magnetic fields in the GRB ejecta, and indicating that the multi-wavelength approach could be fruitful, even if there is currently no consensus on the common origin of the γ-ray and optical emission in the prompt phase of GRBs (e.g. Vestrand *et al.* 2005, Stratta *et al.* 2009, Götz *et al.* 2011, Guidorzi *et al.* 2011, Kopač *et al.* 2013).

Acknowledgements

This work has been supported by ASI grant I/004/11/2, by the French Space Agency (CNES) and by the UnivEarthS Labex program at Sorbonne Paris Cité (ANR-10-LABX-0023 and ANR-11-IDEX-0005-02).

References

Bersier, D., *et al.* 2003, *ApJL*, 583, L63
Coburn, W. & Boggs, S. E. 2003, *Nature*, 423, 415
Covino, S., *et al.* 1999, *A&A*, 348, L1
Covino, S., *et al.* 2003, *A&A*, 400, L9
Covino, S. & Götz, D. 2016, *Astronomical and Astrophysical Transactions*, 29, 205
Ghisellini, G. & Lazzati, D. 1999, *MNRAS*, 309, L7
McGlynn, S., *et al.* 2007, *A&A*, 466, 895
Götz, D., *et al.* 2009, *ApJL*, 695, L208
Götz, D., *et al.* 2011, *MNRAS*, 413, 2173
Götz, D., *et al.* 2014, *MNRAS*, 695, L208
Götz, D., *et al.* 2014, *MNRAS*, 444, 2766
Granot J. 2003, *MNRAS*, 596, L17
Granot, J. & Königl, A. 2003, *ApJL*, 594, L83
Granot, J. & Taylor, G. B. 2005, *ApJ*, 625, 263
Gruzinov, A. & Waxman, E. 1999, *ApJ*, 511, 852
Guidorzi, C., *et al.* 2011, *MNRAS*, 417, 2124
Harrison, F. A., *et al.* 1999, *ApJL*, 523, L121
Israel, G. L., *et al.* 1999, *A&A*, 348, L5
Kopač, *et al.* 2013, *ApJ*, 772, 73
Lazzati, D., *et al.* 2004, *MNRAS*, 347, L1
Lazzati, D. & Begelman, M. C. 2009, *ApJL*, 700, L141
Loeb, A. & Perna, R. 1998, *ApJ*, 495, 597
Lyutikov, M., *et al.* 2003, *ApJ*, 597, 998
Masetti, N., *et al.* 2003, *A&A*, 404, 465
Metzger, B. D., *et al.* 2011, *MNRAS*, 413, 2031
Mundell, C. G., *et al.* 2013, *Nature*, 504, 119
Nakar, E., Piran, T., & Waxman, E. 2003, *JCAP*, 10, 005
Nava, L., *et al.* 2016, *MNRAS*, 455, 1594
Rossi, E. M., *et al.* 2004, *MNRAS*, 354, 86
Rutledge, R. E., Fox, D. B. 2004, *MNRAS*, 350, 1288
Sari, R. 1999, *ApJL*, 524, L43
Stanek, K. Z., *et al.* 1999, *ApJ*, 522, L39
Stratta, G., *et al.* 2009, *A&A*, 503, 783
Toma, K., *et al.* 2008, *ApJ*, 673, L123
Vestrand, W. T., *et al.* 2005, *Nature*, 435, 178
Wang, L. & Wheeler, J. C. 2008, *ARAA*, 46, 433
Waxman, E. 2003, *Nature*, 423, 388
Wiersema, K., *et al.* 2014, *Nature*, 509, 201
Wijers, R. A. M. J., *et al.* 1999, *ApJL*, 523, L33
Wigger, C., *et al.* 2004, *ApJ*, 613, 1088
Yonetoku, D., *et al.* 2011, *ApJL*, 743, L30
Yonetoku, D., *et al.* 2012, *ApJL*, 758, L1

New Frontiers in Black Hole Astrophysics
Proceedings IAU Symposium No. 324, 2016
Andreja Gomboc, ed.

doi:10.1017/S1743921317001156

Generation and decay of the magnetic field in collisionless shocks

Mikhail Garasev and Evgeny Derishev

Institute of Applied Physics of the Russian Academy of Sciences,
46 Ulyanova Str., Nizhny Novgorod 603950, Russia
email: garasev@appl.sci-nnov.ru

Abstract. We present the results of numerical particle-in-cell (PIC) simulations of the magnetic field generation and decay in the upstream of collisionless shocks. We use the model, where the magnetic field in the incoming flow is generated by continuous injection of anisotropic electron-positron pairs. We found that the continuous injection of anisotropic plasma in the upstream of the shock-wave generates the large-scale, slowly decaying magnetic field that is later amplified during the passage of the shock front. In our simulations the magnetic field energy reached ~ 0.01 of the equipartition value, after that it slowly decays on the time scale proportional to the duration of the injection in the upstream. Thus, the magnetic field survives for a sufficiently long time, and supports efficient synchrotron radiation from relativistic shocks, e.g., in GRBs.

Keywords. shock waves, magnetic fields, gamma-ray burst

1. Introduction

Relativistic collisionless shocks are common phenomena in extreme astrophysical sources like AGNs and GRBs. Such shocks are thought to be capable of generating strong magnetic fields and particle acceleration (Achterberg *et al.* (2001), Gruzinov (2001), Medvedev & Loeb (1999)). This makes them natural sources of synchrotron emission. These theoretical predictions have been tested by various numerical simulations (Spitkovsky (2008), Keshet *et al.* (2009), Sironi *et al.* (2013), Sironi *et al.* (2015)).

However, these numerical simulations have discovered some problems in this simple picture. It was found that the magnetic field generated at the shock front via Weibel instability decays very fast on the time scale comparable to the inverse plasma frequency. If the strong magnetic field initially exists in the upstream, then no particle acceleration is observed. So, strong magnetic fields and particle acceleration never meet each other, leaving no room for efficient synchrotron emission.

Solution to this problem could be linked with presence of magnetic turbulence in the upstream flow (Medvedev & Zakutnyaya (2009), Miloslavljevic & Nakar (2006)). Recently a new model of the relativistic collisionless shock was proposed (Derishev & Piran (2016)), where the magnetic field growth is a two-stage process. In the first stage, the magnetic field is generated upstream of the shock as a result of massive production of electron-positron pairs with anisotropic velocity distribution in the fluid comoving frame. The width of this region is about the electron cooling length, which is many orders of magnitude greater than the plasma skin depth, so that the magnetic field is growing very slowly and could be maintained for a very long period of time compared to the plasma time scale. In the second stage, this magnetic field is enhanced at the shock front, and then decays in the downstream. According to the model, the decay length is given by the spatial scale of the turbulent magnetic field formed at the long stage of initial build-up. The goal of this model is to achieve the magnetic field decay length comparable to the electron cooling length.

In this paper we present results of PIC simulations of the magnetic field generation in the upstream of collisionless shocks. The magnetic field build-up results from slowly developing Weibel instability in initially isotropic pair plasma with gradual injection of anisotropic component via generation of electron-positron pairs in collisions of high-energy photons escaping from the shock. In our simulations, we follow the evolution of a small (and, therefore, uniform) region in the comoving frame. Simulations start in the upstream with constant injection rate for anisotropic e^+e^- pairs, passes through a phase of the magnetic field growth, the rate of which depends on the rate of injection, and ends with the observation of the magnetic field decay.

2. Overview of simulation setup

For our simulations we use relativistic PIC-code EPOCH (Arber *et al.* (2015)) in Cartesian 2D geometry (x- and y- axis are in the simulation box, z-axis perpendicular to the simulation box) with periodic boundary conditions. The code was modified to allow injection of new electron-positron pairs (Garasev & Derishev (2016)). During the injection step we randomly distribute particles across the computational domain ensuring that there is no perturbation to the total momenta, charge or current. For this purpose we simultaneously inject two electron-positron pairs with equal but oppositely directed velocities. We take 1000 particles per cell per species in each simulation in order to reduce the numerical noise. The low noise levels are necessary to explore the decay of the magnetic field on long time scales.

Initially our simulation box was filled with isotropic Maxwellian e^+e^- plasma whose distribution function is

$$f_{e,e^+}(p) = \frac{N_0}{\sqrt{(2meT)^3}} \exp\left(-\frac{p_x^2}{2m_eT_x} - \frac{p_y^2}{2m_eT_y} - \frac{p_z^2}{2m_eT_z}\right),$$

where m_e is the electron mass, $T_{x,y,z}$ the temperatures measured in the energy units, p the momentum of a particle and N_0 the number densities of electrons and positrons. The initial temperature was set to $T_{x,y,z} = 50$ keV. We used Cartesian 2D grids with uniform mesh with 1600×1600 cells. The size of a cell in each direction was set to $r_D = \sqrt{kT/(8\pi e 2 N_0)}$. Here e is the elementary charge. Injection of anisotropic pairs starts at the beginning of each simulation at the moment $t_i < 0$, continues with a fixed rate, and then stops at $t = 0$. The injected plasma has elongated two-temperature distribution function with $T_x = 200$ keV and $T_{y,z} = 50$ keV. The total number density of the injected component is δ times the initial number density of particles. We track the evolution of a magnetic field for up to $15000/\omega_{\rm p0}$, where

$$\omega_{\rm p}(t) = \sqrt{8\pi e^2 (N_0 + N_{\rm a}(t))/m_{\rm e}}$$

is the plasma frequency, $\omega_{\rm p0} = \omega_{\rm p}(-t_{\rm i})$ the background plasma frequency and $N_{\rm a}(t)$ is the number density of injected component. It is convenient to use dimensionless time

$$\tau = \int_{-t_{\rm i}}^{t} \omega_{\rm p} {\rm d}t - \tau_{\rm i}, \qquad \tau_{\rm i} = \int_{-t_{\rm i}}^{0} \omega_{\rm p}\, {\rm d}t\,. \qquad (2.1)$$

3. Results

We performed a number of simulations with different injection durations. In agreement with the standard picture of the Weibel instability, the magnetic field filaments start to

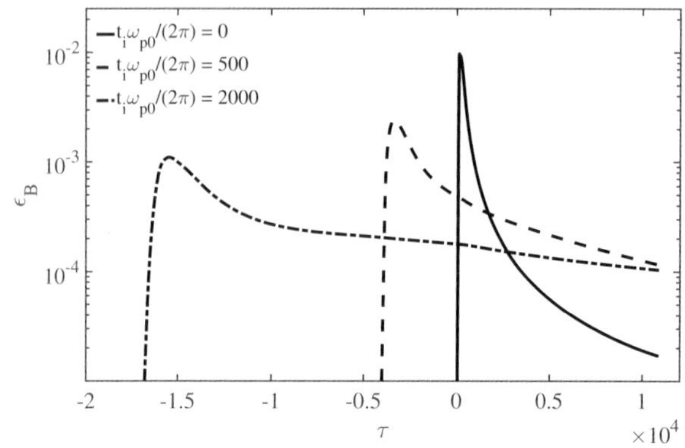

Figure 1. Magnetization parameter (the ratio of the magnetic energy density to the particles energy density) ϵ_B over time for different injections: instantaneous (solid line), $\omega_{p0}t_i/(2\pi) = 500$ (dashed line), and $\omega_{p0}t_i/(2\pi) = 2000$ (dash-dotted line). In all simulations $\delta = 2$.

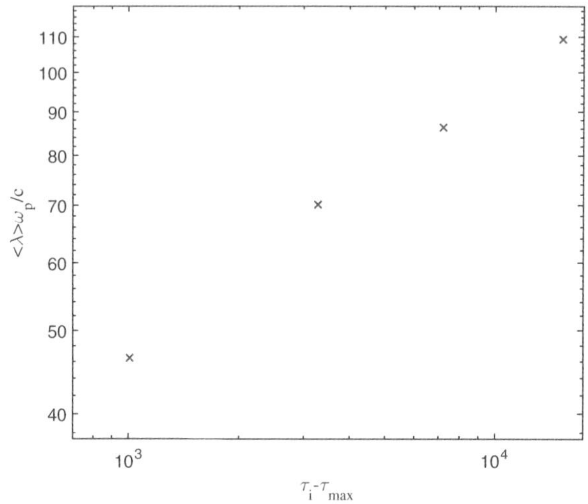

Figure 2. The average wavelength of magnetic field filaments $\langle\lambda\rangle$ at the end of the injection phase as a function of $\tau_i - \tau_{max}$, where τ_{max} is the time of the maximum magnetic field energy.

grow with relatively small spatial scales. The fastest growing modes have their wave vectors nearly perpendicular to the axis of anisotropy. The growing magnetic field deflects the particles and smears out the anisotropy, eventually stopping the magnetic field growth. After that the magnetic filed starts to decay. The decay rate is largest for the short wavelengths. Thus, in the decay phase the spatial scale of the magnetic turbulence grows.

In Fig. 1 we plot three examples of the magnetic field evolution with different injection duration (as observed in our simulations). The maximum value of the observed peak magnetization is about 0.01. This relatively small value is probably due to the small anisotropy of injected component used in our model. It is clearly seen, that the longer injection time corresponds to lower peak magnetization and to longer decay time. As a result, after certain time, the magnetic field generated by long injection overcomes the magnetic field generated by short injection.We attribute this to a larger spatial scale of a magnetic field at the moment of maximum magnetization and to subsequent washing-out of modes with short wavelengths during the rest of injection period (See Fig. 2). From

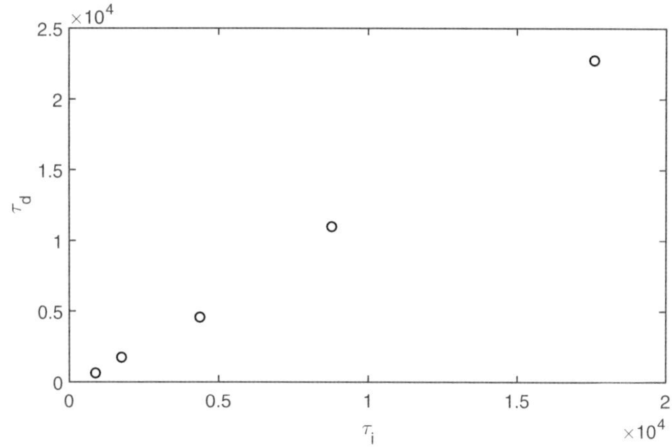

Figure 3. Dissipation timescale τ_d of a magnetic field (it is defined as $\tau_{\rm d} = (\tau_2 - \tau_1)/\ln(\epsilon_{\rm B}(\tau_1)/\epsilon_{\rm B}(\tau_2))$, where $\tau_1 = 0.1\,\tau_{\rm i}$ and $\tau_2 = 0.25\,\tau_{\rm i}$ are two moments of time after injection has stopped) as a function of the injection time.

the simulations we found that the typical spatial scale of the generated magnetic field is much larger than the plasma skin-depth $c/\omega_{\rm p}$ and increases with injection time as

$$\langle\lambda\rangle \propto (\tau_{\rm i} - \tau_{\rm max})^{1/3}.$$

In Fig. 3 we present the measured dependence of the magnetic field dissipation time as a function of the injection time. We observed that the field dissipation time approximately equals to the injection duration a least up to $t_i = 20000\omega_{\rm p}^{-1}$.

We show that the prolonged injection leads to build-up of relatively large-scale magnetic field in the upstream due to development of the Weibel instability. Moreover, this large-scale magnetic field could be amplified at the shock front preserving its scale and then could survive for a long time in the downstream, explaining efficient synchrotron emission from relativistic shocks, e.g. in GRBs.

This work was supported by the Russian Science Foundation grant No 16-12-10528.

References

Achterberg, A., Gallant, Y. A., Kirk, J. G., & Guthmann, A. W. 2001, *MNRAS*, 328, 393
Gruzinov, A. 2001, *Astrophys. Jour. Lett.*, 563, L15
Medvedev, M. V. & Loeb, A., 1999, *ApJ*, 526, 697
Spitkovsky, A. 2008 *Astrophys. Jour. Lett.*, 682, L5
Keshet, U., Katz, B., Spitkovsky, A., & Waxman, E., 2009, *ApJ*, 693, L127
Sironi, L., Spitkovsky, A., & Arons, J., 2013, *ApJ*, 771, 54
Sironi L., Keshet, U., & Lemoine, M., 2015, *Space Sci. Rev.*, 191, 519
Medvedev, M. V. & Zakutnyaya, O. V., 2009, *ApJ*, 696, 2269
Milosavljevic, M. & Nakar, E., 2006, *ApJ*, 651, 979
Derishev, E. & Piran, T. 2016, *MNRAS*, 460, 2036
Arber, T. D., Bennett, K., Brady, C. S., Lawrence-Douglas, A., Ramsay, M. G., Sircombe, N. J., Gillies, P., Evans, R. G., Schmitz, H., Bell, A. R., & Ridgers, C. P., 2015, *Plasma Physics and Controlled Fusion*, 57, 113001
Garasev, M. & Derishev, E. 2016, *MNRAS*, 461 (1), 641

New Frontiers in Black Hole Astrophysics
Proceedings IAU Symposium No. 324, 2016
Andreja Gomboc, ed.

doi:10.1017/S174392131601228X

Low-Γ jets from Compact Binary Mergers as Candidate Electromagnetic Counterparts to Gravitational Wave Sources

Gavin P. Lamb and Shiho Kobayashi

Astrophysics Research Institute, Liverpool John Moores University, L3 5RF, UK
email: g.p.lamb@2010.ljmu.ac.uk

Abstract. Compact binary mergers, with neutron stars or neutron star and black-hole components, are thought to produce various electromagnetic counterparts: short gamma-ray bursts (GRBs) from ultra-relativistic jets followed by broadband afterglow; semi-isotropic kilonova from radioactive decay of r-process elements; and late time radio flares; etc. If the jets from such mergers follow a similar power-law distribution of Lorentz factors as other astrophysical jets then the population of merger jets will be dominated by low-Γ values. The prompt gamma-rays associated with short GRBs would be suppressed for a low-Γ jet and the jet energy will be released as X-ray/optical/radio transients when a shock forms in the ambient medium. Using Monte Carlo simulations, we study the properties of such transients as candidate electromagnetic counterparts to gravitational wave sources detectable by LIGO/Virgo. Approximately 78% of merger-jets result in failed GRB with optical peaks 14-22 magnitude and an all-sky rate of 2-3 per year.

Keywords. gamma rays: bursts, jets and outflows, gravitational waves.

1. Introduction

The merger of a binary system due to gravitational wave (GW) emission where the binary components are either neutron stars (NS) or stellar mass black holes (BH) is a potential source of GWs detectable by current gravitational wave detectors (e.g. LIGO/Virgo). In 2015, advanced LIGO made two unambiguous detections at 5-σ (GW150914 and GW151226) and a third possible detection at 87% confidence (LVT151012) of the inspiral and merger of binary stellar mass BH systems (The LIGO Scientific Collaboration *et al.* 2016). An electromagnetic (EM) counterpart to the inspiral and merger of a binary BH system is not generally expected and no EM counterpart was detected for any of the observed GW signals (Copperwheat *et al.* 2016).

The next GW breakthrough will be the detection of the inspiral and merger of NS-BH or NS-NS systems. For such mergers we expect various EM counterparts, where the nature of the counterpart depends on the viewing angle (Metzger & Berger 2012). When a NS-BH or NS-NS system merge, the merger ejecta forms an accretion disk, tidal tail and disk wind. The rapid accretion from a disk onto a newly formed BH is thought to power bipolar ultrarelativistic jets responsible for short GRBs and their afterglow, the typical timescale for a short GRB is seconds and hours to days for the associated afterglow (Woosley & Bloom 2006; Nakar 2007; Berger 2014). The more isotropic ejecta can power r-process nucleosynthesis where radioactive decay gives rise to macronova emission, these occur on a day to weeks timescale (Tanvir *et al.* 2013; Berger *et al.* 2013; Tanaka 2016). At later times, approximately a month, the merger ejecta can interact with the ambient medium giving rise to radio flares (Nakar & Piran 2011; Hotokezaka *et al.* 2016).

2. Gamma-ray Bursts and Failed GRBs

Except for two cases (Cenko *et al.* 2013, 2015) we currently detect GRBs and their afterglow by a high energy γ-ray trigger. The variability timescale of the prompt γ-ray emission indicates a dependence for the emiting region, $R_d \propto \Gamma^2$, where R_d is the dissipation radius and Γ the bulk Lorentz factor. The photospheric radius $R_\star$, the point at which the fireball becomes optically thin, can be conservatively estimated by considering the density of electrons associated with a baryonic outflow, $R_\star \propto E^{1/2}\Gamma^{-1/2}$. These conditions require GRBs to be produced in ultrarelativistic jets with $\Gamma \sim 100$.

For jets with $\Gamma << 100$ the dissipation radius falls below the photoshere and the prompt γ-rays would be suppressed. Such a failed GRB will still have a broadband orphan afterglow. The observation of afterglows associated with the relativistic jets from binary NS mergers is currently biased by the requirement of a high energy γ-ray trigger, GW triggered searches may reveal a hidden population of low-Γ merger jets by their on-axis orphan afterglow. Astrophysical jets from other accreting BH systems (e.g. AGN, blazars) follow a simple power law distribution with a negative index, $N(\Gamma) \propto \Gamma^{-a}$, where a is ~ 2 for blazars (Saikia *et al.* 2016). If the jets from mergers follow a similar distribution then failed GRBs would outnumber short GRBs.

3. Monte Carlo

To test whether the on-axis orphan afterglow of a failed GRB from a compact stellar merger would be observable, given a GW detection, a Monte Carlo of 2×10^5 events was generated. The events followed the Wanderman & Piran (2015) redshift and luminosity distribution for short GRBs. The bulk Lorentz factor followed a simple power law with index $a = 1.75$ in the limits $3 \leqslant \Gamma \leqslant 10^3$, and independent of a bursts energy. For details of the Monte Carlo and radiative transfer see Lamb & Kobayashi (2016).

4. Results

We found that for a sample in the range < 300 Mpc, the LIGO/Virgo detection limit for face-on NS-NS mergers, that events below a line given by $\Gamma \sim 16(E_{\rm K}/10^{50}{\rm erg})^{0.15}$ always result in failed GRBs, where $E_{\rm K}$ is the isotropic equivalent kinetic energy. For the parameters used we found that 78% of mergers resulted in failed GRBs; the peak flux and time for the on-axis orphan afterglows from the failed GRBs is shown in figure 1. The model parameters used for the afterglows are: $n = 10^{-1}$ protons cm^{-3}, microphysical parameters $\epsilon_B = 10^{-2}$ and $\epsilon_e = 10^{-1}$, the index of the power-law distribution of random electrons accelerated at shock $p = 2.5$, and the jet half-opening angle $\theta_j = 20°$ ensuring that the jet break time is later than the deceleration time for our sample and is within the limits $16 \pm 10°$ found by Fong *et al.* (2015) for short GRBs. The jet opening angle plays a role only when we estimate the jet break time.

5. Discussion & Conclusion

GW emission is strongest on-axis from a merger event (Kochanek & Piran 1993), thus the probability of an on-axis merger is higher, given a GW detection, than if we consider only the isotropic case (see figure 2). If the Lorentz factor of a jet is correlated with the opening angle, where lower Γ gives a wider half-opening angle (as indicated for long GRBs by Ghirlanda *et al.* 2013), then the rates of on-axis orphan afterglow following a GW trigger will be higher than those given here. By using the rate of short GRBs

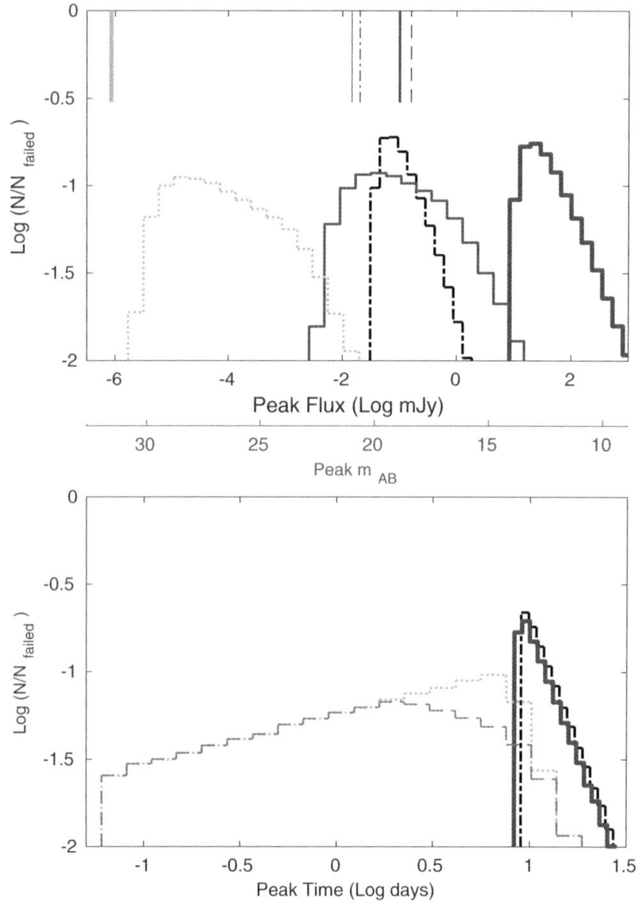

Figure 1. The peak-flux (top panel) and peak-time (bottom panel) distribution of on-axis orphan afterglow from failed GRB events within 300 Mpc. The distributions are normalized by the total number of failed GRBs. X-ray (dotted green line), optical (thick solid red line), radio 10 GHz (thick solid blue line) and radio 150 MHz (thick dash-dotted black line). The vertical lines in the top panel indicate the sensitivity limits of telescopes (thick green XRT, thin red optical ~ 2 m, dash-dotted SKA1-Low, and dashed 48 LOFAR), and the dashed magenta line in the bottom panel shows the distribution of bright events $m_g \leqslant 21$ (Lamb & Kobayashi 2016).

within the LIGO detection volume predicted from the Swift short GRB rate by Metzger & Berger (2012), 0.03 per year, and considering the all sky rate - we find approximately 2.6(26) on-axis orphan afterglows from NS-NS(NS-BH) mergers per year when assuming a jet half-opening angle of 20°.

EM counterparts to GW emission from NS-NS and NS-BH mergers include radio flares, macronova, short GRBs, and on and off axis (orphan) afterglow. A GW triggered search for EM counterparts could reveal a hidden population of low-Γ merger jets from such events and on-axis orphan afterglow of failed GRBs are therefore a strong candidate for EM follow-up searches. The (non)detection of such sources will help constrain the Lorentz factor distribution of such jets (e.g. clustered at high-Γ, a power-law, a log-normal, or multiple populations), and provide constraints on the acceleration process of relativistic jets.

This research was supported by STFC grants. GPL was supported by an IAU travel grant.

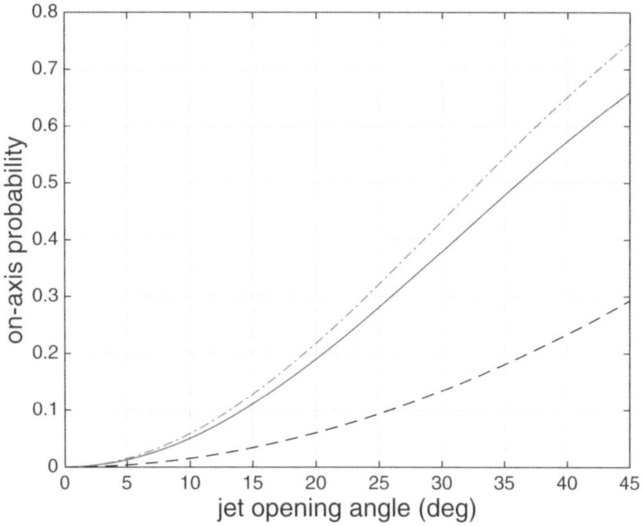

Figure 2. On-axis probability as a function of a jet half-opening angle θ_j. The beaming factor $f_b = 1 - \cos\theta_j$ (black dashed line), given a GW detection the simple approximation $A^3 f_b$ (blue dash-dot line), and the numerical results (red solid line).

References

The LIGO Scientific Collaboration, the Virgo Collaboration, Abbott, B. P., *et al.* 2016, arXiv:1606.04856

Copperwheat, C. M., Steele, I. A., Piascik, A. S., *et al.* 2016, *MNRAS*, 462, 3528

Metzger, B. D. & Berger, E. 2012, *ApJ*, 746, 48

Woosley, S. E. & Bloom, J. S. 2006, *ARA&A*, 44, 507

Nakar, E. 2007, *Phys. Rep.*, 442, 166

Berger, E. 2014, *ARA&A*, 52, 43

Tanvir, N. R., Levan, A. J., Fruchter, A. S., *et al.* 2013, *Nature*, 500, 547

Berger, E., Fong, W., & Chornock, R. 2013, *ApJL*, 774, L23

Tanaka, M. 2016, *Advances in Astronomy*, 2016, 634197

Nakar, E. & Piran, T. 2011, *Nature*, 478, 82

Hotokezaka, K., Nissanke, S., Hallinan, G., *et al.* 2016, arXiv:1605.09395

Cenko, S. B., Kulkarni, S. R., Horesh, A., *et al.* 2013, *ApJ*, 769, 130

Cenko, S. B., Urban, A. L., Perley, D. A., *et al.* 2015, *ApJL*, 803, L24

Saikia, P., Körding, E., & Falcke, H. 2016, *MNRAS*, 461, 297

Wanderman, D. & Piran, T. 2015, *MNRAS*, 448, 3026

Lamb, G. P. & Kobayashi, S. 2016, *ApJ*, 829, 112

Fong, W., Berger, E., Margutti, R., & Zauderer, B. A. 2015, *ApJ*, 815, 102

Kochanek, C. S. & Piran, T. 1993, *ApJL*, 417, L17

Ghirlanda, G., Ghisellini, G., Salvaterra, R., *et al.* 2013, *MNRAS*, 428, 1410

New Frontiers in Black Hole Astrophysics
Proceedings IAU Symposium No. 324, 2016
Andreja Gomboc, ed.

doi:10.1017/S1743921317001442

Search for High Energy emission from GRBs with MAGIC

Alessio Berti[1] for the MAGIC GRB group

[1]INFN Trieste, University of Trieste
via Valerio 2, I-34127, Trieste, Italy
email: Alessio.Berti@ts.infn.it

Abstract. Gamma-Ray Bursts (GRBs) are the most violent explosions in the Universe, releasing a huge amount of energy in few seconds. While our understanding of the prompt and the afterglow phases has increased with *Swift* and *Fermi*, we have very few information about their High Energy (HE, $E \lesssim 100\,\mathrm{GeV}$) emission components. This requires a ground-based experiment able to perform fast follow-up with enough sensitivity above $\sim 50\,\mathrm{GeV}$. The MAGIC (Major Atmospheric Gamma-ray Imaging Cherenkov) telescopes have been designed to perform fast follow-up on GRBs thanks to fast slewing movement and low energy threshold ($\sim 50\,\mathrm{GeV}$). Since the beginning of the operations, MAGIC followed-up 89 GRBs in good observational conditions. In this contribution the MAGIC GRBs follow-up campaign and the results which could be obtained by detecting HE and Very High Energy (VHE, $E \gtrsim 100\,\mathrm{GeV}$) γ-rays from GRBs will be reviewed.

Keywords. Gamma-Ray Bursts, MAGIC, high-energy emission

---

1. Introduction

Gamma-Ray Bursts are transient sources caused by explosive events with a relevant release of energy in the form of HE radiation. Even after almost 50 years since their discovery, there are still open questions about them (see Kumar and Zhang (2015) for a review). One of these concerns the possible emission by GRBs of VHE photons. Observations at lower energies by the *Large Area Telescope* (LAT) instrument onboard the *Fermi* satellite prove that is relatively common for GRBs to emit photons with GeV energies, with the highest energy photon (94 GeV) detected from GRB130427A (Ackermann *et al.* (2014)). Two peculiar features of the GeV emission are the delay with respect to the lower energies and the long lasting duration (see for example Abdo *et al.* (2009)), even up to one day again in the case of GRB130427A. Usually the GeV spectrum can be fitted by an extension of the power law describing the low energy emission, but in some GRBs a second component is clearly present in the LAT data. The physical processes underlying this kind of emission could be better understood with the detection of a possible VHE signal from GRBs with ground-based instruments, like Cherenkov telescopes.

Among the current Imaging Atmospheric Cherenkov Telescopes (IACTs), MAGIC has a low energy threshold and a fast repositioning speed which allow it to follow-up GRBs in an efficient way. In this contribution we describe the possible models of VHE emission and the MAGIC follow-up program for GRBs, focusing on its performance after the new automatic GRB observation procedure was implemented. No VHE signal has yet been detected from GRBs by IACTs experiments.

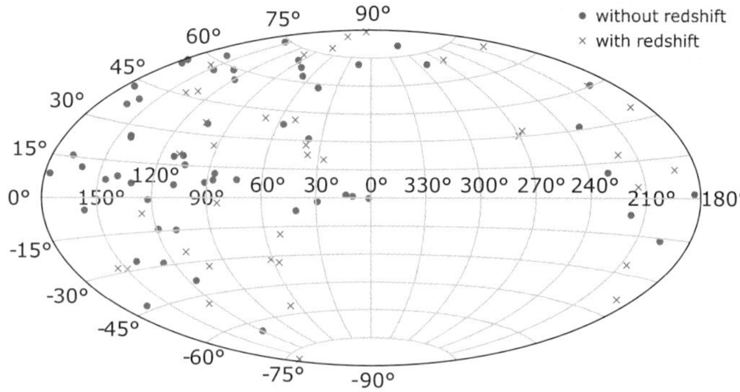

Figure 1. Galactic Aitoff projection of the positions of the GRBs followed up by MAGIC.

2. VHE emission models for GRBs

Possible VHE emission from GRBs could be explained in first approximation as an extension of the low-energy processes. The most simple case is the synchrotron emission by relativistic electrons in the forward shock of the afterglow. As shown by Kumar and Duran (2010), in this case the HE synchrotron emission can be used to predict very well the flux of low energy emission (e.g. optical and X-ray band). Synchrotron models can have a leptonic (see Zhang and Meszaros (2001)) or hadronic origin (see Böttcher and Dermer (1998) and Pe'er and Waxman (2004)). A second model is the Self-Synchrotron Compton (SSC) emission in the afterglow (Sari and Esin (2001)), which however fails in explaining the MeV-GeV delayed emission. More exotic processes include the upscattering of prompt γ-rays by electrons (Meszaros and Rees(1994); Belobodorov (2005); Fan And Wei (2005)), upscattering of CMB photons (Plaga (1995)) or Compton scattering by electrons of photospheric thermal emission (Toma *et al.* (2011)).

Moreover, the detection of the VHE emission of GRBs could answer some pending questions like their possible role in the production of UHECRs or neutrinos (see Waxman (1995) and Paczynsky and Xu (1994)).

3. GRBs follow-up by MAGIC

MAGIC is a system of two Cherenkov telescopes, each with a diameter of 17 m. They are located in the Canary island of La Palma (28.8° N, 17.8° W), almost at the top of the Roque de Los Muchachos mountain at an altitude of 2200 m a.s.l. (Aleksic *et al.* (2016a)). MAGIC observes gamma-ray sources in stereoscopic mode, reaching a sensitivity of less than 0.7% of the Crab Nebula flux for energies above 220 GeV in 50 h and an energy threshold at trigger level of ~50 GeV at zenith (Aleksic *et al.* (2016b)), or even less using the so called *sum trigger*. GRBs are one of the primary targets for MAGIC, being one of its Key Observation Programs (KOP). MAGIC was designed from the beginning to be able to react very fast to external alerts coming from the Gamma-Ray Burst Coordinate Network (GCN†).

Alerts from satellite-based experiments like *Swift* and *Fermi* are sent to MAGIC through TCP/IP socket and a specific program receives and processes them in order to decide if they comply with predefined observation criteria. When this happens, the sky coordinates of the event are sent to the central control of the MAGIC telescopes, which are promptly repointed to the GRB position for data taking. Given their transient nature and their short duration, in particular as far as the prompt emission is concerned,

† `http://gcn.gsfc.nasa.gov/`

the follow-up of GRBs should be done as fast as possible after the onset. The MAGIC telescopes were built with a lightweight carbon-fiber structure which allows them to repoint within few tens of seconds after the alerts.

Since 2013, a new automatic procedure was adopted in order to reduce the number of failures (most of them hardware related), namely without stopping the Data Acquisition (DAQ) system during the slewing. The observation of the GRB is then carried on up to 4 h after the onset taking into account the observational conditions. In the last year, the MAGIC GRB group considered to change the GRB observation strategy, extending the standard four hours of observation in the case of particularly bright events. We refer to this new approach as "late time observations". The primary targets that could be the aim of the new strategy are LAT detected GRBs, since they are likely to present GeV emission extended in time in the afterglow phase. Although the alert system was not yet updated for this kind of targets, there were few events (GRB160310A, GRB160509A, GRB160623A, GRB160625B) which were followed at late times and whose analysis is ongoing. Even if only flux upper limits could be calculated, they can give important informations about the possible VHE emission, in particular in the case of brighter events:

(*a*) constrain the phase space of the several parameters describing the emission in the VHE band;

(*b*) discard or validate emission models proposed to be at the origin of the possible VHE emission;

(*c*) put limits on SSC emission if it dominates over the synchrotron component (see Beniamini *et al.* (2015));

(*d*) obtain lower limits on the intergalactic magnetic field if VHE emission comes from pair echoes (Takahashi *et al.* (2008)).

An example of the use of flux upper limits is the observation of the afterglow of GRB090102 by MAGIC (Aleksic *et al.* (2014)). Flux upper limits for this burst were derived for the GeV-TeV emission both from MAGIC and Fermi-LAT observations (see Figure 2). The use of the sum trigger allowed for MAGIC an analysis energy threshold of ~ 30 GeV. The results were compared with the simulated afterglow emission by different processes, e.g. synchrotron by electrons or protons and SSC by electrons. Although low-energy observations, especially in the optical, constrained well the parameters characterizing the burst, it was not possible to discriminate between a leptonic or hadronic emission process.

4. Conclusions and prospects

The scientific reward which could be obtained by detecting a VHE signal from a GRB is considerable. Beside understanding the physical processes behind this kind of emission, the detection could improve our understanding of other science cases related to GRBs.

First of all, given the cosmological origin of GRBs, their spectrum at very high energies is attenuated by the Extragalactic Background Light (EBL). The characterization of the attenuated spectrum, with a reasonable assumption on the unabsorbed one, can be used to constrain the intensity of the EBL in the infrared region. It is crucial to have a low energy threshold and enough sensitivity in that energy band, since also for moderate redshift GRBs the attenuation is significant.

A second possibility is the discrimination between leptonic and hadronic emission models for the afterglow phase of GRBs. As shown in Aleksic *et al.* (2014), hadronic models can dominate over leptonic ones in some regions of the phase space of the GRBs parameters. However, a good knowledge of the GRBs parameters could be obtained by lower energies observations, which points out to the strong necessity of multi-wavelength observations of GRBs.

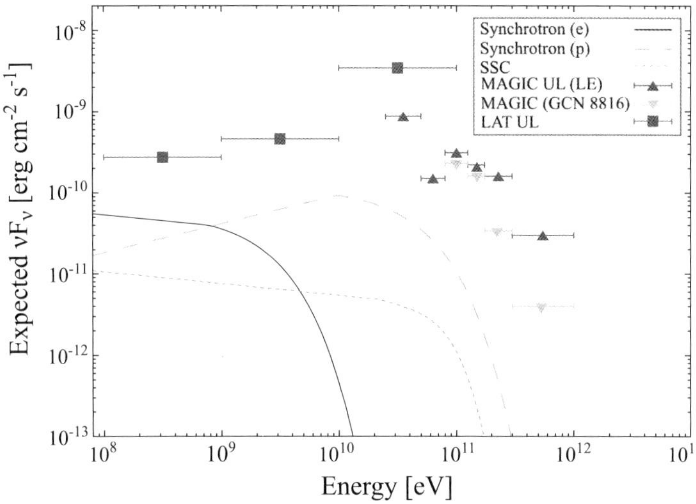

Figure 2. Modelled emission for the afterglow of GRB090102. Up-pointing triangles are MAGIC flux upper limits (95% CL) with optimized low-energy analysis (regular ones are down-pointing triangles). Squares represent LAT 95% CL ULs. The solid and short-dashed lines show the energy flux expected by electron synchrotron and SSC processes, while the long-dashed curve corresponds to the synchrotron emission from protons. From Aleksic *et al.* (2014)

The MAGIC GRB group is now investigating ways to improve the analysis at low energies and the observation strategy. Another improvement could be achieved by using in stereoscopic mode the sum trigger, which is being developed and is supposed to lower routinely the MAGIC energy threshold down to $\sim 30\,\mathrm{GeV}$.

Finally, a forthcoming publication comprising the analysis of all the GRBs observed in good conditions with the new automatic procedure by MAGIC is in preparation.

References

Abdo, A. A., *et al.* 2009, *ApJ* (Letters) 706, L138
Ackermann, M., *et al.* 2014, *Science*, 343, 6166, 42
Aleksic, J., *et al.* 2014, *MNRAS* 30, 490
Aleksic, J., *et al.* 2016, *Astroparticle Physics* 72, 61
Aleksic, J., *et al.* 2016, *Astroparticle Physics* 72, 76
Beloborodov, Andrei M. 2005, *ApJ* 618, L13
Beniamini *et al.* 2015, *MNRAS* 454, 1073
Böttcher, M. & Dermer, C. D. 1998, *ApJ* 499, L131
Fan, Y. Z. & Wei, D. M. 2005, *MNRAS* 364, L42
Kumar, P. & Duran, R. Barniol 2010, *MNRAS* 409, 226
Kumar, P. & Zhang, B. 2015, *Physics Reports* 561, 1-109
Meszaros, P. & Rees, M. J. 1994, *MNRAS* 269, L41
Paczynski, B. & Xu, G. 1994, *ApJ* 427, 708
Pe'er, A. & Waxman, E. 2004, *ApJ* 603, L1
Plaga, R. 1995, *Nature* 374, 430
Sari, R. & Esin, Ann A. 2001, *ApJ* 548, 787
Takahashi *et al.* 2008, *ApJ* (Letters) 687, L5
Toma, K., *et al.* 2011, *MNRAS* 415, 1663
Waxman, E. 1995, *Phys. Rev. Lett.* 75, 386
Zhang, B. & Meszaros, P. 2001, *ApJ* 559, 110

New Frontiers in Black Hole Astrophysics
Proceedings IAU Symposium No. 324, 2016
Andreja Gomboc, ed.

doi:10.1017/S1743921317003295

Extremely Bright GRB160625B with Short-Soft Precursor and long-hard extended emission: Hints for long-term evolution of the GRB Ejecta

Shu-Qing Zhong, Jing Lü, Hou-Jun Lü*, Hai-Ming Zhang, Xiao-Li Huang and En-Wei Liang

GXU-NAOC Center for Astrophysics and Space Sciences, Department of Physics, Guangxi University, Nanning 530004, China
Guangxi Key Laboratory for the Relativistic Astrophysics, Nanning 530004, China
email: lhj@gxu.edu.cn

Abstract. In this paper, we focus on studying the high energy emission of GRB 160625B. The lightcurve of prompt emission is composed of three episodes: short-soft precursor, hard main burst, and possible long extended emission. The spectra of first and third episode can be fitted by a multi-color blackbody and cutoff power-law model, respectively. However, the spectrum of second episode was contributed by both multi-color blackbody and cutoff power-law. One can estimate the Lorenz factor of jet of first two episodes by invoking photosphere model as $\Gamma_0 \sim 175$ and 1694, respectively. It suggests that the ejecta of this case evolved from photosphere dominated initially to internal shock later. On the other hand, the optical emission is very bright during the second episode, which is likely a prompt optical emission. Finally, a more shallower normal decay segment appeared, which is consistent with standard external shock model.

Keywords. gamma rays: bursts; individual: GRB 160625B.

1. Introduction

The origin of gamma-ray burst (GRB) prompt emission is very important to understand and it is related with the ejecta composition, energy dissipation, particle acceleration, and radiation mechanism of GRB jet (Zhang 2011). However, its physical origin remains unclear even after more than forty years of study, e.g. what are the central engine and progenitors of GRBs? How does the ejecta evolve during prompt emission phase? What is the radiation mechanism of prompt emission and later afterglow. Several observational evidences support the existence of two types (black hole or magnetar) central engine and three types (photospheric, leptonic synchrotron or hadronic emission) of emission mechanisms in the relativistic jet are being actively discussed. From observational point of view, a small fraction of GRBs have a soft precursor emission before main burst, and tens seconds quiet phase between precursor and main burst. This was used to constraint the properties of central engine and progenitors. On the other hand, the thermal emission was detected during the prompt emission in several *Fermi* GRBs, such as GRB090902B (Ryde *et al.* 2010). The thermal emission in the prompt phase can be used to constrain the Lorenz factor of GRB jet. Moreover, the optical emission during the prompt emission (or roughly tracking gamma-ray emission) can constrain the emission site by requiring that the synchrotron self-absorption frequency is below the optical band if we assume that this optical emission is from the same emission region as gamma-rays.

GRB 160625B, one interesting case, was detected by *Fermi* Gamma-Ray Burst Monitor (GBM; Burns 2016), the Large Area Telescope (LAT), Konus-Wind, as well as *Swift* X-ray telescope (XRT). The optical and radio afterglows were also detected. More interesting, the optical emission is very bright, and can be compared to GRB 990123 and GRB 080319B. Multi-wavelength observations of this case could provide the opportunity to explore the physical origin of the GRB. Especially, the spectral properties of its prompt emission are quite similar to those of GRB 090902B, which was claimed to show evidence of thermal emission (Ryde *et al.* 2010). In this work, we perform the time-integrated spectral analysis for each part of the prompt emission of GRB 160625B, to obtain the characteristic of the thermal component and the evolution of ejecta, as well as the afterglow modeling analysis. In §2, we give the observations report and data analysis. The spectra fitting result with multi-color blackbody and afterglow model are presented in §3. Conclusions are drawn in §4.

2. Observations and Data Analysis

GRB 160625B triggered two times and one time during the prompt emission phase by *Fermi*/GBM (Burns 2016) and LAT (Dirirsa *et al.* 2016), respectively. The prompt emission lightcurve of this burst is composed of three parts: a short soft-precursor emission with duration about 1 second initially (part I), then a main extremely bright episode with a duration of about several ten seconds appeared after about 200 seconds since GBM first trigger (Batsch *et al.* 2016; Svinkin *et al.* 2016; part II), finally, one long weak emission is following the part II after 11 minutes (part III). Moreover, the *Swift* X-ray telescope (XRT) also observed the burst, and the X-ray lightcurve is from the UK *Swift* Science Data Centre at the University of Leicester. The optical telescopes (Pi of the Sky; Batsch *et al.* 2016) and Global MASTER-Crimea (Lipunov *et al.*2016) also observed this case. All optical observations are extracted from GCN reports (Oates 2016; Kuroda *et al.* 2016; Troja *et al.* 2016; Mazaeva *et al.* 2016; Moskvitin 2016; Mazaeva *et al.* 2016; Guidorzi *et al.* 2016; Valeev *et al.* 2016; Bikmaev *et al.* 2016; Mazaeva *et al.* 2016). The optical lightcurve has a bright flare during the second episode, which is likely an optical prompt emission. Then, following a steep decay ($t^{-3.5}$), it transforms to normal decay segment after 10^4 second (Fig. 1). The X-shooter spectrograph observed the optical spectrum to measure the redshift $z = 1.406$ (Xu *et al.* 2016). The lightcurve is shown in Figure 1.

3. Spectra fitting and afterglow model result

We analyze the spectra of prompt emissions for the time slices of $-0.3 \sim 1.1$ s, $180 \sim 189$ s and $522 \sim 890$ s, which are corresponding to the parts I, II and III in Figure 1. Following Ryde *et al.*(2010), we used multi-color blackbody (mBB) model to fit the spectra of each parts. The first spectrum of part I can be well fitted by mBB model with $T_{\max} = 25.2 \pm 1.1$ keV, $q = 1.63 \pm 0.20$ and a goodness of reduced $\chi^2 = 387/120$. For part II, its spectrum is fit by multi-color blackbody combined with a cutoff power-law, one has $T_{\max} = 967 \pm 49$ keV, $q = 0.92 \pm 0.01$, photon index $\Gamma_{cut} = 1.33 \pm 0.02$ and a high-cut energy 14989 ± 1154 keV. For the third part, cutoff power-law and additional power-law model were used to fit the data. Best fit values are $\Gamma_{cut} = 1.64 \pm 0.04$ and a high-cut energy 692 ± 577 keV, and photon index $\Gamma_{pow} = 1.98 \pm 0.50$. The fitting results are shown in Fig. 2. One can estimate the Lorenz factor of jet of first two episodes by invoking photosphere model as $\Gamma_0 \sim 175$ and 1694, respectively. On the other hand, we used standard afterglow model to fit both X-ray and optical data, and search for the best parameters distributions. We obtain $\epsilon_e = 0.14 \pm 0.06$, $\epsilon_B = (1.98^{+1.53}_{-1.13}) \times 10^{-7}$,

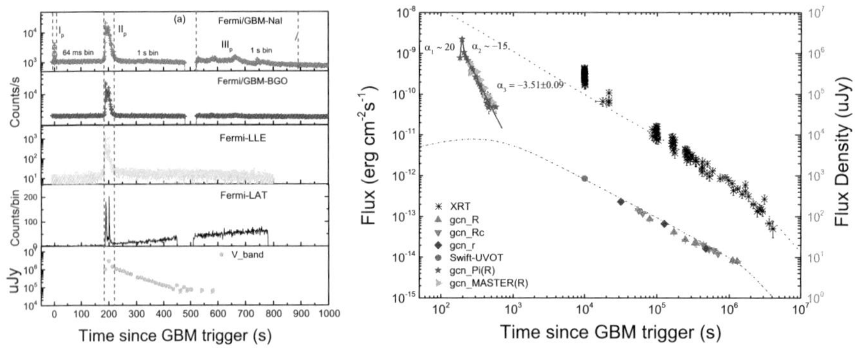

Figure 1. The *left panel*: illustrates the *Fermi* and optical observations of GRB 160625B. *Right panel*: The results of modeling the optical and X-ray afterglow light curves with the standard forward shock model.

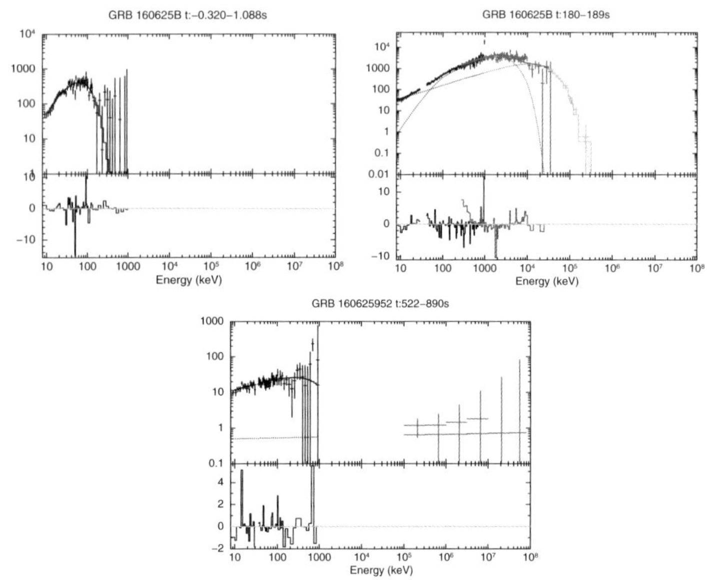

Figure 2. The spectra fitting by multi-color blackbody (part I), multi-color blackbody and cutoff power-law model (part II), and cutoff power-law model (part III). The unit of Y axis is erg cm^{-2} s^{-1}.

$n = 37^{+14}_{-15}$ cm^{-3}, $E_{K,iso} = (1.33^{+0.40}_{-0.44}) \times 10^{55}$ erg, $p = 2.17^{+0.08}_{-0.06}$. Here, we fixed the Lorenz factor as $\Gamma_0 \sim 1694$ in the second episode of prompt emission.

4. Discussion & Conclusion

GRB 160625A is an interesting case. Firstly, it triggered two times and one time during the prompt emission phase by *Fermi*/GBM and LAT, as well as Konus-Wind and other ground optical telescopes. It is the brightest case observed by Konus-Wind in the prompt emission phase, and has the highest isotropic energy ($E_{\gamma,\rm iso} \sim 5 \times 10^{54}$erg) measured by Konus-Wind for more than 21 years of its GRB observations. Secondly, the lightcurve of prompt emission is composed of three episodes: initial short-soft precursor emission (part I), hard main burst after about 200 seconds since GRB trigger (part II), and a weak long emission (part III). Thirdly, the optical emission is as bright as 8.04 magnitude starting from second episode of prompt emission. We find that the spectra of the first and third episode can be fitted by a multi-color blackbody and cutoff power-law model, respectively. However, the spectra of the second episode shows contributions of both multi-color blackbody and cutoff power-law. One can estimate the Lorenz factor of jet of the first two episodes by invoking photospheric model as $\Gamma_0 \sim 175$ and 1694, respectively. This suggests that the photospheric emission is dominating initially, but the ejecta is not in interaction with interstellar medium, therefore there is no afterglow emission during this phase. Then, the photosphere emission is already present in the second episode, but the contribution of internal shock is more and more important. Finally, the internal shock is dominating in the third episode, no thermal component emission is needed. Moreover, the Lorenz factor of first short-soft emission is much lower than during the main burst phase. It is also supported that the photosphere emission is dominating earlier, and the internal shock is more and more important staring from the second episode. On the other hand, the optical emission is very bright during the second episode, which is likely a prompt optical emission. Then followed by a steep decay phase, which may be a tail emission of prompt optical, or due to reversed shock contribution. Finally, a more shallower normal decay segment appeared, which is consistent with standard external shock model.

5. Acknowledgement

This work is supported by the National Basic Research Program of China (973 Program, grant No. 2014CB845800), the National Natural Science Foundation of China (Grant No. 11603006, 11533003), the Strategic Priority Research Program The Emergence of Cosmological Structures of the Chinese Academy of Sciences (grant XDB09000000), the Guangxi Science Foundation (2016GXNSFCB380005, 2013GXNSFFA019001), the One-Hundred-Talents Program of Guangxi colleges, and Scientific Research Foundation of GuangXi University (Grant No XGZ150299). We also acknowledge the use of the public data from the *Fermi* and *Swift* data archive.

References

Batsch T., *et al.*, 2016, *GCN*, 19615, 1
Bikmaev, I., *et al.*, 2016, *GCN*, 19651, 1
Burns, E., 2016, *GCN*, 19587, 1
Dirirsa, F., *et al.*, 2016, *GCN*, 19580, 1
Guidorzi, C., *et al.*, 2016, *GCN*, 19640, 1
Kuroda, D., *et al.*, 2016, *GCN*, 19599, 1
Lipunov, V., *et al.*, 2016, *GCN*, 19683, 1
Mazaeva, E., *et al.*, 2016, *GCN*, 19605, 1
Mazaeva, E., *et al.*, 2016, *GCN*, 19619, 1
Mazaeva, E., *et al.*, 2016, *GCN*, 19680, 1
Moskvitin, A., 2016, *GCN*, 19611, 1
Oates, S., 2016, *GCN*, 19589, 1
Ryde, F., *et al.*, 2010, *ApJ*, 709, L172
Svinkin, D., *et al.*, 2016, *GCN*, 19604, 1
Troja, E., *et al.*, 2016, *GCN*, 19588, 1
Valeev, A. F., *et al.*, 2016, *GCN*, 19642, 1
Xu, D., *et al.*, 2016, *GCN*, 19600, 1
Zhang, B., 2011, CRPhy, 12, 206

New Frontiers in Black Hole Astrophysics
Proceedings IAU Symposium No. 324, 2016
Andreja Gomboc, ed.

doi:10.1017/S1743921317001260

Lorentz Factor Evolution Patterns within Relativistic Jets of GRBs and AGNs

Hai-Ming Zhang, Da-Bin Lin*, Ting-Ting Lin, Bao-Rong Liu, Xiao-Li Huang, Shu-Qing Zhong, Rui-Jing Lu and En-Wei Liang*

Guangxi Key Laboratory for Relativistic Astrophysics, School of Physical Science & Technology, Guangxi University, Nanning 530004, China
GXU-NAOC Center for Astrophysics and Space Sciences, Guangxi University, Nanning 530004, China
email: lindabin@gxu.edu.cn; lew@gxu.edu.cn

Abstract. The Lorentz factor (Γ) is an important parameter related to the relativistic jet physics. We study the evolution patterns of Γ within gamma-ray burst (GRB) and active galactic nuclear jets for individual GRB 090168, GRB 140508A, and 3C 454.3. By estimating the Γ values for well-separated pulses in GRBs 090618 and 140508A with an empirical relation derived from typical GRBs, we find that the Γ evolution pattern in the two GRBs are different. The increasing-to-coasting evolution pattern of Γ in GRB 090618 likely indicates that the GRB fireball is still being accelerated in the prompt phase. The clear decrease evolution pattern of Γ in GRB 140508A suggests the deceleration of the fireball components. By deriving the Γ value through fitting their spectral energy distribution in different flares of 3C 454.3, a pattern of Γ-tracking-γ-ray flux is clearly found, likely indicating that the observed gamma-ray flares are being due to the Doppler boosting effect to the jet emission.

Keywords. gamma rays: bursts (individual: GRBs 090168 & 140508A), active galactic nuclei: individual (3C 454.3), jets and outflows.

1. Introduction

Gamma-ray bursts (GRBs) are the most explosive events in the universe. It is believed that the prompt emission of GRBs is produced in an ultra-relativistic jet. A relativistic jet is also widely found in active galactic nuclei (AGNs). For a relativistic jet, its Lorentz factor Γ is an important parameter to understand its physics, such as the jet composition and acceleration process, etc. In this work, we focus on the Lorentz factor of GRB pulses or gamma-ray flares of AGN jets.

Several methods have been proposed to infer Γ values of GRBs. The most popular method is by taking the peak time of the early afterglow light curve as the deceleration time of the external forward shock (e.g., Sari & Piran 1999). One may also estimate the Γ value with the high-energy cutoff observed in the prompt gamma-ray spectrum based on the "compactness problem" constraint (Fenimore *et al.* 1993; Piran 1999) or with the blackbody emission component (Pe'er *et al.* 2007). Interestingly, a tight relation of Γ_0, $L_{p,\rm iso}$ and $E_{p,z}$ is found with a sample of GRBs whose early optical afterglow onset peaks are clearly detected (Liang *et al.* 2015). To estimate the Γ value of an AGN jet, one needs to measure its Doppler boosting factor (δ) and the viewing angle θ_v to the jet axis. The δ value may be estimated with very long baseline interferometry observations (e.g., Ghisellini *et al.* 1993), the variability brightness temperatures (Lähteenmäki & Valtaoja 1999; Hovatta *et al.* 2009), or the variable timescale of sources (Jorstad *et al.* 2005). Theoretically, the Doppler factors may also be constrained by modeling the observed broadband spectral energy distributions (e.g., Zhang *et al.* 2012). For blazars, it is generally believed

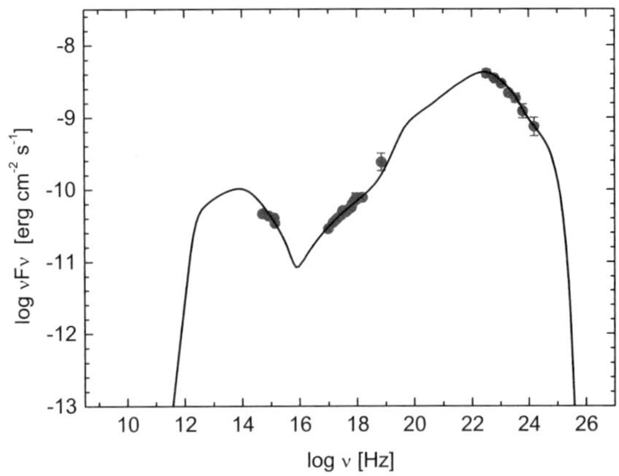

Figure 1. Spectral fitting of observations (points) of 3C 454.3 at MJD 55168.0. The solid line is our spectral fitting result.

that θ_v is very small and $\delta = \Gamma$ is usually adopted in theoretical modeling. This paper studies the evolution patterns of Γ within individual GRBs and blazars.

2. Data

We select GRBs 090618 ($z = 0.45$) and 140508A ($z = 1.03$) for case study. They have well-separated gamma-ray pulses with different peak fluxes. Their *Fermi*/GBM light curves are showed in Figure 2. We use the empirical $L_p - E_{p,z} - \Gamma_0$ relation relation (Liang *et al.* 2015), i.e., $\log \Gamma_0 = 3.33 + 0.46 \log L_{p,\mathrm{iso},52} - 0.43 \log E_{p,z}$, to estimate the Γ values for these pulses. We extract the spectra of these pulses around the peak times ($[t_p - 0.5\mathrm{s}, t_p + 0.5\mathrm{s}]$) and fit them with the Band function to derive their $E_{\mathrm{p},z} = E_p(1+z)$ and $L_{p,\mathrm{iso},52}$.

3C 454.3 (z=0.859, also named PKS 2251+158) is a blazar. It is the brightest gamma-ray source in the gamma-ray sky. *Fermi Large Area Telescope* has detected many bright flares for this source since it was launched in 2008 (Abdo *et al.* 2009). We extract its 8 years' LAT lightcurve and select some bright flares occurred in the period of [MJD55100, MJD55240] and [MJD55480, MJD55600] for our analysis, as shown in Figure 3, where only photons with energy $100\mathrm{MeV} < E < 200\mathrm{GeV}$ are considered. The Γ values of these flares are estimated based on the broad band SED (spectral energy distribution) fitting with the simple one-zone leptonic models (Bonnoli *et al.* 2011, Pacciani *et al.* 2010, Vercellone *et al.* 2011, Cerruti *et al.* 2013) as shown in Figure 1 for an example.

3. Results

Our results are shown in Figures 2 and 3. Different evolution features are found in the two GRBs. An increasing-to-coasting Γ evolution pattern is found in GRB 090618, likely indicating that its fireball is still being accelerated in the initial prompt gamma-ray phase, then kept at almost constant speed. A clear decrease evolution pattern of Γ in GRB 140508A is found. This likely suggests the deceleration of the fireball components for different pulses. These results may imply that there is no universal evolution pattern among GRBs. For 3C 454.3, a pattern of Γ-tracking-γ-ray flux is clearly found in different flares.

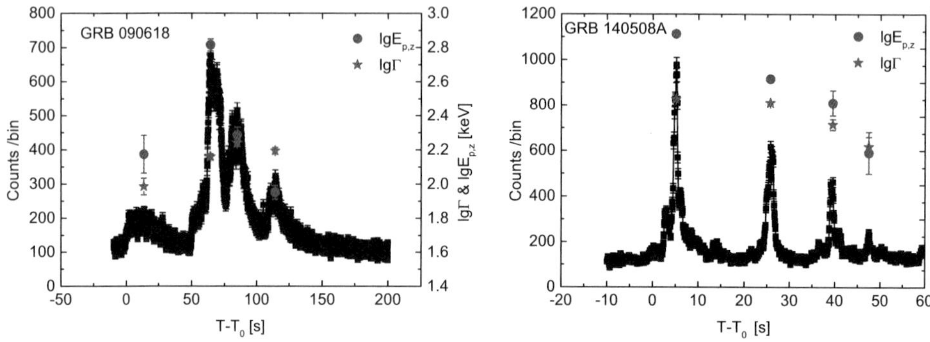

Figure 2. Evolution of photon counts (squares), Γ (stars), and $E_{p,z}$ (dots) in GRB 090618 (left panel) and GRB 140508A (right panel). T_0 is the trigger time of the burst.

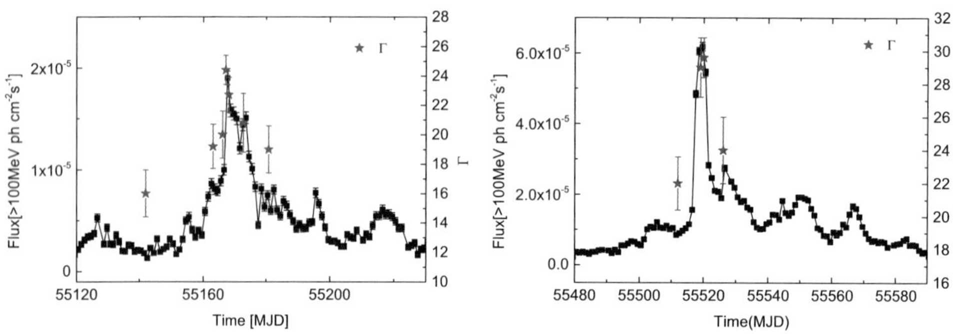

Figure 3. Evolution of flux and Γ (stars) in 3C 454.3 for observations in December 2009 (left panel) and November 2010 (right panel). The error bars of flux are statistical only.

4. Discussion & Conclusion

We have studied the Γ evolution within GRBs 090618 and 140508A as well as within 3C 454.3. It is found that the Γ evolution patterns in the two GRBs are different. This may suggest that there is no universal evolution pattern among GRBs. The widely accepted GRB model is the fireball model. Within the framework of this model, the prompt gamma-ray emission spectrum is expected to be the superposition of a quasi-thermal photosphere emission component and a non-thermal component in the optically thin internal shock region. It was proposed that the spectrum of the first pulse in GRB 090618 is superimposed by such a thermal component (Campana *et al.* 2006). This likely suggests that the emission of this pulse may come from the region around the photospheric radius of the fireball. The pure non-thermal emission in the next three pulses may be attributed to the emission from pure internal shock region. Therefore, the increase of Γ would suggest that the fireball is still accelerated near the photospheric radius. The Γ value keeps almost constant in the next three pulse which suggests that the fireball is within the coasting state. For GRB 140508A, the clear decrease of Γ may be due to the decay of the GRB central engine activities since a random collision process for fireball shells has difficulties in explaining this feature. For 3C 454.3, a pattern of Γ-tracking-γ-ray flux is clearly found. If this is a universal feature, it may suggest that the observed gamma-ray flares would be due to the Doppler boosting effect to the emission from ejecta within a jet powered by the central black hole. Gamma-ray flares of AGNs thus may be a signal of the central engine activity.

5. Acknowledgements

We acknowledge the use of the public data from the Fermi data archive. This work is supported by the National Basic Research Program of China (973 Program, grant No. 2014CB845800), the National Natural Science Foundation of China (Grant No. 11533003, 11403005, 11463001, 11363002), the Strategic Priority Research Program The Emergence of Cosmological Structures of the Chinese Academy of Sciences (grant XDB09000000), the Guangxi Science Foundation (2016GXNSFDA380027, 2014GXNSFAA118011, 2014GXNSFBA118004).

References

Abdo, A. A., Ackermann, M., Ajello, M., *et al.* 2009, *ApJ*, 699, 817
Bonnoli, G., Ghisellini, G., Foschini, L., *et al.* 2011, *MNRAS*, 410, 368
Campana, S., Mangano, V., Blustin, A. J., *et al.* 2006, *Nature*, 442, 1008
Cerruti, M., Dermer, C. D., Lott, B., Boisson, C., & Zech, A. 2013, *ApJL*, 771, L4
Fenimore, E. E., Epstein, R. I., & Ho, C. 1993, *A&AS*, 97, 59
Ghisellini, G., Padovani, P., Celotti, A., & Maraschi, L. 1993, *ApJ*, 407, 65
Hovatta, T., Valtaoja, E., Tornikoski, M., & Lähteenmäki, A. 2009, *A&A*, 494, 527
Jorstad, S. G., Marscher, A. P., Lister, M. L., *et al.* 2005, *AJ*, 130, 1418
Lähteenmäki, A. & Valtaoja, E. 1999, *ApJ*, 521, 493
Liang, E.-W., Lin, T.-T., Lü, J., *et al.* 2015, *ApJ*, 813, 116
Pacciani, L., Vittorini, V., Tavani, M., *et al.* 2010, *ApJL*, 716, L170
Pe'er, A., Ryde, F., Wijers, R. A. M. J., *et al.* 2007, *ApJ*, 664, L1
Piran, T. 1999, *Phys. Rep.*, 314, 575
Sari, R. & Piran, T. 1999, *ApJ*, 520, 641
Vercellone, S., Striani, E., Vittorini, V., *et al.* 2011, *ApJL*, 736, L38
Zhang, J., Liang, E.-W., Zhang, S.-N., & Bai, J. M. 2012, *ApJ*, 752, 157

New Frontiers in Black Hole Astrophysics
Proceedings IAU Symposium No. 324, 2016
Andreja Gomboc, ed.

doi:10.1017/S1743921317001090

Doppler Boosting Effect on the Jet Radiation of Gamma-Ray Bursts and Active Galactic Nuclei

Xiao-Li Huang, Hai-Ming Zhang, Shu-Qing Zhong and En-Wei Liang
Guangxi Key Laboratory for Relativistic Astrophysics, Department of Physics, Guangxi University, Nanning 530004, China
email: lew@gxu.edu.cn

Abstract. High energy photon radiations of gamma-ray bursts (GRBs) and active galactic nuclei (AGNs) are dominated by their jet radiations. It was suggested that relativistic jets powered by different mass-scale black holes may share the same physical laws. A tight relation among the peak luminosity, the peak photon energy in the νf_ν spectrum, and the initial Lorentz factor is found for GRBs. With samples of GeV-TeV BL Lacs, FSRQs, and NLS1 galaxies, we show that these sources do not follow this relation. This may be attributed to the jet geometry and continuous/episodic jet as well as radiation physics for different kinds of sources.

Keywords. Gamma-Ray Bursts; Active Galactic Nuclei; Relativistic Jets; Non-thermal Radiation

1. Introduction

It is believed that gamma-ray bursts (GRBs) are produced by an ultra-relativistic jet powered by stellar black holes (BHs) from core collapses of massive stars or mergers of two compact stars (e.g., Kumar & Zhang 2015), and the high energy photon radiations of active galactic nuclei (AGNs) are dominated by radiations from a mildly relativistic jet fed by accretion of their central super-massive BHs (e.g. Ghisellini *et al.* 2009; Zhang *et al.* 2012). Some comparative studies show evidence of similar jet properties between AGNs and GRBs (e.g., Wang *et al.* 2011; Nemmen *et al.* 2012; Zhang *et al.* 2013). Liang *et al.* (2015) found a tight correlation among the peak luminosity ($L_{\rm p}$), the peak energy ($E_{\rm p,z}$) of the νf_ν spectrum in the GRBs rest frame, and the initial Lorentz factor (Γ_0) of GRBs jets. This paper investigates whether or not the synchrotron radiations from AGNs jets follow the same relation with GRBs jets.

2. Data

Zhang *et al.* (2012, 2015) and Sun *et al.* (2015) derived the jet properties of GeV-TeV selected BL Lacs, flat spectrum radio quasars (FSRQs), and narrow-line Seyfert 1 (NLS1) galaxies by fitting their SEDs with the lepton models. We collected the peak luminosity ($L_{\rm syn}$) and photon energy ($E_{\rm syn,z}$) of the synchrotron radiations and the Doppler boosting factors (δ) of these objects from their analysis. Since they assumed $\delta = 2\Gamma$ in their fits, where Γ is the bulk Lorentz factor of the jets, we calculate the Γ values with $\Gamma = \delta/2$.

3. Results

Firstly, we test if the synchrotron radiations of the AGNs follow the $L_p - E_{p,z} - \Gamma_0$ relation of GRBs, i.e., $L_{p,52} \propto E_{p,z}^{1.34\pm0.14}\Gamma_0^{1.32\pm0.19}$ (Liang *et al.* 2015). We calculate the

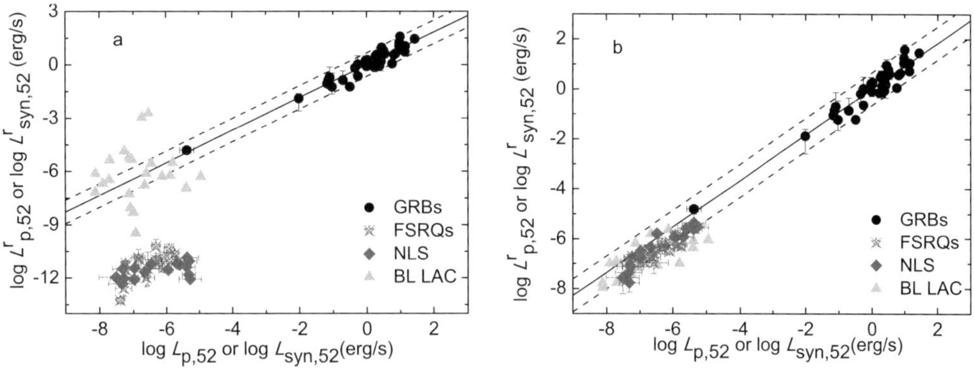

Figure 1. Distributions of the AGNs in the $\log L_{\rm p}^{\rm r} - \log L_{\rm p}$ plane, where $\log L_{\rm p}^{\rm r}$ is calculated by assuming that the AGNs follow the $L_p - E_{p,z} - \Gamma_0$ relation of GRBs (panel a; the solid and dashed lines; Liang *et al.* 2015) or by taking the relations for different type AGNs (panel b).

synchrotron peak luminosity ($L_{\rm syn}^{\rm r}$) for these AGNs with this relation by using their $E_{\rm syn,z}$ and Γ. Figure 1(a) shows $L_{\rm syn}^{\rm r}$ as a function of $L_{\rm syn}$ in comparison with the GRBs from Liang *et al.* (2015). It is found that BL Lac objects are distributed at the low $L_{\rm p}$ end of this relation, but the scatter is very large. Both FSRQs and NLS1 galaxies, on the other hand, heavily deviate from this relation. We then separately make multiple regression analysis to the BL Lacs, FSRQs and NLS1 galaxies to derive their $L_{\rm syn}^{\rm r}(E_{\rm syn,z}, \Gamma)$ relations, and obtain $L_{\rm syn,52} \propto E_{\rm syn,z}^{0.01\pm0.01}\Gamma^{2.35\pm0.52}$, $L_{\rm syn,52} \propto E_{\rm syn,z}^{0.57\pm0.15}\Gamma^{2.58\pm0.81}$, and $L_{\rm syn,52} \propto E_{\rm syn,z}^{0.51\pm0.27}\Gamma^{4.07\pm0.39}$, respectively. Figure 1(b) shows the AGNs and GRBs in the $L_{\rm p}^{\rm r} - L_{\rm p}$ plane by using their $L_{\rm syn}^{\rm r}(E_{\rm syn,z}, \Gamma)$ relations for different types of sources.

4. Conclusions

We have shown that the AGNs jet radiation does not follow the same $L_{\rm syn}^{\rm r}(E_{\rm syn,z}, \Gamma)$ relations as the $L_{\rm p,z}^{\rm r}(E_{\rm p,z}, \Gamma_0)$ relation of GRBs. The dependence of the luminosity on the Lorentz factor is dramatically different among these sources. This may be due to different jet geometry and continuous/episodic jets (e.g., Ghisellini *et al.* 1993). The different dependence of the luminosity on $E_{\rm p,z}$ may hint at different radiation physics among these sources (e.g., Lyu *et al.*2014).

This work is supported by the National Basic Research Program of China (973 Program, No. 2014CB845800), the National Natural Science Foundation of China (No. 11533003), the Guangxi Science Foundation (No. 2013GXNSFFA019001). XLH was supported by an IAU travel grant.

References

Ghisellini, G., Padovani, P., Celotti, A., & Maraschi, L., 1993, *ApJ*, 407, 65
Ghisellini, G., Tavecchio, F., & Ghirlanda, G., 2009, *MNRAS*, 399, 2041
Kumar, P. & Zhang, B., 2015, *PhR*, 561, 1
Liang, E.-W., Lin, T.-T., Lü, J., Lu, R.-J., Zhang, J., & Zhang, B., 2015, *ApJ*, 813, 116
Lyu, F., Liang, E.-W., Liang, Y.-F., Wu, X.-F., *et al.*2014, *ApJ*, 793, 36
Nemmen, R. S., Georganopoulos, M., Guiriec, S., *et al.* 2012, Sci, 338, 1445
Sun, X.-N., Zhang, J., Lin, D.-B., Xue, Z.-W., Liang, E.-W., & Zhang, S.-N., 2015, *ApJ*, 798, 43

Wang, J. & Wei, J. Y., 2011, *ApJ*, 726, L4
Zhang, J., Liang, E.-W., Sun, X.-N., Zhang, B., Lu, Y., & Zhang, S.-N., 2013, *ApJ*, 774, L5
Zhang, J., Liang, E.-W., Zhang, S.-N., & Bai, J. M., 2012, *ApJ*, 752, 157
Zhang, J., Xue, Z.-W., He, J.-J., Liang, E.-W., & Zhang, S.-N., 2015, *ApJ*, 807, 51

New Frontiers in Black Hole Astrophysics
Proceedings IAU Symposium No. 324, 2016
Andreja Gomboc, ed.

doi:10.1017/S1743921317001259

Untriggered search for rapid optical transients with Mini-MegaTORTORA wide-field monitoring system

S. Karpov[1,3], G. Beskin[1,3], A. Biryukov[4], S. Bondar[2], E. Ivanov[2], E. Katkova[2], N. Orekhova[2], A. Perkov[2] and V. Sasyuk[2]

[1]Special Astrophysical Observatory of Russian Academy of Sciences, Russia.
[2]Research and Production Corporation "Precision Systems and Instruments", Russia.
[3]Kazan Federal University, Russia.
[4]Moscow State University, Russia.

Abstract. Here we report ongoing efforts for an untriggered search of rapid optical transients of various astrophysical and non-astrophysical origins on time scales down to fractions of a second with Mini-MegaTORTORA. Mini-MegaTORTORA is a novel 9-channel wide-field optical monitoring system in operation since 2014 at Special Astrophysical Observatory on Russian Caucasus.

Keywords. gamma rays: bursts, instrumentation: miscellaneous

Mini-MegaTORTORA (Biryukov *et al.*2015 and references therein) is a novel robotic instrument developed to detect and characterize fast optical transients of various origins, both cosmological, galactic and near-Earth. It is a 9-channel wide-field (~900 square degrees) monitoring system with temporal resolution of 0.1 seconds and limit down to V~11 mag. Every channel of Mini-MegaTORTORA has 10x10 deg field of view and is equipped with installable photometric and polarimetric filters and coelostat mirror for a rapid repointing in a limited range. It allows to re-configure the system on the fly in order to rapidly follow-up the transients just detected.

Mini-MegaTORTORA started its operation in Jun 2014, and since then continuously monitors the sky looking for fast optical transients. Its real-time transient detection system routinely extracts various kinds of transient from the data stream – rapid flashes, meteors, satellites etc. Data on detected transients are published online.

Transient detection software matches rapid flashes with stellar catalogues to exclude scintillation effects and with NORAD database of satellite orbits to filter out satellite flashes. Some fraction of detected flashes are not identified with satellites, but most probably are of satellite origin. Immediate follow-up observations using Mini-MegaTORTORA rapid reaction mode (see Figure 1) typically reveal faint satellite trails. Durations and shapes of such flashes (Figure 2) are also consistent with satellite ones. Therefore we may conclude that no bright rapid flashes of astrophysical origin are detected in 2.5 years of Mini-MegaTORTORA operation.

Mini-MegaTORTORA also performs follow-up of Swift, Fermi and LIGO-Virgo triggers, including the ones with poor localization accuracy due to its large field of view allowing for simultaneous observations in ~900 sq.deg. sky fields. For the triggers with better localizations, multicolor and/or polarimetric follow-up is performed. Since mid-2015, 4 of 89 Swift GRBs have been followed up in polarimetric mode in 30 to 60 seconds since trigger distribution through GCN network, with no optical emission detections. 9 of 250 Fermi GBM triggers have been also followed up in wide-field mode in 20 to

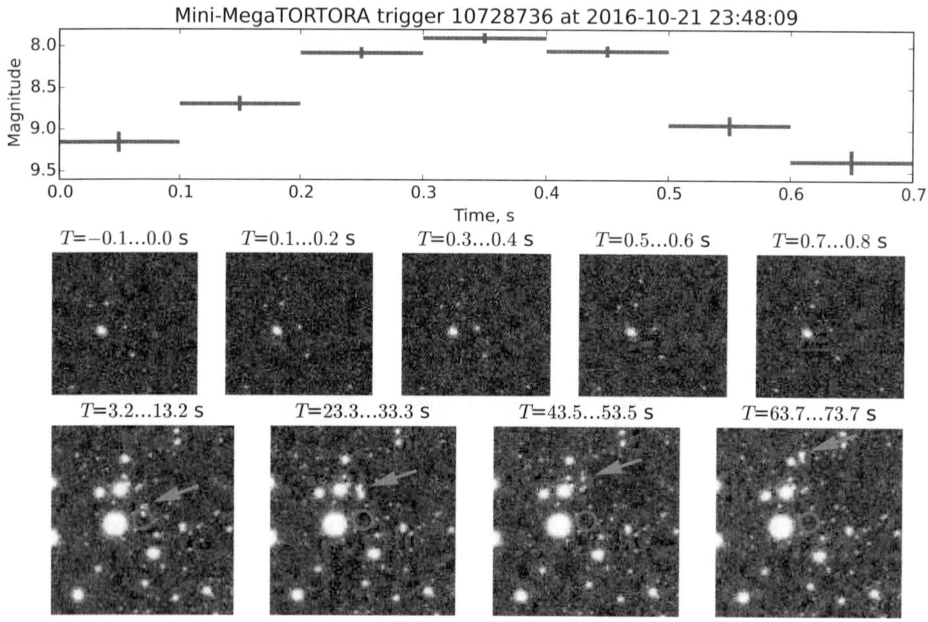

Figure 1. Example of a rapid optical flash independently detected and followed-up by Mini-MegaTORTORA and not identified with satellites from NORAD database. Upper panel – light curve with 0.1 s temporal resolution, middle panel – corresponding detection images (50′ x 50′ around the event), lower panel – follow-up images with 10 s exposures that clearly reveal the satellite, marked by arrow, slowly moving away from the flash position, marked by circle.

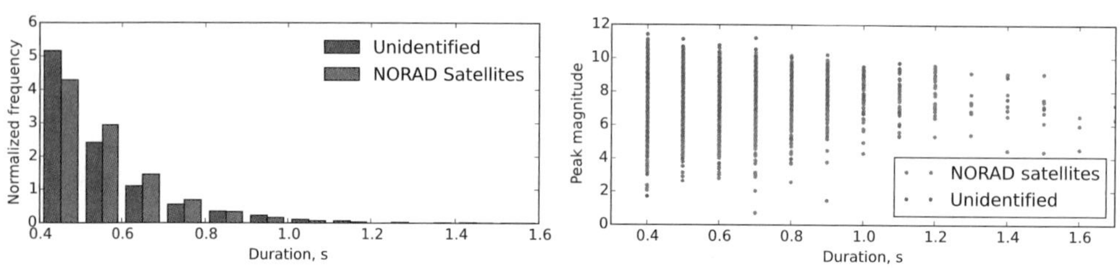

Figure 2. Comparison of durations (left panel) and brightness (right panel) of rapid flashes detected by Mini-MegaTORTORA and identified/non-indentified with NORAD satellites.

90 seconds from the trigger, with bright optical flash of GRB 160625B clearly detected during the gamma activity. All other events were either below the horizon or occurred in bad weather conditions.

Acknowledgements

Mini-MegaTORTORA belongs to Kazan Federal University and the work is performed according to the Russian Government Program of Competitive Growth of Kazan Federal University. Observations on Mini-MegaTORTORA are supported by the Russian Science Foundation grant No. 14-50-00043.

Reference

Biryukov, A., Beskin, G., Karpov, S., Bondar, S., Ivanov, E., Katkova, E., Perkov, A., & Sasyuk, V. 2015, *Baltic Astronomy*, 24, 100

New Frontiers in Black Hole Astrophysics
Proceedings IAU Symposium No. 324, 2016
Andreja Gomboc, ed.

doi:10.1017/S1743921317002046

Synergetic Growth of the Rayleigh–Taylor and Richtmyer–Meshkov Instabilities in the Relativistic Jet

Jin Matsumoto[1] **and Youhei Masada**[2]

[1]Astrophysical Big Bang Laboratory, Riken, Saitama, Japan
email: jin.matsumoto@riken.jp

[2]Department of Physics and Astronomy, Aichi University of Education, Aichi, Japan
email: ymasada@auecc.aichi-edu.ac.jp

Abstract. We investigate the growth of the Rayleigh–Taylor and Richtmyer–Meshkov instabilities at the interface of the relativistic jet using three-dimensional hydrodynamic simulations. The propagation of the relativistic jet that is continuously injected from the boundary of the calculation domain into a uniform ambient medium is solved. We find that the interface of the jet is deformed by a synergetic growth of the Rayleigh–Taylor and Richtmyer–Meshkov instabilities regardless of the launching condition, such as the specific enthalpy of the jet or the effective inertia ratio between the jet and ambient medium. The material mixing between the jet and external medium due to these instabilities causes the deceleration of the jet.

Keywords. galaxies: jets, shock waves, instabilities, methods: numerical

1. Introduction

The relativistic jet is ubiquitous phenomenon among astrophysical systems consisting of a black hole surrounded by an accretion disk, e.g., active galactic nuclei, microquasars, and the central engine of gamma-ray bursts. The dynamics and stability of the relativistic jet is important in order to understand the emission properties of the jet.

The radially oscillating motion of the jet is naturally excited due to the pressure mismatch between the jet and surrounding medium, that is, cocoon when the jet propagates through an ambient medium and is confined by the pressure of the cocoon (Daly & Marscher 1988; Matsumoto *et al.* 2012). In the rest frame of the decelerating jet interface that expands radially, an effective inertia force acts on the interface and is directed outward. Therefore the jet medium is on top of the cocoon medium in the effective gravity in this frame and the Rayleigh–Taylor instability (RTI) is expected to grow at the jet interface (Matsumoto & Masada 2013).

Since the pressure mismatch can not be alleviated immediately without some damping processes, it repeatedly excites the reconfinement shock inside the jet. The convergence of the reconfinement shock produces an outward propagating shock at the center of the jet. The collision of the outward propagating shock with the jet interface is the origin of the Richtmyer–Meshkov instability (RMI) and thus it is expected to be quasi-periodically excited whenever this type of collision occurs (Matsumoto & Masada 2013).

We investigate the growth of the RTI and RMI at the interface of the jet using three-dimensional (3D) hydrodynamic simulations in the cylindrical coordinate system. The propagation of the relativistic jet that is continuously injected from the boundary of the calculation domain ($z = 0$ and $r < r_{jet}$) into a uniform ambient medium is solved considering launching conditions. A small-amplitude (1%) random pressure perturbation is introduced to the injected relativistic flow in order to break the symmetry of the jet.

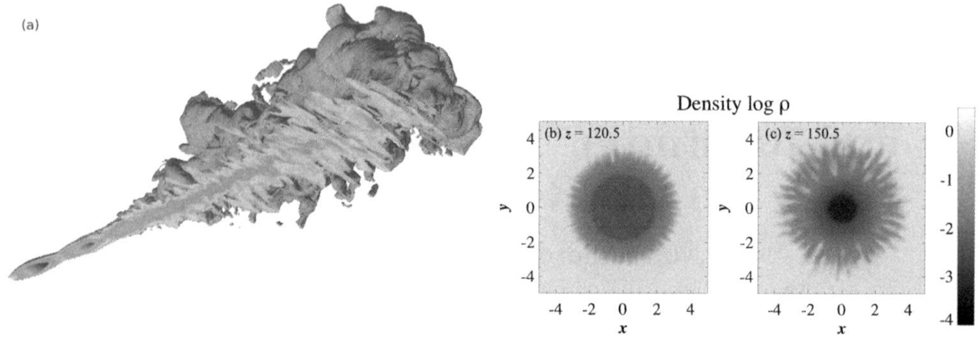

Figure 1. 3D structure of the jet in the large inertia and hot model when the relativistic jet reaches the boundary of the calculation domain ($z = 1000$). Panel (a) shows the density map of the jet in the half simulation box. Panel (b) and (c) show 2D cuts of the density distribution in the x-y (r-ϕ) plane where $z = 120.5$ and $z = 150.5$, respectively.

2. Results

Four-types of models are simulated with focusing on two parameters: One is the effective inertia ratio of the jet to the ambient medium (large or low inertia model), which controls the morphology and dynamics of the propagating jet (Marti *et al.* 1997), and the other is the relativistic hotness of the jet (hot or cold case). Fig. 1 shows the 3D structure of the jet in the large inertia and hot model when the relativistic jet reaches the boundary of the calculation domain ($z = 1000$). The jet–cocoon medium interface becomes unstable for the RTI. The amplitude of the corrugated interface of the jet at $z = 150.5$ is large compared to the interface at $z = 120.5$ due to the RTI. Unlike the RTI, the RMI grows impulsively when the outward propagating shock collides with the jet interface. The material mixing due to the deformation of the interface between the jet and cocoon medium leads to the deceleration of the jet. The flow velocity of the jet in the region where $z \sim 900$ is roughly $0.7c$ although that of the injected relativistic flow at the lower boundary is $0.99c$.

The propagation velocity of the jet head is slow and the jet is surrounded by a thick cocoon in the low inertia model compared to the large inertia model. On the other hand, the pressure of the cocoon is higher in the cold jet case than the hot jet case because the large kinetic energy of the cold jet is converted into the thermal energy of the cocoon at the jet head. Although the relativistic jet shows a rich variety of the propagation dynamics depending on its launching condition, the interface of the relativistic jet is deformed by a synergetic growth of the oscillation–induced RTI and secondary RMI for all models.

Acknowledgements

Numerical computations were carried out on Cray XC30 at CfCA, National Astronomical Observatory of Japan and on Cray XC40 at YITP, Kyoto University.

References

Daly, R. A. & Marscher, A. P. 1988, *ApJ*, 334, 539
Marti, J. M. A., Mueller, E., Font, J. A., Ibanez, J. M. A., & Marquina, A. 1997, *ApJ*, 479, 151
Matsumoto, J., Masada, Y., & Shibata, K. 2012, *ApJ*, 751, 140
Matsumoto, J. & Masada, Y. 2013, *ApJ* (Letters), 772, LL1

New Frontiers in Black Hole Astrophysics
Proceedings IAU Symposium No. 324, 2016
Andreja Gomboc, ed.

doi:10.1017/S1743921317001235

Spatial Distribution of the Gamma-ray Bursts and the Cosmological Principle

Attila Mészáros

Charles University, Faculty of Mathematics and Physics, Astronomical Institute, V Holešovičkách 2, Prague 8, Czech Republic, email: meszaros@cesnet.cz

Abstract. The Cosmological Principle claims that in the large scale average the visible parts of the universe are isotropic and homogeneous. In year 1998 the author, together with his two colleagues, discovered that the BATSE's short gamma-ray bursts are not distributed isotropically on the sky. This first discovery was followed by other ones confirming both the existence of anisotropies in the angular distribution of bursts and the existence of huge Gpc structures in the spatial distribution. All this means that these anisotropies should reject the Cosmological Principle, because the large scale averaging hardly can be provided. This was claimed in year 2009. The aim of this contribution is to survey these publications since 1998 till today.

Keywords. Cosmology: large-scale structure of Universe, observations, theory, miscellaneous

---

1. Cosmological Principle and the Large Scale Averaging

The Cosmological Principle requires that the Universe be spatially homogeneous and isotropic on scales larger than the size of any structure (Peebles 1993). Exactly, it is said on the page 15 of this book that "...in the large scale average the visible parts of our universe are isotropic and homogeneous". The observable part of the Universe has the size of $\sim (10 - 20)$ Gpc, if one uses the so called "proper-motion distance" (for the relevant formulas see, e.g., Weinberg (1972) and Carroll *et al.* (1992)), and thus the averaging should happen far below these scales. In other words, there should exist a transition scale not larger than, say, ~ 1 Gpc, and above this one no structures should exist. Yadav *et al.* (2010) means that this transition scale is $\simeq 260h^{-1}$ Mpc, where h is the Hubble-constant in unit 100 km/(sMpc), and the Cosmological Principle holds.

But, rarely, there are publications claiming supports for the structures with $\sim$ Gpc sizes. For example, Collins & Hawking (1973) and Birch (1982) speak about a possible global rotation in the observable part of the Universe. Other observations (Rudnick *et al.* 2007) claim the existence of structure with size ~ 140 Mpc, but at redshift around 1. A recent publication about the spatial distribution of quasars (Clowes *et al.* 2013) claims the existence of a structure with a scale > 1 Gpc.

The angular distribution of the gamma-ray bursts (GRBs) allows to test this Principle too, because GRBs should be distributed isotropically on the sky. For this test GRBs are especially useful, because they are seen in the gamma-band also in the Galactic plane, too. In year 1996 it was declared that the isotropic angular distribution of BATSE's bursts was fulfilled (see Tegmark *et al.* (1996) and the references therein).

2. Anisotropies in the sky distribution of the gamma-ray bursts

The article Balázs *et al.* (1998) claimed first that the sky distribution of short BATSE's GRBs was not isotropic. This proclaim was then confirmed by several other articles of the author and his collaborators (see Balázs *et al.* (1999), Mészáros *et al.* (2000a), Vavrek

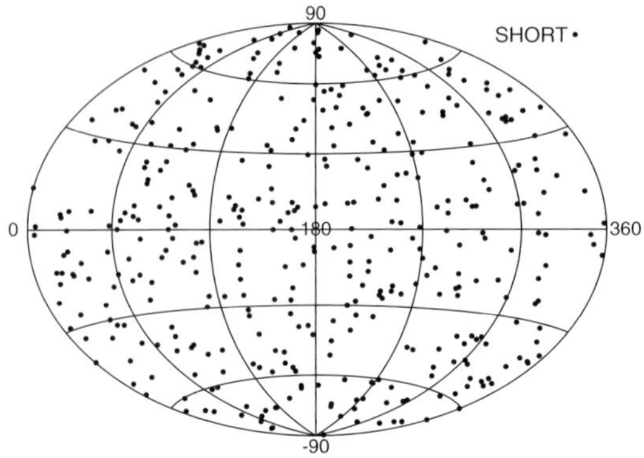

Figure 1. Sky distribution of BATSE's short GRBs in equatorial coordinates.

et al. (2008)). The anisotropic sky distribution of the short BATSE's GRBs is shown on the Figure 1. In addition, both the BATSE's intermediate and long subclasses were found to be distributed also anisotropically (see Mészáros *et al.* (2000b), Mészáros & Štoček (2003)) and Vavrek *et al.* (2008) for more details). After Vavrek *et al.* (2008) the existence of the Gpc structures and thus the problems of the Cosmological Principle were declared by Mészáros *et al.* (2009a) and Mészáros *et al.* (2009b).

Recently, the coauthors of the author, having large enough sample of measured redshifts and with which the study of 3D structures became possible, also found structures on the Gpc scales (Horváth *et al.* 2014, Horváth *et al.* 2015, Balázs *et al.* 2015). Also other authors found similar remarkable results (Sokolov *et al.* 2015, Verkhodanov *et al.* 2015, Ukwatta & Woźniak 2016).

References

Balázs, L. G., Mészáros, A., & Horváth, I. 1998, *A&A*, 339, 1
Balázs, L. G., *et al.* 1999, *A&AS*, 138, 417
Balázs, L. G., *et al.* 2015, *MNRAS*, 452, 2236
Birch, P. 1982, *Nature*, 298, 451
Carroll, S. M., Press, W. H., & Turner, E. L. 1992, *ARAA*, 30, 499
Clowes, R. G., *et al.* 2013, *MNRAS*, 429, 2910
Collins, C. B. & Hawking, S. W. 1973, *MNRAS*, 162, 307
Horváth I., Hakkila, J., & Bagoly, Z. 2014, *A&A*, 561, L12
Horváth, I., *et al.* 2015, *A&A*, 584, A48
Mészáros, A., Bagoly, Z, & Vavrek, R. 2000a, *A&A*, 354, 1
Mészáros, A., *et al.* 2000b, *ApJ*, 539, 98
Mészáros, A. & Štoček, J. 2003, *A&A*, 403, 443
Mészáros A., *et al.*, (2009a) *Baltic Astronomy*, 18, 293
Mészáros A., *et al.*, (2009b) *Sixth Huntsville GRB Symposium, AIP Conf. Proc.*, 1133, 483
Peebles, P. J. E. 1993, *Principles of Physical Cosmology* (Princeton University Press, Princeton)
Rudnick, L., Brown, S., & Williams, L. R. 2007, *ApJ*, 671, 40; Errata: *ApJ*, 678, 1531 (2008)
Sokolov, I. V., *et al.* 2015, *Quark Phase Transition in Compact Objects, SAO RAS, Russia*, 111
Tegmark, M., *et al.* (1996), *ApJ*, 468, 214
Ukwatta, T. N. & Woźniak, P. R. 2016, *MNRAS*, 455, 703
Vavrek, R., *et al.* 2008, *MNRAS*, 391, 1741
Verkhodanov, O. V., Sokolov, V. V., & Khabibullina, M. L. 2015, *Quark Phase Transition in Compact Objects, SAO RAS, Russia*, 142
Weinberg, S. 1972, *Gravitation and Cosmology* (J. Wiley, New York - London - Sydney - Toronto)
Yadav, J. K., Bagla, J. S., & Khandai, N. 2010, *MNRAS*, 405, 2009

Topic 3: Tidal Disruption Events

New Frontiers in Black Hole Astrophysics
Proceedings IAU Symposium No. 324, 2016
Andreja Gomboc, ed.

doi:10.1017/S1743921317001764

Observational Progress in Identifying and Characterizing Tidal Disruption Flares

S. Bradley Cenko

Astrophysics Science Division, NASA Goddard Space Flight Center, Mail Code 661, 8800 Greenbelt Road, Greenbelt, MD 20771, USA

Abstract. I present an overview of observational efforts across the electromagnetic spectrum to identify and study tidal disruption flares (TDFs), when a star wanders too close to a super-massive black hole and is torn apart by tidal forces. In particular I will focus on four unexpected surprises that challenge the most basic analytic picture of these events: 1) large inferred radii for the optical/UV-emitting material; 2) the ubiquity of outflows, detected at radio, X-ray, and UV wavelengths, ranging from speeds of 100 km/s to near the speed of light; 3) the peculiar atomic abundances observed in the UV and optical spectra of these objects; and, 4) the preference for these events to occur in post-starburst galaxies.

Keywords. accretion, black hole physics

---

1. Introduction

When a star passes too close to a massive black hole, tidal forces will overcome self-gravity, unbinding the star. For Sun-like stars and sufficiently low-mass black holes ($M_{\rm BH} \lesssim 10^8\,{\rm M}_\odot$), the disruption will occur outside the event horizon (Hills 1975). Approximately half† of the stellar debris will return to the black hole on highly elliptical orbits (Evans & Kochanek 1989). The resulting emission from this returning bound debris is known as a tidal disruption flare (TDF; Rees 1988).

TDFs have gained increasing interest in the astronomical community in recent years, in large part due to their promise to address a number of key outstanding questions. First and foremost, TDFs provide a means to identify and probe super-massive black holes in distant galaxies *that are otherwise inactive*. As such, they may ultimately be used to extend BH mass measurements and the corresponding studies of black hole and galaxy evolution (e.g., the M-sigma relation; Ferrarese & Merritt 2000; Gebhardt *et al.* 2000) well beyond the current reach of kinematic studies. In addition, TDFs serve as valuable probes of *the process of accretion*, providing insight into such fundamental questions as how accretion disks form, how disks behave at dramatically different Eddington ratios (e.g., De Colle *et al.* 2012), and how relativistic jets are generated in these systems (e.g., Levan *et al.* 2011; Bloom *et al.* 2011; Zauderer *et al.* 2011; Burrows *et al.* 2011; Cenko *et al.* 2012).

Early theoretical and numerical models of the tidal disruption process suggested that the resulting emission should be largely thermal in nature, with temperatures peaking at soft X-ray or extreme ultraviolet wavelengths (Ulmer 1999). Another early prediction was a characteristic $t^{-5/3}$ power-law decay for the mass return rate to pericenter (Phinney 1989), though it was not obvious that the observed emission in any one band would necessarily track the mass return rate (Strubbe & Quataert 2009; Lodato & Rossi 2011). The peak luminosity was anticipated to be roughly the Eddington luminosity for the corresponding black hole (e.g., $10^{44}\,{\rm erg\,s^{-1}}$ for a $10^6\,{\rm M}_\odot$ black hole).

† The remainder is unbound and ejected from the system with high velocity.

The earliest TDF discoveries were made largely in the X-ray band, with the ROSAT all-sky survey leading the way (see, e.g., Komossa 2015 for a review). These flares were identified largely by three properties: 1) large changes in amplitude (in some cases up to factors of thousands); 2) soft X-ray spectra (photon power-law indices $\gtrsim 4$, or blackbody spectra with $kT \gtrsim 50$ eV); and, 3) coincident with the nucleus of a galaxy not previously known to be an AGN. The resulting luminosities and decay rates were largely consistent with early predictions; however it is worth noting that the sampling of these objects (both temporally in the X-rays and contemporaneous multi-wavelength observations) were relatively sparse. Since the completion of the ROSAT mission, TDFs have continued to be discovered at X-ray wavelengths though programs such as dedicated searches of galaxy clusters and via the wide-area coverage of the XMM Slew Survey (see, e.g., Saxton 2017 for an overview).

More recently, wide-field optical transient surveys such as the Palomar Transient Factory (Law *et al.* 2009; Rau *et al.* 2009), Pan-STARRS (Kaiser *et al.* 2010), and the All-Sky Automated Survey for SuperNovae (ASAS-SN; Shappee *et al.* 2014) have reinvigorated the field. While designed largely to discover supernovae by repeatedly imaging large areas of the sky on timescales of $\sim$ days, these surveys have now uncovered ≈ 10 TDF candidates at optical wavelengths. And because these surveys identify their transient candidates *in real-time*, this has provided the opportunity for a broadband multi-wavelength perspective (e.g., optical spectroscopy, radio through X-ray imaging) on TDFs for the first time.

Not surprisingly, these optically discovered events have challenged our understanding of the tidal disruption process on a number of different fronts. Here I describe four recent "observational puzzles" that have arisen in recent years:

- Low and constant temperatures inferred from the optical/UV emission;
- Ubiquity of outflows across a broad range of velocity scales;
- Unusual abundance patterns in optical and UV spectra; and,
- A strong preference for post-starburst host galaxies.

Understanding all of these in the coming years will be critical to turning TDFs into viable probes of black hole mass in the upcoming era of the Large Synoptic Survey Telescope (LSST; Ivezic *et al.* 2008).

2. Blackbody Temperature in Optical/UV TDFs

As discussed above, the first TDF discoveries made at X-ray wavelengths had measured blackbody temperatures $T \lesssim 10^6$ K. For a source with a peak luminosity $L \approx 10^{44}$ erg s^{-1}, the implied blackbody radius is roughly comparable to the tidal radius (and several times the Schwarzchild radius for a 10^6 $M_\odot$ black hole). These results are nicely consistent with models where the returning debris is able to circularize efficiently and form a radiatively efficient, optically thick (i.e., Shakura-Sunyaev) accretion disk.

However, nearly all TDFs discovered at optical and UV wavelengths exhibit very different spectral energy distributions. While the optical/UV emission is reasonably well characterized by a blackbody, the measured temperatures are significantly lower, typically $\approx 3 \times 10^4$ K. As a result, the implied emission region is significantly larger than expected (i.e., much larger than the size anticipated for a newly formed accretion disk), with a characteristic size scale of $\approx 10^{15}$ cm.

Even more puzzling, in most events the spectral temperature does not appear to vary significantly over the evolution of the flare, even though the luminosity may drop by orders of magnitude. Such evolution is dramatically different from supernovae, however, allowing for a powerful discriminant to identify TDFs amongst their much more numerous transient contaminants.

An illustrative example to consider is the nearby (and hence well-studied) ASASSN-14li ($d \approx 90$ Mpc; Holoien *et al.* 2016; Van Velzen et al. 2016). ASASSN-14li is one of the few TDFs to exhibit both luminous optical/UV and X-ray emission, *and they appear to derive from two physically distinct regions.* The implied radius from fitting the X-ray spectrum to a blackbody is consistent with the emission arising from the immediate environs of the tidal radius ($\approx 10^{13}$ cm). But the observed optical/UV emission is significantly more luminous than would be anticipated from extrapolating the X-ray spectrum to these wavelengths. Much like other optically discovered events, the spectral temperature and luminosity suggest a much larger emission region for the optical/UV emission.

Two competing models have been put forth to explain the small observed temperatures from optically discovered events. In the first, a "reprocessing" layer at large radii is responsible for reshaping the emission from the (inner) accretion disk, analogous to the photosphere in a supernova (Guillochon *et al.* 2014; Roth *et al.* 2016). The nature of this reprocessing layer remains somewhat mysterious, although it is natural to associate it with the outflowing material commonly seen in TDFs (see below). Varying the covering fraction and/or geometry of the reprocessing material can presumably alter the ratio of X-ray (accretion disk) to optical/UV (reprocessing) emission that is observed.

Alternatively, it may be that for some tidal disruptions the process of circularization (thought to be driven by stream-stream collisions) may be quite inefficient at generating an accretion disk (Piran *et al.* 2015; Krolik *et al.* 2016). Numerical simulations suggest that shocks may form very efficiently in these stream intersection regions, and thus it may be natural to expect emission directly from these locations (Shiokawa *et al.* 2015). While such a model would naturally explain the large inferred blackbody radii derived from optical/UV events, the wide range of observed X-ray-to-optical ratios in TDFs may be more challenging to accomodate.

3. The Ubiquity of Outflows

Accretion onto black holes, either in the guise of X-ray binaries (i.e., stellar mass black holes) or active galactic nuclei (i.e., super-massive black holes) can often give rise to outflows. The most dramatic example of these are the relativistic jets observed as microquasars and blazars, respectively. Given the similar physics involved in TDFs, it should not be surprising, then, to see analogous behavior.

It has been a bit surprising, however (to the author, at least) to see the large range of velocities over which outflows have been observed in only the handful of well-studied events to date. The most extreme examples are the newly formed relativistic jets that have been associated with tidal disruption events discovered by the *Swift* satellite: Sw J1644+57 (Levan *et al.* 2011; Bloom *et al.* 2011; Zauderer *et al.* 2011; Burrows *et al.* 2011), Sw J2058+05 (Cenko *et al.* 2012), and Sw J1112-82 (Brown *et al.* 2015). These sources are clearly very different beasts than the previously discussed tidal disruptions: extremely luminous ($L \approx 10^{48}$ erg s^{-1}) non-thermal X-ray emission is detected from these sources, lasting for roughly a year. Radio observations provide compelling evidence, mostly clearly in the case of Sw J1644+57, for a newly formed relativistic ($\Gamma_0 \sim 2$–10) jet powering these objects. Despite extensive searches (Bower *et al.* 2013; Van Velzen *et al.* 2013), only marginal evidence in one case (Irwin *et al.* 2015) has yielded potential emission from an off-axis jet, suggesting that this phenomena is relatively rare amongst the TDF population.

While relativistic outflows may be relatively rare following the tidal disruption process, this does not suggest that all TDFs are radio quiet. As one of the most nearby events, ASASSN-14li once again stands as a valuable test case. Faint radio emission (orders of magnitude less luminous than Sw J1644+57) was detected from this source immediately following discovery (Alexander *et al.* 2016; Van Velzen *et al.* 2016). While the interpretation is somewhat complicated by the presence of a pre-outburst radio source, the most plausible interpretation appears to be that the radio emission is powered by an outflow with $v \approx 2 \times 10^4\,\mathrm{km\,s^{-1}}$. A relatively modest fraction of energy is coupled to material moving with this speed ($\approx 10^{48}$ erg), bringing to mind an analogy with supernovae and gamma-ray bursts (Alexander *et al.* 2016). Such outflows would likely be missed in more distant events due to their faintness.

Evidence for outflows with comparable velocity has also been observed from optical spectra of the TDF PS1-11af (Chornock *et al.* 2014). This source exhibited a strong absorption line in spectra near peak that seems most consistent with highly blueshifted ($\Delta v = -13,000\,\mathrm{km\,s^{-1}}$) Mg II $\lambda\lambda$ 2796, 2803. Interestingly, this absorption line is not present in early spectra of PS1-11af, possibly indicating that the observed outflow is associated with a specific accretion state (c.f., black hole X-ray binaries).

Perhaps most surprising of all, high-resolution X-ray spectroscopy of ASASSN-14li revealed a highly ionized, blueshifted absorption system with a remarkably low outflow speed: $\Delta v \approx -200\,\mathrm{km\,s^{-1}}$ (Miller *et al.* 2015). These absorption lines were observed to vary (both in strength and blueshift) on a time scale of a few days, indicating that they likely arose from very close to the black hole. However, the escape velocity at these small radii is significantly larger than the observed blueshift, raising a conundrum – how could such seemingly stable features arise from so close to the black hole?

Absorption lines from high-ionization material were also observed in the UV spectra of ASASSN-14li (Cenko *et al.* 2016). Given the similar measured blueshift, the simplest interpretation is that these lines arise from the same absorption seen in the X-rays. One potential clue to their origin may come from these UV spectra, which conspicuously lack common low-ionization absorption features such as Mg II and Fe II. Given the complex effects caused by photoionization in these systems (see below), one potential explanation is that we are directly observing stellar debris.

4. Peculiar Abundance Measurements

Optical spectra of the first optically selected TDF discovered in real-time, PS1-10jh, revealed a most peculiar abundance pattern: strong, broad emission lines of He II λ 4686 and λ3203, but no corresponding Balmer emission lines. Given that the ionization energy of He II is significantly larger than that of H I, Gezari *et al.* (2012) argued that the lack of H in the spectra of PS1-10jh indicated a H-poor star was disrupted (see also Strubbe & Murray 2015). If indeed the case, this would be a very surprising result, as He core stars are much rarer than typical main sequence (H-rich) stars. However, it would also imply that TDFs could be used as quite remarkable probes to study the chemical composition of individual stars in distant galaxies.

Subsequent to the discovery of PS1-10jh, a number of additional TDF candidates were identified by wide-field optical transient surveys, nearly all of which exhibited the characteristic broad He II emission lines (Arcavi *et al.* 2014). However, Balmer H emission (in particular Hα) was observed in a number of these sources, and the H:He ratio has been observed to vary as a function of time even in the same source. A growing consensus appears to be emerging that photoionization effects play a pivotal role in the observed spectra of TDFs (Guillochon *et al.* 2014), and codes that make the common assumption in AGN environments that the gas is optically thin in the continuum are unlikely to be applicable in TDF environments. For instance, Roth *et al.* (2016) were able to match the observed H:He ratio of <1:5 in PS1-10jh using a H-rich star when properly solving for radiation transport.

The first UV spectrum of a TDF (ASASSN-14li) further reinforced the complicated radiation transport mechanisms in play in these systems (Cenko *et al.* 2016). Like AGN, broad nuclear emission lines are observed in the spectrum of ASASSN-14li. However, these lines display a remarkable abundance pattern: N III], N IV], and He II are quite prominent, while the common quasar emission lines of C III] and Mg II are weak or entirely absent.

The lack of C III could be explained if the gas were above the critical density of $n_{\rm crit} = 10^{9.5}$ cm^{-3} (Osterbrock 1989); constraints from the X-ray spectrum place a lower limit of $n \gtrsim 2 \times 10^9$ cm^{-3}. However, the strong N emission lines are reminiscent of a rare sub-class of quasars known as "N-rich" QSOs. Kochanek (2016) argues that these N-rich QSOs are in fact the tidal disruption of a $\gtrsim 1\,{\rm M}_\odot$ star, for which the high N abundance results from significant CNO processing in the stellar core (pre-disruption).

5. A Preference for Post-Starburst Host Galaxies

As has been the case for both gamma-ray bursts and all sub-classes of supernovae, the host galaxies of transients can often provide important diagnostics. In the case of tidal disruption flares, their hosts appear to exhibit a strong over-abundance of "post-starburst" galaxies: i.e., those galaxies with a strong red continuum from an old stellar population, but also exhibiting absorption lines of Balmer H, indicating the presence of a recent "burst" of star-formation (Arcavi *et al.* 2014; French *et al.* 2016). Post-starburst galaxies are often invoked as relics of recent galaxy mergers. This has been suggested to explain their over-abundance, as galaxy mergers may alter the gravitational potential of the central black hole in a manner that rapidly fills its loss cone.

Understanding observational biases is important to make further progress in studies of TDF hosts. Most ground-based optical surveys select against nuclear transients in blue host galaxies, as these will suffer from significantly more contamination from AGN flares and core-collapse supernovae (potential TDF impostors). It is difficult to come up with a selection effect that would favor post-starburst galaxies over "ordinary" ellipticals; however, more detailed studies of rates are necessary to further understand this result.

6. Future Prospects

With convergence of both theoretical insight (see, e.g., Ramirez-Ruiz 2017 for an overview) and observational capabilities, TDFs will continue to experience exciting breakthroughs in the coming years. In the near-term, the advent of the Zwicky Transient Facility in 2017 should increase the optical TDF discovery rate by an order of magnitude, and eROSITA (Khabibullin *et al.* 2014) may uncover hundreds of TDFs at X-ray wavelengths (though less well-sampled).

However, there is still much work to be done to understand these remarkable transients. In the longer term, LSST is expected to uncover thousands of TDFs each year. But if these objects are to become viable probes of black hole mass in the distant universe, we will need a much better understanding of the physical processes at work in the interim.

References

Alexander, K. D., Berger, E., Guillochon, J., Zauderer, B. A., & Williams, P. K. G. 2016, *ApJL*, 819, L25

Arcavi, I., *et al.* 2014, *ApJ*, 793, 38

Bloom, J. S., *et al.* 2011, *Science*, 333, 203

Bower, G. C., Metzger, B. D., Cenko, S. B., Silverman, J. M., & Bloom, J. S. 2013, *ApJ*, 763, 84

Brown, G. C., Levan, A. J., Stanway, E. R., Tanvir, N. R., Cenko, S. B., Berger, E., Chornock, R., & Cucchiaria, A. 2015, *MNRAS*, 452, 4297

Burrows, D. N., *et al.* 2011, *Nature*, 476, 421

Cenko, S. B., *et al.* 2016, *ApJL*, 818, L32

—. 2012, *ApJ*, 753, 77

Chornock, R., *et al.* 2014, *ApJ*, 780, 44

De Colle, F., Guillochon, J., Naiman, J., & Ramirez-Ruiz, E. 2012, *ApJ*, 760, 103

Evans, C. R. & Kochanek, C. S. 1989, *ApJL*, 346, L13

Ferrarese, L. & Merritt, D. 2000, *ApJL*, 539, L9

French, K. D., Arcavi, I., & Zabludoff, A. 2016, *ApJL*, 818, L21

Gebhardt, K., *et al.* 2000, *ApJL*, 539, L13

Gezari, S., *et al.* 2012, *Nature*, 485, 217

Guillochon, J., Manukian, H., & Ramirez-Ruiz, E. 2014, *ApJ*, 783, 23

Hills, J. G. 1975, *Nature*, 254, 295

Holoien, T. W.-S., *et al.* 2016, *MNRAS*, 455, 2918

Irwin, J. A., Henriksen, R. N., Krause, M., Wang, Q. D., Wiegert, T., Murphy, E. J., Heald, G., & Perlman, E. 2015, *ApJ*, 809, 172

Ivezic, Z., *et al.* 2008, arXiv e-print (astro-ph/0805.2366)

Kaiser, N., *et al.* 2010, in Society of Photo-Optical Instrumentation Engineers (SPIE) Conference Series, Vol. 7733, Society of Photo-Optical Instrumentation Engineers (SPIE) Conference Series

Khabibullin, I., Sazonov, S., & Sunyaev, R. 2014, *MNRAS*, 437, 327

Kochanek, C. S. 2016, *MNRAS*, 458, 127

Komossa, S. 2015, Journal of High Energy Astrophysics, 7, 148

Krolik, J., Piran, T., Svirski, G., & Cheng, R. M. 2016, *ApJ*, 827, 127

Law, N. M., *et al.* 2009, *PASP*, 121, 1395

Levan, A. J., *et al.* 2011, *Science*, 333, 199

Lodato, G. & Rossi, E. M. 2011, *MNRAS*, 410, 359

Miller, J. M., *et al.* 2015, *Nature*, 526, 542

Osterbrock, D. E. 1989, Astrophysics of gaseous nebulae and active galactic nuclei

Phinney, E. S. 1989, in IAU Symposium, Vol. 136, The Center of the Galaxy, ed. M. Morris, 543

Piran, T., Svirski, G., Krolik, J., Cheng, R. M., & Shiokawa, H. 2015, *ApJ*, 806, 164

Rau, A., *et al.* 2009, *PASP*, 121, 1334

Rees, M. J. 1988, *Nature*, 333, 523

Roth, N., Kasen, D., Guillochon, J., & Ramirez-Ruiz, E. 2016, *ApJ*, 827, 3

Shappee, B. J., *et al.* 2014, *ApJ*, 788, 48

Shiokawa, H., Krolik, J. H., Cheng, R. M., Piran, T., & Noble, S. C. 2015, *ApJ*, 804, 85

Strubbe, L. E. & Murray, N. 2015, *MNRAS*, 454, 2321

Strubbe, L. E. & Quataert, E. 2009, *MNRAS*, 400, 2070

Ulmer, A. 1999, *ApJ*, 514, 180

van Velzen, S., *et al.* 2016, *Science*, 351, 62

van Velzen, S., Frail, D. A., Körding, E., & Falcke, H. 2013, *AAP*, 552, A5

Zauderer, B. A., *et al.* 2011, *Nature*, 476, 425

New Frontiers in Black Hole Astrophysics
Proceedings IAU Symposium No. 324, 2016
Andreja Gomboc, ed.

doi:10.1017/S174392131700179X

Relativistic loss cone dynamics: Infall and inspiral rates and branching ratios

Tal Alexander

Department of Particle Physics & Astrophysics, Weizmann Institute of Science,
234 Herzl St, Rehovot 76100, Israel
email: tal.alexander@weizmann.ac.il

Abstract. I describe recent advances in the formulation and modeling of relativistic stellar dynamics around a massive black hole, and the implications for the rates of infall (e.g. tidal disruption) and inspiral (e.g. gradual decay by gravitational wave emission) processes, and their branching ratios.

Keywords. stellar dynamics, black hole physics, Galaxy: center, galaxies: active, gravitational waves

1. Introduction

There are many pathways to the tidal disruption of a star by a massive black hole (MBH) (Figure 1), which assume very different conditions (steady vs transient), dynamics (collisional vs non-collisional) or configurations (e.g. single vs binary MBH). Each further bifurcates into different outcomes (e.g. whether the interacting stars end up bound or unbound to the MBH). Thus, the practical question for the planning and execution of observations: What is the theoretically predicted rate and properties of tidal disruption events (TDEs)? is too general for a useful answer. Theory takes initial and boundary conditions and projects a prediction. However, the initial and boundary conditions themselves are external to the logical chain of theoretical arguments, and unless introduced "by hand", must be supplied by observations. In that sense, all theoretical predictions are conditional.

Here I focus on a specific set of conditions, not because these are necessarily more realistic, but because the theory that describes them is more predictive and / or simpler and better developed. I will assume that the galactic nucleus is (1) in steady state, (2) that the dynamical processes responsible for supplying stars for TDEs are collisional, and (3) that only a single MBH exists in the nucleus.

The dynamics leading to such TDEs are a specific instance of the long-standing general stellar dynamical "loss-cone problem": How do stars in a galactic nucleus get to interact strongly with a MBH, and / or fall into it, and at what rates? (Frank & Rees 1976). The term loss-cone refers to the cone of velocity vectors that take a star from position $\boldsymbol{r}$ relative to the MBH to a periapse distance smaller than a critical value r_{lc}, where some strong interaction of interest, usually destructive, takes place (e.g. direct horizon crossing or stellar disruption). All stars that are initially on such orbits are quickly removed from the system. From that time on the loss rate is determined by the rate at which new stars are deflected into loss-cone orbits. Thus the loss-cone problem corresponds to the question: How do galactic nuclei randomize? (see Alexander 2017 for an extended review).

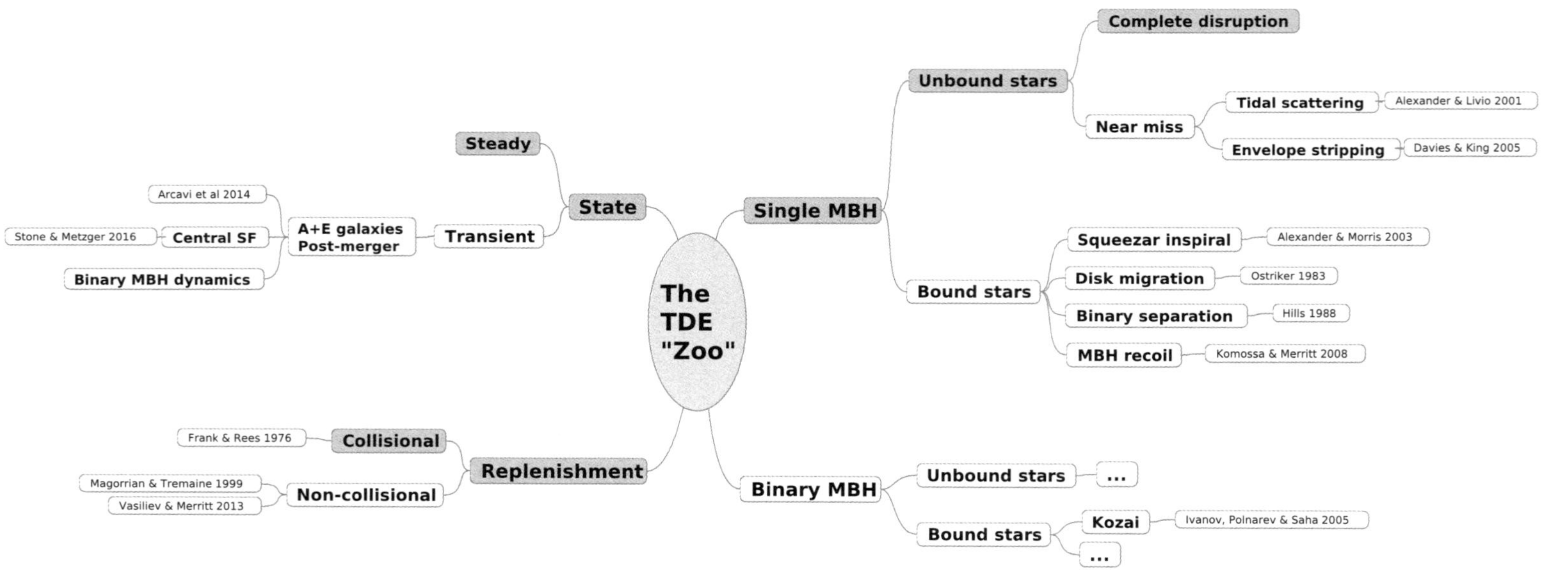

Figure 1. The theoretical TDE "zoo". A schematic of some of the many possibilities, permutations, configurations and dynamical processes that lead to TDEs and related phenomena. The options marked by a darker background are the prompt TDEs discussed here.

2. The classical loss-cone

The classical analysis of the loss-cone problem (Lightman & Shapiro 1977) focused on plunge processes, which require only a single approach to the MBH, such as a TDE, and on scattering by incoherent 2-body interactions. Close to the MBH, inside a critical radius r_d, the accumulated scatter per orbit is small relative the the angular opening of the loss-cone. Stars diffuse slowly into it, and are promptly lost. At any given time the number of stars on loss-cone orbits is negligible. This dynamical regime is called the empty or diffusive loss-cone, where the plunge rate is governed by the slow 2-body relaxation rate. Further out at $r > r_d$, the angular opening of the loss-cone is small enough that the scatter per orbit can cause a star to skip all across the loss-cone before it reaches periapse. This is the full loss-cone regime, where the plunge rate is set by the orbital period. A general analysis of the contributions to the total rate from both regimes (Syer & Ulmer 1999) indicates that the total rate can be conservatively estimated (to within a factor of 2) by taking the smaller of r_d and the radius of the MBH influence $r_h = GM_\bullet/\sigma^2$, where σ is the velocity dispersion in the bulge of the host galaxy, $r_e = \min(r_d, r_h)$ and using the empty loss-cone plunge rate, $\Gamma_p = N_\star(< r_e)/\left[\log(J_c(r_e)/J_{lc})T_E(r_e)\right]$, where $N_\star(< r)$ is the number of stars inside r, $J_{lc} = \sqrt{2GM_\bullet r_{lc}}$ is the angular momentum of the loss-cone, J_c is the maximal (circular) angular momentum, and T_E the 2-body energy relaxation time.

A dissipative dynamical process can extract energy from the orbit (e.g. by gravitational wave (GW) emission (e.g. Amaro Seoane et al 2007), tidal heating (Alexander & Morris 2003), or hydrodynamic interactions with a massive disk (Ostriker 1983)). The dissipation rate is usually a strong function of radius, and occurs mostly near periapse. A gradual inspiral into the MBH requires many repeated periapse passages. This can happen only on tight orbits with short periods, where orbital decay by dissipation is faster than scattering by the background stars, which can abort the inspiral by deflecting the star from its eccentric orbit to a nearly radial plunge orbit, or to a wider orbit, where dissipation is negligible. Figure (2a) shows two possible trajectories of stars in $(\log j, \log a)$ phase-space, where $j = J/J_c = \sqrt{1-e^2}$ is the normalized angular momentum, and a the orbital semi-major axis (sma) in the Keplerian approximation. Plunges occur with high probability for stars with sma $a_c < a < a_h$ (typically $a_c \ll a_h = r_h$), while inspirals occur with high probability for stars with $a < a_c$. The inspiral to plunge event rate ratio reflects the ratio of the small number of stars in the real-space volume inside a_c, relative to the very large number, $N_\star \sim \mathcal{O}(Q)$ ($Q = M_\bullet/M_\star$), in the volume interior to a_h, $\Gamma_i/\Gamma_p \sim N_\star(< a_c)/N_\star(< a_h) \sim \mathcal{O}(0.01)$ (Alexander & Hopman 2003).

3. The relativistic loss-cone: GR precession and resonant relaxation

Non-resonant (incoherent) two-body relaxation is inherent to any system of interacting discrete particles, and provides a minimal randomization rate. The point-like nature of gravitational interactions between stars boosts the relaxation timescale by the logarithmic Coulomb factor, which for a nearly Keplerian system around a MBH scales as $T_{NR} \propto [Q^2 P/N_\star]/\log Q$, where P is the orbital period. Non-resonant relaxation (NR) can change both the orbital energy and angular momentum.

There is however a restricted but much faster form of angular momentum relaxation, resonant relaxation (RR) (Rauch & Tremaine 1996), which operates in nearly spherical potentials, like the one around a MBH, where symmetry restricts orbital evolution (e.g. fixed ellipses in a Keplerian potential). A test star then moves in a static background potential generated by the $N_\star$ orbitally-averaged elliptical "mass wires", which conserve

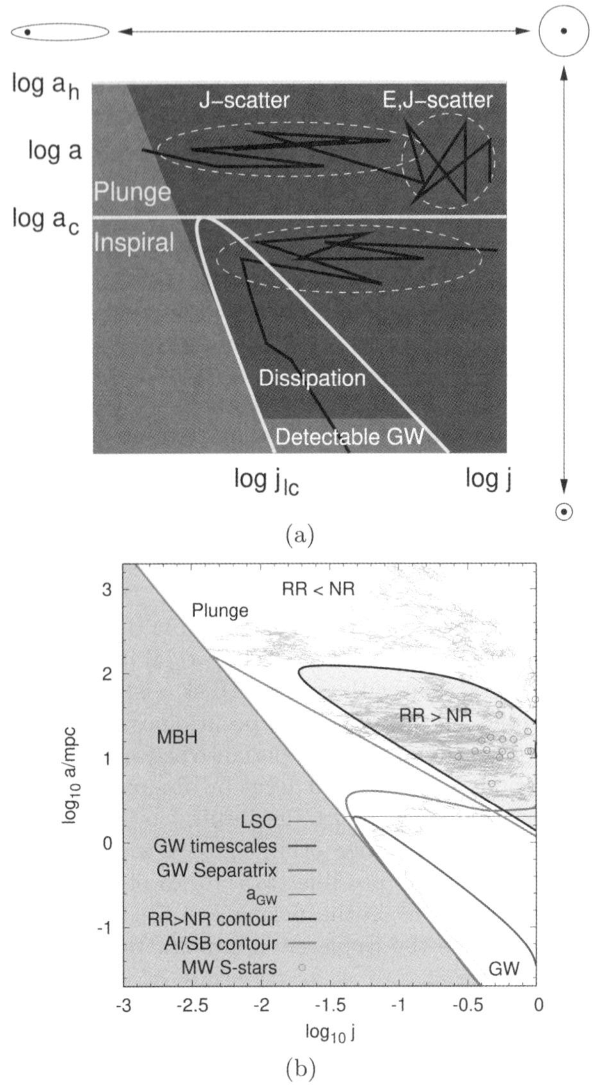

Figure 2. **(a)** A schematic of the "classical" loss-cone, where only 2-body incoherent relaxation is considered (see text). The time to change the angular momentum by order of itself, T_J, is related to the time to change the energy by order of itself, T_E, by $T_J = j^2 T_E$. Stars on nearly circular orbits (upper right corner) are scattered equally in energy and angular momentum. However, once the star is scattered by chance to a somewhat eccentric orbit, scattering in angular momentum becomes much faster. When the orbit lies above the critical semi-major axis (sma) a_c defined by the upper tip of the arc where GW dissipation is faster than the 2-body scatter, it will plunge with high probability directly into the MBH (the wedge of GR-unstable orbits on the left). However, when it lies below a_c, it will cross with high-probability through the GW dissipation region, where it is diverted to an inspiral orbit. **(b)** The phase space of the loss-cone as derived from the η-formalism. The triangular region to the left is the region of unstable orbits that plunge into the MBH. The rounded region on the right is where RR dominates the dynamics. The diagonal line just below it is the SB/AI line, where RR is completely quenched by GR precession. The arc contour is the region where GW dissipation is faster than NR. The critical sma a_c is at the tip of the arc. The curved line that overlaps with part of the arc is the exact separatrix between flow-lines that plunge and those that inspiral. Two stellar tracks are displayed as examples. The S-stars discovered around the Galactic MBH Sgr A* (circles) are observed to lie in the RR-dominated region (see text). (Bar-Or & Alexander 2016).

its energy, but exert a residual torque $\tau_N \sim \sqrt{N_\star} GM_\star/r$ that persists for a coherence time T_c until the small deviations from the Keplerian potential, due to the distributed stellar mass ("mass precession"), to GR precession, or to the mutual resonant interactions between the background stars ("self-quenching") grow to $\mathcal{O}(1)$. The large coherent change in the test star's angular momentum over the coherence time, δJ, then becomes the mean free path of a random walk in phase space on timescales t$\gg T_c$: $\Delta J/J_c \sim (\delta J/J_c)\sqrt{t/T_c} \equiv \sqrt{t/T_{RR}}$. The extended mass wires cannot approach each other as point masses do. However, the RR relaxation timescale is boosted by the long coherence time, $T_{RR} \propto [Q^2 P/N_\star]P/T_c$.

Close to a MBH where mass precession is slow, but far enough that GR is still negligible, the coherence time can be so long that $T_{RR}/T_{NR} \sim \log Q(P/T_c) \ll 1$, and RR dominates the dynamics. Since efficient RR torques can rapidly change $J \to J_{lc}$ and lead to a strong interaction with the MBH, RR is of special relevance to the loss-cone problem. In fact, unchecked RR can drain all the stars from the inner parts of a stellar cusp by efficiently pushing them to plunge orbits (Bar-Or & Alexander 2016). This would have dramatic effects: it would completely suppress GWs from extreme mass ratio inspirals (EMRIs), and would leave no stars or pulsars on relativistic orbits for testing strong gravity (e.g. Will 2008).

However, stars on eccentric orbits undergo substantial in-plane (Schwarzschild) GR precession. In their analysis of the implications of RR for EMRIs, Hopman & Alexander (2006) made the "fortunate coincidence" conjecture: since the GR in-plane precession is a PN1 effect (i.e. an $\mathcal{O}(\beta^2)$ effect), whereas GW emission is a PN2.5 effect, the precession should kick in well before GW dissipation is important, and reverse the sign of the RR torques every half cycle, so that on average the effect of the RR torques should cancel out. In that way, by the time GW emission becomes substantial (it is unaffected by precession), the EMRI orbit will already be completely decoupled from the stellar background, and the EMRI's quasi-periodic GW emission should proceed unimpeded.

The first ever fully-self consistent post-Newtonian N-body simulation (Merritt *et al.*2011) revealed an unanticipated result. Not only did GR precession indeed quench RR well before the GW inspiral began, there appeared to be a barrier in phase space, denoted the "Schwarzschild Barrier" (SB), where the stars' RR-driven random walk in angular momentum toward $j \to 0$ was blocked (Figure 2b). Near the SB, the orbital parameters oscillate at the GR precession frequency for about a coherence time, and then the stars appear to "bounce" back from the barrier to higher j, instead of evolving to a plunge or EMRI orbit, as expected. Only occasionally does a star "jump" across the barrier and reach the MBH. This behavior was explained in a preliminary analysis (Alexander 2010) as related to precession under the influence a residual dipole-like force. However, the SB phenomenon, its nature, implications and very existence remained controversial.

4. The η-formalism for relativistic loss-cone dynamics

The difficulty in incorporating RR in a formal description of dynamics around a MBH is due to the fact that this phenomenon lies in the interface between stochastic and deterministic processes. The standard description of collisional dynamics in terms of Markovian ("memory-less") diffusion is fundamentally inconsistent with RR's long coherence times.

The η-formalism (Bar-Or & Alexander 2014) bridges this gap by expanding the full orbitally averaged Hamiltonian of the test star and background, which expresses the interactions between mass ellipses, to 1st order in a generalized multipole expansion (the description of the ellipse field around the MBH require 3 angles, as opposed to the 2 of the standard multipole expansion of a scalar density field). The resulting expression for the background term is still too complex to solve analytically, but its symmetries can be analyzed, and reveal that to 1st order, the torques by the stellar background can be described as a vector $\boldsymbol{\eta}$ in angular momentum space. Since the change in the angular momentum of the test star is relatively small over T_c, this vector can be approximated as a function of time only. Since the residual torque by the background is due to a superposition of many ellipses, $\boldsymbol{\eta}(t)$ can be further approximated as a Gaussian random noise field. At this level of treatment the temporal behavior of the background noise has to be introduced in a non-self-consistent way (in contrast, the magnitude of the torque, τ_N, is known to good approximation, see Sec. 3).

A key insight is that the nature of the resulting dynamics depends critically on the smoothness (differentiability) of the noise. The power spectrum of a smooth noise must have a maximal frequency above which the power decreases rapidly. For RR, this frequency is $\nu_c \sim \mathcal{O}(T_c^{-1})$. When the precession frequency of the test star, $\nu_{GR} = \nu_0 j^{-2}$ is higher than ν_c, the star is decoupled from the noise: it's orbit is adiabatically invariant under the external perturbations. The SB is therefore the adiabatic invariance (AI) locus $j_{AI}(a) = \sqrt{\nu_0 T_c(a)/2\pi}$ (Figure 2b). Beyond this line, RR is rapidly quenched by GR precession, and only NR affects the dynamics.

The first-order, orbit-averaged Hamiltonian can be integrated in time with a randomly generated realizations of the noise $\boldsymbol{\eta}(t)$, to produce realizations of stellar trajectories in phase space. This is useful for confirming that the various approximations that go into the $\boldsymbol{\eta}$ formalism reproduce the properties of the SB that are observed in the N-body simulations. However, the major practical step forward that is provided by the η-formalism is the derivation of *effective* diffusion coefficients that incorporate AI (Bar-Or & Alexander 2014). These then allow to evolve in time the stellar distribution function by integrating the Fokker-Planck (diffusion) equation by the Monte Carlo method (Bar-Or & Alexander 2016), which corresponds to modeling the realistic $N_\star \to \infty$ limit. By comparison, direct N-body (e.g. Merritt *et al.*2011) is currently limited to $\mathcal{O}(100)$ stars with an unrealistically small mass ratio Q. Another advantage of the Monte Carlo method is that it is simple to add NR, GW, tidal heating and other physical processes, and to switch on or off various physical mechanisms. The $\boldsymbol{\eta}$-formalism Monte Carlo results reproduce the low-$N_\star$ N-body plunge and inspiral rates, and also the behavior observed near the SB/AI line (Bar-Or & Alexander 2016) .

5. Results

Figure (2b) shows the various critical lines in the phase-space of the loss-cone for a Milky Way-like Galaxy with $M_\bullet = 4 \times 10^6\, M_\odot$ and $r_h = 2$ pc, overlayed with two Monte Carlo-generated stellar trajectories, one a plunge and the other a GW inspiral. RR dominates the dynamics only in a limited region that is away from the loss-lines (the last stable orbit or the GW line), and therefore RR does not affect much the loss-rates. Figure (3c) shows the dependence of the loss rates on $M_\bullet$, with and without RR. The difference is small—this confirms the fortunate coincidence conjecture (Sec. 3).

Over long timescales $\sim \mathcal{O}(T_{NR})$, NR completely randomizes the distribution function in phase-space, and it approaches the expected maximal entropy configuration, irrespective of the short timescale behavior near the SB/AI line, since the steady state should not depend on the nature of the randomization process. This is shown in Figure (3a). A slight enhancement in the local plunge rate is seen where the RR-dominated region extends closest to the loss-line (Figure 3b).

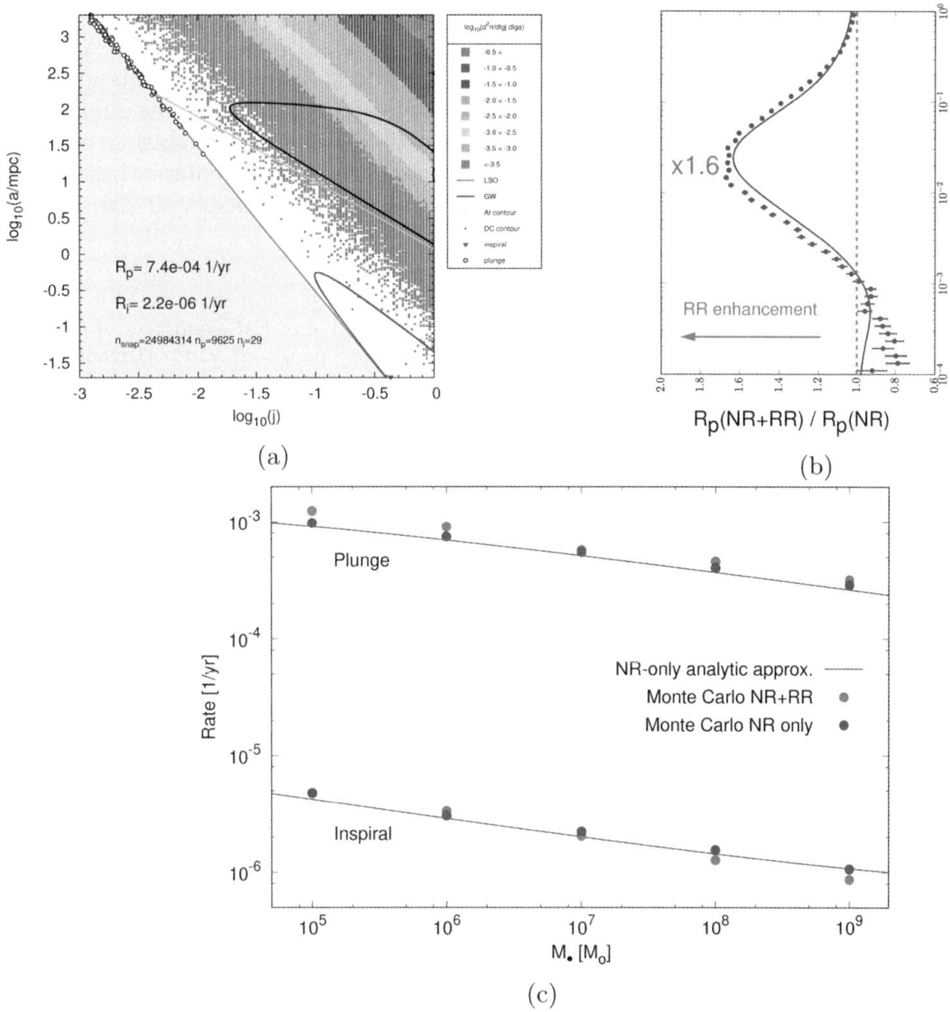

Figure 3. **(a)** An η-formalism Monte Carlo simulation of a Milky Way-like galaxy, taking into account NR, RR, GR and mass precession, and assuming a smooth noise. The color scale shows the steady state stellar distribution function, the end-points of some of the stellar trajectories are shown as black circles, and the resulting steady-state loss rates are quoted on the left. **(b)** The small relative differential enhancement due to RR, as function of sma a, scaled to correspond to panel (a). **(c)** The steady state plunge and inspiral rates as function of MBH mass $M_\bullet$, with and without RR. (Adapted from Bar-Or & Alexander 2016).

While RR does not affect much the steady state loss rates, it can have a strong effect on stars that happen to orbit in the RR-dominated region. The S-stars observed around the MBH of the Milky Way Sgr A* (main sequence B-stars) are such stars (Gillessen et al 2009). There are two leading scenarios for explaining their presence so close to the MBH. One is that they were tidally captured by the Hills mechanism (Hills 1988) (a 3-body interaction between the MBH and an incoming binary, where one star is captured on a tight eccentric orbit and the other is ejected as a hyper-velocity star), in which case their initial eccentricity was very high, and the other that they migrated from a stellar disk such as the one observed today around Sgr A* (Levin 2007), in which case their initial eccentricity was low. Their presently observed eccentricity distribution is thermal (random), which requires post-formation / capture evolution of angular momentum. η-formalism Monte Carlo experiments show (Sabsovich *et al.*2017) that for either scenarios to work, one must assume the existence of a dense "dark" cusp of remnants and low-luminosity stars that can exert the rapid RR torques on the S-stars. The simulations further indicate that the Hills mechanism is favored over the disk migration (see also Perets et al 2009). RR is likewise relevant for modeling the dynamics of relativistic stars and pulsars near Sgr A* for the purpose of testing strong gravity (e.g. Will 2008).

References

Alexander, T. 2010 in: P. Amaro-Seoane & E. Porter (eds.), *GW and EM signatures of MBH binaries and EMRIs*, online proc. `http://www.aei.mpg.de/ pau/conf_vid4/Alexander.pdf`
Alexander, T. 2017, *ARAA* in press.
Alexander, T. & Livio, M. 2001, *ApJ* (Letters) 560, L143
Alexander, T. & Morris, M. 2003, *ApJ* (Letters) 590, L25
Alexander, T. & Hopman, C. 2003, *ApJ* (Letters) 590, L29
Amaro Seoane, P., *et al.* 2007, *CQG* 24, 113
Arcavi, I., *et al.* 2014, *ApJ* 793, 38
Bar-Or, B. & Alexander, T. 2014, *ApJ* 820, 129
Bar-Or, B. & Alexander, T. 2016, *CQG* 31, 244003
Davies, M. & King, A. 2005, *ApJ* (Letters) 624, L25
Frank, J. & Rees, M. J. 1976, *MNRAS* 176, 633
Gillessen, S., *et al.* 2009, *ApJ* 692, 362
Hills, J. G. 1988, *Nature* 331, 687
Hopman, C. & Alexander, T. 2006, *ApJ* 645, 1152
Ivanov, P. B., Polnarev, A. G. & Saha, P. 2005 *MNRAS* 358, 1361
Komossa, S. & Merritt, D. 2008 *ApJ* (Letters) 683, L21
Levin, Y. 2007 *MNRAS* 374, 515
Lightman, A. P. & Shapiro, S. L. 1977, *ApJ* 211, 244
Magorrian, J., *et al.* 1998 *AJ* 115, 2285
Merritt, D., Alexander, T., Mikkola, S., & Will, C. 2011, *PRD* 84, 044024
Ostriker, J. P. 1983, *ApJ* 273, 99
Perets, H. B., *et al.* 2009, *ApJ* 702, 884
Rauch, K. P. & Tremaine, S. 1996, *New Astron.* 1, 149
Sabsovich, D., Alexander, T. & Bar-Or, B. 2017 *in prep.*
Stone, N. C. & Metzger, B. D. 2016 *MNRAS* 455, 859
Syer, D. & Ulmer, A. 1999, *MNRAS* 306, 35
Vasiliev, E. & Merritt, D. 2013, *ApJ* 774, 87
Will, C. M. 2008, *ApJ* (Letters) 674, L25

New Frontiers in Black Hole Astrophysics
Proceedings IAU Symposium No. 324, 2016
Andreja Gomboc, ed.

doi:10.1017/S1743921316013119

The role of electromagnetism in tidal disruption events

Andrej Čadež

University of Ljubljana
email: andrej.cadez@fmf.uni-lj.si

Abstract. Tidal disruption events are characterized by the prompt release of very high energy electromagnetic radiation, which rapidly carries away a significant fraction of mass-energy. The time scale od prompt release is quite comparable to the time scale of gravitational collapse. We suggest that generation of very high energy radiation from pulsar nebulae may be an example of the relativistic coupling between gravity and electromagnetism. The main ingredient of our picture comes from the observation that the electron - ion energy exchange time scale is much longer than the electron ion energy exchange time scale, which leads to thermodynamic decoupling of electron and ion gases, and thus to their different temperatures. Under such conditions the dominating hotter component pushes the colder component further out and thus generates a global electric field which constrains electrons and ions by their mutually generated electric field, yet allows them to reach very high kinetic energies

Keywords. plasmas, relativity, cosmic rays

1. Introduction

High energy phenomena in astrophysics take many forms, from gamma ray bursts exploding in high energy jets, to steady accretion X-ray emitting disks and finally to plerionic nebulae, such as Crab, which emit gamma rays at TeV energies. The emission of a significant portion of energy by a relatively small number of photons, which require a relatively small number of very high energy charged particles to produce, is the common characteristic of all these phenomena. Jets are another common trait. Since the Crab nebula and its pulsar have been studied extensively, providing a wealth of data on its behaviour, it might be of interest to discuss the many chalenges that these data present.

A closely studied property of the Crab pulsar is the history of its slow down by the Jodrell Bank Radio Observatory. After 23 years of monitoring Lyne *et al.*(1993) were able to state: "Between glitches, the rotational slowdown of the Crab pulsar is well described by a power law with braking index 2.51±0.01.", while 22 years later the same authors (Lyne *et al.* (2015)) came to the conclusion: "Overall, a low mean braking index of 2.342(1) is measured for the whole period, compared with values close to 2.5 in intervals between glitches". Using the published Jodrell Bank data since 1988, together with insight from scarce but much more precise optical observations, Čadež *et al.* (2016) come to a different conclusion. Instead of analyzing the frequency and frequency derivative behaviour, they represent the rotational phase of the pulsar during the mentioned period by ten episodes delimited by nine mayor glitches. Each episode, lasting a few years, can be fitted to a specific braking index law, with the braking index taking values between ~2.2 and ~2.6. The phase difference from the pure braking index law always follows a simple relaxation decay with an amplitude no larger than 17 turns†. This result is interpreted as a hint

† With the total phase change during the episode of 3×10^9 turns.

that differences between braking episodes are caused by abrupt changes in electromagnetic coupling of the rotating magnetic field of the pulsar to the surrounding nebula. Considering the relaxation time scales and the durations of episodes, we explore this as a hint that coupling to the braking mechanism is caused by changes in the structure of the surrounding nebula rather than in the interior structure of the pulsar. This interpretation eventually relates to other observed, yet unexplained phenomena such as the formation of a global magnetic field, production of extremely high energy radiation, formation of jets, accelerated expansion of the nebula etc.

2. How to heat the nebula?

Let us consider the pulsar as a generator of a rotating electromagnetic field at the center of the nebula, a medium with a certain permittivity ε and conductivity σ. The propagation of electromagnetic wave in such a medium is characterized by the dispersion relation $k^2 = \varepsilon\left(\frac{\omega}{c}\right)^2 + i\mu_0\sigma\omega$, where omega is the angular frequency of the propagating wave – the angular rotation frequency of the pulsar in our case. The solution of Maxwell's equations can be expanded in multipoles of damped propagating waves, which dissipate energy along the way. Considering the solution which at $r = r_p$ limits to a rotating magnetic dipole ($|kr_p| \ll 1$) with a dipole moment p_m, we find the following expressions for the energy outflow through the sphere with radius r_p:

$$\varepsilon > 0 : L = \frac{\mu_0}{6\pi}\omega^2\left(\varepsilon^{3/2}\frac{\omega^2}{c^3} + \frac{\mu_0\sigma}{r_p}\right)p_m^2 \qquad n = 3 - \frac{2}{1 + \varepsilon^{3/2}\left(\frac{\omega}{c}\right)^2\frac{r_p}{\mu_0\sigma c}} \tag{2.1}$$

$$\varepsilon < 0 : L = \frac{\mu_0}{6\pi}\omega^2\frac{\mu_0\sigma}{r_p}p_m^2 \qquad n = 1 \tag{2.2}$$

Considering that energy outflow is fed by rotational energy of the pulsar ($\frac{d}{dt}\left(\frac{1}{2}J\omega^2\right) = -L$), one can calculate the corresponding braking index $n = \frac{\ddot{\omega}\omega}{\dot{\omega}^2}$, which is also listed above. This simple model explains the average observed braking index $n \sim 2.5$ if $\varepsilon > 0$ and $\sigma = \frac{1}{3}\varepsilon^{3/2}\left(\frac{\omega}{c}\right)^2\frac{r_s}{\mu_0 c} = 3.5\times10^{-12}\left[\frac{A}{Vm}\right]\varepsilon^{3/2}$. On first sight both conditions seem difficult to satisfy: 1. considering the often quoted electron density $\sim 10^3 cm^{-3}$, one finds that the plasma frequency should be a few kHz, which would imply that below a few kHz ε should be negative. 2. Even if the value of ε should be found positive, the required plasma conductivity is unusually low. Yet, the conditions at the center of the Crab nebula may be sufficiently unusual to allow both obstacles to be surmounted. First consider the conductivity: $\sigma = \frac{n_e e^2}{m_e \nu_c}$, where ν_c is the collision frequency of an electron with all other particles and fields present. Because of the evidently low density of plasma in question, electron-electron, electron-ion and electron-atom collision frequencies are many orders of magnitude too low to satisfy our condition, but the remaining candidate — the exchange of energy-momentum with electromagnetic field through synchrotron radiation — can do the job. So we derive the electron collision frequency ν_c as the inverse of synchrotron lifetime to write: $\nu_c = \frac{\sigma_T}{m_e c}\frac{B^2}{\mu_0}\gamma_e$, where σ_T is the Thomson cross section, B the strength of the static component of the magnetic field, and γ_e the relativistic Lorentz gamma factor of electrons. When this expression is used to calculate the conducivity inserted in eq.(2.1), one obtains the following expression for the braking power: $L = \int \sigma_T c\gamma_e^2 n_e \frac{B^2}{\mu_0}dV \approx 2\langle{\gamma_e}^2 n_e\rangle\sigma_T c\int\frac{B^2}{2\mu_0}dV$. Here $\langle\rangle$ stands for the suitable average over the volume with high magnetic field near the pulsar; the magnetic energy, represented by the above integral, is taken to be the energy belonging to a static magnetic dipole with a strength 10^8T at the surface of the pulsar (at r_p = 10km). In this way the braking power becomes

$L = 3.3 \times 10^{38} [\mathrm{erg/s}] \frac{n_e}{1000\mathrm{cm}^{-3}} \left(\frac{\gamma_e}{10^4}\right)^2$, which should be compared to the invoked slow down luminosity ($L_{Crab} \sim 2 \times 10^{38}$ $[erg/s]$). If one further relates the high energy peak at $\sim 10^{12}$eV in the energy spectrum of the Crab (Kohri *et al.* (2012)) to the energy of electrons in the heating region, one comes to the conclusion that extremely low electron density in the central region suffices to couple pulsar rotation to the nebula. However, it is quite clear that pulsar's gravitational field can not bind plasma of correspondingly high temperature to the region where the electromagnetic field can heat it. Thus, this proposal can only work if large scale electric and magnetic fields can provide suitable potential wells to sustain the conversion of energy in the central heating region.

We start with a very simple scenario, considering the central 7×10^5km region as the heater where at least electrons can be accelerated to very high kinetetic energy by the central electromagnetic source (the pulsar). The heater is surrounded by a tenuous high energy synchrotron nebula, visible almost exclusively in X-rays (Chandra image), which merges with the ejecta shell at the distance R roughly 1pc from the pulsar. The electron-ion energy echange in the synchrotron nebula is extremely slow – of order millions of years. Thus, no mechanism within the nebula enforces equipartition of energy between electrons and ions. The average energy of electrons is regulated mainly by the balance of energy input from the heater and cooling by synchrotron radiation. Ions are only weakly coupled to radiation but couple thermaly to the relatively cold ejecta boundary. Thus we expect electrons in the synchrotron nebula to have higher kinetic temperature than ions. The phase space distribution of constituents of both gassses can be described by Boltzmann distribution functions $f_e(\vec{r},\vec{p}) = e^{-\alpha_e - \beta_e H_-}$ and $f_i(\vec{r},\vec{p}) = e^{-\alpha_i - \beta_+ H_+}$, where the single particle Hamiltonians $H_\pm = \alpha(r)\sqrt{m_\pm^2 c^2/\psi^4(r) + m_\pm^2 c^2} \pm e\phi(\vec{r})$ govern the motion of electrons ("-") and ions ("+") in the gravitational field of the pulsar and the electric field they generate. Here $\alpha(r) = \left(1 - \frac{M^*}{2r}\right) / \left(1 + \frac{M^*}{2r}\right)$ and $\psi = 1 + \frac{M^*}{2r}$ are the relevant Schwarzschild metric coefficients in isotropic coordinates, $M^* = GM_p/c^2$ is the gravitational radius of the pulsar and $\phi(\vec{r})$ is the electric potential generated by the charge distribution. The Lagrange multipliers α_e, α_i, β_e and β_i are determined by the numbers and energies of ions and electrons. Introducing $\tau_e = m_e c^2/\beta_e$, $\tau_i = m_p c^2/\beta_i$, $F(\vec{r}) = \frac{e\phi(\vec{r})}{m_e c^2 \tau_e}$, $\tilde{F}(\vec{r}) = \frac{e\phi(\vec{r})}{m_e c^2 \tau_p}$, $\tilde{\alpha}(r) = m_i\alpha(r)/m_e$, $B_n(x) = e^x \mathrm{BesselK}[n,x]$ and denoting Compton wavelength as λ_c and classical electron radius as r_{cl}, these can be written as :

$$N_e = \int \frac{4\pi}{\lambda_c^3 \alpha} \tau_e e^{F-\alpha_e} B_2\left(\frac{\alpha}{\tau_e}\right) dV \tag{2.3}$$

$$N_i = \int \frac{4\pi}{\lambda_c^3 \tilde{\alpha}} \tau_i e^{\tilde{F}-\alpha_e} \left(\frac{m_p}{m_e}\right)^2 B_2\left(\frac{\tilde{\alpha}}{\tau_e}\right) dV \tag{2.4}$$

$$E_e = \int \frac{m_e c^2}{\lambda_c^3 \alpha^2} e^{F-\alpha_e} \tau_e^3 \left[(3-F)\frac{\alpha}{\tau_e} B_0\left(\frac{\alpha}{\tau_e}\right) + \left(6 - 2F + \frac{\alpha^2}{\tau_e^2}\right) B_1\left(\frac{\alpha}{\tau_e}\right)\right] dV \tag{2.5}$$

$$E_i = \int \frac{m_p c^2}{\lambda_c^3 \tilde{\alpha}^2} e^{\tilde{F}-\alpha_e} \tau_e^3 \frac{m_p}{m_e} \left[(3-\tilde{F})\frac{\tilde{\alpha}}{\tau_p} B_0\left(\frac{\tilde{\alpha}}{\tau_p}\right) + \left(6 - 2\tilde{F} + \frac{\tilde{\alpha}^2}{\tau_p^2}\right) B_1\left(\frac{\tilde{\alpha}}{\tau_p}\right)\right] dV \tag{2.6}$$

Here $dV = 4\pi r^2 \psi^6(r) dr$ and the electric potential equation turns into:

$$F''(r) + \frac{2}{r\psi(r)} F'(r) = -16\pi^2 \frac{r_{cl}}{\lambda_c^3} \frac{\psi^4(r)}{\alpha(r)} \left[-e^{F-\alpha_e} B_2\left(\frac{\alpha(r)}{\tau_e}\right) + \frac{m_p^2 \tau_i}{m_e^2 \tau_e} e^{-\tilde{F}-\alpha_i} B_2\left(\frac{\alpha(r) m_i}{m_e \tau_i}\right)\right], \tag{2.7}$$

with boundary conditions $N_e = N_i$ and $F(r_{ej}) = 0$ at the outer boundary of ejecta shell. An insight into eq.(2.7) can be obtained from the low energy ($\tau_e, \tau_i \ll 1$), low gravity

$(r_p \gg M^*)$, $\tau_e \neq \tau_i$ solution $F(r) \approx \frac{15}{8}\left(\frac{M^*}{r} - \frac{M^*}{r_{ej}}\right)\left(\frac{1}{\tau_i} - \frac{1}{\tau_e}\right)^{-1} + \mathcal{O}\left(\frac{M^{*2}}{r^2}\right)$, which makes the right hand side of eq.(2.7) vanish†, while the remainder is $\mathcal{O}\left(M^{*2}/r^4\right)$. The star and the ejecta shell act like a spherical capacitor charged with the hotter componet at the outer electrode. The potential difference $U = \frac{m_e c^2}{e}\tau_e\left(F(r_{ej}) - F(r_s)\right)$ is obviously quite small in the nonrelativistic limit. In higly relativistic case equation (2.7) must be solved numerically. Solutions $F(r)$ again only slightly differ from the capacitor field form $F \approx \frac{M^* q}{r}$, only the value of q must be determined numerically. In the regime $10^4 < \tau_e < 5 \times 10^4$, which corresponds to electrons capable of emitting at $\sim 10^{12}$eV, characteristic of Crab high energy radiation, and $\tau_i \sim 1300$, q is found to have an almost constant value $q \approx 21$. With this solution one can go back to the braking power calculation and refine it by using the radial dependence of electron number density, expressed as $n_e = n_0 e^{qM^*/r}$, to better evaluate the volume integral. Neglecting the dependence of α and ψ on r, realizing that $\gamma_e = 3\tau_e$ and defining $x = \frac{M^* q}{r_p} = 3.15$, the result can be written in the form

$$
\begin{aligned}
L &= 27 \times 3.3 \times 10^{35}\,[\mathrm{erg/s}]\left(\frac{n_e}{\mathrm{cm}^{-3}}\right)\left(\frac{\gamma_e}{10^4}\right)^2\left[e^x\left(\frac{1}{x} - \frac{2}{x^2} + \frac{2}{x^3}\right) - 2\frac{1}{x^2}\right] \\
&= 3.7 \times 10^{37}\,[\mathrm{erg/s}]\left(\frac{n_e}{\mathrm{cm}^{-3}}\right)\left(\frac{\tau_e}{10^4}\right)^2 = 0.18 L_{Crab}\left(\frac{n_e}{\mathrm{cm}^{-3}}\right)\left(\frac{\tau_e}{10^4}\right)^2 . \qquad (2.8)
\end{aligned}
$$

3. Conclusion

We have identified the tenuous region between the shock front and the compact remnant of a supernova-like explosion as a region where gravity and electromagnetism strongly couple through tenuous relativistic plasma. Applying these ideas to the tenuous plasma of inner Crab nebula, equations (2.2) and (2.8) plausibly connect pulsar braking power to high electron temperature and low density of the inner nebula. In the proposed mechanism gravity and electromagnetism couple directly, since the main component of the electric field is produced by charges on the capacitor between the central compact object and outer ejecta layer. The plasma inside the capacitor does not importantly contribute to electric field by polarization (as long as it is collisionless), however the polarization is sufficient to concentrate the hotter component (electrons) in the inner region and deplete the colder component there. This eliminates the plasma frequency limit in the inner region, and allows electrons to couple strongly and rapidly to electromagnetic field of the central object. This may be the the fastest way to turn electromagnetic or gravitational energy transfer into observable radiation.

References

Lyne, A. G., Pritchard, R. S., & Graham-Smith, F. 1993, *MNRAS*, 265, 1003
Lyne, A. G., Jordan, C. A., Graham-Smith, F., *et al.* 2015, *MNRAS*, 446, 857
Čadež, A., Zampieri, L., Barbieri, C., *et al.* 2016, *A&A*, 587, A99
Kohri, K., Ohira, Y., & Ioka, K. 2012, *MNRAS*, 424, 2249

† It is worth noting that the right hand side of equation (2.7) does not vanish for $F = 0$, only because $B_2(x)$ deviates from a pure power law. In the classical nonrelativistic treatment $B_2\left(\frac{\alpha(r)}{\tau_e}\right)$ would be replaced by $\sqrt{\frac{2\pi\tau_e}{\alpha(r)}}$ and similarly $B_2\left(\frac{\alpha(r)m_i}{m_e\tau_e}\right)$. In this case $\alpha^{-1/2}(r)$ would be the common factor of the two terms on the right hand side of eq.(2.7), which could be made to vanish for $F = 0$ at all r with a proper choice of α_i and α_e. Thus, the emergence of electric field should be considered as a relativistic effect.

New Frontiers in Black Hole Astrophysics
Proceedings IAU Symposium No. 324, 2016
Andreja Gomboc, ed.

doi:10.1017/S1743921316012898

Star Formation Close to Sgr A* and Beyond the Nuclear Cluster

Farhad Yusef-Zadeh[1] **and Mark Wardle**[2]

[1]CIERA, Department of Physics and Astronomy, Northwestern University, 2145 Sheridan Rd, Evanston, IL 60208, US
email: zadeh@northwestern.edu

[2]Department of Physics and Astronomy and Research Center for Astronomy, Astrophysics & Astrophotonics, Macquarie University, Sydney NSW 2109, Australia
email: mark.wardle@mq.edu.au

Abstract. Two modes of star formation are involved to explain the origin of young stars near Sgr A*. One is a disk-based mode, which explains the disk of stars orbiting Sgr A*. The other is the standard cloud-based mode observed in the Galactic disk. We discuss each of these modes of star formation and apply these ideas to the inner few parsecs of Sgr A*. In particular, we focus on the latter mode in more detail. We also discuss how the tidal force exerted by the nuclear cluster makes the Roche density approaching zero and contributes to the collapse of molecular clouds located tens of parsecs away from Sgr A*.

Keywords. Galaxy: nucleus, stars: formation, ISM: clouds

1. Introduction

The environment of Sgr A*, the 4 million Solar mass black hole at the center of the Galaxy, provides a window to close-up study of the center of galaxies and is a potential Rosetta Stone for understanding star formation under extreme physical conditions. A critical question is whether tidal shear in the vicinity of supermassive black holes (SMBHs) is able to completely suppress or induce star formation.

Star formation is modified by tidal effects due to Sgr A* as well as high external pressure due to variety of processes close to Sgr A*. There is empirical evidence for three different regimes of star formation in this region of the Galaxy, as described below.

First, close to Sgr A*, there is a concentration of over 100 OB stars within 0.04 and 0.4 pc of Sgr A* providing strong evidence for formation of stars within 0.5 pc of Sgr A* in the last few million years (Paumard *et al.* 2006; Lu *et al.* 2009). To explain star formation in a disk, a plausible suggestion is that gravitational collapse took place in a gaseous disk captured by the black hole. Star formation in a captured disk around Sgr A* explains the compactness of the observed stellar disk. This implies that the raw material for the disk was captured from a cloud as it temporarily engulfed Sgr A* several million years ago while passing through the central parsec of the Galaxy. Due to the cancellation of angular momentum of the captured cloud material that passes on opposite sides of the black hole, it naturally produces a compact, gravitationally unstable disk (Wardle and Yusef-Zadeh 2008, 2014). One of the key feature of this mechanism is that clouds of gas can be brought in very close to supermassive black holes. This is mainly because of the loss of angular momentum of the initial cloud, thus the captured cloud generates not only a high rate of star formation but also provides the fuel necessary to feed SMBHs.

Second, massive stars such as IRS 8 and about ten bow-shock early type stars are distributed within the ionized streamers 0.5-2 pc from Sgr A* (Geballe *et al.* 2006; Sanchez-

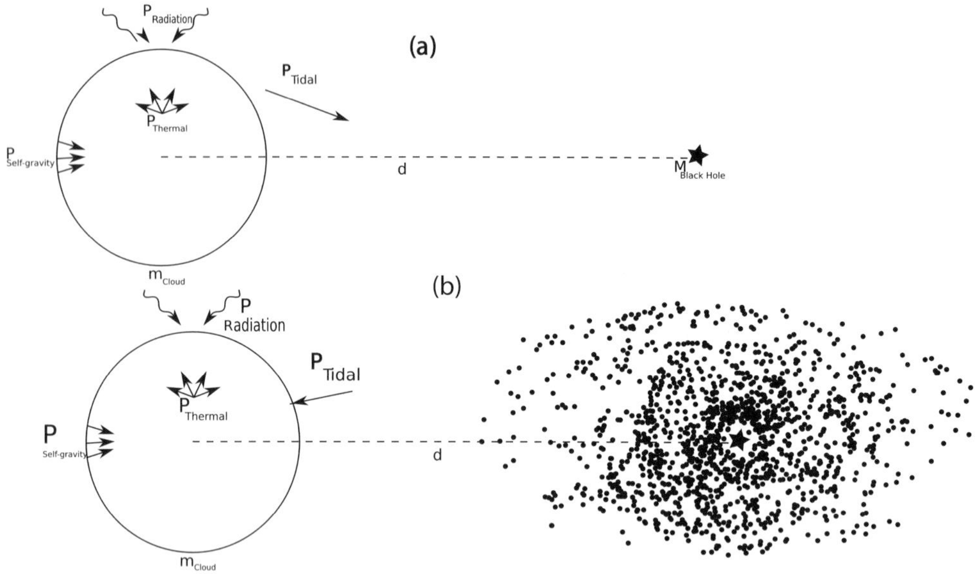

Figure 1. *(a) Top* A Schematic diagram showing the four forces that a cloud experiences as it approaches Sgr A* with its point mass gravitational potential. *(b) Bottom* Same as (a) except that an extended concentration of mass (the nuclear cluster) serves as the gravitational potential and the tidal force on the cloud becomes compressive.

Bermudez *et al.* 2014). On a scale of tens of parsecs, there are also clusters of massive young stars (e.g., the Arches and Quintuplet star clusters) with similar ages to those of the central cluster near Sgr A* (Stolte *et al.* 2015). The question is how star formation proceeds in these three different regions.

In addition to the population of young massive stars near Sgr A*, several lines of evidence point to on-going star formation within two parsecs of Sgr A*. First, the discovery of water masers with multiple and single velocity components (Yusef-Zadeh *et al.* 2015a). Second, SED modeling of 64 infrared excess sources in the inner pc of Sgr A* are consistent with their being YSOs (Yusef-Zadeh *et al.* 2015a). Third, 44 partially resolved compact sources with size scales ranging between 400 and 1600 AUs have been identified (Yusef-Zadeh *et al.* 2015b). The bow-shock appearance of these sources facing the direction of Sgr A* suggests a population of photoevaporative protoplanetary disks (proplyds) associated with newly formed low mass stars. The disks are externally illuminated by strong Lyman continuum radiation from the 100 OB and WR massive stars distributed within 0.5 pc of Sgr A*. We have recently detected millimeter emission from five proplyd candidates supporting the idea that these sources have cool disks (Yusef-Zadeh *et al.* 2016; submitted). Fourth, several SiO (5-4) sources within a pc of Sgr A* are suggestive of YSOs with protostellar outflows. Their SiO (5-4) luminosities and their velocity widths are similar to protostellar outflows in star forming clouds in the disk of the Galaxy (Yusef-Zadeh *et al.* 2013). Finally, a recent ALMA study of this region (Tsuboi *et al.* 2016) suggests molecular clumps of 10-100 Solar mass, distributed in the streamers. Altogether the evidence for on-going star formation implies that stellar disk of OB stars is formed close to Sgr A* rather than being migrated from large distances. Here, we focus on the mode of star formation beyond the inner 0.5 pc of Sgr A* and summarize the results of a recent work, details of which can be found in Wardle and Yusef-Zadeh (2016, in preparation).

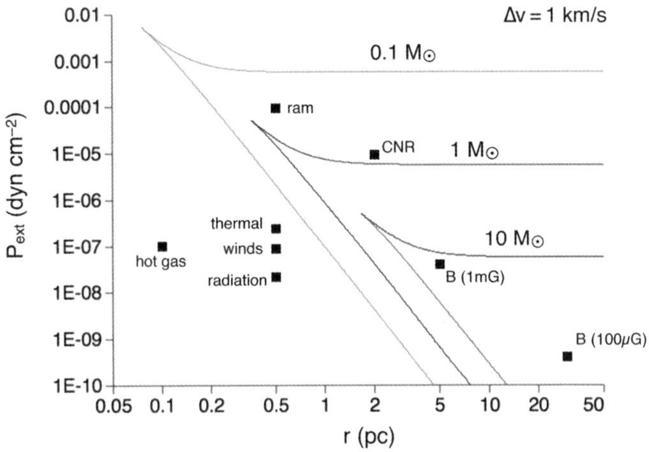

Figure 2. The external pressure experienced by a Galactic center cloud as a function of distance. The plots in the top, middle and bottom correspond to 0.1, 1 and 10 Solar mass cloud masses, respectively.

2. Cloud-based Mode of Star Formation

Although cloud capture model can explain the origin of the stellar disk orbiting Sgr A*, a different process is needed to explain the formation of isolated stars or of clusters beyond 0.5 pc of Sgr A*. The gravitational potential within the inner $\sim$5 pc is dominated by the mass of Sgr A*. The Roche density is 10^8 cm^{-3} at 1 pc from Sgr A* and the gas density inferred from a number of measurements is lower by at least 1.5 orders of magnitude. However, the gravitational stability of a cloud also depends on the internal and external pressures (Chen *et al.* 2016; Wardle and Yusef-Zadeh 2016). Figure 1a demonstrates the disruptive forces due to the tidal force from Sgr A* and the expansive internal pressure. These are opposed by confining forces, self-gravity and compression by external pressure such as radiation, cosmic rays and shocks. Thus, the stability and the evolution of a cloud clearly depends on the balance between the internal and external forces.

Figure 2 shows the range of external pressures for which clouds with 1 km s^{-1} line width are stable as a function of distance from Sgr A*. There are different types of pressure forces acting on a cloud at different distance from Sgr A*. The calculated value of pressure on each curve should be compared to the actual pressure at each radius, which is contributed by a variety of different mechanisms such as the ram pressure associated with motion through the external medium or the pressure exerted by a jet or outflow from Sgr A* that impinges upon the cloud. The circumnuclear molecular ring with its high gas density, the hot X-ray gas, stellar winds from massive stars, the UV radiation from massive stars, the ionized thermal gas in the mini-spiral (the streamers) and the magnetic field contribute within a few parsecs of Sgr A*.

By way of illustration consider the fate of a one Solar mass cloud. Figure 2 shows that this cloud could collapse if it is at least 0.5 pc from Sgr A*. If the cloud is closer than this, it will be torn apart by tidal effects. In the region between the diagonal and horizontal boundaries, the cloud remains stable and neither collapses nor is disrupted. One of the important implications is that lower mass clouds can approach Sgr A* more closely than high mass clouds. Given sufficient external pressure, lower mass stars are preferentially formed closer to the black hole. This implies that a population of proplyds associated with newly formed low mass stars should be in the vicinity of Sgr A* (see Yusef-Zadeh *et al.* 2015b).

3. Star Formation Beyond the Nuclear Cluster

SMBHs in external galaxies with a mass $\sim 100 - 1000$ times Sgr A*, dominate the nuclear stellar potential within 100 pc to 1 kpc. Star formation in the innermost regions of galaxies is generally presumed to be suppressed by strong tidal forces. In the case of Sgr A*, the potential on 10-200 pc scales is dominated by the nuclear cluster. It is generally presumed (e.g. Morris & Serabyn 1996) that tidal forces can only be overcome if the gas density exceeds the Roche density based on the interior mass. However, this condition may never arise since the Roche density drops to zero wherever the gravitational potential is not Keplerian, as described below.

It turns out that tidal forces respond differently when a cloud is experiencing a potential dominated by a point mass, Sgr A*, vs an extended mass concentration (e.g., the nuclear cluster (Jog 2013; Wardle & Yusef-Zadeh 2016). The schematic diagram in Figure 1b presents the distribution of the nuclear cluster centered on Sgr A*. A cloud within a couple of parsecs experiences mainly the Keplerian potential of Sgr A* and at large distances beyond a few parsecs, the potential of the nuclear cluster becomes significant. The rotation curve is Keplerian in the inner parsec, where the enclosed mass is dominated by Sgr A*, flattens between 1 and 10 pc as the stellar cluster starts to dominate the enclosed mass, and slowly rises beyond that. The consequence of such a change in the mass distribution profile is that clouds latteral compression by the tidal force of the nuclear cluster dominates radial stretching. Another consequence of this external force on a cloud beyond the nuclear cluster is that the Roche density approaches zero. The reason for such a drop in Roche density is that the differential force across a cloud is not significant in the radial direction as long as the rotation curve is flat or rising. Thus, the acceleration across the radial direction becomes insignificant. This situation is different for a cloud embedded in a potential of a point mass where the rotation curve drops rapidly as a function of distance from Sgr A*. This result is different than what has generally been assumed that the gas density must be greater than the Roche density. It is conceivable that prominent young star clusters such as the Arches and Quituplet clusters benefited from the contribution of the nuclear cluster in compressing the gas cloud, thus the cloud becomes gravitationally unstable. If the potential is dominated by the Dark matter in the early universe, then star formation may be induced in a cloud as long as clouds of gas do not follow a point mass potential.

References

Chen, X., Amaro-Seoane, P., & Cuadra, J. 2016, *ApJ*, 819, 138
Geballe, T. R., Najarro, F., Rigaut, F., & Roy, J.-R. 2006, *ApJ*, 652, 370
Jog, C. J. 2013, *MNRAS*, 434, L56
Lu, J. R., Ghez, A. M., Hornstein, S. D., *et al.* 2009, *ApJ*, 690, 1463
Paumard, T., Genzel, R., Martins, F., *et al.* 2006, *ApJ*, 643, 1011
Sanchez-Bermudez, J., Alberdi, A., Schödel, R., *et al.* 2014, *A&AP*, 572, L1
Stolte, A., Brandner, W., Grebel, E. K., Lenzen, R., & Lagrange, A.-M. 2005, *ApJ*, 628, L113
Tsuboi, M., Kitamura, Y., Miyoshi, M., *et al.* 2016, *PASJ*, 68, L7
Wardle, M. & Yusef-Zadeh, F. 2008, *ApJ*, 683, L37
Wardle, M. & Yusef-Zadeh, F. 2014, *ApJ*, 787, L14
Yusef-Zadeh, F., Royster, M., Wardle, M., *et al.* 2013, *ApJ*, 767, L32
Yusef-Zadeh, F., Roberts, D. A., & Wardle, M., *et al.* 2015b, *ApJ*, 801, L26
Yusef-Zadeh, F., Wardle, M., Sewilo, M., *et al.* 2015a, *ApJ*, 808, 97

New Frontiers in Black Hole Astrophysics
Proceedings IAU Symposium No. 324, 2016
Andreja Gomboc, ed.

doi:10.1017/S1743921317001089

What can Fermi LAT observation of the Galactic Centre tell us about its active past?

Gabrijela Zaharijas[1,*], Jovana Petrović[2] and Pasquale Serpico[3]

[1]Laboratory for Astroparticle Physics, University of Nova Gorica, Vipavska 13, SL-5000, Slovenia, email: gabrijela.zaharijas@ung.si (* Speaker)

[2]University of Novi Sad, Department of Physics, Novi Sad, Serbia *and* University of Belgrade, Department of astronomy and astrophysics, Belgrade, Serbia

[3]LAPTh, Univ. de Savoie, CNRS, B.P.110, Annecy-le-Vieux F-74941, France

Abstract. The Fermi-LAT gamma-ray data in the inner Galaxy region show several prominent features possibly related to the past activity of the Milky Way's super massive black hole. At a large, 50 deg scale, the Fermi LAT revealed symmetric hour glass structures with hard energy spectra extending up to 100 GeV (and dubbed 'the Fermi bubbles'). More recently and closer to the Galactic centre, at the 10 deg scale, several groups have claimed evidence for excess gamma-ray emission that appears symmetric around the Galactic center and has an energy spectrum peaking at few GeVs. We explore here the possibility that this emission originates in inverse Compton emission from high-energy electrons produced in a short duration, burst-like event injecting $10^{52} - 10^{53}$ erg, roughly 10^6 yrs ago. Several lines of evidence suggest that a series of 'burst like' events happened in the vicinity of our black hole in the past and gamma-ray observations may offer a new view of that scenario.

Keywords. Gamma rays, Galactic centre

1. Introduction

The center of our Galaxy represents one of the most interesting targets for astroparticle physics, as the observations of the non-thermal processes in the region offer the 'front seat view' on dense environments surrounding supermassive black holes and active star forming regions. In addition, the Galactic Centre (GC) is expected to be the brightest spot in terms of dark matter (DM) annihilation emission and is therefore an attractive target which could help resolve the long standing mystery of the nature of these particles which make up 85% of the matter density in the Universe (Ade *et al.* (2015)).

In this work we focus on gamma rays, which are copiously produced in interactions of cosmic rays with the interstellar medium and fields. They are also expected to be emitted in the self-annihilation or decay of dark matter particles in one of the most popular Weakly Interacting Massive Particle (WIMP) DM models.

The Fermi LAT satellite was launched in June 2008. It detects gamma rays in energy range 30 MeV to $\gtrsim$300 GeV†. Its data are public and actively used by the community‡. The diffuse gamma-ray emission from the cosmic ray population of the Milky Way constitutes majority of photons received by the LAT. In the crowded region of the Galactic center this emission is specially hard to model and to distinguish from the numerous non-thermal astrophysical sources, in part due to the limited angular resolution of the LAT. This challenges make this one of the hardest-to-model regions.

† $https://www.slac.stanford.edu/exp/glast/groups/canda/lat_Performance.htm$
‡ $http://fermi.gsfc.nasa.gov/ssc/data/access/lat/$

Over the course of Fermi's eight year mission many groups have performed an analysis of the GC region, using the public Fermi LAT data (see e.g. Hooper *et al.* (2010), Calore *et al.* (2014), Ajello *et al.* (2015)). Most of the studies reached similar conclusion, that the commonly used ('pre Fermi') models of diffuse emission are not sufficient to explain the observation and that residual emission, dubbed the *'Galactic Centre Excess' (GCE)*, is present in this region at GeV energies.

The most important properties of the claimed residuals are i) the emission is extended, reaching up to few kilo-parsecs away from the GC with a $\sim r^{-2.4}$ profile (where r s the distance from the GC), ii) spectrum of the GCE can be modelled by a power law with an exponential cut-off (PLexp), of the type $E^{-\Gamma}\ \exp\left[-E/E_{\rm cut}\right]$ with parameters in the range $\Gamma = 0.5 \div 1$, $E_{\rm cut} \sim 2 \div 3$ GeV† and iii) Their total flux at 1-3 GeV, integrated within 1° of Galactic Center, is $\sim 10^{-10}$ erg cm^{-2} s^{-1}.

The claim that the GCE properties are exactly those expected from the annihilation of WIMP DM sparked significant attention. Among conventional astrophysical sources, cumulative emission from a population of unresolved milli-second pulsars (MSP) represent one of the most concrete possibilities (see e.g. Abazajian *et al.* (2012)). During the Fermi-LAT mission, numbers of detected gamma ray MSP discovered sky rocketed. Their spectra is intriguingly similar to the one inferred for the residuals, being a PLexp with $\Gamma \sim 1.5$, $E_{\rm cut} \sim 3.3$ GeV and they could naturally be produced during one of the past star burst periods in the region or possibly deposited there by the in-falling globular clusters in the past (Brandt *et al.* (2015)).

All the above mentioned contributions are by hypothesis *steady state*. We relax that assumption in Petrović *et al.* (2014) notting that the high energy sky is however significantly time-dependent, and some nuclear regions in external galaxies do show major signs of activity. We explore the possibility that the GCE might be an 'echo' of a past transient event at the GC, which injected high energy electrons in the medium. The GCE emission we observe today could then be due to the inverse Compton emission produced in the propagation of those electrons emitted in the past. Properties of the excess, in particular its i) energy cut-off, ii) spatial extension and iii) overall flux would then determine the age and total energy injected by the transient event. Similar idea was studied in Carlson *et al.* (2014), but in that case interactions of cosmic ray protons with the gas in the disk were considered as the source of the observed gamma rays.

2. Transient event in the Galactic Centre past?

The time-dependent spectrum $Q(E)$ of cosmic ray electrons injected in a bursting episode (a delta function in time and position), propagating (via diffusion and energy-losses) in a homogenous medium is well-known from classical literature on cosmic ray astrophysics. Here we closely follow Atoian *et al.* (1995), which provides a transparent and general analytical solution, allowing us to illustrate the main physical effects.

The energy distribution function of particles at time t post-burst and distance r from the source is given by

$$\frac{dn}{dE_e}(r,t,\gamma) = \frac{N_0\ \gamma^{-\alpha}}{\pi^{3/2}\ r_{\rm diff}^3}\ (1 - b\,t\,\gamma)^{\alpha-2}\ \ e^{-(r/r_{\rm diff})^2} \tag{2.1}$$

where $\gamma = E_e/m_e c^2$ parametrizes the energy E of electrons, α is the index of the electron injection spectrum, b accounts for inverse Compton and synchrotron energy losses as $d\gamma/dt = b\gamma^2$. As we are interested in a GC region extending up to 10 degrees we choose

† More recent data analyses which used larger data sets also detect emission above 10 GeV.

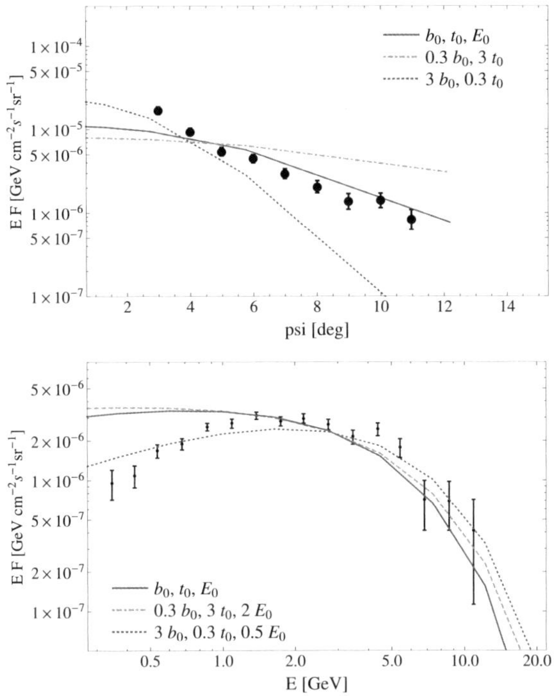

Figure 1. *Top Panel:* Latitude profile of the IC emission from an electron population injected t_0 (Red, Solid), 0.3 t_0 (Orange, Dashed) and 3 t_0 (Blue, Dotted) years ago (where $t_0 = 1$ Myr). *Bottom Panel:* The spectra of the IC emission (the same color scheme) at 5° away from the Galactic plane. The overall energetics is given in units of $E_0 = 4 \times 10^{52}$ erg.

values for the energy loss parameter which are higher than the locally measured ones, but appropriate for the Inner Galaxy.

Energy losses determine the *energy cut-off* in the electron spectrum set by the cooling in time t of electrons with formally infinite injection energy as $\gamma_{\rm cut} = (b\ t)^{-1}$.

The *spatial extension* of the electron flux is determined by the diffusion length $r_{\rm diff}$,

$$r_{\rm diff} = 2\left(D(\gamma)\ t\ \frac{1-(1-\gamma/\gamma_{\rm cut})^{1-\delta}}{(1-\delta)\gamma/\gamma_{\rm cut}}\right)^{1/2} \tag{2.2}$$

where the diffusion coefficient D is taken to be $D(\gamma) \simeq D_0\ (\gamma/\gamma_*)^{\delta}$, with $E_* = 3$ GeV, $\delta = 0.6$ and $D(10\text{ GeV}) = 6\ 10^{28}\ \text{cm}^2/\text{s}$. As $r_{\rm diff}$ explicitly depends on the age of the source t, it breaks a degeneracy between the energy loss parameter b and age, which determine the spectral cut-off. The function $r_{\rm diff}$ also changes the spectrum in the sense of depleting the low-energy part of dn/dE_e the farther one is from the origin, at a given time, since less energetic electrons diffuse more slowly.

In order to calculate the Inverse Compton (IC) gamma-ray fluxes from this electron population we follow Colafrancesco *et al.* (2005). The IC emissivity can be written as

$$J_{IC}(E_\gamma) = \int dE_e\ \frac{dn_e}{dE_e}(E_e)\ P_{IC}(E_\gamma, E_e) \tag{2.3}$$

where dn_e/dE_e is given by Eq. (2.1) and P_{IC} is the inverse Compton power which depends on the energy and density of the Inter Stellar Radiation Field (ISRF) and the differential Klein-Nishina cross section.

If we keep the spectral injection and diffusion indices fixed to their fiducial values, in principle one has **three observables:** spectral shape, angular shape and normalization with **four major parameters:** E_{TOT} (the total energy output of the source), D_0, b_0, t_0. The *angular shape* is controlled by $r_{\rm diff}$, which in turn can be altered via $(D_0\ t_0)$, see Eq.(2.2). This singles out a *Myr timescale* for t_0, for fiducial value of D_0. Once this parameter is fixed, only minor spectral slope adjustments are possible by varying $\alpha \in 2.1 - 2.4$, and $D_0(4\ {\rm GV}) = 2\ 10^{28} - 10^{29}{\rm cm}^2{\rm s}^{-1}$, but the key spectral parameter, the cutoff energy $E_{\rm cut} = m_e(b\,t_0)^{-1}$ is determined by the same parameter, t_0. It is remarkable that the observed cutoff in the spectrum is fully consistent with this estimate, as shown in Fig. 1. These analytic results have later been confirmed and studied in detail by a dedicated numerical simulation (Cholis *et al.* (2015)), which found that data are best described by a series of (two to three) leptonic bursts.

3. Summary

We show that a bursting event, injecting $\sim 10^{52}$ ergs of energy in a standard power-law cosmic ray electron spectrum some Myrs ago can reproduce naturally the spectra and angular features of the claimed GeV excess in the inner Galaxy. Intriguingly, there are many hints that the GC may have experienced an active past, with the most spectacular manifestation provided by the "Fermi-LAT bubbles" (Su *et al.* (2010), Ackermann *et al.* (2014)). The main goal of our calculations was to raise awareness on the importance of accounting for transient events when dealing with extended excesses at the GC.

References

Ade, P. A. R., *et al.* [Planck Collaboration], *Astron. Astrophys.* 594, A13 (2016) doi:10.1051/0004-6361/201525830 [arXiv:1502.01589 [astro-ph.CO]].

Hooper, D., & L. Goodenough, *Phys. Lett. B* 697, 412 (2011) doi:10.1016/j.physletb.2011.02.029 [arXiv:1010.2752 [hep-ph]].

Calore, F., Cholis, I., & Weniger, C., *JCAP* 1503 (2015) 038 doi:10.1088/1475-7516/2015/03/038 [arXiv:1409.0042 [astro-ph.CO]].

Ajello, M., *et al.* [Fermi-LAT Collaboration], *Astrophys. J.* 819 (2016) no.1, 44 doi:10.3847/0004-637X/819/1/44 [arXiv:1511.02938 [astro-ph.HE]].

Abazajian, K. N. & Kaplinghat, M., *Phys. Rev. D* 86, 083511 (2012) Erratum: [Phys. Rev. D 87, 129902 (2013)] doi:10.1103/PhysRevD.86.083511, 10.1103/PhysRevD.87.129902 [arXiv:1207.6047 [astro-ph.HE]].

Brandt, T. D. & Kocsis, B., *Astrophys. J.* 812, no. 1, 15 (2015) doi:10.1088/0004-637X/812/1/15 [arXiv:1507.05616 [astro-ph.HE]].

Petrović, J., Serpico, P. D., & Zaharija, G., *JCAP* 1410, no. 10, 052 (2014) doi:10.1088/1475-7516/2014/10/052 [arXiv:1405.7928 [astro-ph.HE]].

Atoian, A. M., Aharonian, F. A., & Volk, H. J., *Phys. Rev. D* 52, 3265 (1995). doi:10.1103/PhysRevD.52.3265

Colafrancesco, S., Profumo, S., & Ullio, P., *Astron. Astrophys.* 455, 21 (2006) doi:10.1051/0004-6361:20053887 [astro-ph/0507575].

Daylan, T., Finkbeiner, D. P., Hooper, D., Linden, T., Portillo, S. K. N., Rodd, N. L., & Slatyer, T. R., *Phys. Dark Univ.* 12, 1 (2016) doi:10.1016/j.dark.2015.12.005 [arXiv:1402.6703 [astro-ph.HE]].

Cholis, I., Evoli, C., Calore, F., Linden, T., Weniger, C., & Hooper, D., *JCAP* 1512 (2015) no.12, 005 doi:10.1088/1475-7516/2015/12/005 [arXiv:1506.05119 [astro-ph.HE]].

Su, M., Slatyer, T. R., & Finkbeiner, D. P., *Astrophys. J.* 724, 1044 (2010) doi:10.1088/0004-637X/724/2/1044 [arXiv:1005.5480 [astro-ph.HE]].

Ackermann, M., *et al.* [Fermi-LAT Collaboration], *Astrophys. J.* 793, no. 1, 64 (2014) doi:10.1088/0004-637X/793/1/64 [arXiv:1407.7905 [astro-ph.HE]].

New Frontiers in Black Hole Astrophysics
Proceedings IAU Symposium No. 324, 2016
Andreja Gomboc, ed.

doi:10.1017/S1743921317001107

Swift J1644+5734: the EVN view

Z. Paragi[1], J. Yang[1,2,3], S. Komossa[4], A. van der Horst[5], L. I. Gurvits[1,6], R. M. Campbell[1], D. Giannios[7] and T. An[3,8]

[1]Joint Institute for VLBI ERIC (JIVE), Postbus 2, NL-7990 AA Dwingeloo, the Netherlands
email: zparagi@jive.eu

[2]Dept. of Earth and Space Sciences, Chalmers Univ. of Technology, Onsala Space Observatory SE-439 92 Onsala, Sweden

[3]Shanghai Astronomical Observatory, Chinese Acad. of Sciences, 200030 Shanghai, P.R. China

[4]QianNan Normal University for Nationalities, Longshan Street, Duyun City of Guizhou Province, China

[5]Dept. of Physics, The George Washington University, 725 21st Street NW, Washington, DC 20052, USA

[6]Dept. of Astrodynamics and Space Missions, Delft University of Technology, NL-2629 HS Delft, the Netherlands

[7]Dept. of Physics and Astronomy, Purdue University, 525 Northwestern Avenue, West Lafayette, IN 47907, USA

[8]Key Lab. of Radio Astronomy, Chinese Academy of Sciences, 210008 Nanjing, P.R. China

Abstract. A small fraction of Tidal Disruption Events (TDE) produce relativistic jets, evidenced by their non-thermal X-ray spectra and transient radio emission. Here we present milliarcsecond-resolution imaging results on TDE J1644+5734 with the European VLBI Network (EVN). These provide a strong astrometric constraint on the average apparent jet velocity $\beta_{\rm app} < 0.27$, that constrains the intrinsic jet velocity for a given viewing angle.

Keywords. techniques: interferometric, astrometry, radio continuum: galaxies, galaxies: active, X-rays: bursts, black hole physics

---

1. Introduction

Tidal Disruption Events (TDE) may help us understand two outstanding questions of black hole astrophysics: how black holes form and grow, and what impact they have on the evolution of their host galaxies (see a recent review by Komossa 2015). The first question can be addressed through constraining the low-mass end of the supermassive black hole mass-function (Stone & Metzger 2016). The idea here is that TDEs "light-up" dormant black holes (Giannios & Metzger 2011) that have not shown an active galactic nucleus (AGN) signature before. As the size of both the black hole and the disrupted star can be estimated well (e.g. Gezari *et al.* 2012, Guillochon *et al.* 2014), TDEs provide means to weigh black holes down to $10^4 - 10^5 M_{\odot}$, the majority of which has remained undetected to date. The second question can be addressed through observations of newly formed radio jets (triggered by a TDE) on milliarcsecond scales. Very long baseline interferometry (VLBI) is a powerful tool to localise the origin of H I outflows and to reveal the dynamics of atomic gas swept-up by the radio jet, catching AGN jet feedback in action (Morganti *et al.* 2013); while VLBI of jetted TDEs provide us with a unique new chance of observing and understanding the very early phases of jet formation and evolution in an otherwise pristine environment, without any past AGN activity.

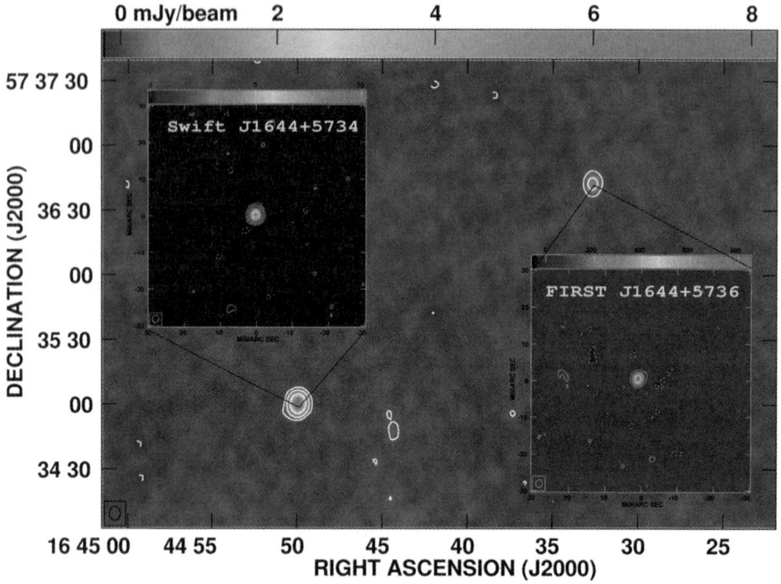

Figure 1. EVN images of the radio counterpart to Swift J1644+5734 and the in-beam reference source (insets), with a WSRT image as background. From Yang *et al.* (2016), Fig. 1.

Future blind radio surveys will potentially detect a large number of TDEs, but it has been recognised that a multi-band approach is essential for their identification. The role of VLBI may be two-fold: very high resolution observations of strong candidates may reveal their nuclear origin, supporting the TDE identification at early times. On the other hand, securely identified TDEs (from X-ray properties and/or optical spectra) that occur off-nucleus may point to peculiar systems such as a white dwarf disrupted by an intermediate-mass black hole (Shcherbakov *et al.* 2013), a TDE from a recoiling black hole (Komossa & Merritt 2008), or it may reveal a secondary galactic nucleus in a merger (Liu & Chen 2013). The latter is potentially interesting since the TDE rate is apparently higher in a certain type of post-starburst galaxies that have recently undergone a major merger event (French, Arcavi & Zabludoff 2016). It has been shown that jetted TDEs will be more efficiently detected in large-scale radio surveys than with current or near-future X-ray missions, therefore TDEs will be primary targets of synchrotron transients with the Square Kilometre Array (SKA); the expected numbers however depend strongly on the fraction of jetted events, and their relativistic beaming factors (Donnarumma & Rossi 2015). For the first time, we obtained a strong constraint on the average apparent jet speed for J1644+5734, briefly described below. More details of this result can be found in Yang *et al.* (2016).

2. The first jetted TDE

The transient Swift J1644+5734 was extremely luminous from radio to the γ rays, that has been widely interpreted as a non-thermal TDE with a relativistic jet (Levan *et al.* 2011; Bloom *et al.* 2011; Burrows *et al.* 2011; Zauderer *et al.* 2011). The variability properties in the different bands indicated that the X-ray and the radio emission were most likely produced in different regions, the latter related to a decelerating external shock in an initially ultrarelativistic outflow (Giannios & Metzger 2011). Similar models have been invoked to explain multi-band properties of γ-ray afterglows (e.g. Zauderer *et al.* 2013). The VLBI technique has the power to directly reveal relativistic ejecta by for example resolving them. In case of J1644+5734 Berger *et al.* (2012) predicted that marginally resolving the source may be possible at high frequencies with current VLBI instrumentation; at 22 GHz and above however the emission is expected to fade quickly. Another way, often successfully applied in quasars up to even very high redshifts (see Frey *et al.* 2015), is to detect the proper motion of relativistic ejecta. We followed this alternative route.

3. European VLBI Network (EVN) observations

We observed Swift J1644+5734 at six epochs between 12 April 2011 and 9 March 2015 with the EVN at 5 GHz. Detection of targets at the few-mJy level requires frequent visits to a bright, compact calibrator, a technique called phase-referencing (see Reid & Honma 2014). This technique also provides the relative position between the target and the calibrator. For this purpose we used one of the International Celestial Reference Frame (ICRF) defining sources, J1638+5720 as primary calibrator. The precision of the derived relative position however depends strongly on the separation between these sources (in our case 55 arcminutes). The larger the angular distance, the larger the systematic errors are. Therefore our strategy was to identify a more nearby compact source in our very first observation, that could be used as a reference point in the following epochs. We did find a candidate in the FIRST catalogue that was unresolved on milliarcsecond scales, FIRST J1644+5736, with a separation of 2.9 arcminutes. Swift J1644+5734 and FIRST J1644+5736 were within the primary beam of the smaller telescopes, and only the large dishes had to regularly nod between these sources (see Fig. 1 for the EVN images).

4. Constraints from VLBI

Using the technique described above, we achieved very high precision relative astrometry between our in-beam reference source and the radio "afterglow" of Swift J1644+5734. The standard deviation of our five relative positions spanning three years were 13 μas in Right Ascension and 11 μas in Declination, just a tiny fraction (1/160th) of the synthesized beam. We see no detectable proper motion in Swift J1644+5734 (relative to FIRST J1644+5736). The 3-σ upper limit to the average (over 3 years) apparent jet velocity is $\beta_{\rm app} < 0.27$. The implied limit on the intrinsic jet speed as a function of viewing angle is shown on Fig. 2.

If the radio emission is produced by a forward shock analogous to gamma-ray burst jets with a Lorentz-factor $\Gamma = 2$ (blue line in Fig. 2) or higher (cf. Zauderer *et al.* 2013), the implied viewing angle is less than 3 degrees. A lower average Lorentz factor (e.g. due to a dense interstellar medium) would allow larger viewing angles. In this case, the observations would indicate that significant deceleration took place within about three years after relativistic ejecta were triggered in Swift J1644+5734, and this may be expected in other non-thermal (jet-forming) TDEs as well. Ultra-precise VLBI astrometry commenced at a closer epoch to the trigger event may give stronger constraints for other TDEs in the future, possibly for systems that are less-beamed (i.e. have a larger viewing angle, but not completely de-boosted).

Alternatively, the radio jet in Swift J1644+5734 might have been mildly relativistic ($\beta_{\rm int} < 0.3$) from the beginning, in accordance with what is expected for purely radiatively driven jets (Sadowski & Narayan 2015). While Kara *et al.* (2016) detected relativistic reverberation in the accretion flow of the system, and argued for a mildly relativistic wind

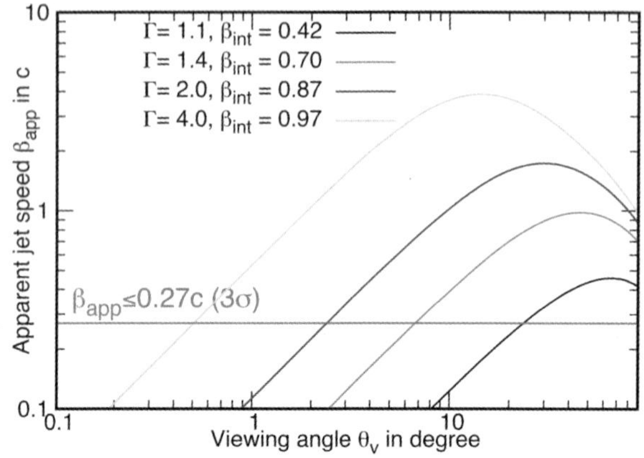

Figure 2. Constraints on Swift J1644+5734 average intrinsic jet velocity $\beta_{\rm int}$ and Lorentz factor for various viewing angles, from the proper motion upper limit (horizontal line). From Yang *et al.* (2016), Fig. 3.

as the source of X-rays. These models however do not naturally explain the broad-band properties of Swift J1644+5734, that imply a jet with an initially high Lorentz factor.

Relativistic-jet TDEs remain in the forefront of transient research with the EVN, and the first result claiming a resolved jet in ASASSN-14li has just been published by Romero-Cañizales *et al.* (2016). In the future, we intend to launch a SKA-VLBI (Paragi *et al.* 2015) follow-up program for the most promising candidates found by the SKA.

References

Komossa, S. 2015, *J. High Energy Astrophys.*, 7, 148
Stone, N. C. & Metzger, D. E. 2016, *Mon. Not. R. Astron. Soc.*, 455, 859
Giannios, D. & Metzger, D. E. 2011, *Mon. Not. R. Astron. Soc.*, 416, 2102
Gezari, S., Chornock, R., Rest, A., *et al.* 2012, *Nature*, 485, 217
Guillochon, J., Manukian, H., & Ramirez-Ruiz, E. 2012, *ApJ*, 783, 23
Morganti, R., Fogasy, J., Paragi, Z., Oosterloo, T., & Orienti, M. 2013, *Science*, 341, 1082
Shcherbakov, R. V., Pe'er, A., Reynolds, C. S., *et al.* 2013, *ApJ*, 769, id. 85.
Komossa, S. & Merritt, D. 2008, *ApJ* (Letters), 683, L21
Liu, F. K. & Chen, X. 2013, *ApJ*, 767, id. 18
French, K. D., Arcavi, I., & Zabludoff, A. 2016, *ApJ* (Letters), 818, L21
Donnarumma, I., & Rossi, E. M. 2015 *ApJ*, 803, 36
Yang, J., Paragi, Z., van der Horst, A. J., *et al.* 2016 *Mon. Not. R. Astron. Soc.*, 462, L66
Levan, A. J., Tanvir, N. R., Cenko, S. B., *et al.* 2011, *Science*, 333, 199
Bloom, J. S., Giannios, D., Metzger, B. D., *et al.* 2011, *Science*, 333, 203
Burrows, D. N., Kennea, J. A., Ghisellini, G., *et al.* 2011, *Nature*, 476, 421
Zauderer, B. A., Berger, E., Soderberg, A. M., *et al.* 2011, *Nature*, 476, 425
Zauderer, B. A., Berger, E., Margutti, R., *et al.* 2013, *ApJ*, 767, 152
Berger, E., Zauderer, A., Pooley, G. G., *et al.* 2012 *ApJ*, 748, 36
Frey, S., Paragi, Z., Fogasy, J. O., & Gurvits, L. I. 2015, *Mon. Not. R. Astron. Soc.*, 466, 2921
Reid, M. J. & Honma, M. 2014, *ARA&A*, 52, 339
Sadowski, A. & Narayan, R. 2015, *Mon. Not. R. Astron. Soc.*, 453, 3213
Kara, E., Miller, J. M., Reynolds, C., & Dai, L. 2016 *Nature*, 535, 388
Romero-Cañizales, C., Prieto, José, L., Chen, X., *et al.* 2016, *ApJ* (Letters), 832, L10
Paragi, *et al.* 2015, PoS(AASKA14), id. 143

New Frontiers in Black Hole Astrophysics
Proceedings IAU Symposium No. 324, 2016
Andreja Gomboc, ed.

doi:10.1017/S1743921317001776

Tidal disruption events seen in the XMM-Newton slew survey

Richard Saxton[1], S. Komossa[2], Andrew Read[3], Paulina Lira[4], Kate D. Alexander[5] and Iain Steele[6]

[1]XMM SOC, ESAC, Apartado 78, 28691 Villanueva de la Cañada, Madrid, Spain
email: rsaxton@sciops.esa.int

[2]QianNan Normal University for Nationalities, Longshan Street, Duyun City of Guizhou Province, China

[3]Dept. of Physics and Astronomy, University of Leicester, Leicester LE1 7RH, U.K.

[4]Universidad de Chile, Observatorio Astronomico Nacional Cerro Calan, Santiago, Chile

[5]Harvard-Smithsonian Center for Astrophysics, 60 Garden St., Cambridge, MA 02138, USA

[6]Astrophysics Research Institute, Liverpool John Moores University, Liverpool CH41 1LD, UK

Abstract. XMM-*Newton* performs a survey of the sky in the 0.2-12 keV X-ray band while slewing between observation targets. The sensitivity in the soft X-ray band is comparable with that of the ROSAT all-sky survey, allowing bright transients to be identified in near real-time by a comparison of the flux in both surveys. Several of the soft X-ray flares are coincident with galaxy nuclei and five of these have been interpreted as candidate tidal disruption events (TDE). The first three discovered had a soft X-ray spectrum, consistent with the classical model of TDE, where radiation is released during the accretion phase by thermal processes. The remaining two have an additional hard, power-law component, which in only one case was accompanied by radio emission. Overall the flares decay with the classical index of $t^{-5/3}$ but vary greatly in the early phase.

Keywords. X-rays: galaxies, galaxies: active

1. Introduction

A tidal disruption event (TDE) occurs when a stellar object is gravitationally destroyed and subsequently accreted by a super-massive black hole (SMBH). It represents the onset of a brief phase of accretion activity in an otherwise dormant galactic nucleus and is characterised by a sudden dramatic rise in luminosity, and a decline to quiescence on a timescale of months to years.

Early work predicted that stellar debris would return into a tightly bound orbit about the SMBH and form an optically thick accretion structure close to the SMBH (e.g. Rees 1988). This would emit thermal radiation of temperature $kT_{eff} \approx 40 M_6^{-1/4}$ eV (Ulmer 1999) whose high energy tail would be visible in the soft X-ray band. Pioneering observations by ROSAT identified several X-ray flares in non-active galaxies (Komossa & Bade 1999; see Komossa 2015 for a recent review) characterised by a very soft X-ray spectrum. Explanations such as variable AGN, blazars, ULX, supernova or supernova remnants were excluded by the amplitude of the variability (factor $>$ 1000 for NGC 5905 and RXJ 1242-1119; Komossa *et al.* 2004) and high luminosities of up to several $\times 10^{44}$ ergs/s. As their light curves have fallen consistently with the predicted rate, $t^{-5/3}$ (Rees 1988, Guillochon & Ramirez-Ruiz 2013), TDEs are the most likely explanation for these events.

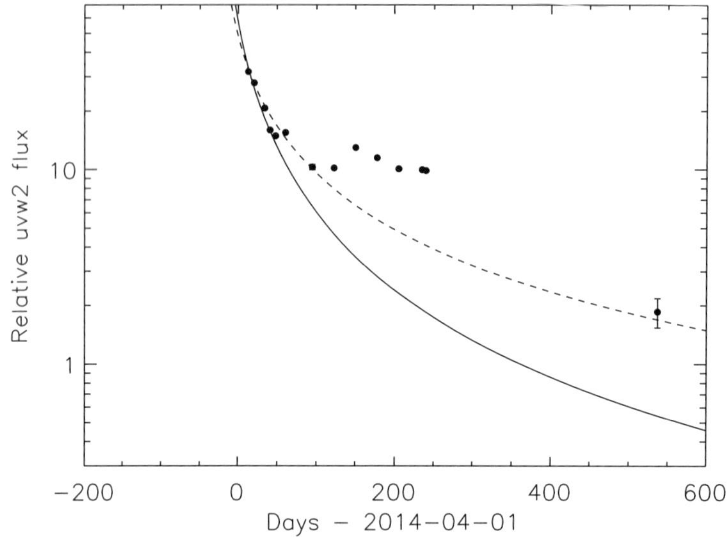

Figure 1. The Swift-uvw2 filter (1928Å) light curve of XMMSL1 J0740-85 fit with indices of -5/3 (solid line) and -1.2 (dashed).

2. XMM-*Newton* slew survey

The XMM-Newton slew survey collates the data taken by the EPIC-pn camera while the satellite moves between targets. It has a sensitivity in the 0.2–2 keV energy band which is well matched with the ROSAT all-sky survey. Comparisons between these two surveys have led to the detection of several transient sources, including five candidate TDEs (Tab. 1). NGC 3599 and SDSS J1323+48 were found in archival data (Esquej *et al.* 2008) while SDSS J1201+30 (Saxton *et al.* 2012), 2MASX J0619-65 (Saxton *et al.* 2014) and XMMSL1 J0740-85 (Saxton *et al.* 2017) were found at the peak of their emission and had their decays monitored with Swift and XMM-*Newton*. XMMSL1 J063045.9-603110, a bright slew transient from 2011, has been proposed as a TDE in a dwarf galaxy (Mainetti *et al.* 2016). However, a contemporary optical spectrum localised the flare to the local group (B. Cenko, p.comm.) and so this remains unconfirmed.

The initial estimate of the tidal disruption rate from the XMM-*Newton* slew data was given as 2.3×10^{-4} gal^{-1} yr^{-1} (Esquej *et al.* 2008) based on two TDE. Scaling this calculation for the five which have now been found in data taken up until the end of 2014, gives a somewhat lower value of 1.4×10^{-4} gal^{-1} yr^{-1}.

2.1. *Temporal characteristics*

The XMM-*Newton* slew survey TDEs have faded in X-rays over periods of 6 months and more with an index compatible with $t^{-5/3}$ in line with the ROSAT TDE and the original expectations. In XMMSL1 J0740-85 the UV flux fell with a flatter index of ~ 1.2 (Fig. 1). This could be due to more flux entering the UV band as the flare cools. In detail though, the UV curve is seen to recover after about 100 days; a behaviour seen in other TDE (e.g. LeLoudas *et al.* 2016.). In analogy with the "supernova hump" in GRB this may imply a secondary emission mechanism; possibly circularisation and infall (LeLoudas *et al.* 2016; Krolik *et al.* 2016) or expelled unbound debris shocking against the ambient medium (Khokhlov & Melia 1996).

There is a lot of variability in the X-ray light curves. About 20 days after discovery, SDSS J1201+30 dropped in X-ray flux by a factor 50 within 1 week (Fig. 2). A similar

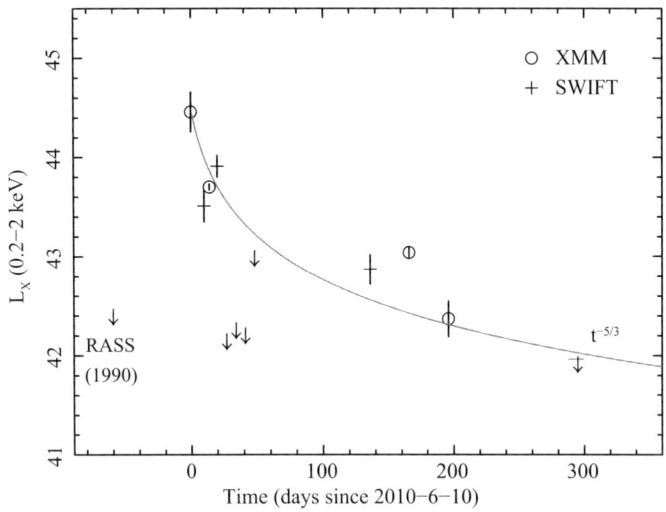

Figure 2. The 0.2–10 keV light curve of SDSS J1201+30 fit.

drop after ~ 30 days was seen in XMMSL1 0740-85. The spectrum actually softens during the drop in 0740-85 excluding variations in neutral absorption and leaving ionised absorption or a change in the intrinsic emission as possible causes. Liu, Li & Komossa (2014) interpret the dip as evidence for a close binary black hole system, where the fall back of material onto the primary black hole is interrupted by the presence of the secondary.

Simulations have shown that the development of a TDE light curve is dependent on when the streams of tidal debris intersect each other (Hyashaki *et al.* 2013; Guillochon & Ramirez-Ruiz 2015; Shiokawa *et al.* 2015). Early interactions appear to be rare and in the majority of cases circularisation occurs late and at a large distance from the SMBH, leading to a *delayed* TDE with a longer rise to peak and a plateau that lasts months or even years. NGC 3599 with a plateau of $\geqslant$ 18 months may have been a *delayed* TDE (Saxton *et al.* 2015). Optically-selected TDEs have either no X-ray emission or flat X-ray light curves with L_X of a few$\times 10^{41-42}$ ergs/s (e.g. ASSASN 14li, Holohein *et al.* 2016; ASSASN 15lh, Margutti *et al.* 2016). We are investigating a TDE recently found in the slew survey with similar behaviour. It could be that these TDEs are all delayed in some way, however there is a similarity between the luminosity and X-ray light curves of these events and those of extremely luminous young supernova remnants such as SN 2006jd (Stritzinger *et al.* 2012) and SN 2010jl (Chandra *et al.* 2012). The featureless TDE spectra, nevertheless, argue against the X-ray emission being produced by ISM shocks.

2.2. *Spectral characteristics*

Material falling back to the inner orbits around a $10^{6-7} M_\odot$ black hole will heat up and emit radiation with an equivalent temperature of a few tens of eV. While this broadly matches the emission seen in the soft X-ray spectrum in no case have we been able to join the X-ray and UV emission with a single thermal multi-colour disk model.

In two of the slew sources, XMMSL1 J0740-85 and 2MASX J0619-65, a hard component is visible, stretching out to at least 10 keV. A similar component was seen in NGC 4845 (Nikolajuk & Walter 2013) and ASASSN-15oi (Holoien *et al.* 2016b). Weak radio emission was seen in NGC 4845 and XMMSL1 J0740-85 (Alexander *et al.*2016) consistent with an outflow or a weak relativistic jet. This implies that a source of high-energy comptonising

Table 1. Candidate TDE found in the XMM-*Newton* slew survey.

Name	z	Log(L_{bol})	Flare?[1] Soft-X	Hard-X	Radio
NGC 3599	0.002	42	Y	N	?
SDSS 1323+48	0.07	44	Y	N	?
SDSS 1201+30	0.146	45	Y	N	N
2MASX 0619-65	0.07	44	Y	Y	N
XMMSL1 0740-85	0.019	44	Y	N	Y

Notes:
[1] Was there a flare at this wavelength?

electrons, analogous to the comptonising corona of a steady-state AGN, can be produced during disruptions.

2.3. *Host galaxy*

A disproportionate fraction of optically-selected TDE are found in post-starburst galaxies (Arcavi *et al.* 2014), a feature that is shared by XMMSL1 J0740-85. Arcavi *et al.* (2014) suggested that starbursts triggered by galaxy mergers could produce binary black holes that dynamically enhance the TDE rate. Alternatively, starbursts could create central stellar overdensities where enhanced encounter (two body relaxation) rates lead to more TDE flares (Stone & Metzger 2016).

References

Alexander, K. D., *et al.* 2016, arXiv:1610.03861
Chandra, P., *et al.* 2012, *ApJ Letters*,750, L2
Esquej, P., Saxton, R., Komossa, S., Read, A., & Freyberg, M. J., 2008, *A&A*, 489, 543
Guillochon, J. & Ramirez-Ruiz, E. 2013, *ApJ*, 767, 25
Guillochon, J. & Ramirez-Ruiz, E. 2015, *ApJ*, 809, 166
Hayasaki, K., Stone, N., & Loeb, A. 2013, *MNRAS*, 434, 909
Holoien, T., Kochanek, C., Prieto, J., Stanek, K., Dong, S., *et al.* 2016, *MNRAS*, 455, 2918
Holoien, T., Kochanek, C., Prieto, J., Grupe, D., Chen, P., *et al.* 2016, arXiv:1602.01088
Khokhlov, A. & Melia, F. 1996, *ApJ* 457, L61
Komossa, S. & Bade, N. 1999, *A&A*, 343, 775
Komossa, S., Halpern, J., Schartel, N., *et al.* 2004, *ApJ*, 603, L17
Komossa, S. 2015, *JHEAp* 7, 148
Krolik, J., *et al.*, 2016, *ApJ*, 827, 127
Leloudas, G., Fraser, M., Stone, N., van Velzen, S., Jonker, P. G., *et al.* 2016, arXiv:160902927
Liu, F., Li, S., & Komossa, K. 2014, *ApJ*, 786, 103
Mainetti, D, Campana, S. & Colpi, M. 2016, *A&A*, 592, 41
Margutti, R., *et al.* 2016, arXiv:1610.01632
Nikolajuk, M. & Walter, R., 2013, *A&A*, 552, 75
Rees, M. 1988, *Nature*, 333, 523
Saxton, R., Read, A., Esquej, P., *et al.* 2012, *A&A*, 541, 106
Saxton, R., Read, A., Komossa, S., *et al.* 2014, *A&A*, 572, 1
Saxton, R., *et al.* 2015, *MNRAS*, 454, 2798
Saxton, R., Read, A., Komossa, S., *et al.* 2017, *A&A*, 598, A29
Shiokawa, H., Krolik, J., Cheng, R., Piran, T., & Noble, S. 2015, *ApJ*, 804, 85
Stone, N. & Metzger, B. 2016, *MNRAS*, 455, 859
Stritzinger, M, Taddia, F. Fransson, C., *et al.* 2012, *ApJ*, 756, 173
Ulmer, A. 1999, *ApJ*, 514, 180

New Frontiers in Black Hole Astrophysics
Proceedings IAU Symposium No. 324, 2016
Andreja Gomboc, ed.

doi:10.1017/S1743921317001636

Tidal Disruption Events and stellar-mass black holes in OGLE and Gaia surveys

Łukasz Wyrzykowski, A. Hamanowicz and K. A. Rybicki

Warsaw University Astronomical Observatory
Al. Ujazdowskie 4, Warszawa 00-478, Poland
email: `lw@astrouw.edu.pl`

Abstract. Tidal Disruption Events (TDE) allow to probe the super massive black holes (SMBH) in the cores of galaxies and could be a source of black hole mass growth. We present the search for candidates for TDEs conducted within OGLE and Gaia surveys. Our preliminary results indicate that TDEs can occur in cores of galaxies exhibiting different levels of activity, from quiescent, through weak-AGNs to highly active QSOs. We also present how Gaia can help study the mass distribution of Milky Way single black holes via microlensing.

Keywords. super massive black holes, tidal disruption events, OGLE, Gaia

---

Introduction

Tidal Disruption Events (TDE) offer an opportunity to detect and study Super Massive Black Holes (SMBH) at lower-end of their mass distribution. However, the number statistics are still too low what makes it difficult to employ TDEs as a useful tool. Moreover, there is still no consensus about the nature of some nuclear flares being found both in real-time and in archives, e.g., changing-look QSOs/AGNs (Shappee *et al.* 2014, LaMassa *et al.* 2015, MacLeod *et al.* 2016, Ruan *et al.* 2016) or slow-blue hypervariables (Lawrence *et al.* 2016). New, well observed and studied examples are necessary and here we present our search for nuclear flares from OGLE and Gaia surveys. Additionally, we describe how Gaia will help measure the mass distribution of stellar-mass black holes in the Milky Way via gravitational microlensing.

OGLE

OGLE† is a Polish large-scale photometric monitoring survey, operating since 1992, using a dedicated 1.3m telescope in Las Campanas Observatory, Chile. Since 2010 (OGLE-IV) it utilises the 32-chip CCD mosaic camera with field-of-view of 1.4 sq.deg (Udalski *et al.* 2015). OGLE-IV monitors regularly the Galactic Bulge, Southern Galactic Disk and 700 sq.deg. region around the Magellanic Clouds with depth of 22 mag in I-band per single image and $\sim$24 mag on stacks.

Within OGLE we search for TDEs in both real-time data‡ (Wyrzykowski *et al.* 2014) and archival data. State-of-the-art Difference Imaging Analysis and well sampled images of galaxies allow for very precise astrometry of centres of galaxies. We have located about 6 million galaxies and conducted subtraction at their very centres, yielding light curves covering years 2010-2016. We found candidates for AGNs, supernovae and couple of long-term events, likely TDEs. Their detailed analysis is on-going. These data will be used to derive first unbiased TDE-rates. We have also identified some very short-duration nuclear flares, which could be related to low-mass black hole TDEs (Wyrzykowski *et al.* 2014, MacLeod, M. *et al.* 2016).

† http://ogle.astrouw.edu.pl
‡ http://ogle.astrouw.edu.pl/ogle4/transients/

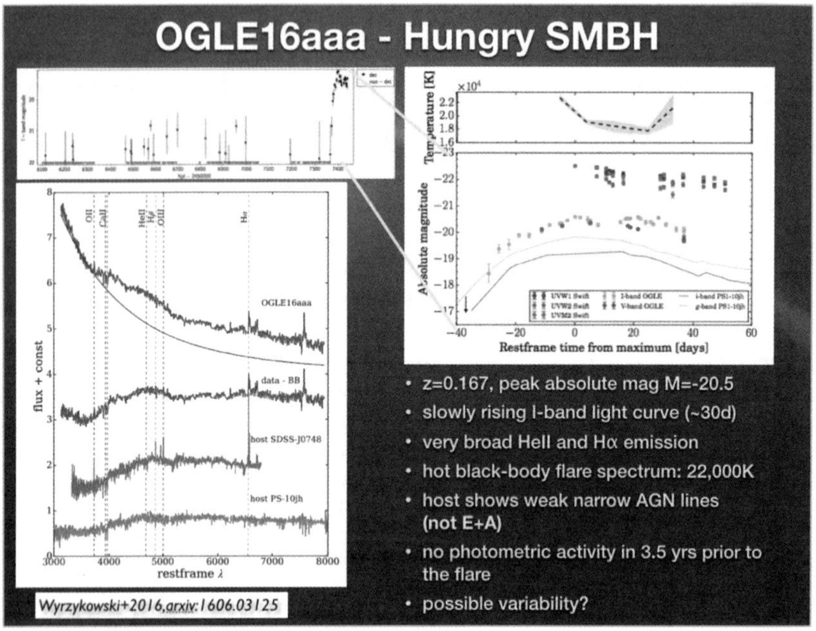

Figure 1. OGLE16aaa: first TDE candidate from OGLE (Wyrzykowski *et al.* 2016b)

OGLE16aaa – Hungry SMBH

OGLE16aaa, found in January 2016, was classified as OGLE's first TDE detected in real-time (Wyrzykowski *et al.* 2016b). Its spectral and multi-colour follow-up revealed high temperatures (~20,000K) remaining at that level over the duration of the event. Detection of very broad HeII and Hα emission features, along with absolute magnitude above -20.5 mag, were consistent with a current picture of TDEs (Arcavi *et al.* 2014).

However, the host galaxy of OGLE16aaa did not belong to a canonical and rare E+A type of galaxies (French *et al.* 2016), but exhibited narrow forbidden emission lines, consistent with a weak AGN, similarly to another TDE candidate, SDSS J0748. This hints that TDEs could have been notoriously missed in previous searches, requiring a quiescent galaxy with no sign of AGN activity, leading to biases in TDE rates. The weak-AGN would also be a sign of previous episodes of stellar consumption by the SMBH, hundreds or thousands years ago.

Gaia

Gaia is the ESA's space astrometric mission in operation since mid-2014 with its main aim to derive parallaxes and proper motions of 1 billion stars of the Milky Way. It scans the entire sky with its two 1.4m telescopes, therefore it allows for fairly rapid detection of transient astrophysical phenomena in the Gaia Photometric Science Alerts system†. It reaches around 20.5 mag in its broad G-band and a typical cadence is 30 days.

Since Gaia is an astrometric space mission, it yields very precise positional measurements. Currently, the positional accuracy reaches about 100 mas, however, it should be possible soon to improve these measurements to 100 μas, i.e., of order of 100 pc in physical distance for typical observed galaxies.

Moreover, each Gaia's photometric measurement is accompanied with low-dispersion (R<100) red and blue spectra (BP/RP). The spectra alone can help distinguish major

† http://gsaweb.ast.cam.uk/alerts

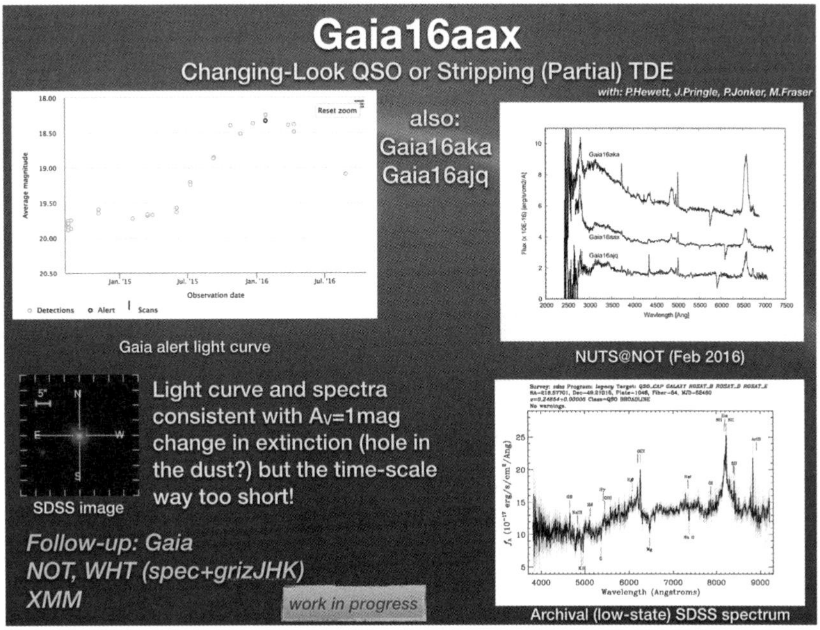

Figure 2. Gaia16aax and other nuclear flares from Gaia. Candidates for changing-look AGNs or partial TDEs.

types of transients, in particular, the most common SN Type Ia can be recognised based on their broad spectral lines. This feature of Gaia allows to immediately pick potentially interesting transients of rare type, in particular Tidal Disruption Events (TDE), which spectra typically show hot black body continuum.

Among the candidates for nuclear flares, so far Gaia has detected a handful, however, none of them were obvious cases of TDEs. However, Gaia16aax, along with Gaia16aka and Gaia16ajq, were found to be nuclear flares, and their NOT spectra showed blue continuum with broad Hα and Hβ and narrow AGN lines. Among possible explanations we consider a "changing-look" AGN or Stripping (Partial) TDE.

Stellar mass black holes via microlensing with OGLE and Gaia

Mass distribution of Galactic black holes is a key issue connected with our understanding of the late stages of the stellar evolution at its low-mass end and with recent detections of very massive products of black hole mergers via gravitational waves, at its heavy-mass end. Gravitational microlensing is probably the only channel which can increase the number of candidate black holes, both single and in binaries.

There is more than 15000 microlensing events discovered so far, however vast majority have unknown mass/distance/velocity due to degeneration of these parameters in the microlensing model. Gould (2000) predicted that 0.8% of the events are due to black holes, that means that about a dozen of events detected every year have a black hole as a lens.

In order to uniquely derive the mass and distance of the lens, and hence to conclude on its nature by comparing to the observed light, one requires to measure two observables: microlensing parallax, π_E (from photometry), and angular size of the Einstein Ring, θ_E (from astrometry). Combination of ground-based (e.g. OGLE) photometry and Gaia astrometric time-series, would allow to measure masses and distances of nearly all microlensing events occurring on the sky, hence recognising lensing black holes.

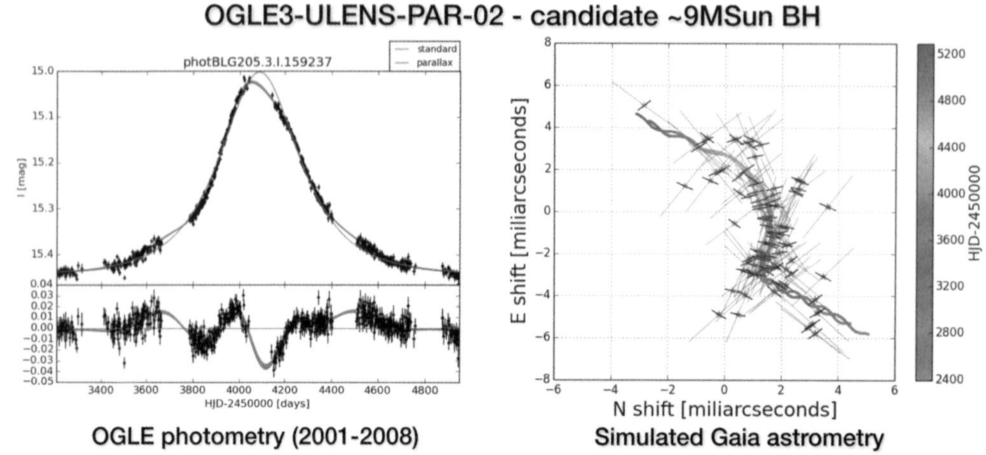

Figure 3. Combination of ground-based photometry and Gaia's astrometry time-series will yield masses and distances of microlensing events, hence the confirmation of lensing black holes (Wyrzykowski *et al.* 2016a). Left panel shows an example of an old microlensing event, most likely due to a black hole. Right panel shows simulated Gaia astrometric time-series with best fit model (solid line).

OGLE3-ULENS-PAR-02, which occurred in years 2002-2008. is the best candidate for a $\sim$9 $M_\odot$ black hole lens (Wyrzykowski *et al.* 2016a). A similar event, discovered now during the operation of Gaia, would produce a microlensing astrometric deviation to the parallax+proper motion curve of the source, as shown in Fig.3.

Summary

- OGLE and Gaia are well suited for finding transients in galaxy centres
- Advantages: superb astrometry (Gaia, OGLE); instantaneous low-resolution spectra (Gaia)
- TDEs can be found also around active (or formerly active) black holes (OGLE16aaa) - most TDE searches so far excluded AGN hosts - is there a bias in TDE rates computation?
- Follow-up programme: spectroscopy (VLT, SALT, NOT, WHT); photometry (OPTICON Time-Domain network)
- Gaia will help find Galactic single/binary stellar mass black holes via microlensing

Acknowledgments

We thank for help at various stages of this work from numerous people, in particular Drs. Zuzanna Kostrzewa-Rutkowska, Nikolay Britavskiy, Peter Jonker, Jochen Greiner, Iair Arcavi, James Guillochon, Morgan Fraser, Nadejda Blagorodnova. We also acknowledge work of the OGLE group, PESSTO group, ESA Gaia DPAC and Gaia Photometric Science Alerts Team.

References

Arcavi, I., *et al.* 2014, *ApJ*, 793, 38
Gould, A. 2000, *ApJ*, 535, 928
French, K. D., Arcavi, I., & Zabludoff, A. 2016, *ApJ*, 818, L21
LaMassa, S. M., *et al.* 2015, *ApJ*, 800, 144
Lawrence, A., *et al.* 2016, *MNRAS*, 463, 296

MacLeod, C. L., *et al.* 2016, *MNRAS*, 457, 389
MacLeod, M, *et al.* 2016, *ApJ*, 819, 3
Ruan, J. J., *et al.* 2016, *ApJ*, 826, 188
Shappee, B. J., *et al.* 2014, *ApJ*, 788, 48
Udalski, A., Szymański, M. K., & Szymański, G. 2015, *AcA*, 65, 1
Wyrzykowski, Ł., *et al.* 2014, *AcA*, 64, 197
Wyrzykowski, Ł., *et al.* 2016a, *MNRAS*, 458, 3012
Wyrzykowski, Ł., *et al.* 2016b, *arXiv:1606.03125v2*, accepted for *MNRAS Letters*

New Frontiers in Black Hole Astrophysics
Proceedings IAU Symposium No. 324, 2016
Andreja Gomboc, ed.

doi:10.1017/S1743921317001016

Tidal disruption events from different kinds of astrophysical objects: a preliminary analysis

Aurora Clerici and Andreja Gomboc

Laboratory for Astroparticle Physics, University of Nova Gorica,
Vipavska 11c, 5270 Ajdovščina, Slovenia
email: aurora.clerici@ung.si

Abstract. Tidal disruption events are a powerful tool to study quiescent massive black holes residing in the centre of galaxies. Occasionally, astrophysical objects such as stars, planets and smaller bodies are captured and tidally disrupted by the massive black hole, giving rise to a luminous flare. A detailed study of disruption parameters and the emitted radiation can give important insights on the black hole and its surroundings.

Keywords. black hole physics, Galaxy: nucleus, methods: numerical

---

1. Introduction

Tidal disruption events (TDEs) are useful tools to study quiescent massive black holes (MBHs). In the last two decades, the interest for these astrophysical phenomena has grown due to the availability of more observational data (for a recent review see Komossa 2015). Furthermore, with the growing power of computers, hydrodynamical simulations of TDEs are accessible, providing new means to test theoretical models and to model observational data. Even though a lot of work has been done on TDEs, there are important aspects which are not yet fully understood, for example, the main mechanisms governing the circularization process of tidal debris, how is energy released and where are the radio, optical/UV and X-ray radiations produced (for a review on recent developments see Lodato *et al.* 2015; Krolik *et al.* 2016).

MBHs can be found in the centre of galaxies. The nearest MBH is the one residing in our Galactic Centre (GC), which, thanks to its proximity, provides us the possibility to study the MBH environment in detail. In addition, the GC environment harbours a variety of astrophysical objects (Alexander 2005). While the study of the bound debris of a TDE can give us direct information regarding the disrupted object and the MBH, the unbound debris can provide us data about the surrounding environment. In this respect, we're investigating possible outcomes of the tidal disruption of stars, planets and asteroids. The main goal is to highlight the different physical parameters governing the tidal disruptions and obtain a better understanding of the MBH in the GC.

2. Theory and motivation

A star is disrupted when it is brought too close to a MBH: when the tidal forces overcome the star's self-gravity, the star is torn apart. In the process, about half of the stellar material is unbound and expelled, while the other half is bound and returns to the MBH as a debris stream (Rees 1988). The latter then accretes and produces a luminous flare, which carries information on the MBH and stellar properties. TDEs are particularly useful in the study of otherwise quiescent MBHs.

Specific conditions need to be satisfied in order to tidally disrupt an object. In the Newtonian picture, if a star of mass m_* and radius r_* orbits a MBH of mass m_{BH} with orbital pericentre r_p smaller than the tidal radius $r_t = r_*(m_{BH}/m_*)^{1/3}$, we may expect the tidal disruption of the star. However, there is a critical mass of the BH above which the star crosses the horizon before being disrupted: for example, a white dwarf can be disrupted by a $\sim 10^3 M_\odot$ BH, but is swallowed by a $\sim 10^6 M_\odot$ BH, which instead can disrupt solar-type stars. In the relativistic picture, also the BH spin, which affects the radius of the innermost stable circular orbit, must be taken into account.

The study of the tidal disruption of objects by a MBH can provide information about the BH itself and the surroundings. For instance, it has been shown that the final accretion of dense objects with mass of the order of 10^{20} g onto the Sgr A* BH can explain the flares detected at various wavelengths, which are characterised by short duration and quasi periodic oscillations (Kostić *et al.* 2009). Furthermore, there is a debate regarding the nature of the object called G2 (i.e. a gas cloud, a planetary embryo). In this respect, it has been considered that planets initially bound to S-stars could be captured on eccentric orbits by the MBH and their orbital properties may be compatible with those of G2 (Trani *et al.* 2016). With these considerations in mind, we want to investigate the tidal disruption parameters of different astrophysical objects such as stars, planets and asteroids.

3. On going work and future perspectives

We are performing simulations using a smoothed particle hydrodynamics (SPH) code, where an object is represented by a set of particles. When considering a star, the initial conditions are obtained by placing the SPH particles within a close packing sphere configuration according to a polytropic model. A similar procedure is used for a planet. After checking the stability of the object, it is set on an orbit which brings it within the tidal radius of the BH in order to be disrupted.

We start investigating the tidal disruption parameters of relatively simple objects. We plan to follow the evolution of their debris and to build the expected light curves using ray tracing codes. We will then extend the study to include increasingly detailed objects. For example, given $m_{BH} = 10^6 M_\odot$, we can study first the disruption of solar-type stars with hydrodynamics, then the evolution of the star's magnetic field during the disruptive process with magnetohydrodynamics (Guillochon *et al.* 2016); for planets we have to take into account the mantle+core structure, and the possibility that they are still bound to the parent star. In addition, we can consider relativistic effects affecting the disruption due to a spinning MBH. This work's first aim is to shed light on the dynamics and the evolution of the objects near the MBH in the GC.

References

Alexander, T. 2005, *Phys. Rep.*, 419, 65
Guillochon, J. & McCourt, M. 2016, eprint arXiv:1609.08160
Komossa, S. 2015, *J. High Energy Astrophys.*, 7, 148
Kostić U., Čadež A., Calvani M. & Gomboc, A. 2009, *A&A*, 496, 307
Krolik, J., Piran, T., Svirski, G., & Cheng, R. M. 2016, *ApJ*, 827, 127
Lodato, G., Franchini, A., Bonnerot, C., & Rossi, E. M. 2015, *J. High Energy Astrophys.*, 7, 158
Rees, M. J, 1988, *Nature*, 364, 1105
Trani, A., Mapelli, M., Spera, M., & Bressan, A. 2016, eprint arXiv:1608.07230

New Frontiers in Black Hole Astrophysics
Proceedings IAU Symposium No. 324, 2016
Andreja Gomboc, ed.

doi:10.1017/S1743921317002228

Accretion and wind dynamics in tidal disruption events

T. Mageshwaran[1] and A. Mangalam[2]
Indian Institute of Astrophysics,
Bangalore, India 560034,
email: [1]mageshwaran@iiap.res.in, [2]mangalam@iiap.res.in

Abstract. We have constructed self similar models of time dependent and non relativistic accretion disks in both sub and super-Eddington phase of TDEs with wind outflows for a general viscosity prescription which is a function of surface density of the disk Σ_d and radius r. The physical parameters are black hole (BH) mass $M_\bullet$, specific orbital energy E and angular momentum J, star mass $M_\star$ and radius $R_\star$. We have considered an accretion disk where matter is lost due to accretion by black hole and out flowing wind (in case of super-Eddington) and added through fallback of the disrupted debris. We have simulated the light curve profiles in various spectral bands and fit them to the observations to determine the above mentioned physical parameters.

Keywords. Galaxy: nucleus, black hole physics, accretion, accretion disks

1. Introduction

A star is tidally disrupted if its pericenter $r_p \leqslant r_t$ where $r_t \sim R_\star(M_\bullet/M_\star)^{1/3}$ is the tidal radius and these events are called as tidal disruption events (TDEs) (Rees (1988)). Mageshwaran & Mangalam (2015) have constructed a detail stellar dynamical model of TDE using parameters that include black hole (BH) mass $M_\bullet$, specific orbital energy E and angular momentum J, star mass $M_\star$ and radius $R_\star$. We define the dimensionless energy $\bar{e} = E/E_m$ and angular momentum $\ell = J/J_{lc}$ where J_{lc} is the loss cone angular momentum and $E_m = GM_\bullet/r_t$ is the maximum orbital energy such that $J_{lc} = 0$.

We have considered a self similar model of a time dependent and non-relativistic accretion disk for both sub-Eddington and super-Eddington disk with wind outflows due to strong radiative pressure. We have also considered a general viscosity prescription given by $\Pi_{r\phi} = -K\Sigma_d^b r^d$ where Σ_d is the surface density of the disk, r is the radius, K, b and d are constants (Mangalam (2001)). We assume a α viscous model for sub-Eddington disk with gas pressure dominating such that $b = 5/3$ and $d = -1/2$ and the radiative viscosity for super-Eddington disk with radiation pressure given by $P_r \propto \rho^\gamma$ where $\gamma = 4/3$ such that $b = 1/3$ and $d = -5/6$ (Mangalam (2003)). The non relativistic disk equations in the limit of $u_r \ll u_\phi$, where u_r is the radial velocity and u_ϕ is the azimuthal velocity, are given by $\dot{\Sigma}_d = -(1/r)\partial_r(ru_r\Sigma_d) - \dot{\Sigma}_w$; $\omega^2(r) = (1/r)\partial_r\Phi(r)$; $u_r\Sigma_d\partial_r(r^2\omega) + \dot{\Sigma}_w r^2\omega(r) = -(1/r)\partial_r(r^2\Pi_{r\phi})$, where $\dot{\Sigma}_d$ and $\dot{\Sigma}_w$ are time derivative of the surface density of disk and wind components respectively and $\dot{\Sigma}_w$ is zero for sub-Eddington disk and $\Phi(r) = -GM_\bullet/r$. For a super-Eddington disk, assuming hydrostatic equilibrium up to the height $Z = Z_{ph}$ we construct the vertical density structure ρ which is then used in the Eddington approximation to derive Z_{ph}. Using vertical momentum equation at Z_{ph}, implies $\dot{\Sigma}_w \propto \Sigma_d^e r^f \sqrt{1 - \eta/\eta_{ph}}$, where $e = 0.12$, $f = -3/2$, η and η_{ph} is the ratio of radiation pressure to total pressure at the mid plane and photosphere of the disk respectively. The self similar form of the accretion disk is

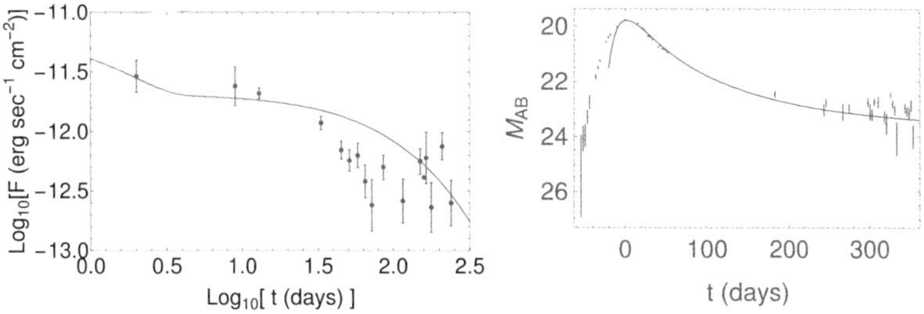

Figure 1. (Left) The sub-Eddington model fit to the XMMSL1 J061927.1-655311 (Saxton (2014)) in X-ray band. (Right) The super-Eddington model fit to the PS1-10jh (Gezari *et al.*(2012)) observations in optical g band.

taken to be $\Sigma_d = \Sigma_0 (t/t_0)^{\beta} A \xi^p$ and $\xi = r/r_0(t/t_0)^{-\alpha}$, where Σ_0, r_0, t_0, A, α, β, p are the constants derived using disk equations and the mass of disk $M_d(t_0) = \int_{t_m}^{t_0} \dot{M}_{fb}(t)\, \mathrm{d}t$, where $\dot{M}_{fb}$ and t_m are the infall rate of disrupted debris after disruption and orbital period of innermost debris respectively as given in Mageshwaran & Mangalam (2015).

We assume a seed accretion disk formed by the circularization of the debris with initial mass $M_d(t_0)$ and the derived self similar structure of Σ_d and $r_0 = q\ r_{in}$, where r_{in} is the inner radius taken to be innermost stable circular orbit r_{ISCO} which is a function of black hole spin j. This seed disk evolves in effect of mass addition by the infall of disrupted debris and mass loss due to accretion by black hole and wind such that the structure of Σ_d remains same. Applying mass conservation, $\dot{M}_d = \dot{M}_{fb} - \dot{M}_a - \dot{M}_w$ where $\dot{M}_d$, $\dot{M}_a$ and $\dot{M}_w$ are the rate of the change of disk mass, accretion rate by the black hole and mass loss due to out flowing wind respectively, we calculate the evolution of outer radius $r_{out}(t)$. The temperature of the sub-Eddington disk is then given by $\sigma T_e^4 = (3/8)\ \omega \Pi_{r\phi}$ where $\omega = \sqrt{GM_{\bullet}/r^3}$ and for super-Eddington disk, the temperature is derived using radiative Eddington approximation at Z_{ph}. We now discuss the light curve profiles in various spectral bands fit to the observations and estimated physical parameters.

2. Conclusions

We have obtained the self similar structure of a time dependent and non relativistic accretion disk. The parameters $\{\beta,\ \alpha,\ p\}$ are $\{-8/7,\ 5/21,\ 3/2\}$ for sub-Eddington disk and $\{-0.95,\ 1.22,\ -0.77\}$ for super-Eddington disk. The bolometric luminosity for sub-Eddington disk is $L \propto t^{-1.9}(\xi_{out}^{5/2}(t) - \xi_{in}^{5/2}(t))$ and for super-Eddington disk is $L \propto t^{1/2}(\xi_{out}^{0.4}(t) - \xi_{in}^{0.4}(t))$. The Fig 1 (Left) shows the sub-Eddington model fit to the observations in X-rays for XMMSL1 J061927.1-655311 and the derived parameters are $\bar{e} = 0.00316$, $\ell = 0.9$, $M_6 = 3.15$, $m = 1.77$, $q = 2$ and black hole spin $j = 0.15$. The Fig 1 (right) shows the super-Eddington model fit to the PS1-10jh observations in optical g band and the derived parameters are $\bar{e} = 0.0001$, $\ell = 1$, $M_6 = 36$, $m = 1.56$, $q = 2$, $j = 0.5$.

References

Gezari *et al.* 2012, *Nature*, 485, 217
Mageshwaran, T. & Mangalam, A. 2015, *ApJ*, 814, 141
Mangalam, A. 2001, *A&A*, 379, 1138
Mangalam, A. 2003, *Bulletin of the Astronomical Society of India*, 31, 207
Rees, Martin J. 1988, *Nature*, 333, 523
Saxton *et al.* 2014, *A&A*, 572
Strubbe, Linda E. & Quataert, Eliot. 2009, *MNRAS*, 400, 2070

New Frontiers in Black Hole Astrophysics
Proceedings IAU Symposium No. 324, 2016
Andreja Gomboc, ed.

doi:10.1017/S1743921317001491

Explosive nucleosynthesis in tidal disruption events of massive white dwarfs, and their debris

Ataru Tanikawa[1,2], Yushi Sato[1,3], Ken'ichi Nomoto[4,5], Keiichi Maeda[6,4], Naohito Nakasato[7] and Izumi Hachisu[1]

[1]Department of Earth Science and Astronomy, College of Arts and Sciences, The University of Tokyo, 3-8-1 Komaba, Meguro-ku, Tokyo 153-8902, Japan; tanikawa@ea.c.u-tokyo.ac.jp
[2]RIKEN Advanced Institute for Computational Science, 7-1-26 Minatojima-minami-machi, Chuo-ku, Kobe, Hyogo 650-0047, Japan
[3]Department of Astronomy, Graduate School of Science, The University of Tokyo, 7-3-1 Hongo, Bunkyo-ku, Tokyo 113-0033, Japan
[4]Kavli Institute for the Physics and Mathematics of the Universe (WPI), The University of Tokyo, 5-1-5 Kashiwanoha, Kashiwa, Chiba 277-8583, Japan
[5]Hamamatsu Professor
[6]Department of Astronomy, Kyoto University, Kitashirakawa-Oiwake-cho, Sakyo-ku, Kyoto 606-8502, Japan
[7]Department of Computer Science and Engineering, University of Aizu, Tsuruga Ikki-machi Aizu-Wakamatsu, Fukushima 965-8580, Japan

Abstract. We perform SPH simulations coupled with nuclear reactions to follow tidal disruption events (TDEs) of white dwarfs (WDs) by intermediate mass black holes (IMBHs). We consider an oxygen-neon-magnesium (ONeMg) WD with $1.2M_\odot$ as well as a helium (He) WD with $0.3M_\odot$, and a carbon-oxygen (CO) WD with $0.6M_\odot$. Our WD models have different numbers of SPH particles, N, up to a few 10 million. We find that nucleosynthesis does not converge against N even for $N > 10^7$. For all the WDs, the amount of radioactive nuclei, such as ^{56}Ni, decreases with increasing N. Nuclear reactions might be extinguished for infinitely large N. Our results show that these kinds of TDEs, if solely powered by radioactive decays, are much dimmer optical transients similar to Type Ia supernovae as previously suggested.

Keywords. hydrodynamics, nuclear reactions, nucleosynthesis, white dwarfs

---

1. Introduction

A number of tidal disruption events (TDEs) has been found. Some of them are possibly TDEs of a WD (WD TDEs). We focus on the WD TDE as an optical transient resulting from its thermonuclear explosion. In a WD TDE, the WD is heated by compression in the direction perpendicular to the orbital plane. Then, the WD may undergo explosive nuclear burning, yielding radioactive nuclei such as ^{56}Ni. Rosswog *et al.* (2008) and Rosswog, Ramirez-Ruiz & Hix (2009) studied WD TDEs and their nucleosynthesis signatures. A WD TDE needs not only an adiabatic compression but also a shock heating in order to reach high temperature enough to start explosive nuclear reactions. However it is not clear whether previous simulations have high enough resolution to detect the shock wave.

In this paper, we perform SPH simulations with various numbers of SPH particles from a few 0.01 million up to a few 10 million. We aim to assess whether the shock wave is generated, and whether nucleosynthesis is converging against N even if the shock wave is not generated.

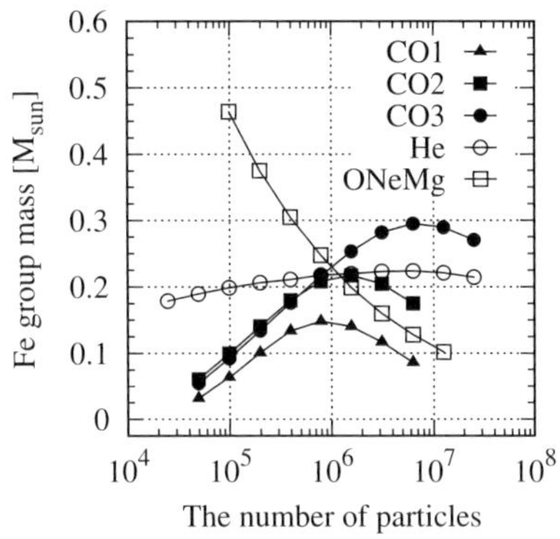

Figure 1. Mass fractions of Fe group elements as a function of N at the time just after the nuclear reactions have ceased

2. Method

We follow the evolution of WD TDEs by means of our SPH code coupled with a nuclear reaction network. We adopt SPH modeling for the WDs, and use a fixed potential to model the IMBH gravity. Our SPH code solves the vanilla ice SPH equations with self-gravity among SPH particles with adaptive gravitational softening. We use the Helmholtz equation of state with (or without) Coulomb corrections (Timmes & Swesty 2000), and include nuclear reactions with Aprox13 (Timmes, Hoffman, & Woosley 2000).

We adopt a He WD with $0.3M_{\odot}$, a CO WD with $0.6M_{\odot}$, and a ONeMg WD with $1.2M_{\odot}$. We set an IMBH with $500M_{\odot}$ for the He WD and CO WD, and an IMBH with $100M_{\odot}$ for the ONeMg WD. For the CO WD, we model the IMBH gravity as Newtonian potential, the Paczyński-Wiita (PW) potential and a generalized Newtonian potential obtained by Tejeda & Rosswog (2013). The model names are "CO1", "CO2", and "CO3". For the He WD and ONeMg WD, we assume PW potential. These model names are "He" and "ONeMg". In all the models, we change the number of particles for the WDs, N, from a few 0.01 million up to a few 10 million. The ratios of pericenter distances to tidal radii are 5.0 for models COs and He, and 3.0 for model ONeMg.

3. Results

Figure 1 shows that the nucleosynthesis does not converge against N even for $N > 10^7$. For model ONeMg, the amount of these elements decreases monotonically. For the other models, these elements also decrease at $N > 10^7$. Nuclear reactions might be extinguished for infinitely large N. Our results show that these kinds of TDEs, if solely powered by radioactive decays, are much dimmer optical transients similar to Type Ia supernovae as previously suggested.

References

Rosswog, S., Ramirez-Ruiz, E., & Hix, W. R. 2009, *ApJ*, 695, 404

Rosswog, S., Ramirez-Ruiz, E., Hix, W. R., & Dan, M. 2008, *Computer Physics Communications*, 179, 184

Tejeda, E., & Rosswog, S. 2013, *MNRAS*, 433, 1930

Timmes, F. X., Hoffman, R. D., & Woosley, S. E. 2000, *ApJS*, 129, 377

Timmes, F. X. & Swesty, F. D. 2000, *ApJS*, 126, 501

Topic 4: Active Galactic Nuclei

New Frontiers in Black Hole Astrophysics
Proceedings IAU Symposium No. 324, 2016
Andreja Gomboc, ed.

doi:10.1017/S1743921317001922

The remarkable AGN jets

Serguei Komissarov
School of Mathematics, University of Leeds,
Leeds LS29JT, United Kingdom
email: s.s.komissarov@leeds.ac.uk

Abstract. The jets from active galactic nuclei exhibit stability which seems to be far superior compared to that of terrestrial and laboratory jets. They manage to propagate over distances up to a billion of initial jet radii. Yet this may not be an indication of some exotic physics but mainly a reflection of the specific environment these jets propagate through. The key property of this environment is a rapid decline of density and pressure along the jet, which promotes its rapid expansion. Such an expansion can suppress global instabilities, which require communication across the jet, and hence ensure its survival over huge distances. At kpc scales, some AGN jets do show signs of strong instabilities and even turn into plumes. This could be a result of the flattening of the external pressure distribution in their host galaxies or inside the radio lobes. In this regard, we discuss the possible connection between the stability issue and the Fanaroff-Riley classification of extragalactic radio sources. The observations of AGN jets on sub-kpc scale do not seem to support their supposed lack of causal connectivity. When interpreted using simple kinematic models, they reveal a rather perplexing picture with more questions than answers on the jets dynamics.

Keywords. black hole physics, relativity, galaxies: jets, hydrodynamics, MHD, radio continuum: galaxies

1. Introduction

The terrestrial and laboratory gas and plasma jets are know to be susceptible to instabilities, which tend to destroy them over relatively short distances. This is particularly true for subsonic jets but also applies to supersonic jets. Although, supersonic jets are more resilient, the typical laboratory jets slow down and dissipate most of its kinetic energy over the distances which do not exceed a hundred initial jet radii. In contrast, the AGN jets appear to be much more robust. Indeed, the initial radius of an AGN jet is expected to be comparable to the gravitational radius of the central black hole, which is $r_i \sim 10^{14}$ cm for $10^9 M_{\odot}$, and these jets can often be traced up to the distance of 100 kpc, which is about one billion of the initial jet radius. At first glance, this remarkable contrast between the AGN jets and their terrestrial counterparts suggests a dramatically different physics. After all, these jets are relativistic and produced by black holes. We cannot reproduce them in a physical laboratory. However, the results of numerous theoretical and computational studies suggest that this may not be the right conclusion.

The properties of relativistic fluid jets are found to be quite similar to those of non-relativistic jets, when using proper parametrisation. For example, initially it was claimed that high-Lorentz-factor jets are less unstable. However, both the relativistic Mach number and the specific inertia (mass density in the lab frame) depend on the Lorentz factor (Konigl 1980, Komissarov & Falle 1996) and hence the results of the instability studies are well in line with the known dependence of growth rates on the Mach number and density ratio for non-relativistic jets. Moreover, the jets from young stars, which are non-relativistic, show a similar ability to propagate over huge distances, not much different from AGN jets when measured in the units of initial jet radius. The magnetic field can suppress some non-magnetic instabilities but introduces new current-driven instabilities, which could be even more destructive given its predominantly toroidal (azimuthal) structure expected far away from the magnetic central engine.

The laboratory jets are approximately cylindrical (albeit with oscillations due to initial difference in pressure with the external gas) and most theoretical and numerical stability studies have been focused on cylindrical jets. This symmetry assumption allows to simplify the problem and helps to make it treatable by standard analytic techniques. The AGN jets are not cylindrical, with the half-opening angle in the region of few degrees. This may not seem a lot and at first suggests that the cylindrical approximation may be suitable. On the other hand, over the long length of the jets, this amounts to a huge increase in jet radius. The same applies to the stellar jets. Although the stability of expanding jets is not well studied, it has been argued that such an expansion has a stabilising effect (e.g. Moll *et al.* 2008).

2. Causality and jet stability

It helps to distinguish between local and global instabilities of jets. Global modes imply coherent motion on the scale of the jet radius. One typical example of such a global instability is the kink mode, where the whole jet moves sideways attaining a snake-like or a helical shape. Another is the pinch mode, where the jet radius varies independently of the azimuthal angle. Such instabilities are recognised as the most threatening to the jet survival, as they lead to strong interaction with the surrounding gas in the non-linear regime. Local instabilities are small scale motions. They may lead to local dissipation and mixing but do not endanger the jet integrity. Clearly, only the global instabilities are relevant to the issue of the astrophysical jets ability to propagate over very large distances.

The very nature of global instabilities implies a causal communication across the jet. Without such a communication it is impossible to support such highly-organised motions. In fluids, the communication speed is limited by the fundamental wave speeds, the sound speed in gas dynamics and the fast magnetosonic speed in MHD. For supersonic (super-fast-magnetosonic) flows, these waves cannot propagate upstream. In fact, they are confined to a downstream-pointing cone known as the Mach cone. A higher Mach number corresponds to a narrower Mach cone and hence a longer distance over which the jet becomes aware of an upstream perturbation. This explains why the growth rate of instabilities in cylindrical jets decreases with the Mach number. In expanding supersonic jets, the Mach number monotonically increases and hence the growth rate decreases along the jet. The instabilities become suppressed altogether when the typical opening angle of the Mach cone becomes smaller than that of the jet (This argument applies equally well to both relativistic and non-relativistic flows.).

Consider for example a freely-expanding non-relativistic unmagnetised supersonic jet. Such a jet is almost conical and propagates at constant speed. Its density $\rho \propto z^{-2}$, pressure $p \propto \rho^{\gamma}$, sound speed $a^2 \propto p/\rho$ and the Mach angle $\theta_m \propto a$. From these we find $\theta_m \propto z^{1-\gamma}$ and hence the Mach angle is bound to eventually become less than the opening angle of the jet, leading to its global stability. Presumably, this simple argument was behind the assumption made in some early studies of kpc AGN jets that they emerge from AGN as cold conical flows (e.g. Falle 1991).

If a jet is not freely expanding but instead it is confined by an atmosphere with the power law pressure $p_{ext} \propto z^{-\kappa}$ then the ratio of the Mach angle and the jet opening angle $\theta_m/\theta_j \propto z^{(2-\kappa)/2}$ (Porth & Komissarov 2015). This indicates that $\kappa = 2$ is a critical value, with jets becoming causally-disconnected for $\kappa > 2$. Such causally-disconnected

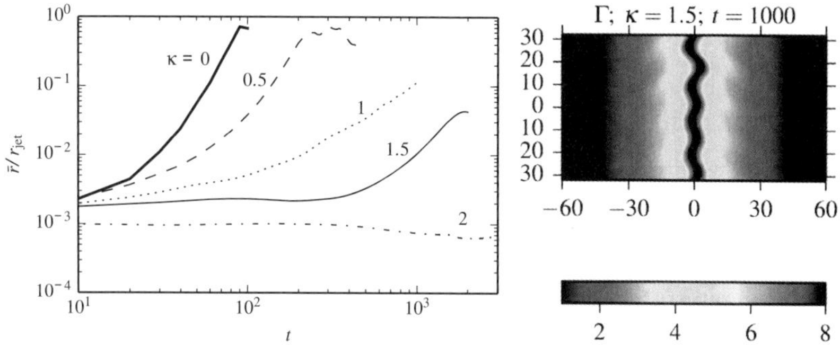

Figure 1. *Left panel*: Amplitude evolution of a kink-mode instability as observed in the simulations by Porth & Komissarov (2015). The amplitude is defined as the sideways displacement of the jet barycentre measured in the units of the instantaneous jet radius. The time is measured in the units of r_0/c, where r_0 is the initial jet radius. *Right panel*: The unstable core of the jet in the atmosphere with $\kappa = 3/2$. The image shows the Lorentz factor distribution.

jets are not necessarily freely expanding conical flows. Indeed, in a conical jet its pressure decreases as $p \propto z^{-2\gamma}$, which is faster than that of the external gas if $\kappa < 2\gamma$. Hence for $2 < \kappa < 2\gamma$ the jets must still be confined by the external pressure, which is now communicated to the jet interior by shock waves. However, these are high amplitude waves and not the small perturbations associated with instabilities, which suggests that for $\kappa > 2$ jets are still globally stable. Relativistic and magnetic effects do not seem to alter this conclusion (Porth & Komissarov 2015).

The central engines of AGN and stellar jets are dominated by compact objects, whose gravitational field has a strong impact on the distribution of surrounding gas. The overall effect is a rapid decrease of both its density and pressure away from the centre. For a polytropic atmosphere of central mass, one has $\kappa = \gamma/(\gamma - 1)$, which is higher than 2 when $1 < \gamma < 2$. For a spherical adiabatic wind, $\kappa = 2\gamma$, which is also steeper than the critical one. A self-collimating magnetic wind can in principle deliver $\kappa < 2$. For the Bondi accretion $\kappa = 3\gamma/2$, which is still larger than 2 for $\gamma > 4/3$. Thus, steep gradients of external pressure, bordering the critical value, are expected to be quite common close to the central engine. Unfortunately, we are still unable to measure the parameters of the external gas on the scales below 1 kpc. Instead, we have to rely mostly on indirect model-dependent estimates. Taken together, this circumstantial evidence suggests that $\kappa \simeq 2$ is indeed quite typical for the AGN environment (e.g. Phinney 1983, Begelman *et al.* 1984).

In order to study the effect of jet expansion on its stability, Porth & Komissarov (2015) used periodic box simulations where the pressure of external gas was forced to decrease as a power function of time, similarly to what would have been seen in a frame moving radially at constant speed through a power-law atmosphere. The left panel of Fig. 1 shows the growth of a kink-mode perturbation in a moderately relativistic magnetised jet with a purely azimuthal magnetic field. One can see, that the growth rate decreases with κ and the instability is totally suppressed for $\kappa = 2$, in line with the expectations.

One characteristic property of magnetised jets is the development of a magnetically-confined core which refuses to expand with the same rate as the rest of the jet and remains causally connected. Hence, this core can develop instabilities and gets destroyed even when the rest of the jet remains stable. In fact, this is an example of a local instability at work. The right panel of Fig. 1 shows how this instability develops in the simulations by Porth & Komissarov (2015). Such local instabilities may power emission from jets without endangering their integrity.

3. The Fanaroff-Riley division

The issue of jet stability is often discussed in connection to the global division of extragalactic radio sources into two main classes according to the classification by Fanaroff & Riley (1974). The main motivation behind this connection is based on the remarkable similarity in appearance between the kpc-scale jets (and tails) of some FR1-type sources (e.g. 3C31) and the unstable transonic jets produced in laboratory. These sources do not show any cocoons separating their jets from the external gas and in some cases this gas is observed in X-rays (e.g. Schreier *et al.* 1982 for M87). Furthermore, the theoretical model based on the idea of a turbulent flow confined by the external pressure and entraining significant amount of surrounding gas has successfully reproduced a number of the observational results (e.g. Bicknell 1984, Komissarov 1990, Laing & Bridle 2002). Thus, we need to take seriously the possibility that some AGN jets become reconfined by the external gas pressure at the distances of $0.1 \div 1\,\mathrm{kpc}$ from the central engine and develop instabilities. In fact, the optical observations of giant elliptical galaxies indicate that their stellar distribution flattens near the centre. The size of this central core depends on the galactic mass and is scattered around 1 kpc. At such distances, the gravitational potential of the central black hole is sub-dominant and the gas distribution is expected to reflect that of the stars. In fact, it is known that outside of the core the gas distribution does mimic the stellar one (Mathews & Brighenti 2003). Hence, the X-ray gas of galactic coronas is likely to have a central core with flat pressure distribution which could be sufficiently high to reconfine the FR1-type jets.

The reconfinement process involves a strong shock, driven into the jet by the external pressure. It turns out (see Porth & Komissarov 2015) that for a uniform external gas of pressure p_0 and an initially conical jet of kinetic power L_j, the reconfinement shock reaches the jet axis at the distance $z_{r,kpc} \approx 0.3(L_{j,44}/p_{0,-9})^{1/2}$, where the units are 1 kpc, 10^{44} erg/s and 10^{-9} dyn/cm^2. Since $p_{0,-9} \approx 1$ is typical for the central parts of the X-ray coronas of elliptical galaxies and for FR1 sources $L_{j,44} < 1$ (Cavagnolo *et al.* 2010), the reconfinement of FR1 jets inside the galactic cores seems very likely. On the other hand, the kinetic power of FR2 is significantly higher and the reconfinement process cannot be completed inside the cores. Outside, the pressure of the X-ray gas decreases rapidly again, $p \propto z^{-1.5\pm0.25}$ (Mathews & Brighenti 2003). This slows down not only the reconfinement process but also the development of instabilities that follows. So it is reasonable to conclude that jets which manage to escape the core will remain as stable supersonic flows for much longer.

In fact, the observations show that FR1 and FR2 sources belong to different regions in the $P - L_O$ plane, where P is the source radio power and L_O is the optical luminosity of the host galaxy, with the dividing line $P^{FR} \propto L_O^2$ (Owen & Ledlow 1994). Assuming that the critical jet power, L_j^{FR}, separating FR1 and FR2 sources is set by the condition $z_r = z_c$, where z_c is the X-ray core radius, Porth & Komissarov (2015) obtain $L_j^{FR} \propto p_0 z_c^2$. Thus, the Owen-Ledlow result can be explained if the parameters of X-ray cores depend on the optical luminosity (mass) of the parent galaxy in an agreeable way. Unfortunately, the X-ray cores are rarely resolved. If, however, their radius scales in the same way as that of the stellar distribution then $z_c \propto L_O^{1.1}$ and if the gas temperature scales as the stellar velocity dispersion then $p_0 \propto L_O^{-0.15}$ (Porth & Komissarov 2015). This leads to $L_j^{FR} \propto L_O^{2.05}$, which is basically the same as observed, provided $P \propto L_j$. Overall, this is a very encouraging result.

Contrary to being often described as the prototype FR1 source, 3C31 represents only a small sub-class of FR1 sources. Most of them have jets submerged into radio lobes, which makes them more like FR2 sources. The difference in that in FR2 sources the jets remain well collimated all the way and terminate in a very bright and compact hot spot near the outer edge of the radio lobe, whereas the jets of FR1 sources seem to "disappear" halfway through the lobe (de Ruiter *et al.* 1990). These observations show that the model of a naked jet has its limitations and a better model of FR-1 sources has to incorporate not only the conditions close to the AGN, where their jets are naked, but also the distant regions where these jets inflate radio-lobes.

The widely accepted current model of FR2 radio sources involves two jets surrounded by a light cocoon, which shields them from direct contact with the interstellar/intergalactic gas (Falle 1991, Komissarov & Falle 1998, Kaiser & Alexander 1997). This cocoon is filled with plasma which has been supplied by the jets during their previous activity and passed through the jet termination shocks, associated in the model with the leading hot spots of a FR2 source. This cocoon is over-pressured with respect to the external gas and drives a strong shock through it. The expansion of this cocoon is slow compared to its sound speed and hence its pressure distribution is more or less uniform.

Falle (1991) has shown that the cocoon pressure is always sufficient to reconfine the initially free conical AGN jet and turn it into a quasi-cylindrical flow whose length/radius ratio increases with time. Falle (1991) argued that eventually the cylindrical section becomes long enough for the jets to develop instabilities and become turbulent, now inside the cocoon. He proposed that this may result in the change in morphology from the FR2 to the FR1 type. However his 2D axisymmetric simulations could not confirm this prediction. The 2D simulations of jets with slab symmetry by Komissarov & Falle (2003) did exhibit a kink-mode instability but the lobe structure still remained more or less unchanged. The expected transition has been achieved only in the very recent 3D hydrodynamic simulations by Massaglia *et al.* (2016) and RMHD simulations by Tchekhovskoy & Bromberg (2016). It turns out that the 3D effects are critical in this problem. While an interesting and promising idea, it remains to be seen if it can explain the separation between FR classes in the Owen-Ledlow plane.

The analytical model of FR2 sources by Falle (1991) deals only with power-law density distributions of the external gas as this assumption leads to a self-similar solution. In the solution the whole jet is surrounded by a melon-shaped expanding cocoon. This naturally leads to a picture where the interstellar gas is evacuated from the parent galaxy. However, this is likely to be an over-simplification. From the theoretical viewpoint, an external gas distribution with flat central core introduces a characteristic length-scale to the problem and this is likely to alter the dynamics. Observationally, the two-lobed structure of FR-2 radio-sources, even at low radio frequency where the synchrotron losses are week (e.g. Kassim *et al.* 1996), suggests that the melon-shape of their cocoons is not generic. This conclusion is strengthened by the X-ray observations which show the presence of hot gas in the gap between the lobes of FR2 sources (Nulsen *et al.* 2015). Thus, it looks more likely that close to AGN the FR2 jets are "naked" and only further out they are cocooned.

4. Complications

The idea that the rapid expansion of AGN jets delays the onset of global instabilities until kpc scales offers a simple and attractive explanation to the most basic observed properties of extended extragalactic radio sources within the framework of well known physics. Yet we need to learn more about the properties of ISM and the dynamics of AGN jets on sub-kpc scales in order to see if these jets are indeed largely causally-disconnected on these scales.

Using the basic steady-state theoretical models, one may expect such flows to have a relatively simple structure. Not affected by global instabilities, they should be more or less straight. In the absence of strong internal dissipation of kinetic energy, their acceleration should be quite efficient and so we expect them to be highly supersonic. Even for the rather slow magnetic collimation-acceleration mechanism, the acceleration should be completed by the distances corresponding to the blazar zone, $z \lesssim 1\,\mathrm{pc}$ (Komissarov *et al.* 2007). Their shape is not that well determined. Although, highly-magnetised causally-disconnected jets are almost conical (Komissarov *et al.* 2009), the AGN jet are not expected to remain highly-magnetised on the distances above one parsec. As to weakly-magnetised jets, we have already noted that they cannot be conical unless $\kappa > 2\gamma$†. Assuming that the dissipation associated with the shocks which adjust the jet pressure to that of the external gas with $< 2\kappa < 2\gamma$ is weak and they can still be treated as adiabatic, we find that the jet radius $r_j \propto z^{\kappa/2\gamma}$. So the shape is still parabolic. Weak oscillations associated with stationary shocks may be superimposed on top of this parabolic shape throughout the jet length. We also expect to find $\theta_j/\theta_m > 1$. For magetically-dominated flows $\theta_m \approx \sigma^{1/2}/\Gamma$, where $\sigma \gg 1$ is the relativistic magnetisation parameter, whereas for particle-dominated plasma ($\sigma \ll 1$) $\theta_m \approx \beta_s/\Gamma$, where $\beta_s = a/c$ is the sound speed. Hence the condition of causal disconnection reads as $\theta_j\Gamma > \sigma^{1/2}$ and $\theta_j\Gamma > \beta_s$ for these two regimes respectively.

The properties of AGN jets on sub-kpc can be studied using the VLBI method; normally down to about one parsec from the central engine. These observations allow to measure the jet apparent opening angle and the apparent velocity of individual components in the plane of the sky. To convert these into the actual values one has to know the angle between the jet and the line of sight. A number of more or less reasonable assumptions can be made to estimate the viewing angle. For individual sources they may lead to significant errors but for a large sample one expects to obtain statistically reliable results. Analysis of large samples has become the main approach of the current studies of VLBI-jets. Several groups studied the distributions of Γ and θ_j using somewhat different models and techniques and they all agreed that $\theta_j\Gamma = 0.1 \div 0.2$ (Jorstad *et al.* 2005, Pushkarev *et al.* 2009, Clausen-Brown *et al.* 2013). This means that the jets can be causally disconnected only if they are particle-dominated and cool, with the sound speed $\beta_s \leqslant 0.1$. This implies that no matter what is the main energy type of the jets near the central source by the distance of about one parsec it is converted into the kinetic energy of bulk motion. This is consistent with the conclusion reached by Komissarov *et al.* (2007) that the magnetic acceleration of AGN jets should be completed inside the blazar zone. The alternative mechanism of thermal acceleration is even faster, with $\Gamma \propto r_j$.

However, Lister *et al.* (2013) have found that in the MOJAVE sample the jet components identified at larger projected distance z_p from the core tend to have higher apparent speed. Moreover, Homan *et al.*(2015) have shown that individual components move with predominantly positive acceleration at $z_p < 4\,\mathrm{pc}$ and predominantly negative accelerations at $z_p > 30\,\mathrm{pc}$. If these proper motions are indicative of the bulk flow velocity then in stark conflict with the current theoretical models the jet speed keeps increasing even on the scales well above $z = 1\,\mathrm{pc}$ and hence even at such large distances there is still plenty of free energy in the magnetic or thermal form to power this acceleration.

The M87 jets stands out from the rest by the fact that this galaxy is very close and the

† For very hot jets with $p \gg \rho c^2$ this condition reads $\kappa > 4$ but jets which have reached their asymptotic speed cannot remain that hot.

jet can be traced almost all the way from the event horizon to the kpc scales. According to the observations, there exist three distinct zones. In the *near zone*, which extends up to about $z = 0.1\,\mathrm{pc}$ ($100\,r_s$), the jet shape is approximately parabolic with $r_j \propto z^{0.75}$ (Hada *et al.* 2013). Here the jet is still inside the AGN. The *middle zone* extends up $z = 300\,\mathrm{pc}$. In this zone, the jet shape is a different parabola with $r_j \propto z^{0.56}$ (Hada *et al.* 2013). This shape change may indicate a change in the jet environment. Presumably, in the middle zone the jet is already surrounded by ISM of the parent galaxy‡. The *far zone* extends into kiloparsecs and the jet shape is almost conical there (e.g. Nakamura &Asada 2014).

The apparent speed of individual features is seen to increase with distance in the middle zone from the Lorentz factor $\Gamma \approx 1$ at $z = 50\,\mathrm{pc}$ to $\Gamma \approx 6$ at $z = 300\,\mathrm{pc}$ (Asada *et al.* 2014). Kovalev *et al.*(2007) report even smaller speeds, below few percent of the speed of light, for $1\,\mathrm{pc} < z < 50\,\mathrm{pc}$. Optical observations indicate that in the far zone the apparent speed of proper motion decreases with distance. However, the velocity pattern is extremely complicated, with very large local variations within individual knots (Meyer *et al.* 2013). If taken as measurements of the bulk speed of a quasi-steady jet, the results by Asada *et al.* (2014) and Kovalev *et al.*(2007) indicate that up to the distance $z \approx 50\,\mathrm{pc}$ the jet is still subsonic! Indeed, in the magnetic model the fact that the jet eventually reaches $\Gamma \approx 6$ means that at the start of the acceleration zone the flow magnetisation parameter $\sigma > 6$. This corresponds to the Alfvén speed $c_a > 0.91c$. In the thermal model, this means that at the start of the zone the jet plasma is ultra-relativistically hot with the sound speed $a \approx c/\sqrt{3}$. Both the numbers are high enough to imply a subsonic motion at $z < 50\,\mathrm{pc}$, which is also in stark conflict with our current understanding of AGN jets. The low speeds also imply the Doppler factor of order unity, which suggests that the M87 jet is intrinsically one-sided and hence rather exceptional. Interestingly and in conflict with the above results, Walker *et al.* (2008) report the apparent speed $\approx 2c$ at $z \approx 1\,\mathrm{pc}$. Perhaps the interpretation of VLBI observations is not that straightforward and the observed proper motions are poor indicators of the jet speed.

Idealised steady-state theoretical models of jets have a very simple structure, either entirely featureless or with few clearly identifiable reconfinement shocks. Even a quick glance at the high-resolution images of AGN jets reveal that their structure is much richer. Variable central engine, cloudy environment and instabilities are the most obvious factors which can be responsible for this. All these factors can strongly modify the jet dynamics and make the idealised models redundant. More adequate models can be based on advanced computer simulations which have to accommodate the 3D nature of AGN jets. In these models, the issue of non-thermal particle acceleration and emission has to be addressed with rigour for the sake of direct comparison with the observations. Although some may think that we have captured the essence of the AGN jets, observations keep asking us challenging questions about this remarkable astrophysical phenomenon and force us to have another look at the basics again and again. We may well be just at the beginning of a long and complicated journey.

References

Asada, K., Nakamura, M., Doi, A., Nagai, H., & Inoue, M., 2014, *ApJ Letters*, 781, 2

Begelman, M. C., Blandford, R. D., & Rees, M. J., 1984, *Reviews of Modern Physics*, 56, 255

Bicknell, G. V., 1984, *ApJ*, 286, 68

‡ Using the theoretical shape $r_j \propto z^{\kappa/2\gamma}$ and assuming $\gamma = 5/3$, we find $\kappa = 2.5$ and 1.87 for the near and middle zones respectively. These values seem quite reasonable.

Cavagnolo, K. W., McNamara, B. R., Nulsen, P. E. J.., Carilli, C. L., Jones, C., & Brzan, L., 2010, *ApJ*, 720, 1066

Clausen-Brown, E., Savolainen, T., Pushkarev, A. B., Kovalev, Y. Y., & Zensus, J. A., 2013, *A&A*, 558, 144

Falle, S. A. E. G., 1991, *MNRAS*, 250, 581

Fanaroff, B. L. & Riley, J. M., 1974, *MNRAS*, 167, 31

Hada, K., Kino, M., Doi, A., Nagai, H., Honma, M., Hagiwara, Y., Giroletti, M., Giovannini, G., & Kawaguchi, N., 2013, *ApJ*, 775, 70

Homan, D. C., Lister, M. L., Kovalev, Y. Y., Pushkarev, A. B., Savolainen, T., Kellermann, K. I., Richards, J. L., & Ros, E., 2015, *ApJ*, 798, 134

Jorstad, S. G., Marscher, A. P., Lister, M. L., *et al.*, 2005, *AJ*, 130, 1418

Kaiser, C. R. & Alexander, P., 1997, *MNRAS*, 286, 215

Kassim, N., Perley, R. A., Carilli, C. L., Harris, D. E., & Erickson, W. C., 1996, in *Cygnus A – Studay of a Radio Galaxy, Proceedings of the Greenbank Workshop*, edited by. C.L. Carilli and D.E. Harris, Cambridge University Press, p.182

Kharb, P., Lister, M. L., & Cooper, N. J., 2010, *ApJ*, 710, 764

Konigl, A., 1980, *PhFl*, 23, 1083

Komissarov, S. S., 1990, *Ap&SS*, 171, 105

Komissarov, S. S. & Falle, S. A. E. G., 1996, *ASPC*, 100, 173

Komissarov, S. S. & Falle, S. A. E. G., 1998, *MNRAS*, 297, 1087

Komissarov, S. S. & Falle, S. A. E. G., 2003, *MNRAS*, 343, 1045

Komissarov, S. S., Barkov, M. V., Vlahakis, N., & Königl, A., 2007, *MNRAS*, 380, 15

Komissarov S. S., Vlahakis, N., Königl A., & Barkov, M. V., 2009, *MNRAS*, 394, 1182

Kovalev, Y. Y., Lister, M. L., Homan, D. C., & Kellermann, K. I., 2007, *ApJ Letters*, 668, 27

Laing, R. A. & Bridle, A. H., 2002, *MNRAS*, 336, 1161

Lister, M. L., Aller, M. F., Aller, H. D., *et al.*, 2013, *AJ*, 146, 120

Massaglia, S., Bodo, G., Rossi, P., Capetti, S., & Mignone, A., 2016, *arXiv:1609.02497*

Mathews, W. G. & Brighenti, F., 2003, *Ann.Rew.A&A*, 41, 191

Meyer, Eileen T., Sparks, W. B., Biretta, J. A., *et al.*, 2013, *ApJ Letters*, 774, 21

Moll, R., Spruit, H. C., Obergaulinger, M., 2008, *A&A*, 492, 621

Nakamura, M. & Asada, K., 2014, in *The Innermost Regions of Relativistic Jets and Their Magnetic Fields, EPJ Web of Conferences*, ed. Jos L. Gmez, v61, id.01004

Nulsen, P. E. J.., Young, Andrew, J. K., Ralph, P., McNamara, B. R., & Wise, M. W., 2015, *IAU Symposium Proceedings*, 313, 236

Owen, F. N. & Ledlow, M. J., 1994, *ASP Conference Series*, 54, 319

Phinney, E. S., 1983, *PhD Thesis, University of Cambridge*

Porth, O. & Komissarov, S. S., 2015, *MNRAS*, 452, 1089

Pushkarev, A. B., Kovalev, Y. Y., Lister, M. L., & Savolainen, T., 2009, *A&A*, 507, L33

de Ruiter, H. R., Parma, P., Fanti, C., & Fanti, R., 1990, *A&A*, 227, 351

Schreier, E. J., Gorenstein, P., & Feigelson, E. D., 1982, *ApJ*, 261, 42

Tchekhovskoy, A. & Bromberg, O., 2016, *MNRAS Letters*, 461, 46

Walker, R. C., Ly, C., Junor, W., & Hardee, P. J., 2008 , *JPhCS*, 131, 012053

New Frontiers in Black Hole Astrophysics
Proceedings IAU Symposium No. 324, 2016
Andreja Gomboc, ed.

doi:10.1017/S1743921317001272

Observational View of Magnetic Fields in Active Galactic Nuclei Jets

Talvikki Hovatta

Tuorla Observatory, Department of Physics and Astronomy, University of Turku
Väisäläntie 20, 21500 Kaarina, Finland
email: talvikki.hovatta@utu.fi

Abstract. According to the currently favored picture, relativistic jets in active galactic nuclei (AGN) are launched in the vicinity of the black hole by magnetic fields extracting energy from the spinning black hole or the accretion disk. In the past decades, various models from shocks to magnetic reconnection have been proposed as the energy dissipation mechanism in the jets. This paper presents a short review on how linear polarization observations can be used to constrain the magnetic field structure in the jets of AGN, and how the observations can be used to constrain the various emission models.

Keywords. polarization, galaxies: active, galaxies: jets, galaxies: magnetic fields

1. Introduction

Active galactic nuclei (AGN) are among the most extreme objects in the universe. They are the centers of distant galaxies hosting a supermassive black hole, billions of times the mass of the Sun. About 10% of AGN produce collimated, relativistic outflows or plasma jets that shine brightly over the entire electromagnetic spectrum from radio to gamma-ray energies. The formation and stability of these jets are not yet fully understood. It is believed that energy is extracted from the central black hole by large-scale magnetic fields (Blandford & Znajek 1977) while rotating inner regions of the accretion disk form an outflowing wind that helps to collimate the jet (Blandford & Payne 1982). In this picture, the jet is launched as a magnetically dominated outflow with strong magnetic fields accelerating the flow to relativistic velocities. General relativistic magnetohydrodynamic simulations show that the efficiency of the jet generation can be very high if the black hole is threaded by a dynamically important magnetic field (Tchekhovskoy *et al.* 2011).

While the simulations suggest that the jets are likely highly magnetized near the black hole, their composition and magnetic field structure in the parsec scales, where they produce most of the emission, is much less clear. A link between the magnetic fields near the black hole and the parsec-scale jets was established by Zamaninasab *et al.* (2014) who showed that there is a correlation between the magnetic flux of the jet and the accretion disk luminosity, as predicted by the magnetically arrested disk (MAD) model (Tchekhovskoy *et al.* 2011). This result illustrates how studying the emission regions in the parsec-scale jets may help us gain knowledge about the jet formation processes.

One of the main open questions is the nature of the emission region. Traditionally, the flaring behavior of AGN jets has been well explained by shock-in-jet models, especially in the radio bands (e.g., Hughes *et al.* 1985, Marscher & Gear 1985). These models assume the jet to be kinetically dominated at parsec scales. Shock models struggle to explain the very fast variability detected at high-energy gamma-ray bands, and models involving magnetic reconnection have been invoked to explain the high-energy emission (e.g., Giannios *et al.* 2009). In this case the jet flow should be magnetically dominated,

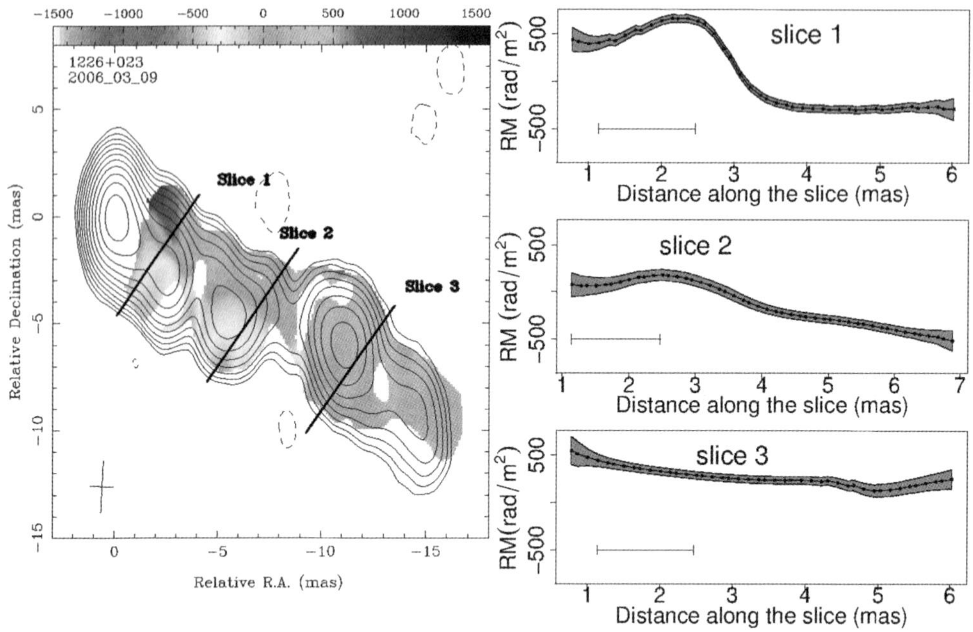

Figure 1. Faraday rotation map of 3C 273 obtained with the VLBA at 8 - 15 GHz. A rotation measure gradient is seen extending the full length of the jet. This can be interpreted as a signature of a helical magnetic field. Figure adapted from Hovatta *et al.* (2012).

although recent particle-in-cell simulations show that even in reconnection models, the magnetic field and particle energies can be in equipartition at the emission region (Sironi *et al.* 2015).

Another major open question is the structure of the magnetic field at the emission site. According to the simulations, the magnetic field is helical close to the black hole but whether it preserves its order in the parsec scales is debated upon. For example, in the model of Marscher *et al.* (2008), the magnetic field is helical at the acceleration and collimation zone of the jet but gets disrupted and turbulent beyond a standing shock in the jet. Polarization observations of some jets, however, indicate that the magnetic field could be helical even on the parsec scales (e.g., Asada *et al.* 2002, Gabuzda *et al.* 2004, Hovatta *et al.* 2012). Figure 1 shows an example adapted from Hovatta *et al.* (2012).

In this contribution, I will discuss how linear polarization observations in the radio and optical bands can be used to constrain jet emission models and the magnetic field structure of the jets. This is not intended as a comprehensive review of all possible models, but more to give some examples on how observations are used to study magnetic fields in AGN jets.

2. Polarization observations

The optical and radio emission of AGN jets is synchrotron emission, which is intrinsically highly polarized. In an optically thin emission region with a uniform magnetic field, the polarization degree is up to 70% (e.g., Pacholczyk 1970), although such high polarization degree values are not typically seen in AGN jets in the radio (e.g., Aller *et al.* 2003, Lister & Homan 2005) or optical (e.g., Angel & Stockman 1980, Pavlidou *et al.* 2014) bands. This has been taken as evidence for disordered magnetic fields. The emission in AGN jets is often described with the Stokes parameters I (for total intensity), Q and U (for linear polarization) and V (for circular polarization). I will only discuss linear polarization here. Using the Stokes parameters, the polarization degree and the electric vector position angle (EVPA) can be defined as $m = (\sqrt{Q^2 + U^2})/I$ and EVPA$= 1/2\tan^{-1}(U/Q)$. In the simplest case, in an optically thin jet, the polarization position angle is perpendicular to the magnetic field direction and one can use the EVPA observations to infer the direction of the magnetic field. However, one should note that in relativistic jets viewed at a small angle to the line of sight of the observer, the situation is more complicated (Lyutikov *et al.* 2005).

Another way to obtain information about magnetic fields in blazar jets is via Faraday rotation observations. When synchrotron radiation passes through magnetized plasma, Faraday rotation of the EVPAs proportional to the line-of-sight magnetic field and electron density occurs (e.g. Burn 1966). In the simplest case, the effect can be described by a linear dependence between the observed electric vector position angle (EVPA; $\chi_{\rm obs}$) and wavelength squared (λ^2), given by

$$\chi_{\rm obs} = \chi_0 + 0.81 \int n_e \mathbf{B} \cdot \mathrm{dl} = \chi_0 + \mathrm{RM}\lambda^2, \qquad (2.1)$$

where χ_0 is the intrinsic EVPA and RM is the rotation measure (in rad/m^2), related to the electron density n_e of the plasma (in cm^{-3}) and the magnetic field component $\mathbf{B}$ (in μG) along the line of sight (in parsecs). The RM can thus be estimated by observing the EVPA at different frequencies. This will give us information on the line-of-sight component of the magnetic field. For example, if the rotation measure is positive, the magnetic field is coming towards the observer, and if negative it is going away from the observer. Therefore it was suggested that a gradient in a Faraday rotation measure transverse to the jet direction could reveal helical magnetic field structures (Blandford 1993).

3. Constraining emission models through polarimetry

Polarization observations of AGN jets in the radio and optical bands have been conducted since the 1970s (see e.g., Saikia & Salter 1988 for a review). At the same time, a large number of theoretical models were developed to explain the total intensity and polarization observations of AGN jets (e.g, Jones & O'Dell 1977, Blandford & Königl 1979, Laing 1980). In the 1980s it was suggested that most of the radio variability, especially in the cm-band, are due to shocks in the relativistic jets where the magnetic field is predominantly turbulent (Hughes *et al.* 1985, Marscher & Gear 1985, Jones *et al.* 1985).

The field of polarization modeling has received a new boost since an exciting connection to high-energy gamma-ray emission was found (Marscher *et al.* 2008). A rotation in the optical polarization angle was coincident with an ejection of a new very long baseline interferometry (VLBI) component from the core, and a very-high-energy (VHE; $>$ 100 GeV) detection of the source BL Lac by the MAGIC telescope. A further connection between the optical polarization and gamma-ray flares was suggested in two quasars 3C 279 (Abdo *et al.* 2010) and PKS 1510-089 (Marscher *et al.* 2010, Aleksić *et al.* 2014). In both of these the optical polarization angle was seen to rotate by more than 180 degrees over a course of 20 to 50 days during which a sharp gamma-ray flare was observed. Following all these observations, a number of new models have been published in the last years (e.g., Marscher 2014, Zhang *et al.* 2014, Hughes *et al.* 2015, Zhang *et al.* 2015). I will go through some of these models with the emphasis on the observations that can be used to constrain them.

3.1. *Turbulence in the jets*

Turbulent magnetic field produces stochastic variations to the polarization angle and degree (Jones *et al.* 1985, Marscher 2014), such as an EVPA rotation of any length, typically accompanied by a low polarization degree. The challenge in constraining these types of models is the stochastic nature of the variations so that it is not possible to fit the individual light curves (Marscher 2014). Instead, one should observe a large number of objects and compare the statistical properties of the variations with the turbulent models.

To study the connection between high-energy emission and optical polarization in a statistical manner, the RoboPol program was initiated in 2013 (Pavlidou *et al.* 2014). The RoboPol instrument is mounted on the 1.3-m telescope in Skinakas Observatory in Crete, where the observing season lasts from April until November. During its first three observing seasons, RoboPol observed about 100 AGN twice per each week in search of optical polarization angle rotations. The main goal was to study the differences in gamma-ray detected and non-detected objects, and to search for a connection between rotations and flaring behavior.

RoboPol has detected 40 optical polarization angle rotations during the first three observing seasons, tripling the number of known rotations (Blinov *et al.* 2015, Blinov *et al.* 2016a, Blinov *et al.* 2016b). Comparing the rotations with random walk simulations showed that while stochastic variability for any individual rotation could not be ruled out, it was highly unlikely that all of them would be of random walk origin. Kiehlmann *et al.* (2016) suggested that the smoothness of the rotation could be used as an additional indicator when comparing the rotations with random walk models. They also illustrated the importance of good sampling in polarization observations by revealing how additional data changed the rotation reported by Abdo *et al.* (2010) in the quasar 3C 279.

Figure 2 shows an example of a EVPA rotation in the source 1ES 1727+502 studied in Hovatta *et al.* (2016). Based on random walk simulations, conducted in the same manner as in Kiehlmann *et al.* (2016), a stochastic origin for the rotation cannot be excluded.

3.2. *Emission in a helical field*

A rotation in the polarization angle could also be due to an emission feature tracing a magnetic field in the jet as suggested by Marscher *et al.* (2008) and Marscher *et al.* (2010). In this model, the magnetic field in the jet is helical in the acceleration and collimation zone, which is probed by the optical band, and a rotation is seen when the emission feature moves along the magnetic field. The rotation may be accompanied by flaring in other bands when the emission feature reaches a standing shock in the jet.

Another alternative is a shock moving down a jet with a helical field, in which case the EVPA rotation would be due to light travel time effects when parts of the shock are seen at different times (Zhang *et al.* 2014). This model was used to successfully fit the EVPA rotation in the quasar 3C 279 (Zhang *et al.* 2015), although one should bear in mind the caveat that with the additional data from Kiehlmann *et al.* (2016), the rotation was not as long as originally presented by Abdo *et al.* (2010).

Whether the magnetic field in the jet is helical, especially in regions beyond the acceleration and collimation zone, is still unclear. As explained in Sect. 2, observations of a Faraday rotation measure gradient across the jet could be an indication of such a field. First signatures of a helical magnetic field were observed in the quasar 3C 273 using multifrequency very long baseline array (VLBA) observations (Asada *et al.* 2002, Zavala & Taylor 2005). Many more claims of such gradients have been made (e.g., Gabuzda *et al.* 2004) but the issue has been controversial due to the limited resolution across the jets in blazars (Taylor & Zavala 2010).

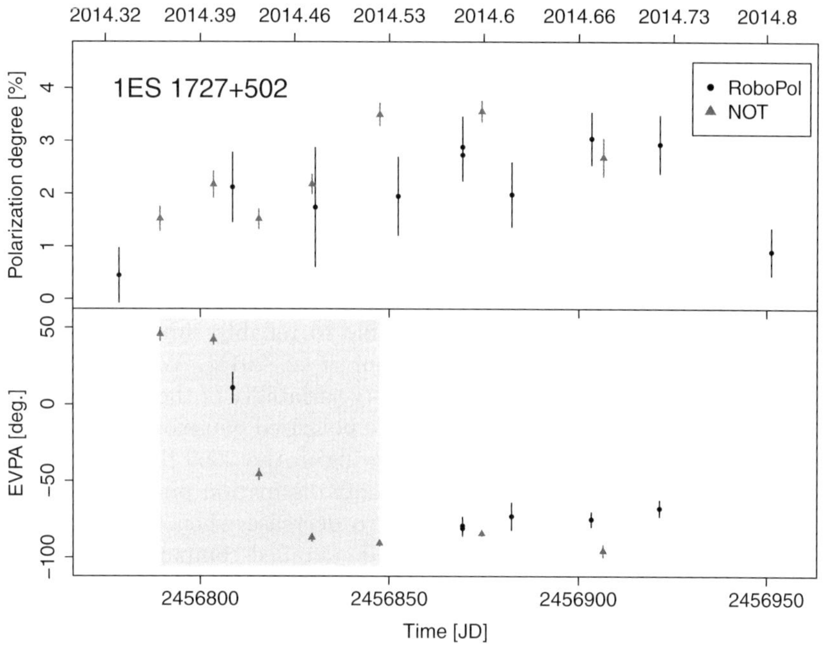

Figure 2. Polarization degree and EVPA of 1ES 1727+502 in 2014 observed with RoboPol and the Nordic Optical Telescope (NOT). The EVPA shows a rotation of nearly 150 degrees in about 60 days. Figure adapted from Hovatta *et al.* (2016).

In Hovatta *et al.* (2012), we performed Monte Carlo simulations to quantify the significance of the rotation measure gradients in the large sample of parsec-scale jets in the MOJAVE (Monitoring of Jets in Active galactic nuclei with VLBA Experiments) sample. Our observations confirmed the gradient in 3C 273 (see Fig. 1), and we also found significant gradients in three other quasars. In Zamaninasab *et al.* (2013) we modelled the gradient of the quasar 3C 454.3 using a magnetic field with both helical and turbulent components. In the other two sources, the gradient span only over a small portion of the jet and detailed modeling was not possible.

Nowadays performing simulations to quantify the significance of the gradients is a common practice (e.g., Murphy & Gabuzda 2013) and the number of significant gradients has steadily increased (e.g., Gabuzda *et al.* 2015). However, rotation measure gradients can also arise from changes in the density of the Faraday rotating material (e.g. Gómez *et al.* 2011) so that detailed modeling of the gradients should be done in order to confirm that they are indeed due to helical fields. It is also unclear whether the helical field is within the jet or in a sheath layer around the jet because Faraday rotation is a propagation effect originating in the plasma outside the emission region.

3.3. *Shock in a jet*

As stated earlier, shock-in-jet models have been successful in reproducing the observed flaring especially in the radio bands. Typically, radiative transfer simulations are generated to estimate the parameters of the shocks (e.g. Hughes *et al.* 2011, Hughes *et al.* 2015) and these are then compared with observations at multiple bands (Aller *et al.* 2014). For example, as shown by Hughes *et al.* 2015, the polarization degree and range of EVPA values can be used to constrain the magnetic field geometry and jet orientation, while total intensity behavior is indistinguishable in different models.

Interestingly, although the magnetic field is assumed to be predominantly turbulent, in some cases an ordered field component (possibly helical) could also be present (Aller *et al.* 2016). This supports the findings of e.g., Zamaninasab *et al.* (2013), Blinov *et al.* (2015), and Kiehlmann *et al.* (2016), who also suggest that there are both turbulent and ordered magnetic field components / deterministic variations in the jets.

3.4. *Magnetic reconnection*

Magnetic reconnection (see Kagan *et al.* 2015 for a review) is a new hot topic in the field. It was originally invoked to explain the fast high-energy variability of blazars through the jet-in-jet model (Giannios *et al.* 2009). In the recent years, particle-in-cell simulations have evolved significantly, and it is now possible to reliably simulate the complicated structure of the reconnection layer (e.g., Sironi *et al.* 2015). While the reconnection models can already be compared to total intensity variability of the sources (Petropoulou *et al.* 2016), there are no explicit models for the polarized emission.

Zhang *et al.* (2015) stated that the EVPA swing in the 2009 flare of 3C 279 was reproduced by a model that "favors magnetic energy dissipation process during the flare" because the magnetic field strength was seen to decrease when the total intensity increased. However, this model did not yet include detailed comparisons to a specific reconnection model. Considering the growing interest in the magnetic dissipation models, it is likely that in the next few years more detailed models with observational predictions will become available.

3.5. *Statistical trends*

From the previous sections, it is clear that polarization observations can be used to constrain various types of emission models and the magnetic field structure in the jets. In order to generalize the findings in individual sources into the AGN jet population, a statistical approach must be used. This is what the RoboPol program aims for by observing a sample of about 100 sources with high cadence. RoboPol has found, for example, that any class of blazars from low to high synchrotron peaking objects can show rotations (see also Jermak *et al.* 2016, Hovatta *et al.* 2016). However, there seems to be a specific class of "rotators" that do so more often, and rotations are more common in the low spectral peaking objects (Blinov *et al.* 2016b).

The low synchrotron peaking objects also have higher optical polarization degree than the high-peaking objects (Angelakis *et al.* 2016), a trend that has also been seen in the smaller sample studied in Jermak *et al.* (2016) and earlier in the radio band (Lister *et al.* 2011). These general trends can be explained with a simple, qualitative model where a shock moves down a jet, which has both helical and turbulent magnetic field components (see Angelakis *et al.* 2016 for details). Any model put forward to explain the polarization behavior in individual flares, should also account for these general trends.

4. Future directions

One challenge in connecting the observations of magnetic fields to the theory of jet formation is that often the observations, especially in the radio bands, probe regions $10^3 - 10^5$ gravitational radii away from the black hole. Optical observations do not suffer from this restriction, but for example, Faraday rotation with λ^2 wavelength dependence, is not typically seen at optical wavelengths. A solution can be provided by going to millimeter-band observations, as demonstrated by Martí-Vidal *et al.* (2015). They used ALMA polarization observations to probe the Faraday rotation at the jet base of the lensed quasar PKS 1830-211, and found extremely high Faraday rotation of $10^8 \mathrm{rad/m}^2$, which is two orders of magnitude higher than previous observations in other sources (e.g., Plambeck *et al.* 2014). They inferred this as a signature of a very high magnetic field at the base of the jet, in support of magnetically launched jet models. Whether similar high RM values are a common property of quasars remains to be seen with future ALMA observations. Especially interesting will be future observations using ALMA as part of the global VLBI array, which may be able to spatially resolve the regions of high magnetic fields.

A major leap forward will come with the future X-ray polarization missions because the amount of polarization for different high-energy emission mechanisms in AGN jets, for example between leptonic and hadronic models, is very different (Zhang & Böttcher 2013), and X-ray polarization observations can be used to distinguish between the models.

5. Summary

Magnetic fields in AGN jets can be probed through polarization observations, which are most easily done at radio and optical bands. Modeling the polarization degree and position angle behavior can be used to test different blazar emission models, while Faraday rotation observations at radio, and now also in the millimeter bands, can be used to probe the line-of-sight magnetic field component. A statistical approach is favored when connecting the results to different AGN jet populations.

References

Abdo, A. A., Ackermann, M., Ajello, M., *et al.* 2010, *Nature*, 463, 18
Aleksić, J., Ansoldi, S., Antonelli, L. A., *et al.* 2014 *A&A*, 569, 46
Aller, M. F., Aller, H. D., & Hughes, P. A. 2003, *ApJ*, 586, 33
Aller, M. F., Hughes, P. A., Aller, H. D., *et al.* 2014, *ApJ*, 791, 51
Aller, M. F., Hughes, P. A., Aller, H. D., *et al.* 2016, *Galaxies*, 4, 35
Angel, J. R. P. & Stockman, H. S. 1980, *ARA&A*, 18, 321
Angelakis, E., Hovatta, T., Blinov, D., *et al.* 2016 *MNRAS*, 463, 3365
Asada, K. Inoue, M., Uchida, Y., *et al.* 2002, *PASJ*, 54, L39
Blandford, R. 1993, in: M. Burgarella, M. Livio, & C. P. O'Dea (eds.) *Astrophysical Jets*, (Astrophysics and Space Science Library, Vol. 103; Cambridge: Cambridge Univ. Press), 15
Blandford, R. D. & Königl, A. 1979, *ApJ*, 232, 34
Blandford, R. D. & Payne, D. G. 1982, *MNRAS*, 199, 883
Blandford, R. D. & Znajek, R. L. 1977, *MNRAS*, 179, 433
Blinov, D., Pavlidou, V., Papadakis, I., *et al.* 2015, *MNRAS*, 453, 1669
Blinov, D., Pavlidou, V., Papadakis, I. E., *et al.* 2016a, *MNRAS*, 457, 2252
Blinov, D., Pavlidou, V., Papadakis, I. E., *et al.* 2016b, *MNRAS*, 462, 1775
Burn, B. J. 1966, *MNRAS*, 133, 67
Gabuzda D. C., Knuettel, S. & Reardon, B. 2015 *MNRAS*, 450, 2441
Gabuzda, D. C., Murray, E., & Cronin, P. 2004, *MNRAS*, 351, L89
Giannios, D., Uzdensky, D. A., & Begelman, M. C. 2009, *MNRAS*, 359, L29
Gómez, J.-L., Roca-Sogorb, M., Agudo, I., *et al.* 2011, *ApJ*, 733, 11
Hovatta, T., Lister, M. L., Aller, M. F., *et al.* 2012, *AJ*, 144, 105
Hovatta, T., Lindfors, E., Blinov, D., *et al.* 2016 *A&A*, 596, 78
Hughes, P. A., Aller, H. D., & Aller, M. F. 1985, *ApJ*, 298, 301
Hughes, P. A., Aller, M. F., & Aller, H. D. 2011, *ApJ*, 735, 81
Hughes, P. A., Aller, M. F., & Aller, H. D. 2015, *ApJ*, 799, 207
Jermak, H., Steele, I., Lindfors, E., *et al.* 2016, *MNRAS*, 462, 4267
Jones, T. W. & O'Dell, S. 1977, *ApJ*, 215, 236
Jones, T. W., Rudnick, L., Aller, H. D., *et al.* 1985, *ApJ*, 290, 627

Kagan, D., Sironi, L. Cerutti, B., & Giannios, D. 2015, *Space Science Reviews*, 191, 545
Kiehlmann, S., Savolainen, T., Jorstad, S. G., *et al.* 2016, *A&A*, 590, A10
Laing, R. A. 1980, *MNRAS*, 193, 439
Lister, M. L. & Homan, D. C. 2005, *AJ*, 130, 1389
Lister, M. L., Aller, M. F., Aller, H. D., *et al.* 2011, *ApJ*, 742, 27
Lyutikov, M., Pariev, V. I., & Gabuzda, D. C. 2005, *MNRAS*, 360, 869
Marscher, A. P. 2014, *ApJ*, 780, 87
Marscher, A. P. & Gear, W. K. 1985, *ApJ*, 298, 114
Marscher, A. P., Jorstad, S. G., D'Arcangelo, F. D., *et al.* 2008, *Nature*, 452, 966
Marscher, A. P., Jorstad, S. G., Larionov, V. M., *et al.* 2010, *ApJ*, 710, L126
Martí-Vidal, I., Muller, S., Vlemmings, W., Horellou, C., & Aalto, S. 2015, *Science*, 348, 311
Murphy E. & Gabuzda D. C. 2013, in *The Inner-most Regions of Relativistic Jets and Their Magnetic Fields*, EPJ Web of Conferences, Volume 61, id.07005
Pacholczyk, A. G. 1970, *Radio Astophysics, Nonthermal Processes in Galactic and Extragalactic Sources* (San Francisco, CA: Freeman)
Pavlidou, V., Angelakis, E., Myserlis, I., *et al.* 2014, *MNRAS*, 442, 1693
Petropoulou, M., Giannios, D., & Sironi, L. 2016, *MNRAS*, 462, 3325
Plambeck, R. L., Bower, G. C., Rao, R., *et al.* 2014, *ApJ*, 797, 66
Saikia, D. & Salter, C. 1988, *Ann. Rev. in Astron. & Astrophys.* , 26, 93
Sironi, L., Petropoulou, M., & Giannios, D. 2015, *MNRAS*, 450, 183
Taylor, G. B. & Zavala, R. T. 2010, *ApJ*, 722, L183
Tchekhovskoy, A., Narayan, R., & McKinney, J. C. 2011, *MNRAS*, 418, L79
Zamaninasab, M., Clausen-Brown, E., Savolainen, T., & Tchekhovskoy, A. 2014, *Nature*, 510, 126
Zamaninasab, M., Savolainen, T., Clausen-Brown, E., *et al.* 2013, *MNRAS*, 436, 3341
Zavala, R. T. & Taylor, G. B. 2005, *ApJ*, 626, L73
Zhang, H. & Böttcher, M. 2013, *ApJ*, 774, 18
Zhang, H. Chen, X., & Böttcher, M. 2014, *ApJ*, 789, 66
Zhang, H. Chen, X., Böttcher, M., *et al.* 2015, *ApJ*, 804, 58

New Frontiers in Black Hole Astrophysics
Proceedings IAU Symposium No. 324, 2016
Andreja Gomboc, ed.

doi:10.1017/S1743921317001739

Ultrafast VHE Gamma-Ray Flares of IC 310

Maxim V. Barkov[1,2], **Felix Aharonian**[3,4] **and Dmitriy V. Khangulyan**[5]

[1]DESY, Platanenallee 6, 15738 Zeuthen, Germany
email: maxim.barkov@desy.de

[2]Astrophysical Big Bang Laboratory, RIKEN, 2-1 Hirosawa, Wako, Saitama 351-0198, Japan
email: maxim.barkov@riken.jp

[3]Dublin Institute for Advanced Studies, 31 Fitzwilliam Place, Dublin 2, Ireland

[4]Max-Planck-Institut für Kernphysik, Saupfercheckweg 1, D-69117 Heidelberg, Germany

[5]Department of Physics, Rikkyo University, Nishi-Ikebukuro 3-34-1, Toshima-ku, Tokyo 171-8501, Japan

Abstract. In 2012 November MAGIC detected a bright flare from IC 310. The flare consisted of two sharp peaks with a typical duration of $\sim 5\,\mathrm{min}$. The energy released during that event has been estimated to be at the level of $2 \times 10^{44}\,\mathrm{erg\,s^{-1}}$.

In this work we derive an upper limit on the possible luminosity of flares generated in black hole (BH) magnetosphere, which depends very weakly on the mass of BH and is determined by disk magnetisation, viewing angle, and pair multiplicity. Since all these parameters are smaller than a unit, the luminosity 2×10^{43} erg s^{-1} can be taken as a strict upper limit for flare luminosity for several minutes variability time. This upper limit appears to be approximately an order of magnitude below the value measured with MAGIC. Thus, we conclude that it seems very unfeasible that the magnetospheric processes can be indeed behind the bright flaring activity recorded from IC 310.

Keywords. gamma rays: theory, galaxies: active, galaxies: active

---

1. Introduction

The gamma-ray emission of blazars is strongly variable with fluxes which match well the sensitivity of the *Fermi*/LAT in the MeV/GeV band and the current arrays of imaging atmospheric Cherenkov telescopes (IACT) like H.E.S.S., MAGIC, and *Veritas* in the VHE band. During the strongest flares of BL Lac objects like Mkn 421, Mkn 501 and PKS 2155-304 the energy fluxes of VHE gamma-rays can exceed $f_{\rm VHE} = 10^{-10}$ erg/cm^{2}s. Such fluxes can be studied with IACT arrays which provide huge, as large as 10^9 cm^2, detection areas, in almost background free regime with a detection rate exceeding 1 Hz. This allows variability studies on timescales of minutes. The fluxes of flares in MeV/GeV gamma-rays from powerful quasars like 3C 279 and 3C 454.3 can be significantly larger, $f_{\rm VHE} = 10^{-8}$ erg/cm^{2}s. But because of the limited detection area of *Fermi*/LAT and *AGILE* ($\simeq 1$ m^2), the variability studies can be performed on hour timescales. It is interesting to compare these timescales with the minimum time that characterises a black hole system as an emitter, namely, the light crossing time of the gravitational radius of the black hole:

$$\tau_0 = r_{\rm g}/c = 5 \times 10^3 M_9\,\mathrm{s}. \tag{1.1}$$

Note that $r_{\rm g} = GM/c^2 = 1.5 \times 10^{14} M_9\,\mathrm{cm}$ is the minimum value for the gravitational radius corresponding to the extreme Kerr black hole (i.e., twice smaller than the Schwarschild radius).

Thus, for a typical mass range of black holes in powerful AGN, $M \geqslant 10^8 M_\odot$, the IACT arrays have a unique potential for exploring the physics of black holes close to the event horizon on timescales significantly shorter than τ_0. Remarkably, such ultra-fast VHE gamma-ray flares events have been already detected from at least three AGN†, PKS 2155−304 (Aharonian *et al.* 2007), Mkn 501 (Albert *et al.* 2007), and IC 310 (Aleksić *et al.* 2014). The GeV gamma-ray flares detected on timescales of minutes from 3C 279 (Ackermann *et al.* 2016) and hours from 3C 279 (Hayashida *et al.* 2015) and 3C 454.3 (Abdo et al. 2011), are close to τ_0 (the black hole masses in these objects are estimated in the intervals $(0.3 - 0.8) \times M_9$ (Woo & Urry 2002; Gu *et al.* 2001; Nilsson *et al.* 2009) and $(0.5 - 4) \times M_9$ (Gu *et al.* 2001; Bonnoli *et al.* 2011), respectively). For comparison, the characteristic timescales of the even shortest (~ 1 ms; Pozanenko & Loznikov 2002; Golkhou *et al.* 2015) gamma-ray burst events which associate, most likely, with solar mass black holes, exceed τ_0 by several orders of magnitude.

The detection of the variable VHE gamma-ray emission from AGN on timescales significantly shorter than τ_0 is an extraordinary result and requires a careful treatment and interpretation. If the emission is produced in a relativistically moving source the variability time-scale seen by an observer is shorten by the Doppler factor $\delta_{\rm em} = 1/\Gamma_{\rm em}(1 - \beta\cos\theta_{\rm em})$, where $\Gamma_{\rm em} = 1/\sqrt{(1-\beta_{\rm em}^2)}$ and $\theta_{\rm em}$ are bulk Lorentz factor of the emitting source and the angle between its velocity and the line of sight, respectively. Thus if we want to increase the proper size of the emitter, R', (i.e. the source size in co-moving reference frame) to a reasonable value of $R' \geqslant r_{\rm g}$, the Doppler factor should exceed, for example, in the case‡ of PKS 2155−304, $\delta_{\rm em} = 25$.

However, there is another issue of conceptual importance which cannot be ignored. The problem is that if the jet perturbations propagate from the central engine, e.g., in the form of sequence of blobs ejected with different Lorentz factors (leading to internal shocks), the size of the emitter in the laboratory frame does not depend on the Doppler factor. Let us define the proper size of the production region as $R' = \lambda\Gamma_{\rm j} r_{\rm g}$, where $\Gamma_{\rm j}$ and λ are the jet bulk Lorentz factor and a dimensionless parameter defining the size of the production region that should exceed 1 ($\lambda \geqslant 1$) for the conventional assumptions. Combining the causality requirement and the limitation on the variability timescale, we have

$$t_{\rm var} \geqslant \tau_0 \frac{\lambda\Gamma_{\rm j}}{\Gamma_{\rm em}}\,. \tag{1.2}$$

The observational requirement of $t_{\rm var} = 0.04\tau_0$ inferred from the VHE flares of PKS 2155−304 (Aharonian *et al.* 2007) requires $\Gamma_{\rm em} \simeq 25\lambda\Gamma_{\rm j}$, i.e., the emitter moves relativistically in the frame of the jet which, in its turn, moves relativistically towards the observer. The "jet in the jet" model suggested by Giannios *et al.* (2009) can be consider as a possible realization of this general scenario. Alternatively, i.e., if the source of the flare does not move relativistically in respect to the jet ($\Gamma_{\rm em} \simeq \Gamma_{\rm j}$), the size of the source should be much less than the black hole's gravitational radius: $\lambda \simeq 0.04$. There are at least two reasons to invalidate the requirement of $\lambda > 1$: (1) the flare might originate in a small fraction of the BH magnetosphere, or (2) the perturbation that leads to the flare might have an *external* origin, i.e. not directly related to the black hole. An analogy for the first scenario could be emission of radio loud pulsars. In these objects the radio pulses are believed to originate in the polar cap region that substitute only a small fraction of the pulsar surface. Note that for the typical pulsar radius $R_{\rm psr}$ of 10 km, $\tau_0 = R_{\rm psr}/c \sim 30\mu$s is too small to be probed through the variability of radio emission.

† For a recent summary of such systems see Vovk & Babić (2015)

‡ This estimate differs by a factor of 2 from the one derive in Aharonian *et al.* (2007) since in that paper one used the Schwarzschild radius for the estimate while we use the Kerr radius.

The second possibility can be realized if a star or a cloud penetrates to the jet from outside (Barkov *et al.* 2010, 2012a,b; Bosch-Ramon *et al.* 2012; Khangulyan *et al.* 2013). Since the star radius $R_* \ll r_{\rm g}$, this scenario may initiate perturbations on scales smaller than the black hole's gravitation radius $r_{\rm g}$ (Barkov *et al.* 2012a).

Note that while all these three scenarios can explain, in principle, the detected time structures on scales $t_{\rm var} < \tau_0$, there are important additional requirements that should be fulfilled: (1) the overall energy budget should be feasible, (2) the emitter should be optically thin, i.e. gamma-rays should be able to escape their production region without significant absorption, (3) the proposed radiation mechanism should be able to explain the detected spectral features of gamma-radiation.

In this paper we discuss the fisibility of the source of the flare which could appear in a magnetospheric gap occupying a small volume in the proximity of the black hole close to the event horizon (Neronov & Aharonian 2007; Levinson & Rieger 2011);

2. Addressing the "sub-horizon" scale variability

Several groups argued that gamma-ray flares can be generated immediately in the magnetosphere of the central BH (see, e.g., Levinson & Rieger 2011). However, in certain cases one does not validate feasibility of such a scenario from the point of the available power (see, e.g., Aleksić *et al.* 2014) or the minimum variability time scale (see, e.g., Hirotani & Pu 2015) . While in the case of flare production in relativistically moving blobs the apparent luminosity of the emission is strongly enhanced by the Doppler boosting and beaming effects, the emission generated in the magnetosphere allows to evaluate directly the true power spent for the flare production. Moreover, since the gap should collapse from both sides (Timokhin 2010), the observed variability time scale provided an upper limit on the size of the gap: $h < t_{\rm var}c$. As it is shown below, these two simple facts make possible a conclusion that the magnetospheric origin for the gamma-ray emission meets energy limitations that clearly exclude production of bright flares in the vicinity of the BH. Importantly, the derived upper limit depends only marginally on the central BH mass, thus detection of such events possible only from the nearby supermassive BH, e.g. from M87 as suggested by Levinson & Rieger (2011) or even from the Galactic center (we note here that due to unfavorable orientation of the line of sight the emission from the Galactic center may be severely attenuated). The later constraint has been ignored by Hirotani & Pu (2015) for explanation of a very short flare detected with MAGIC from IC 310 (see, e.g., Aleksić *et al.* 2014).

The electrical field in the gap can be estimated as

$$E \sim B_{\rm bh}\frac{R\Omega_{\rm F}\sin\theta}{c}\,, \tag{2.1}$$

where $B_{\rm bh}$ is the magnetic field at the BH horizon, $\Omega_{\rm F}$ is the angular velocity of the frame, R is radius, and θ is polar angle. The gap potential drop can be estimated as

$$\Delta V \lesssim hB_{\rm bh}\frac{R\Omega_{\rm F}\sin\theta}{c}\,, \tag{2.2}$$

where $h \sim t_{\rm var}c < r_{\rm g}$ is the thickness of the gap. An estimated difference by a factor h/R from Eq. 2.2 is often used for the gap potential (see, e.g., Blandford & Znajek 1977; Levinson & Rieger 2011). The nature of the difference between these two estimates is related to influence of magnetospheric charges located outside the gap. The estimate given by Eq. 2.2 does not account for this contribution, and since the external charges, even if they are located outside the gap, tend to decrease the electrical field in the gap, Eq. 2.2

provides a strict upper limit on the gap potential (see also Broderick & Tchekhovskoy 2015).

The luminosity of the particles accelerated in the gap can be estimated as (Levinson & Rieger 2011)

$$L_\gamma < 4\pi R^2 c\kappa\rho_{\rm GJ}\Delta V\,, \tag{2.3}$$

where κ and $\rho_{\rm GJ} = \Omega_{\rm F} B_{\rm bh}\sin\theta/(2\pi c)$ are pair multiplicity and Goldreich-Julian charge density, respectively. For an efficient operation of the gap acceleration an obvious condition $\kappa \ll 1$ should be fulfilled.

For a Kerr BH with maximum angular momentum the angular velocity $\Omega_{\rm F}$ can be estimated as:

$$\frac{\Omega_{\rm F}}{c} \simeq \frac{1}{4r_g}\,. \tag{2.4}$$

Thus, taking $R = r_{\rm g}$ (otherwise, one can account for the drop of the magnetic field strength away from the horizon) one obtains

$$L_\gamma < \frac{1}{8}\kappa B_{\rm bh}^2 r_{\rm g} hc\sin^2\theta\,. \tag{2.5}$$

We should note that if $h \to r_{\rm g}$ the luminosity estimate provided by Eq. (2.5) (after averaging over polar angle θ) exceeds by a factor of 2 the luminosity of the Blandford-Znajek mechanism (Blandford & Znajek 1977). This indicates that Eq. (2.5) can be taken as a safe upper limit for the available power.

For a minute variability time-scale, $t_{\rm var} = 5\,t_{\rm var,5}$ min, the gap thickness $h = 10^{13} t_{\rm var,5}$ cm is small as compared to the gravitational radius of supermassive BH with $M_8 > 1$, the estimated gamma-ray luminosity cannot exceed the following value:

$$L_\gamma < 5\times 10^{43}\kappa B_4^2 M_8 t_{\rm var,5}\sin^2\theta\,{\rm erg\,s^{-1}}\,, \tag{2.6}$$

where $B_{\rm bh} = 10^4 B_4$ G.

Equation (2.6) contains two parameters that are determined by the condition in the close vicinity of the BH: pair multiplicity, κ, and the magnetic field strength, B. Importantly, these parameters are essentially defined by the same process: by the accretion rate. Magnetic field at the BH horizon needs to be supported by accretion disk, therefore the field strength is directly determined by the accretion rate. On the other hand, the accretion rate also defines the intensity of photon fields in the magnetosphere, and consequently the multiplicity parameter (see, e.g., Levinson & Rieger 2011). The multiplicity parameter should remain strictly below unity, since otherwise the gap potential will be completely screened, this implies an upper limit on the accretion rate, and consequently on the magnetic field strength.

In previous works by Levinson & Rieger (2011); Aleksić *et al.* (2014) the estimated value of the maximum accretion rate compatible with existence of a vacuum gap in magnetosphere was

$$\dot m < 3\times 10^{-4} M_8^{-1/7}\,, \tag{2.7}$$

where $\dot m$ is accretion rate measured in the Eddington units:

$$\dot M_{\rm edd} = \frac{4\pi m_p G M_{\rm BH}}{\eta c\sigma_t}\,. \tag{2.8}$$

Here m_p and σ_t are mass of proton and Thompson cross-section, respectively. Efficiency of the accretion, η, determines a fraction of gravitational energy that is radiated by the accretion flow. To derive the estimate provided by Eq. (2.7), Levinson & Rieger (2011) adopted a value of $\eta = 0.1$. Levinson & Rieger (2011) also have provided an estimate for magnetic field strength at the BH horizon as

$$B_{\rm bh} = 1.3 \times 10^5 \left(\dot{m}/M_8\right)^{1/2} \ {\rm G}\,, \tag{2.9}$$

where we rescaled the numerical coefficient to the normalization used through out our paper. Therefore, one obtains an estimate on the maximum luminosity achievable at magnetospheric gap as

$$L_\gamma < 3 \times 10^{42} \kappa M_8^{-1/7} t_{\rm var,5} \sin^2\theta \, {\rm erg\,s^{-1}}\,. \tag{2.10}$$

The numerical coefficient in the above estimate differs (100 times smaller) from the value obtained by Levinson & Rieger (2011), which is related to the fact that we used Kerr radius while Levinson & Rieger (2011) provided the estimate for the Schwardschild radius. Such an approach is justified in the case of M87 given the relatively low gamma-ray flux and large overall uncertainties in this estimate. In other cases, e.g., for IC 310, the energy budget is much tighter, therefore below we provide a somewhat more accurate treatment of the case of magnetosphere around a Kerr BH. For example, the accretion efficiency of a flow accreted by a Kerr BH should be rather high $\eta = 0.3$ (see, e.g., Thorne 1974). Theoretically predicted value is, in fact, higher, $\eta = 0.42$ (Bardeen *et al.* 1972), but gravitational redshift and photon aberration should decrease it.

The strength of the magnetic field can be obtained by extrapolating the conditions at the last marginally stable orbit. Let us assume that the magnetic field in the disk is:

$$B_{\rm d} = \sqrt{8\pi\beta_{\rm m} p_g} \tag{2.11}$$

here $\beta_{\rm m}$ and p_g are disk magnetization and gas pressure in accretion disk that supports the magnetic field near the horizon. The gas pressure can be estimated using the solution for radiatively inefficient accretion flow (see Narayan & Yi 1994) (more accurate treatment of accretion flow revile correction less than 20-30% (see Narayan & Yi 1995)) as

$$p_g = \frac{\sqrt{10}\dot{M}\sqrt{GM_{\rm BH}}}{12\pi\alpha_{ss}R^{5/2}}\,, \tag{2.12}$$

here α_{ss} is non-dimensional viscosity parameter in the disk (Shakura & Sunyaev 1973). This reduces Eq. (2.11) to the following form:

$$B_{\rm bh} = 1.5\frac{\beta_{\rm m}^{1/2}(\dot{M}c)^{1/2}}{(\alpha_{ss})^{1/2} r_{\rm g}}\,. \tag{2.13}$$

Then the total gamma-ray luminosity, Eq. (2.5), can be estimated as

$$L_\gamma < \frac{\sqrt{10}}{12}\frac{\beta_{\rm m}\kappa(h/r_g)\sin^2\theta\dot{M}c^2}{\alpha_{ss}}\,. \tag{2.14}$$

For the sake of consistency we will estimate the multiplicity coefficient, κ, at Kerr radius through the density of the photons in the magnetosphere (see, e.g., Levinson & Rieger 2011) as

$$\kappa \equiv \frac{n_\pm}{n_{\rm GJ}} \approx 6 \times 10^6 \frac{\dot{m}^{7/2} M_{\rm BH,8}^{1/2}}{(\eta\alpha_{ss})^{7/2}\beta_{\rm m}^{1/2}}\,. \tag{2.15}$$

Thus, one obtains an upper limit on the accretion rate allowing magnetospheric emission as:

$$\dot{m} < 10^{-2}\frac{\eta\alpha_{ss}\beta_{\rm m}^{1/7}}{M_{\rm BH}^{1/7}}\,. \tag{2.16}$$

Equations (2.16) and (2.8) substituted to Eqs. (2.13) and (2.14) yield estimates for the largest magnetic field that is consistent with vacuum gap

$$B_{\rm bh} < 5 \times 10^3 \left(\frac{\beta_{\rm m}}{M_{\rm BH,8}} \right)^{4/7} \ {\rm G} \tag{2.17}$$

and the maximum luminosity of particles accelerated by the gap potential drop

$$L_\gamma < 2 \times 10^{43} \beta_{\rm m}^{8/7} \kappa t_{var,5} M_{\rm BH,8}^{-1/7} \sin^2\theta \ \ {\rm erg\ s^{-1}}\,. \tag{2.18}$$

This limit is independent on α_{ss} and η. We can see that the derived upper limit on photon luminosity is significantly smaller than the observed VHE luminosity of 2×10^{44} erg s^{-1}†. In addition, several factors can reduce the available power even further: (i) for favorable viewing angles, the luminosity may be damped by a large factor $\sin^2\theta < 0.1$; (ii) photon-photon absorption on local infrared field may be severe in powerful AGNs.

Given the weak dependence of Eq. (2.18) on the BH mass, and possible strong impact of the gamma-gamma absorption in case of large masses of BHs, detection of magnetospheric short flares from nearby lighter BH viewed by angle $\lesssim \pi/4$ (to avoid severe absorption in the accretion disk) should be more feasible.

3. Discussion and Conclusions

Equation (2.18) determines the maximum luminosity of vacuum gaps that can collapse quicker than $t_{\rm var}$. It has been assumed for its derivation that magnetic field is determined by accretion regime that in turn determines the intensity of the photon field in the magnetosphere. In the case of the steady accretion, this seems to be a very feasible approximation, however this may look less certain in the case of rapidly changing accretion rate, since the processes that govern change in the accretion rate and escape of the magnetic field may have different characteristic timescales. The dominant contribution to the photon field is coming from plasma located at distances $r \sim 2r_g$, and the characteristic viscous accretion time (density decay time in the disk) is

$$t_{\rho,{\rm decay}} \simeq \frac{2r_{\rm g}}{c\alpha_{\rm ss}} \simeq 10^4 \alpha_{\rm ss,-1}^{-1} M_8 \ {\rm s}\,. \tag{3.1}$$

When the accretion fades, the decay of the magnetic field is determined by magnetic field reconnection rate (Komissarov 2004), i.e.

$$t_{\rm B,decay} \simeq \frac{\pi r_{\rm g}}{0.3c} \sim 10^4 M_8 \ {\rm s}\,. \tag{3.2}$$

Since these two time-scales are essentially identical, it is natural to expect that the field strength and disk density will decay simultaneously. Thus, Equation (2.18) should also be valid for time-variable accretion regime.

According to the estimate provided by Eq. (2.18) the possible luminosity of flares generated in BH magnetosphere depends very weakly on the mass of BH and is determined by disk magnetization, viewing angle, and pair multiplicity‡. Since all these parameters are smaller than unit, then the numerical coefficient in Eq. (2.18) can be taken as a strict

† In the paper (Hirotani & Pu 2015) the efficiency of BZ mechanism was required on the level 10 000%, unfortunately in 3D simulations McKinney *et al.* (2012) the efficiency is limited by value 300%. We should remark that equation (2.14) with $\alpha_{\rm ss} = 0.1$ gets efficiency on the level 250% that is very close to results of 3D simulations.

‡ Equation (2.18) does not account for relativistic effects that should be small unless the gap is formed close to the horizon. If the vacuum gap is close to the horizon, then gravitational redshift should even strengthen the constraints imposed by the variability time.

upper limit for flare luminosity for the given variability time. This upper limit appears to be approximately an order of magnitude below the value measured with MAGIC (Aleksić *et al.* 2014). Thus, we conclude that it seems very unfeasible that this processes can be indeed behind the bright flaring activity recorded from IC 310.

References

Abdo, A. A., *et al.* 2011, *ApJ*, 733, L26
Ackermann, M., *et al.* 2016, *ApJ*, 824, L20
Aharonian, F., *et al.* 2007, *ApJ*, 664, L71
Albert, J., *et al.* 2007, The Astrophysical Journal, 669, 862
Aleksić, J., *et al.* 2014, Science, 346, 1080
Bardeen, J. M., Press, W. H., & Teukolsky, S. A. 1972, *ApJ*, 178, 347–370
Barkov, M. V., Aharonian, F. A., Bogovalov, S. V., Kelner, S. R., & Khangulyan, D. 2012a, *ApJ*, 749, 119
Barkov, M. V., Aharonian, F. A., & Bosch-Ramon, V. 2010, *ApJ*, 724, 1517
Barkov, M. V., Bosch-Ramon, V., & Aharonian, F. A. 2012b, *ApJ*, 755, 170
Blandford, R. D., & Znajek, R. L. 1977, *MNRAS*, 179, 433
Bonnoli, G., Ghisellini, G., Foschini, L., Tavecchio, F., & Ghirlanda, G. 2011, *MNRAS*, 410, 368
Bosch-Ramon, V., Perucho, M., & Barkov, M. V. 2012, *A&A*, 539, A69
Broderick, A. E., & Tchekhovskoy, A. 2015, ArXiv e-prints
Giannios, D., Uzdensky, D. A., & Begelman, M. C. 2009, *MNRAS*, 395, L29
Golkhou, V. Z., Butler, N. R., & Littlejohns, O. M. 2015, ArXiv e-prints
Gu, M., Cao, X., & Jiang, D. R. 2001, *MNRAS*, 327, 1111
Hayashida, M., *et al.* 2015, *ApJ*, 807, 79
Hirotani, K., & Pu, H.-Y. 2015, ArXiv e-prints
Khangulyan, D. V., Barkov, M. V., Bosch-Ramon, V., Aharonian, F. A., & Dorodnitsyn, A. V. 2013, *ApJ*, 774, 113
Komissarov, S. S. 2004, *MNRAS*, 350, 427
Levinson, A. & Rieger, F. 2011, *ApJ*, 730, 123
McKinney, J. C., Tchekhovskoy, A., & Blandford, R. D. 2012, *MNRAS*, 423, 3083
Narayan, R. & Yi, I. 1994, *ApJ*, 428, L13
—. 1995, *ApJ*, 444, 231–243
Neronov, A. & Aharonian, F. A. 2007, *ApJ*, 671, 85
Nilsson, K., Pursimo, T., Villforth, C., Lindfors, E., & Takalo, L. O. 2009, *A&A*, 505, 601
Pozanenko, A., & Loznikov, V. 2002, in Lighthouses of the Universe: The Most Luminous Celestial Objects and Their Use for Cosmology, ed. M. Gilfanov, R. Sunyeav, & E. Churazov, 194
Shakura, N. I. & Sunyaev, R. A. 1973, *A&A*, 24, 337
Thorne, K. S. 1974, *ApJ*, 191, 507
Timokhin, A. N. 2010, *MNRAS*, 408, 2092
Vovk, I. & Babić, A. 2015, *A&A*, 578, A92
Woo, J.-H. & Urry, C. M. 2002, *ApJ*, 579, 530

New Frontiers in Black Hole Astrophysics
Proceedings IAU Symposium No. 324, 2016
Andreja Gomboc, ed.

doi:10.1017/S174392131700151X

Monitoring of the radio galaxy M 87 at Very High Energy with MAGIC during a low emission state between 2012 and 2015

C. Arcaro[1], P. Bangale[2], M. Manganaro[3], D. Mazin[2,4], P. Colin[2], Ie. Vovk[2], K. Mannheim[5] for the MAGIC Collaboration and K. Hada[6], H. E. Jermak[7], J. P. Madrid[8], F. Massaro[9], S. Richter[10], F. Spanier[10] and R. C. Walker[11]

[1]Università di Padova and INFN, I-35131 Padova, Italy
email: cornelia.arcaro@pd.infn.it

[2]Max-Planck-Institut für Physik, D-80805 München, Germany
[3]Inst. de Astrofísica de Canarias, E-38200 La Laguna, Tenerife, Spain; Universidad de La Laguna, Dpto. Astrofísica, E-38206 La Laguna, Tenerife, Spain
[4]also at Japanese MAGIC Consortium
[5]Universität Würzburg, D-97074, Würzburg, Germany
[6]Mizusawa VLBI Observatory, National Astronomical Observatory of Japan, 2-12 Hoshigaoka-cho, Mizusawa, Oshu, Iwate 023-0861, Japan
[7]Astrophysics Research Institute, Liverpool John Moores University, Liverpool, UK
[8]CSIRO, Astronomy and Space Science, PO Box 76, Epping NSW 1710, Australia
[9]Dipartimento di Fisica, Università degli Studi di Torino, via Pietro Giuria 1, I-10125 Torino, Italy
[10]North-West University, Potchefstroom, South Africa [11]National Radio Astronomy Observatory, Socorro, NM 87801, USA

Abstract. We present the preliminary results from observing the nearby radio galaxy M 87 for 156 hours (between the years 2012 and 2015) with the MAGIC telescopes, which lead to a significant very high energy (VHE, $E > 100\,\mathrm{GeV}$) detection of the source in quiescent states each year. Our VHE analysis combined with quasi-simultaneous data at other energies (from gamma-rays, X-rays, optical and radio) provides a unique opportunity to study the source variability and its broadband spectral energy distribution, which is found to disfavour a one-zone synchrotron/synchrotron self-Compton model. Therefore, other alternative scenarios for the photon emission are explored. We also find that the VHE emission is compatible with being produced close to the source radio core as previous data already indicated. A detailed paper presenting full results of the observing campaign is in preparation.

Keywords. BL Lacertae objects: individual: M 87 - galaxies: active - galaxies: jets - gamma rays: observations - radiation mechanisms: nonthermal

---

Messier 87, commonly known as M 87, is a giant elliptical radio galaxy of Fanaroff-Riley-I-type (FR I; Fanaroff & Riley 1974), located in the nearby Virgo Cluster, at a distance $\sim 16\,\mathrm{Mpc}$ (e.g. Blakeslee *et al.* 2009, Bird *et al.* 2010), and powered by a supermassive black hole.

The relativistic jet of M 87 is misaligned with respect to our line of sight by an angle between $5 - 25°$ (Biretta *et al.* 1999, Acciari *et al.* 2009), allowing the study of the jet during its evolution from the core to the extended lobe. The jet features several knots along his length, the inner knot being HST-1 located at $0.85''$ from the core, that have been resolved in radio, optical and X-ray bands (Perlman *et al.* 2001, Wilson & Yang 2002).

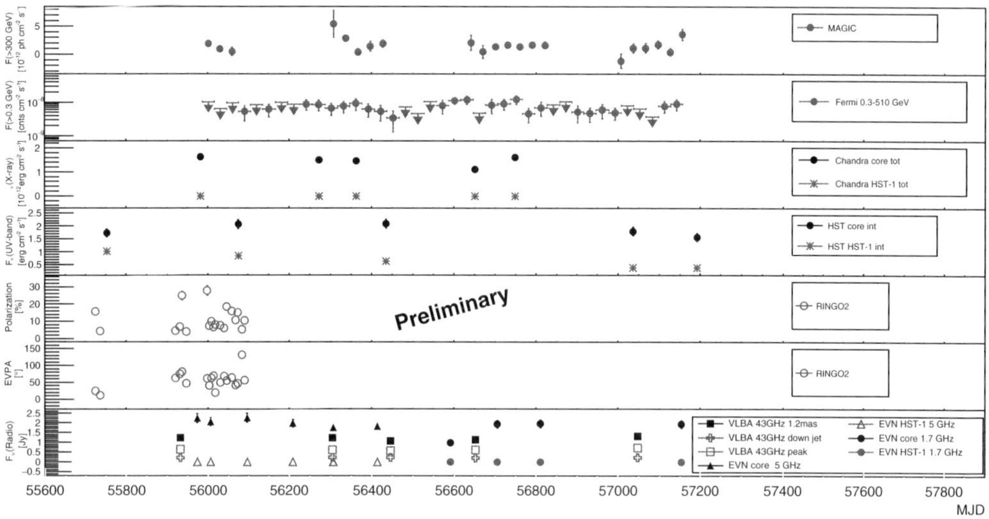

Figure 1. *From top to bottom:* VHE (filled circles) and HE gamma-ray (filled circles) data from MAGIC and *Fermi*-LAT; Galactic absorption corrected X-ray data (core: filled circles; HST-1: asterics) from *Chandra*; optical data in the UV band (core: filled circles; HST-1: asterics) from HST are corrected for extinction (Cardelli *et al.* 1989). Optical polarimetry data (open circles) taken with V+R filter by Liverpool telescope with RINGO2; radio data at 43 GHz by VLBA (1.2 mas: filled squares; peak: open squares; down jet: open cross) and EVN at 1.7 and 5 GHz (core: filled/open triangles; HST-1: filled/open circles). Upper limits of 95% confidence level are indicated by downward arrows. The light curves are daily binned except VHE and HE gamma rays, where a monthly and a 30-day binning were applied, respectively.

Here we present the MAGIC VHE monitoring dataset of M 87 between 2012 and 2015 together with multiwavelength (MWL) data from high energy (HE; $100\,\mathrm{MeV} < E < 100\,\mathrm{GeV}$), gamma rays to radio frequencies both of the core and HST-1. No major flare was detected in that period in any of the observed bands, which allows us to study the source in quiescent flux state. The data quality is sufficient to constrain some emission models and to study the broad band spectral energy distribution (SED) of M87.

MAGIC is a stereoscopic system of two 17-m diameter IACTs situated at the Roque de los Muchachos, on the Canary island of La Palma (28.75°N, 17.86°W) at a height of 2200 m above sea level, with an integral sensitivity of $(0.66 \pm 0.03)\%$ of the Crab nebula flux above 220 GeV in 50 hrs at low zenith angles (Aleksić *et al.* 2016). M87 observations were performed between December and July, in each year from 2012 to 2015 at zenith angles ranging from 15° to 50° during dark time and under Moon light conditions. After quality cuts, a total of 157 hrs of effective observation time of data were used for further analysis.

The MWL dataset presented in Fig. 1 shows no hint for any enhanced activity between 2012 and 2015. We note however that the monitoring in VHE, X-rays, optical and radio was not very frequent, wherefore it cannot be fully excluded that flaring periods took place in the last years.

The positions of the excess of the VHE emission obtained from observations by MAGIC (RA: $12^{\mathrm{h}}30^{\mathrm{m}}49.54^{\mathrm{s}} \pm 0.94^{\mathrm{s}}{}_{\mathrm{stat}} 4.8^{\mathrm{s}}{}_{\mathrm{sys}}$ Dec: $12°23'38.90'' \pm 14.65''_{\mathrm{stat}} \pm 1.2''_{\mathrm{sys}}$) between 2012 and 2015, by H.E.S.S. (RA: $12^{\mathrm{h}}30^{\mathrm{m}}47.2^{\mathrm{s}} \pm 1.4^{\mathrm{s}}{}_{\mathrm{stat}} \pm 1.3^{\mathrm{s}}{}_{\mathrm{sys}}$, Dec: $12°23'51'' \pm 19''_{\mathrm{stat}} \pm 20''_{\mathrm{sys}}$; Aharonian *et al.* 2006) and VERITAS (RA: $12^{\mathrm{h}}30^{\mathrm{m}}46^{\mathrm{s}} \pm 4^{\mathrm{s}}{}_{\mathrm{stat}} \pm 6^{\mathrm{s}}{}_{\mathrm{sys}}$, Dec: $12°23'21'' \pm 50''_{\mathrm{stat}} \pm 1'30''_{\mathrm{sys}}$; Acciari *et al.* 2008) are compatible within the errors. The Very Large Array (VLA) radio telescope image taken at 330 MHz (Owen *et al.* 2000)

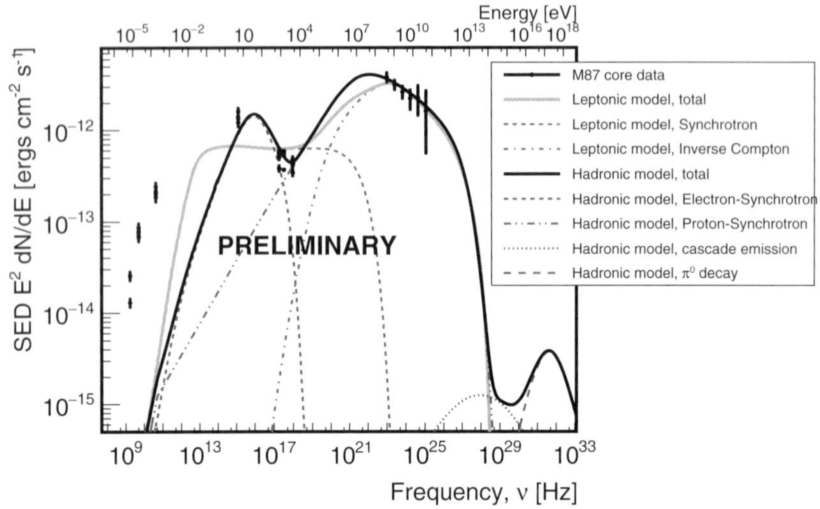

Figure 2. Averaged SED (black points) of the radio core of M87 compiled from quasi-simultaneous 2012-2015 MWL observations: HE gamma-ray data (*Fermi*-LAT); Galactic absorption corrected X-ray (*Chandra*), optical data in the UV band (HST) corrected for extinction (Cardelli *et al.* 1989), radio data at 1.7 GHz and 5 GHz (EVN) and at 43 GHz (VLBA). The models represent two possible scenarios, where the GeV-TeV is either dominated by SSC emission (leptonic case; black solid line) or by the synchrotron radiation of the relativistic protons (hybrid case; gray solid line).

is used as a reference, indicating that the VHE emission is located close to the radio core (RA: $12^{h}30^{m}49.42^{s}$, Dec: $12°23'28.04''$; `http://images.nrao.edu/57`center of radio image taken as reference). The upper limit for an extended VHE emission at 99.9% confidence level of 0.042°, corresponding to 11.5 kpc is shown with previous results obtained from H.E.S.S. observations (0.05°, i.e., 13.7 kpc).

As no clear variability in the TeV regime is detected, the model constraints are not as strict. We assume the relevant parts connected to the TeV gamma-ray emission zone to be the HE gamma-ray and the X-ray data detected for the core region, while radio data from the Very Long Baseline Array (VLBA) and the European VLBI Network (EVN) originate from a larger region. The optical data from Hubble Space Telescope (HST), detected from the core of M 87 presumably originate much closer to the black hole and where the jet is launched. We are, therefore, modelling the X-ray, GeV and TeV data, while the radio to optical data are considered to be upper limits to our models.

The applied model was designed as a hybrid model and can cover the range from purely leptonic scenarios to the full photo-hadronic reaction chain. The fully time-dependent implementation is based on the geometry of Weidinger & Spanier 2010. The acceleration mechanism and the implementation of all leptonic processes was adopted from Richter & Spanier 2016 and the photo-hadronic framework was implemented following Hümmer *et al.* (2010).

In the leptonic model the GeV-TeV emission is dominated by the inverse Compton emission, while radio to X-ray emission is synchrotron emission from the same electron population. In the hybrid model, the parts from radio to soft X-rays also originate from synchrotron emission of electrons. Due to the higher magnetic field and the assumption of protons being injected into the acceleration zone, the higher energies are dominated by synchrotron emission of protons in this case.

We note that *Chandra* data ($10^{17} - 10^{18}$) Hz are in the sensitive area of the transition between the two bumps in the SED. In the leptonic model we did not succeed to properly match the *Chandra* photon index, that seems to be softer than in the model. The reason for that is a rather flat synchrotron self-Compton (SSC) spectrum in the Inverse Compton (IC) regime up to 10 TeV without an apparent cut-off in the data. This requires high maximum energies for the electron population, leaving the IC in the Klein-Nishina regime and resulting in a different, softer photon index in the IC regime. The high HE flux in comparison to the X-ray flux adds to the difficulties and requires a high particle density. The resulting Doppler factor is low, as one might expect so close to the black hole. Summarising, a homogeneous leptonic model is able to describe the five orders in magnitude flat photon spectrum in HE and VHE gamma rays (*Fermi*-LAT and MAGIC) but it must be pushed to extreme limits and still has possible troubles in reproducing the X-ray data.

In the hybrid scenario it is easier to fit the available data. However, the number of parameters is higher than in the leptonic case and the two components (synchrotron from electrons and synchrotron from protons) are basically independent of each other, as their densities are unrelated. On the other hand, the hybrid model has a clear prediction on the flux of π_0 decay at $\sim$10 PeV, which can be probed by instruments like HAWC after a long exposure.

Acknowledgements

We would like to thank the Instituto de Astrofísica de Canarias for the excellent working conditions at the Observatorio del Roque de los Muchachos in La Palma. The financial support of the German BMBF and MPG, the Italian INFN and INAF, the Swiss National Fund SNF, the ERDF under the Spanish MINECO (FPA2012-39502), and the Japanese JSPS and MEXT is gratefully acknowledged. This work was also supported by the Centro de Excelencia Severo Ochoa SEV-2012-0234, CPAN CSD2007-00042, and MultiDark CSD2009-00064 projects of the Spanish Consolider-Ingenio 2010 programme, by grant 268740 of the Academy of Finland, by the Croatian Science Foundation (HrZZ) Project 09/176 and the University of Rijeka Project 13.12.1.3.02, by the DFG Collaborative Research Centers SFB823/C4 and SFB876/C3, and by the Polish MNiSzW grant 745/N-HESS-MAGIC/2010/0.

References

Acciari, V. A., Beilicke, M., Blaylock, G., *et al.* 2008, *ApJ*, 679, 397
Acciari, V. A., Aliu, E., Arlen, T., *et al.* 2009, *Science*, 325, 444
Aharonian, F., Akhperjanian, A. G., Bazer-Bachi, A. R. *et al.* 2006, *Science*, 314, 1424
Aleksić, J., Ansoldi, S., Antonelli, L. A., *et al.* 2016, *Astroparticle Physics*, 72, 76
Blakeslee, J. P., Jordan, A., Mei, S., *et al.* 2009, *ApJ*, 694, 556
Biretta, J. A., Sparks, W. B., & Macchetto, F. 1999, *ApJ*, 520, 621
Bird, S., Harris, W. E., Blakeslee, J. P., & Flynn, C. 2010, *A&A*, 524, A71
Cardelli, J. A., Clayton, G. C., & Mathis, J. S.1989, *ApJ*, 345, 245
Fanaroff, B. L. & Riley, J. M. 1974, *MNRAS*, 167, 31P
Hümmer, S., Rüger, M., Spanier, F., & Winter, W. 2010, *ApJ*, 721, 630
Owen, F. N., Eilek, J. A., & Kassim, N. E. 2000, *ApJ*, 543, 611
Perlman, E. S., Biretta, J. A.,Sparks, W. B., *et al.* 2001, *ApJL*, 551, 206
Richter, R. & Spanier, F. 2016, *ApJ*, 829, 56
Wilson,A. S. & Yang, Y. 2002, *ApJ*, 568, 140
Weidinger, M., Spanier, F. *et al.* 2010, *IJMPhysD*, 19, 887

New Frontiers in Black Hole Astrophysics
Proceedings IAU Symposium No. 324, 2016
Andreja Gomboc, ed.

doi:10.1017/S1743921317001648

The Extremes of AGN Variability

S. Komossa[1], D. Grupe[2], N. Schartel[3], L. Gallo[5], J. L. Gomez[6], W. Kollatschny[7], G. Kriss[8], K. Leighly[9], A. L. Longinotti[10], M. Parker[4], M. Santos-Lleo[3], D. Wilkins[5] and M. Zetzl[7]

[1]QianNan Normal University for Nationalities, Duyun City, China [2]Morehead State University, Morehead, KY 40351, USA [3]XMM-Newton SOC, ESA, ESAC, 28692 Villanueva de la Canãda, Spain [4]Institute of Astronomy, Cambridge CB3 0HA, UK [5]Department of Astronomy and Physics, Saint Mary's University, Halifax, NS B3H 3C3, Canada [6]Instituto de Astrofisica de Andalucia-CSIC, E-18008 Granada, Spain [7]Institut für Astrophysik, Universität Göttingen, 37077 Göttingen, Germany [8]Space Telescope Science Institute, Baltimore, MD, 21218, USA [9]Homer L. Dodge Department of Physics and Astronomy, The University of Oklahoma, Norman, OK 73019, USA [10]Instituto de Astronomia, UNAM, 04510 Mexico D.F., Mexico

Abstract. We present results from our ongoing monitoring programs aimed at identifying and understanding Active Galactic Nuclei (AGN) in extreme flux and spectral states. Observations of AGN in extreme states can reveal the nature of the inner accretion flow, the physics of matter under strong gravity, and they provide insight on the properties of ionized absorbers and outflows launched near supermassive black holes (SMBHs). We present new results from our long-term monitoring of IC 3599, WPVS007, and Mrk 335, multi-wavelength follow-ups of the newly identified changing-look AGN HE 1136–2304, and UV–X-ray follow-ups of the binary SMBH candidate OJ 287 after its 2015 optical maximum, now in a new optical-X-ray–high-state.

Keywords. Black holes, galaxies, blazars, accretion, jets

1. Introduction: AGN in extreme flux and spectral states

The intrinsic X-ray spectra of radio-quiet AGN consist of two main components. (1) Thermal quasi-black-body emission from the accretion disk, contributing to soft X-rays, and (2) a hard X-ray spectral component likely arising from coronal emission above the accretion disk, and due to inverse Compton scattering of disk photons off a population of relativistic particles. The observed spectra are modified by the effects of absorption, reflection and re-emission (see Turner & Miller 2009 and Fabian 2016 for recent reviews). Ionized absorbing material near the SMBH imprints absorption edges and absorption lines on the observed spectra. Matter seen in reflection or emission from the inner accretion disk will inevitable be subject to relativistic effects, including blurring due to the high gaseous velocities, relativistic Doppler beaming, and gravitational redshift, producing characteristic emission-line profiles with blue horns and broad red wings. Further, if the corona is located very close above the inner disk, the effect of relativistic lightbending is strong, causing a strongly reflection-dominated spectrum. These effects therefore provide us with valuable diagnostics of the extreme physics near SMBHs. Spectral signatures are most pronounced in extreme states, especially low flux states. Therefore, we are carrying out programs to identify AGN in extreme flux and spectral states.

Based on our long-term AGN monitoring program with the satellite *Swift* (Gehrels *et al.* 2004), we have followed-up or newly identified some of the most highly variable AGN known today. When these undergo exceptional flaring activity, or go into deep low-states, we then trigger deep follow-up observations including with XMM-Newton and NuSTAR, and at other wavebands.

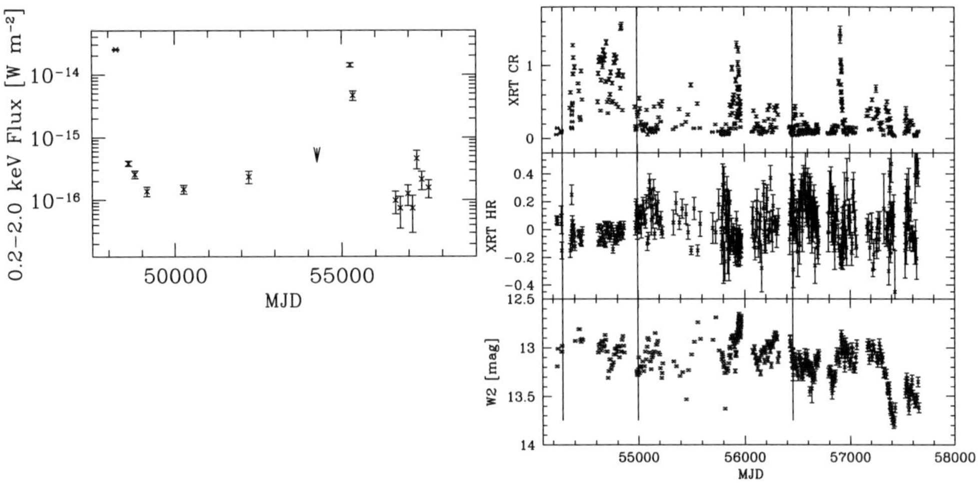

Figure 1. *Left:* Long-term X-ray lightcurve of **IC 3599** with two giant (factors ~ 100) outbursts. Our most recent *Swift* monitoring has shown that the X-rays continue to remain faint, while the UV is brighter again. *Right: Swift* lightcurve of **Mrk 335** (upper panel: X-ray telescope (XRT) countrate, middle panel: hardness ratio, lower panel: UVOT W2 magnitude). While historically X-ray bright, most recently, Mrk 335 has mostly remained in a low X-ray flux state, with occasional flaring.

2. IC 3599

IC 3599 is unique among AGN in its high-amplitude X-ray transience, accompanied by variable emission lines, first seen in 1990. The discovery of a second such (factor ~100) outburst 20 yrs after the first (Komossa *et al.* 2014, Grupe *et al.* 2015), has ruled out an outburst scenario of a single stellar tidal disruption event (TDE) flaring only once. Variants of repeat TDEs in binary SMBH systems do not match observations well or require further testing. An accretion disk instability is well consistent with the data (Grupe *et al.* 2015). If confirmed, this is the first case where such a disk instability has been identified among accreting SMBHs. In 2015 IC3599 reached a third high-state in optical-UV emission, possibly indicating the onset of a third outburst. So far, however, the X-rays have remained faint (Fig. 1), implying that either the UV emission varies independent of the X-rays, or the UV predates the X-ray rise by months-years. Multiple flaring is not expected in scenarios, where both pairs of a binary star are tidally disrupted, producing only two peaks (Mandel & Levin 2015), but is well consistent with the scenario of a disk instability.

3. WPVS007

WPVS007 underwent a unique, giant-amplitude drop in its X-ray emission (factor 400) first detected with ROSAT, not seen in any other AGN. It has remained X-ray faint ever since the 1990s, except occasional rapid flaring. Results from years of multi-wavelength monitoring (e.g., Grupe *et al.* 2013, Komossa *et al.* 2014) along with HST UV spectroscopy (Leighly *et al.* 2015), suggest a rare viewing geometry into the central engine, with our line of sight grazing a clumpy molecular torus and the winds launched from its edge (Leighly *et al.* 2015). Its 2016 lightcurve is overall consistent with a steady decrease in extinction, while X-rays remain undetected.

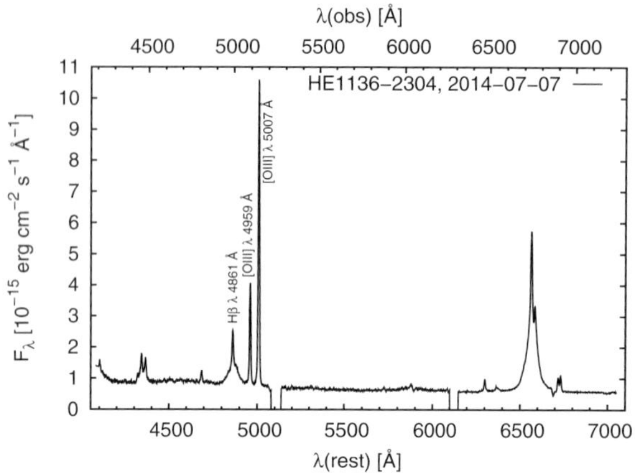

Figure 2. SALT optical spectrum of HE 1136-2304 from 2014, revealing its Seyfert-type change. In 1993, the broad Hα line was barely detected. Lines labelled are Hβ, [OIII]4363 and [OIII]5007.

4. Mrk 335

The NLS1 galaxy Mrk 335 is among the X-ray brightest, most variable AGN known; with pronounced episodes of X-ray flaring, and deep, long-lasting X-ray low-states rich in spectral features (e.g., Grupe *et al.* 2012; our Fig. 1). The broad-band X-ray spectrum is consistent with blurred reflection from an accretion disk around a nearly maximum spinning black hole ($a > 0.9$; Parker *et al.* 2014, Gallo *et al.* 2015, Wilkins *et al.* 2015), or alternatively with an extreme partially covering absorber, with $\log N = 23.3$ at low-state. While the nature of the lamppost illuminating the disk remains unknown, the base of a jet has been considered as a candidate (Gallo *et al.* 2015). Knowing accretion conditions of nearby, low-power radio sources provides us with an independent path to understanding the question, why only ~10% of all AGN are radio-loud. Dedicated X-ray follow-ups including with Suzaku, NuSTAR, and XMM-Newton have been triggered in a deep low-state in 2013, a bright flare state in 2014, and most recently with XMM-Newton and HST in a deep UV low-state in 2015. The latter observation shows broad, fast (-5000 km s^{-1}) UV absorption in Lyα, C IV, and possibly O VI.

5. HE 1136–2304

The Seyfert galaxy HE 1136-2304 was detected in an X-ray flare state in the XMM-Newton slew survey, subsequently confirmed with *Swift* after increasing in X-ray flux by a factor ~ 30. Follow-up observations with NuSTAR, XMM-Newton, and SALT (South Africa Large Telescope), quasi-simultaneous within 3 days, have identified HE 1136-2304 as one of the rare "changing-look" AGN (Parker *et al.* 2016), dramatically changing its optical Seyfert type, from almost type 2 in 1993 to a Seyfert 1.5 with strong Balmer emission lines in 2014 (Fig. 2). The observations are consistent with an intrinsic change in accretion rate.

6. The binary SMBH candidate OJ 287

Galaxy mergers and the binary black holes at their centers offer unique views on galaxy formation and evolution, and on black hole demographics and growth across cosmic times. Coalescing supermassive binaries are the strongest sources of gravitational waves in the universe. An intense search for wide and close systems in all stages of their evolution

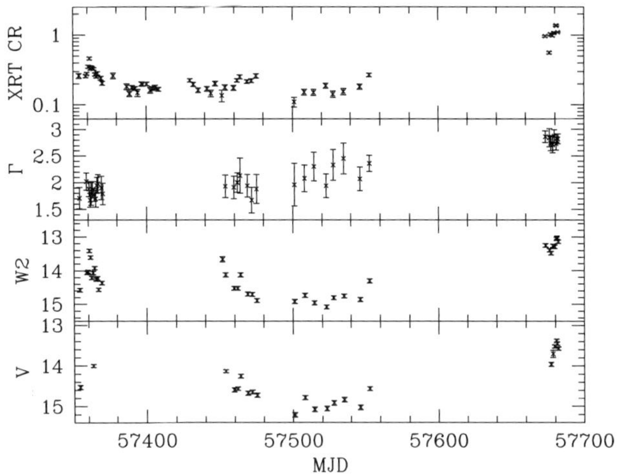

Figure 3. X-ray, UV, and optical lightcurve of OJ 287 taken with *Swift*, between December 2015 and October 2016.

is currently ongoing (review by Komossa & Zensus 2016). OJ 287 is one of the best candidates for hosting a sub-parsec binary SMBH, with quasi-periodic optical maxima every ~12 years. Reports of a high optical peak flux in December 2015 (ATel 8372), along with a suggestion that this may correspond to the next decadal peak of OJ 287, which would then imply strong orbital precession of the secondary SMBH (Valtonen *et al.* 2016), has triggered many follow-up projects. Multi-frequency flux density monitoring with the 100m Effelsberg radio telescope started in December 2015 (Komossa *et al.* 2015). OJ 287 was also observed with *Swift*, first in triggered mode following the optical rise to a maximum in December 2015 (ATel 8401), more recently by us in order to follow the long-term X-ray evolution along with other wavebands (Fig. 3). In X-rays, the highest recent state was detected on October 12, 2016; more than two times brighter than the December 5 high-state, accompanied by bright UV emission, and by a significant softening of the X-ray spectrum ($\Gamma_x = 2.9 \pm 0.1$). Ongoing monitoring of OJ 287, and extremely high-resolution Event Horizon Telescope imaging, provide us with a unique chance of testing the binary SMBH scenario, and understanding one of the most massive blazars known.

References

Fabian, A. C. 2016, *AN*, 337, 375
Gallo, L., *et al.* 2015, *MNRAS*, 446, 633
Gehrels, N., *et al.* 2004, *ApJ*, 611, 1005
Grupe, D., *et al.* 2012, *ApJS*, 199, 28
Grupe, D., *et al.* 2013, *AJ*, 146, 78
Grupe, D., *et al.* 2015, *ApJ*, 803, L28
Komossa, S., *et al.* 2014, *PoS(SWIFT10)*, 143
Komossa, S., *et al.* 2015, *ATel*, 8411, 1
Komossa, S., & Zensus, J. A. 2016, *IAUS*, 312, 13
Leighly, K. M., *et al.* 2015, *ApJ*, 809, L13
Mandel, I. & Levin, Y. 2015, *ApJ*, 805, L4
Parker M., *et al.* 2014, *MNRAS*, 443, 1723
Parker M., *et al.* 2016, *MNRAS*, 461, 1927
Turner, T. J. & Miller, L. 2009, *A&ARv*, 17, 47
Wilkins, D., *et al.* 2015, *MNRAS*, 454, 4440
Valtonen, M., *et al.* 2016, *ApJ*, 819, L37

New Frontiers in Black Hole Astrophysics
Proceedings IAU Symposium No. 324, 2016
Andreja Gomboc, ed.

doi:10.1017/S1743921317003490

X-ray fluctuation timescale and Black Hole mass relation in AGN

Amri Wandel[1] **and Mathew Malkan**[2]

[1]Racah Institute, The Hebrew University of Jerusalem, Jerusalem 91904, Israel
email: amri@huji.ac.il

[2]Dept. of Astronomy, University of California, Los Angeles, California , U.S.A.
email: malkan@astro.ucla.edu

Abstract. We analyze the fluctuations in the X-ray flux of 20 AGN (mainly Seyfert 1 galaxies) monitored by RXTE and XMM-Newton with a sampling frequency ranging from hours to years, using structure function (SF) analysis. We derive SFs over four orders of magnitude in the time domain (0.03-300 days). Most objects show a characteristic time scale, where the SF flattens or changes slope. For 10 objects with published power-spectral density (PSD) the break time scales in the SF and PSD are similar and show a good correlation . We also find a significant correlation between the SF timescale and the mass of the central black hole, determined for most objects by reverberation mapping.

Keywords. AGN, X-ray, Fluctuations, Structure Functions, Power Spectral Density, Black Holes

1. Introduction

The strong, irregular X-ray variability observed in many Seyfert galaxies on timescales as short as hours indicates that these X-rays are emitted from a region close to the central black hole. Doubling time scales were used as the first quantification of variability; characteristic time scales of X-ray variability seemed to correlate with luminosity and black hole mass Wandel & Mushotzky (1986).

For quasi-periodic and stochastic light curves the variability is often characterized through the power spectral density (PSD). The PSD is the variability amplitude per frequency, so it describes the variability power contained within a frequency interval. A similar measure that is sometimes used is the structure function, which describes the variability amplitude as a function of time scale. The variability of quasi-periodic and stochastic light curves may then be characterized by representing the power spectrum in a parametric form. However, irregular sampling or sequences of regular sampling separated by gaps often cause problems in measuring variability features from a light curve, which is the case in all ground-based and often also space-based data. Ideally the variability must be quantified over a broad a range of time scales, to maximize the probability of detecting a characteristic variability time scale and to avoid spurious features caused by an insufficiently long observing time.

The Rossi X-ray Timing Explorer (XTE) has been the first mission to provide sustained monitoring on given objects for time scales up to months and years. High dynamic-range X-ray PSDs were published for approximately 10 AGN, (e.g., Edelson & Nandra (1999), Uttley, McHardy & Papadakis (2002), Markowitz *et al.* (2003), Chatterjee *et al.* (2009;2011), Rothshild *et al.* (2011)). Typically they are parameterized as a broken power law; on long timescales the fluctuation power is flat, scaling as a power of f^{0-1}, and on short timescales it is descending more steeply, with slopes of 2-3. The break between

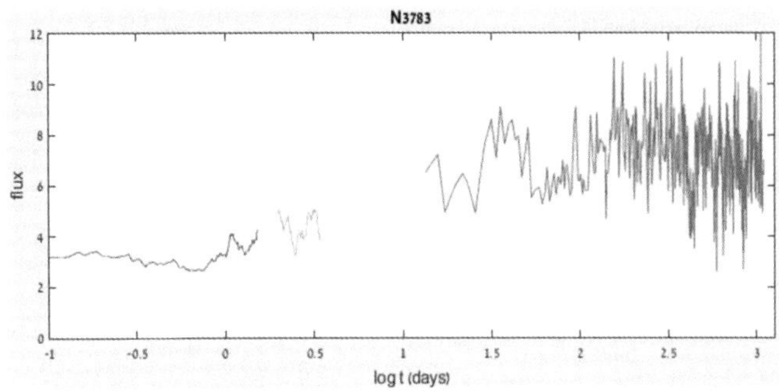

Figure 1. Combined logarithmic light curve of NGC 3783. The X-ray flux vs. log (1/frequency) in units of days. Shown are the light curves of XTE (right curve) and XMM (the two left curves).

these two regimes is in the range of roughly 10^{4-6} Hz, corresponding to time scales t_{PSD} ranging from a few hours up to weeks.

2. Structure functions

The structure function (SF) measures the mean relative amplitudes of pairs of flux variations as a function of their time difference (Simonetti *et al.* (1985)). While the PSD requires strictly even sampling to compute, the SF can be derived for uneven and gappy sampling. The SF has been applied to optical AGN monitoring data (Webb and Malkan (2000); Collier & Peterson (2001)) as well as to X-ray light curves of blazars. As long as the number/size of gaps are not too large, meaningful results can be obtained; however this procedure is not foolproof, and sampling limitations (especially the lack of long enough timescales) can produce spurious features, as demonstrated by Emmanoulopoulos, McHardy & Uttley (2010) (hereafter E10), who give a detailed analyses of the caveats and cautions for SF measurement and interpretation. Recently a new method - continuous-time autoregressive moving average (CARMA) has been suggested (Kelly *et al.* 2014) to handle irregular sampling and gaps.

We define the amplitude-averaged SF as

$$S(\tau)^2 = \sum_k [f(t_k) - f(t_k - \tau)]^2$$

where $f(t_k)$ is the normalized flux at time point t_k. As advised in E10, we work with the log f-log SF. Typically, the fluctuation amplitude is increasing with time difference, as larger fluctuations are present on longer timescales.

To derive errors on each SF bin, we used two methods. The first is an enhanced χ^2 method (Collier & Peterson (2001)). The statistical uncertainties in structure functions are defined by $\sigma_i)/(N_i/2)^{1/2}$ where σ_i is the root mean square (rms) deviations about the mean structure function value in bin i and N_i is the number of pairs of data points in the i-th bin. This ad hoc, but reasonable, error estimates reflect the fact that not all pairs of measurements in a given bin are independent. The second method, which for sparser sampling gives more realistic errors for light curves, is a Monte Carlo procedure (e.g., Kataoka *et al.* (2001) and E10).

For each light curve/SF segment we assume a form for the underlying PSD (using the published PSD parameters if they exist; otherwise we make a PSD analyses with the incomplete data and use the resulting power law (Emmanoulopoulos, private communication). For each curve segment, we simulate 100 light curves using the algorithm of Timmer & Koenig (1995), re-sample them in the same manner as the observed light

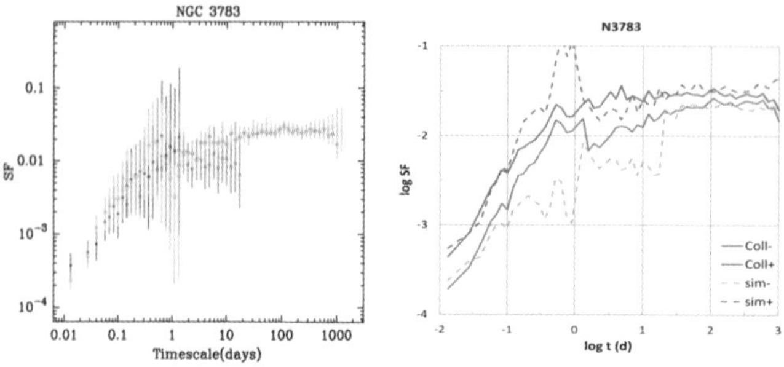

Figure 2. Combined log(SF)-log t diagram for NGC 3783.Upper and lower error limits, calculated for the 95%-th percentile with both methods, MC simulations (dashed) and Collier & Peterson (2001) (solid).

curve, and measured the SF. This yields relatively larger errors towards longer time scales, which reflects systematic uncertainties in the SF introduced by red-noise leak and/or by gaps in the sampling. We adopt the 95% confidence errors on each binned SF point, as a trade-off between 1σ errors for χ^2 -fitting and a higher confidence level, that would reflect more closely the systematic SF behavior for each time scale bin. Finally we combine the log(SF)-log t diagrams for all data segments for each object to produce a combined SF over a broad range of timescales.

Similar SF figs of have been obtained for NGC 3516, 4051, 4151, 5584, 7469, IC4329a, MCG 6-30-15, Fairall 9, Mkn 766. The two last ones do not show an obvious break (Wandel, Markowitz and Malkan, in preparation). Comparing χ^2-fits of the SF to an unbroken power-law and a broken one, we find that for almost all objects with a full range time series (4 orders of magnitude in temporal range) data sources, the improvement in a χ^2 fit when including a break is statistically significant. It is important though to caution that the SF points are not all independent, and unlike the Monte Carlo method applied to PSDs, this method also does not take into account distortion effects such as aliasing or red-noise leak, which could introduce scatter into the determination of the break point t_{SF}.

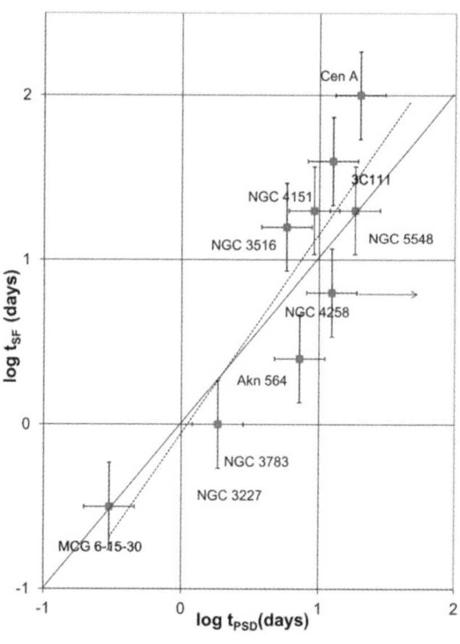

Figure 3. The time scale corresponding to the SF break vs. that of the break in the PSD. The best fit has a slope of 1. Solid line denotes equality, dashed line is the best fit. Typical errors are denoted by the crosses; the vertical error bars would be larger for a higher confidence level.

3. The SF-break timescale and the Black Hole mass relation

It can be shown that for good enough data the PSD and SF power-law slopes are related: when the sampling is of sufficient quality and dynamic range the slope of PSD(ν) minus one equals the slope of SF(τ) (E10, appendix A). If a change in power-law slope exists in the broadband PSD, we should expect to find a change in power-law slope in the broadband SF at a similar time scale. Comparing the SF and the PSD timescales for the objects with published PSD, it appears that a distinct break in the SF corresponds to the break in the PSD.

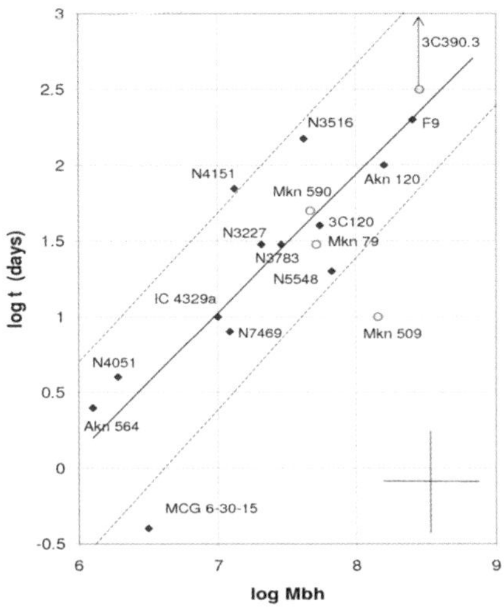

Figure 4. SF break time vs. black hole mass for 13 Seyfert galaxies. Four more objects with weak breaks are denoted by circles. The solid line is the best linear fit, dashed lines are linear extrapolations of the two breaks in the PSD of Cyg X-1. Typical errors are denoted by the cross in the lower right.

The PSD timescales scale approximately linearly with the estimated black hole mass (McHardy *et al.* (2006); Uttley and McHardy (2005)). Using the structure function method enables us to include also objects which do not have sufficient sampling for constructing a PSD. Altogether we have 17 objects. We find a similar relation between the SF break timescale and the BH mass (fig. 3). The black hole masses of most of the Seyferts are estimated by reverberation mapping (Peterson *et al.* 2004). For Akn 564, 3C111 and MCG 6-15-30, which do not have reverberation data, single-epoch Hβ is used with the empirical L-R relation (Wandel, Peterson & Malkan 1999; (Kaspi *et al.* 2000). We note that the timescale of the SF-break is strongly correlated with the black hole mass, the best fit being consistent with a linear relation, $M_{\rm BH} = 1.010^6(\tau_{SF}/\text{day})$. Almost all points coincide, within the errors, with the extrapolated "Cyg-X1 line" marked by the dashed diagonal lines in fig. 3, which shows the extrapolation from the break time - $M_{\rm BH}$ relation for the two breaks in the PSD of Cyg X-1, corresponding to 0.2 sec and 5 sec, respectively. This resultsis consistent with similar trend found in AGN between the black hole mass and the break time in the PSD.

We acknowledge a long lasting collaboration with Alex Markowitz. AW acknowledges the hospitality of the Physics and Astronomy department at UCLA.

References

Chatterjee *et al.* 2009, *ApJ 704, 1689*
Chatterjee et al. 2011, ApJ 734, 43
Collier, S. & Peterson, B. 2001, ApJ 2001,555, 775
Edelson, R. & Nandra, K. 1999, ApJ 514, 682
Emmanoulopoulos, D., & McHardy, I. M. and Uttley, P. 2010, MNRAS 404, 931
Kaspi, S. et al. 2000, ApJ 533, 631
Kataoka et al. . 2001, ApJ 560, 659
Kelly, B. C. et al. 2014, ApJ 788, 33
Markowitz et al. 2003, ApJ 593, 96
McHardy et al. 2005, MNRAS 348, 783
McHardy, I. M. et al. 2006, Nature 444 730
Peterson, B. et al. 2004, ApJ 613, 632
Rothschild, R. E., et al. 2011, ApJ 733, 23
Simonetti et al.. 1985, ApJ 295, 46S
Timmer & Koenig 1995, A & A300, 707
Uttley,P., McHardy, I. M. & Papadakis 2002, MNRAS 332,231
Uttley, P. & McHardy, I. M. 2005, MNRAS 363,586
Wandel, A. & Mushotzky, R. F. 1986, ApJ 306, L61
Wandel, A., Peterson, B.& Malkan, M. 1999, ApJ 526, 579
Webb, W.& Malkan, M. 2000, ApJ 540, 652

New Frontiers in Black Hole Astrophysics
Proceedings IAU Symposium No. 324, 2016
Andreja Gomboc, ed.

doi:10.1017/S1743921317002277

Periodic optical variability of AGN

E. Bon[1], P. Marziani[2], N. Bon[1]

[1]Belgrade Observatory, Serbia
[2]INAF, Osservatorio Astronomico di Padova, Italy
email: ebon@aob.rs

Abstract. Here we present the evidence for periodicity of an optical emission detected in several AGN. Significant periodicity is found in light curves and radial velocity curves. We discuss possible mechanisms that could produce such periodic variability and their implications. The results are consistent with possible detection of the orbital motion in proximity of the AGN central supermassive black holes.

Keywords. quasars: general, black hole physics, accretion, accretion disks

1. Introduction

Different mechanisms have been proposed to explain active galactic nuclei (AGN) emission variability (see e.g. Gaskell 1983, 2009 and the references therein). Besides outflows, jet precession, disk precession, disk warping, spiral arms, flares, and other instabilities, one of the most interesting possibilities involves the existence of a binary system in the core (see eg., Komossa 2006; Bogdanović *et al.* 2008; Gaskell 2009; Bon *et al.* 2012; Bogdanović 2015; Graham *et al.* 2015a, 2015b; Bon *et al.* 16, Charsi *et al.* 2016, Liu *et al.* 2016. and the references therein) and the tidal disruption event (see for e.g. Komossa and Zensus 2015). If black hole (BH) mass grows via major mergers we might expect to see the signature of a binary black hole in some or many active galaxies, if the merger process involves slow coalescence of the two components.

The variability in the nuclei of galaxies was recognized even before AGN were defined as a class. There are light curves showing variability in over 100 years of observations in some objects, for example NGC4151 (from 1906, see e.g., Oknyanskij 2007)), 3c273 (from 1880s Smith & Hoffleit 1963), with long variability timescale of more than one decade. In the 1950s it was reported that the NGC 5548 nucleus appeared to vary by about a magnitude (see more in Gaskell & Klimek 2003).

It is well known that many AGN show variability in different time scales and all wavelengths (Gaskell & Klimek 2003). The variability timescales are affected by speed at which physical processes propagate; most relevant are the speed of light $c \sim 3 \cdot 10^5$ km s^{-1}, the orbital speed $v_{orb} \sim (GM/R)^{1/2}$ and the sound speed $v_s \sim (kT/m)^{1/2}$, where $c > v_{orb} >> v_s$. The shortest timescale corresponds to the light crossing timescale. Reverberation campaigns are based on measuring light travel time between a spectral feature representative of the ionizing continuum and an emission line. If the time delays are measured as a function of radial velocity, it is possible at least in principle to map the line emitting region.

Orbital timescales are longer (Netzer 2013). Measuring them could be very useful for determining the dynamical mass of the central BH, while sound speed timescale can give us information about the property of the disk or the broad line region (BLR).

A question arises: can we identify periodic variations? Such periodic variability should correspond to orbital motion exclusively, while the other two processes could produce only quasi periodic signals. Having an estimate of the BH masses, we can easily filter expected domains of periodicity that could correspond to the orbital motion, for the optical spectral range, where we have the longest observing records. Unfortunately, the brightest active galactic nuclei were identified only about 70 years ago. Therefore observing records are relatively short, and do not allow to trace many orbits in historical light curves. Extensive spectroscopic observations started even later, because the brightest AGN were still faint for the instrumentation available till the mid 1970s. With such limitation in the data, it is really hard to prove that the signal is periodic. If we were able to reveal periodic signals on the orbital timescales, combining results from the light crossing timescales, and measuring the broad emission line shapes (since we know that the low-ionization part of the emitting region is virialized) we could really be able to map the BLR.

2. Light curve variability patterns

AGN variability patterns present in their light curves mainly correspond to red noise, giving spuriously high significance levels to low frequency periods (see Westman *et al.* 2011, Vaughan *et al.* 2016 and Bon *et al.* 2016).

AGN have been since long known to be dominated by red noise-like light curves. Therefore, standard methods, such as the Lomb-Scargle (LS) method (Lomb 1976, Scargle 1982) may show very high peaks but the probability of a false positive computed with conventional statistical approximations may not be valid. Such problems in using LS periodogram tests for the analysis of AGN X-ray light curves has been known for some time, and they appear especially serious for very frequent quasi-periodic variability detections over long time-scales which turned out to be false (e.g., see discussion in Vangham 2006).

AGN light curves can be modeled as damped random walks (DRW) or with autoregressive models (as it has been known since many years; e.g., Fahlman & Ulrych 1975, Gaskell & Peterson 1987). There are some attempts to systematically search for periods in quasars LCs taking this problem into account (e.g. see Charsi *et al.* 2016, Liu *et al.* 2016). These studies show that periodicities of some candidates show very significant power-spectrum peaks compared to expectation for red noise. In some cases there is evidence of even more then one period with very high significance compared to autoregressive (AR) expectations (see e.g. Bahta *et al.* 2016).

Nevertheless, even though activity within one period might look stochastic compared to the DRW tests, repeating patterns could appear on very long light curves, showing the same complex patterns starting to repeat themselves after some periodic interval (see folded light curve presented in Fig. 1, where one can see that noisy pattern repeated after 5676 days in the NGC 5548 optical light curve; more details are available in Bon *et al.* 2016). Periodicity was found using a new method, similar to phase dispersion minimization (PDM, see Stellingwerf 1978), specially developed for the uneven-sampled 43 yr data series of NGC 5548 (Bon *et al.* 2016). Beside light curves, radial velocity curves of NGC 5548 measured on the broad Hβ profile show periodicity of 5550 d, with a peak above 95% AR significance level (of LS false alarm probability), as shown in Fig. 2, (see Table 3 in Bon *et al.* 2016). Using a PDM method and the new compiled light curve from Fig. 1 we found significant peak at 5678 d.

Until now there are only few candidates with the detection of the same periodicity in both light and radial velocity curves: e. g., NGC 4151 (Bon *et al.* 2012) and NGC 5548 (Li *et al.* 2016, Bon *et al.* 2016). Both candidates have been monitored for very long time intervals, with probably the longest and best sampled data series of spectra and light curves, spanning over 43 years for NGC5548 and over 25 yr for NGC4151.

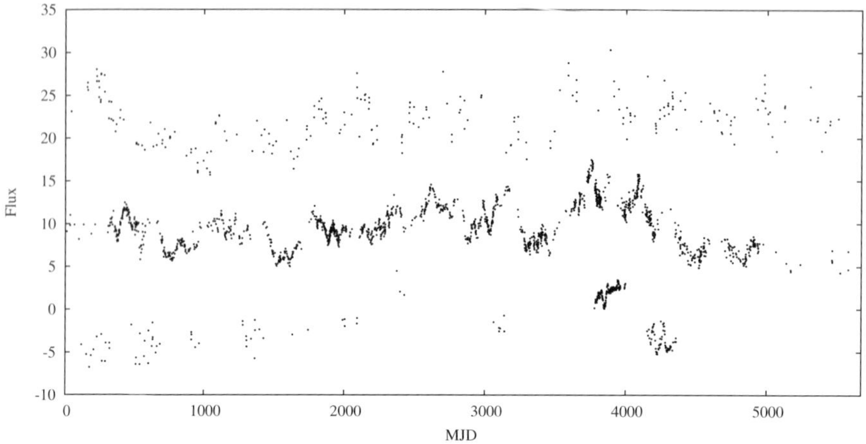

Figure 1. Folded phased diagram of NGC 5548 continuum flux 43 yr long light curve, assuming 5676 days periodicity, here presented to show an example of repeating noisy pattern light curve. Top light curve is pre AGN watch epoch, middle is AGN watch epoch Sergeev *et al.* (2007), and bottom is post AGN watch epoch constructed from new observations presented in Bon *et al.* (2016), Lu *et al.* (2016), Pei *et al.* (2017).

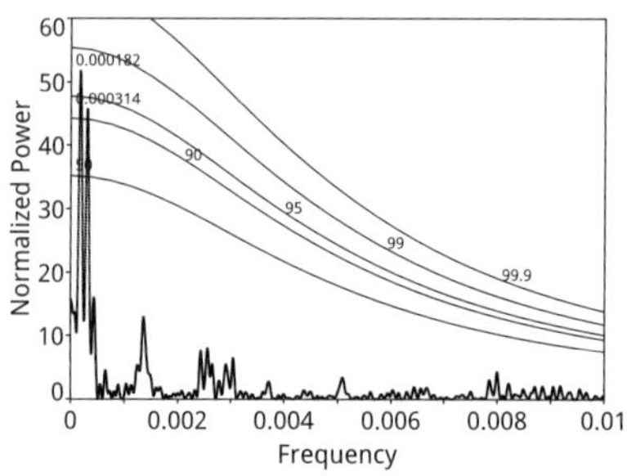

Figure 2. Example of Lomb-Scargle periodogram of NGC 5548 radial velocity curve constructed from the red side broad Hβ half-widths measured at 75% of maximum (see Bon *et al.* 2016). Red noise background spectrum significance levels are marked.

3. Physical interpretations

Probably the most plausible scenario for relatively short periodicity is orbital motion (see eg. Bon *et al.* 2012, 2016; Graham *et al.* 2015a, 2015b). Some mechanisms, are geodetic precession (Begelman *et al.* 1980), accretion disk precession due to a second black hole (Katz 1997), and disk-self warping induced by radiation pressure (Pringle 1996). Some of these periodic AGN are recognized as SMBBH candidates, like NGC4151 (Bon *et al.* 2012), OJ287 (e.g. Pihajoki 2016), PG1302-102 (Graham *et al.* 2015a). Geodetic precession is excluded since it occurs on time scales that are in general much longer than currently available monitoring times. Periodicity could be due to the presence of a long-lived hotspot in the disk (Jovanović *et al.* 2010). We note however that the flare lasted only about a year, and was spiraling outwards in the model of Jovanović *et al.* (2010). Therefore, it would be more likely that the hotspot could be periodically generated by an orbiting body passing through the accretion disk (eg. Kieffer & Bogdanovic 2016, Pihajoki 2016, Bon *et al.* 2016), with an orbital period similar to S0 stars in our Galaxy, or alternatively, a periodic enhancement of the inner part of the disk producing the X-ray emission (Fausnaugh *et al.* 2015, Bon *et al.* 2016).

Graham *et al.* 2015b calculated that warped disks around single black holes are not favored in an AGN context unless very special conditions are satisfied. If the accretion disk is warped, it could lead to precession of the jet, but again with periodicity much longer then current monitored time. Yet another alternative possibility could be an obscuration of the central accretion disk, and of part of the BLR by a moving obscuring object (Bon *et al.* 2016), or moving emitting region, since BLR could consist of two different emitting components (Bon 2008, Bon *et al.* 2009a,2009b, 2015).

References

Begelman, M. C., Blandford, R. D., & Rees, M. J. 1980, Nature , 287, 307

Bhatta, G., Zola, S., Stawarz, Ł., *et al.* 2016, Astrophys. J. , 832, 47

Bogdanović, T., S., Britton, D., Sigurdsson, S., & Eracleous, M., 2008, *ApJS*, 174 , 455

Bogdanović, T., Eracleous, M., & Sigurdsson, S. 2009, New Astronomy Reviews , 53, 113

Bogdanović, T. 2015, *Astrophysics and Space Science Proceedings*, 40, 103

Bon, E., 2008, SerAJ, 177, 9

Bon, E., Popović, L. Č., Gavrilović, N., La Mura, G. , & Mediavilla, E., 2009, Mon. Not. R. Astron. Soc. , 400, 924

Bon, E., Gavrilović, N., La Mura, G., & Popović, L. Č., 2009, *NewAR*, 53, 121

Bon, E., Jovanović, P., Marziani, P., *et al.* 2012, *ApJ*, 759, 118

Bon, E., Zucker, S., Netzer, H., *et al.* 2016, *ApJS*, 225, 29

Bon, N., Bon, E., Marziani, P., & Jovanović, P. 2015, Astrophysics and Space Science , 360, 7

Charisi, M., Bartos, I., Haiman, Z., *et al.* 2016, *MNRAS*, 463, 2145

Fausnaugh, M. M., Denney, K. D., Barth, A. J., *et al.* 2016, *ApJ*, 821, 56

Fahlman, G. G. & Ulrych, T. J. 1975, *ApJ*, 201, 277

Gaskell, C. M. & Peterson, B. M. 1987, *ApJS*, 65, 1

Gaskell, C. M. & Klimek, E. S., 2003 Astronomical and Astrophysical Transactions, 22, 661

Gaskell, C. M.,2009, *New Astronomy Reviews* , 55, 140

Graham, M. J., Djorgovski, S. G., Stern, D., *et al.* 2015a, *Nature*, 518, 74

Graham, M. J., Djorgovski, S. G., Stern, D., *et al.* 2015b, *MNRAS*, 453, 1562

Jovanović, P., Popović, L. Č., Stalevski, M., & Shapovalova, A. I. 2010, Astrophys. J. , 718, 168

Kieffer, T. F. & Bogdanović, T. 2016, *ApJ*, 823, 155

Komossa, S. & Zensus, J. A. 2016, *Star Clusters and Black Holes in Galaxies across Cosmic Time*, 312, 13

Li, Y. R., Wang, J. M., Ho, L. C., *et al.* 2016, *ApJ*, 822, 4

Liu, T., Gezari, S., Burgett, W., *et al.* 2016 ApJ, 833, 6

Lomb, N. R., 1976, Astrophysics and Space Science , 39, 447

Lu, K.-X., Du, P., Hu, C., *et al.* 2016, Astrophys. J. , 827, 118

MacFadyen, A. I. & Milosavljević, M. 2008, *ApJ*, 672, 83

Netzer, H. 2013, *The Physics and Evolution of Active Galactic Nuclei* Cambridge, UK: Cambridge University Press

Pei, L., Fausnaugh, M. M., Barth, A. J., *et al.* 2017, arXiv:1702.01177

Pihajoki, P. 2016, *MNRAS*, 457, 1145

Plavchan, P., Jura, M., Kirkpatrick, J. D., Cutri, R. M., & Gallagher, S. C. 2008, Astrophysical Journal Supplement , 175, 191-228

&Pringle, J. E. 1996, *MNRAS*, 281, 357

Scargle, J. D., 1982, *ApJ*, 263, 835

Sergeev, S. G., Doroshenko, V. T., Dzyuba, S. A., *et al.* 2007, Astrophys. J. , 668, 708

Stellingwerf, R. F. 1978, *ApJ*, 224, 953

Vaughan, S. & Uttley, P. 2006, *Advances in Space Research*, 38, 1405

Vaughan, S., Uttley, P., Markowitz, A. G., *et al.* 2016, *MNRAS*, 461, 3145

Wilson, A. S. & Colbert, E. J. M. 1995, *ApJ*, 438, 62

New Frontiers in Black Hole Astrophysics
Proceedings IAU Symposium No. 324, 2016
Andreja Gomboc, ed.

doi:10.1017/S174392131700237X

Quasi-Periodicities at Year Time Scales in Blazars

Stefano Covino[1], Angela Sandrinelli[1,2] and Aldo Treves[2]
[1]INAF / Brera Astronomical Observatory, via Bianchi 46, 23897, Merate (LC), Italy
[2]Università degli Studi dell'Insubria, Via Valleggio 11, 22100 Como, Italy

Abstract. We examine the 2008-2016 gamma-ray and optical light curves of a number of bright Fermi blazars. In a fraction of them, the periodograms show possible evidence of quasi-periodicities related in the two bands. This coincidence strengthens their physical meaning. Comparing with results from the periodicity search of quasars, the presence of quasi-periodicities in blazars suggests that the basic condition for its observability is related to the relativistic jet in the observer direction, but the overall picture remains uncertain.

Keywords. BL Lacertae objects: general, methods: statistical

1. Introduction

Blazars are active galactic nuclei with a relativistic jet pointing in the observer direction. They are highly variable in all spectral bands (e.g. Falomo *et al.* 2014), and periodicities for these sorces have been hypothesized since the discovery of the first members of the class. Yet, no fully convincing case has been identified, with the possible exception of OJ 287, with a reported 12 year period (e.g. Sillanpää *et al.* 1988) that appears rather robust, but not uncontroversial (e.g. Hudec *et al.* 2013).

In recent years it has become also clear that blazars are the main constituents of the extragalactic γ-ray sky and the *Fermi* mission since its launch has monitored the entire celestial sphere every 3 hours. Therefore, γ-ray light curves of blazars are becoming easily available (Abdo *et al.* 2010), opening the possibility to look for periodicities in multiple and independent bands. In addition, robotic telescopes often developed for the monitorig of high-energy transients as gamma-ray bursts, are now operational since almost a decade, providing a large amount of homogeneous data that can fruitfully be analyzed for identifying long-term oscillations.

One of these small-size robotic facilities is the Rapid Eye Mount (REM, Zerbi *et al.* 2004; Covino *et al.* 2004) telescope (Fig. 1) located at the ESO premise of La Silla since 2004 and since then devoting a large fraction of the available observing time to blazar monitoring in the Southern hemisphere (Fig. 2). Our team derived and analyzed the REM data for six BL Lac objects: PKS 0537-441, PKS 0735+17, OJ 287, PKS 2005-489, PKS 2155-304, and W Comae, and of the flat spectrum radio quasar PKS 1510-089 (Sandrinelli *et al.* 2014a).

2. Quasi Periodicities

For some of the monitored sources, a search for periodicity has been carried out with optical, from REM and other facilities, and high-energy *Fermi* data (Sandrinelli *et al.* 2014b, 2016a,b). We recently also performed analogous analyses on three BL Lac objects in the Northen sky (0716+714, MRK 421 and BL Lac, Sandrinelli *et al.* 2017) and similar investigations have been proposed by other teams (e.g. Ackermann *et al.* 2015). Generally

Figure 1. The REM telescope at the ESO facility of La Silla.

speaking, in a fair fraction of the analyzed sources, possible periodicities at year time-scale have been singled out (Fig. 3).

The proposed quasi-periodicities correspond to peaks in the power density spectra that stand out of the local structured noise rather clearly. However, once a correction for the "trial-factors" due to the large number of frequencies analyzed is applied, the global significance of the periods turns out to be rather low. The whole subject is rather complex and widely discussed in the literature (e.g. Horne & Baliunas 1986; Frescura *et al.* 2008; Süveges 2014; Süveges *et al.* 2015; Vaughan 2005, 2010), and a proper treatment of the statistical significance of the discussed quasi-periodicities will be subject of a forthcoming paper. At this stage it is enough to consider that, beyond any correct evaluation of the chance probabilities for these findings, we observe that in several cases there is a coincidence, within the errors, of the same periods in independent bands as the optical and gamma-rays. This suggests that there could be a physical meaning associated to these periods worth discussing in some detail.

3. Discussion

For a few sources low-significance periodicities simultaneously present in the optical and in the gamma-rays have been identified, enforcing their possible physical relevance. The convincing cases are PKS 2155-304 (Sandrinelli *et al.* 2014b), PG 1553+11 (Ackermann *et al.* 2015), PKS 0537-447 (Sandrinelli *et al.* 2016b) and BL Lac (Sandrinelli *et al.* 2017). These oscillations could be directly related to a real periodicity, as the orbital one of a system of two binary supermassive black holes, or indirectly as being due to an accretion disc/jet precession.

Blazar emissions are characterized by a chaotic component with superimposed a possible periodicity, which is therefore possibly hidden and difficult to single out if, as it is

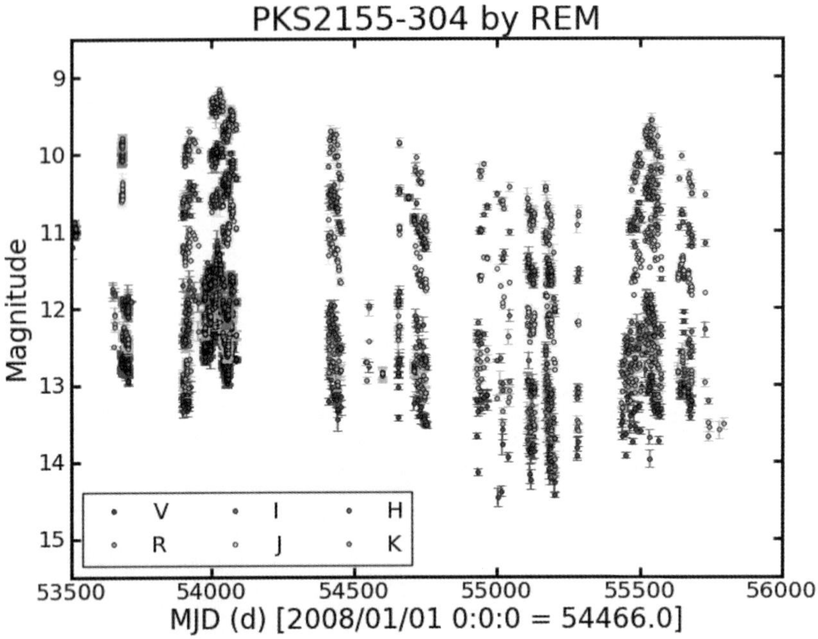

Figure 2. Multicolor light-curve for PKS2155-304.

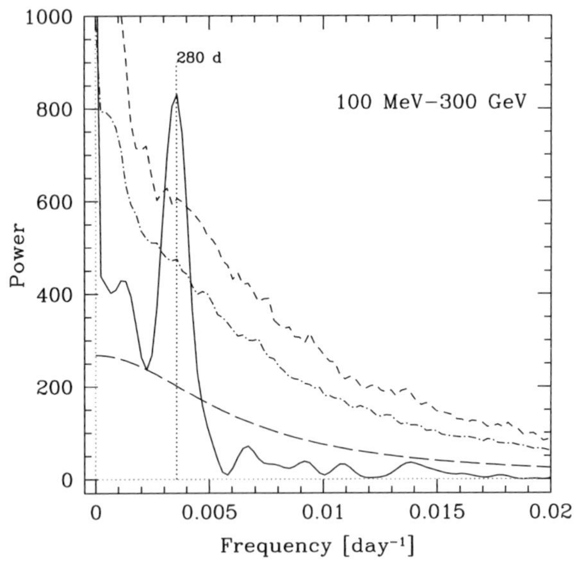

Figure 3. Power density spectrum for the *Fermi* gamma-ray data of an active phase of PKS 0537-441 (black line, from Sandrinelli *et al.* 2016b). A possible periodicity at $T \sim 280$ days is indicated. 95% and 99% significance curves derived from the local noise model (dashed line) are also shown.

typically the case, the data are unevenly sampled and the time-series have of course a finite length. Furthermore, the fact that optical observation campaigns on a blazar are basically activated in case of a high state of the source introduces another complicating factor.

Recently, year-like periodicities have been searched for quasars starting from various surveys, detecting several positive candidate cases among hundred thousands objects (Charisi *et al.* 2016, and references therein). Therefore, there could be a remarkable difference with respect to bright *Fermi* blazars in terms of fraction of quasi-periodicity detections with roughly a year time-scale. If this will be proved to be true it could be related with the main characteristic that distinguishes them from quasars, the presence of a relativistic jet pointing in the observer direction. The net effect would be to obviously amplify the visibility of the oscillations, since it can magnify the variability through known relativistic effects. Furthermore, as already mentioned, it is possible that the oscillations are due to jet instabilities, independent of the presence of a binary companion.

References

Abdo, A. A., Ackermann, M., Ajello, M., *et al.* 2010, *ApJ*, 722, 520
Ackermann, M., Ajello, M., Albert, A., *et al.* 2015, *ApJ*, 813, 41
Charisi, M., Bartos, I., Haiman, Z., *et al.* 2016, *MNRAS*, 463, 2145
Covino, S., Stefanon, M., Sciuto, G., *et al.* 2004, *SPIE* 5492, 1613
Falomo, R., Pian, E., & Treves, A. 2014, *ARAA*, 22, 73
Frescura, F. A. M., Engelbrecht, C. A., & Frank, B. S. 2008, *MNRAS*, 388, 1693
Horne, J. H. & Baliunas, S. L. 1986, *ApJ*, 302, 757
Hudec, R., Bašta, M., Pihajoki, P., & Valtonen, M. 2013, *A&A*, 559, A20
Sandrinelli, A., Covino, S., & Treves, A. 2014a, *A&A*, 562, 79
Sandrinelli, A., Covino, S., & Treves, A. 2014b, *ApJ*, 793, 1
Sandrinelli, A., Covino, S., & Treves, A. 2016a, *AJ*, 151, 54
Sandrinelli, A., Covino, S., & Treves, A. 2016b, *ApJ*, 820, 20
Sandrinelli, A., Covino, S., Treves, A., *et al.* 2017, *A&A*, in press (arXiv:1701.04454)
Sillanpää, A., Haarala, S., Valtonen, M. J., Sundelius, B., & Byrd, G. G. 1988, *ApJ*, 325, 628
Süveges, M. 2014, *MNRAS*, 440, 2099
Süveges, M., Guy, L. P., Eyer, L., *et al.* 2015, *MNRAS*, 450, 2052
Vaughan, S. 2005, *A&A*, 431, 391
Vaughan, S. 2010, *MNRAS*, 402, 307
Zerbi, F. M., Chincarini, G., Ghisellini, G., *et al.* 2004, *SPIE*, 5492, 1590

New Frontiers in Black Hole Astrophysics
Proceedings IAU Symposium No. 324, 2016
Andreja Gomboc, ed.

doi:10.1017/S1743921317001065

Radio and γ-ray loud narrow-line Seyfert 1 galaxies in the spotlight

V. Karamanavis[1,⋆], E. Angelakis[1], S. Komossa[2], I. Myserlis[1], D. Blinov[3] and J. A. Zensus[1]

[1]Max-Planck-Institut für Radioastronomie, Auf dem Hügel 69, D-53121 Bonn, Germany
email: vkaramanavis@mpifr.de

[2]QianNan Normal University for Nationalities, Longshan Street, Duyun, Guizhou, China

[3]Department of Physics and Institute of Theoretical and Computational Physics, University of Crete, 71003 Heraklion, Greece

Abstract. Narrow-line Seyfert 1 (NLS1) galaxies provide us with unique insights into the drivers of AGN activity under extreme conditions. Given their low black hole (BH) masses and near-Eddington accretion rates, they represent a class of galaxies with rapidly growing supermassive BHs in the local universe. Here, we present the results from our multi-frequency radio monitoring of a sample of γ-ray loud NLS1 galaxies (γNLS1s), including systems discovered only recently, and featuring both the nearest and the most distant γNLS1s known to date. We also present high-resolution radio imaging of 1H 0323+342, which is remarkable for its spiral or ring-like host. Finally, we present new radio data of the candidate γ-emitting NLS1 galaxy RX J2314.9+2243, characterized by a very steep radio spectrum, unlike other γNLS1s.

Keywords. galaxies: active, galaxies: jets, galaxies: Seyfert, radio continuum: galaxies

1. Introduction

Narrow-line Seyfert 1 galaxies are a particular class of active galactic nuclei (AGN) that show: (i) small width of their broad optical emission lines, i.e. FWHM(H_β) $\leqslant 2000\,\mathrm{km\,s^{-1}}$ (Osterbrock & Pogge 1985), pointing to a low black hole (BH) mass for these systems ($\sim 10^6 - 10^8\,M_\odot$), (ii) super-strong Fe II emission complexes, (iii) rapid X-ray variability, hinting that the broad-line region (BLR) and accretion disk are directly visible, (iv) near-Eddington accretion rates with ratios L/L_{Edd} between 0.1 and 1 (Boroson & Green 1992), (v) super-soft X-ray spectra, and (vi) a multitude of intriguing multi-wavelength properties (see review by Komossa 2008).

There exists also an interesting small fraction of their population being radio-loud, featuring relativistic jets, and detected at γ rays by *Fermi* (γNLS1s; Komossa *et al.* 2006; Abdo *et al.* 2009a,b). These few sources, the radio- and γ-ray-loud population, are exceptional because they show blazar-like observational attributes such as flat radio spectra, high brightness temperatures ($T_b \sim 10^{10} - 10^{14}$ K), Doppler boosting, γ-ray emission and in few cases, one-sided jets (D'Ammando *et al.* 2013; Angelakis *et al.* 2015; Karamanavis 2015; Fuhrmann *et al.* 2016). However, possessing non-blazar physical properties, especially almost two orders of magnitude lower BH masses and high accretion rates. In this context, they constitute a source of new insights, able to give us clues on the formation of extragalactic jets and their evolution under conditions of high accretion rate and in a regime that is not probed by classical blazars. As possible explanations for their appearance, different scenarios have been proposed. Do they represent a young AGN population that rapidly grow their BHs? And how their orientation, with respect to the observer's line of sight, enters this challenging equation?

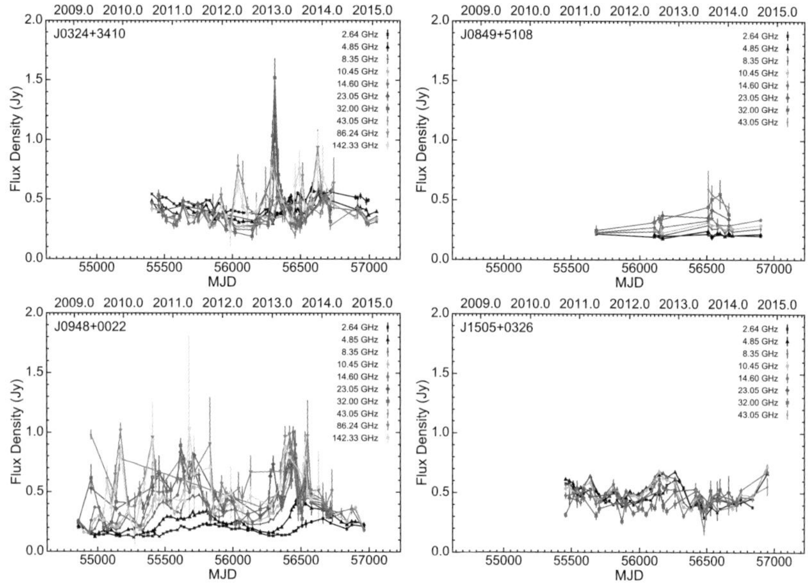

Figure 1. Updated radio light curves of γNLS1s between 2.6 and 42 GHz, until 2015 January. Where available, data at 86 and 142.33 GHz are also shown.

In the following, we present our radio monitoring of γNLS1s using primarily the 100-m telescope at Effelsberg and the IRAM 30-m telescope, while focusing on latest results. Furthermore, we focus on two γNLS1s which have been studied in detail: (i) RX J2314.9+2243 with dedicated multi-wavelength monitoring, and (ii) 1H 0323+342 with the help of high-resolution very-long-baseline interferometry (VLBI).

2. Radio monitoring of NLS1 galaxies

Our monthly monitoring of the jet emission from γ-ray-emitting NLS1 galaxies constitutes the most comprehensive such program at cm and short mm wavelengths (Angelakis *et al.* 2015). Observations with the Effelsberg 100-m telescope cover eight bands between 2.64 and 43.05 GHz. Until 2014, the IRAM 30-m telescope provided coverage at 86.24 and 142.33 GHz. The data collected in more than five years is the longest data set with the widest frequency coverage available for these objects (see Fig. 1). It includes also radio polarisation monitoring at 2.64, 4.85, 8.35, and 10.45 GHz, with measurements of its linear and circular components and EVPAs. The observed sample comprises the following sources: 1H 0323+342 (J0324+3410), SBS 0846+513 (J0849+5108), PMN J0948+0022 (J0948+0022), and PKS 1502+036 (J1505+0326). Recently, three more γNLS1s were included, namely FBQS J1644+2619, SDSS J1222+0413, and B3 1441+476.

All sources show the typical blazar behaviour although with lower flux densities. γNLS1s are variable and flare often, but the events are characterized by smaller amplitude and shorter duration. However, more energetic outbursts can be seen from time to time which dominate a source's light curve. Their spectra exhibit intense evolution with time, and steep, flat or even inverted spectra can be seen at different epochs. This evolution occurs faster than in blazars and can be ascribed to evolving shocks. The estimates of their jet powers are comparable with the output of the least energetic blazars. Furthermore, their moderate T_b and associated Doppler factors ($\delta \lesssim 10$), point to mildly relativistic jets as the driver of their phenomenology (Angelakis *et al.* 2015).

Most sources do not display detectable polarisation. However, 1H 0323+342 is ~5.5% linearly polarized, higher than what is typically observed in blazars (e.g. ~3.5–5% Myserlis 2015). Its EVPA, at ~45°, lies perpendicular to the jet orientation, suggesting a magnetic field orientation almost parallel to the jet. PKS 1502+036 appears to be polarized at 8.35 GHz (~2–3% with its EVPA at ~55°) during flaring periods. In the R band, γNLS1s have been monitored with RoboPol since 2013 (e.g. Angelakis *et al.* 2016). Their mean linear polarisation ranges from less than 1% to up to 20%.

3. The curious case of RX J2314.9+2243

This is a mildly radio-loud (R = 10–20) NLS1 galaxy at a redshift z = 0.1692 with a BH mass of $8\times10^7 M_\odot$ and a ratio $L/L_{\rm Edd} = 0.2$ (Komossa *et al.* 2006). What sets this source apart from the rest of the NLS1 population is its tentative detection at γ rays by *Fermi* (L. Foschini, priv. comm.) and its steep radio spectrum ($\alpha = -0.76$; Komossa *et al.* 2015). The ongoing radio observations verify the steep spectrum with flux densities $S_{4\,\rm GHz} = 10 \pm 2$ mJy and $S_{8\,\rm GHz} = 6 \pm 1$ mJy on two occasions in 2014 October. To date, all γ-ray detected NLS1s have been associated with flat spectrum sources and RX J2314.9+2243 could be the first steep radio spectrum NLS1 to emit γ rays.

Komossa *et al.* (2015) address the emission processes shaping the spectral energy distribution (SED) of RX J2314.9+2243. The source is a luminous infrared emitter with an SED showing a broad emission hump between IR and UV bands, that steepens in the UV. An unusual feature of the source, compared to other NLS1s with steep X-ray spectra (e.g. Boller *et al.* 1996; Grupe 2004; Zhou *et al.* 2006), is the flatness of its X-ray spectrum, hinting at jet activity. In the UV on the other hand, the spectrum is very steep but there is no evidence of optical extinction beyond what is expected from our Galaxy, as would be expected by the dusty environment of luminous infrared galaxies (LIRGs). Therefore the IR to UV emission is most likely of non-thermal (i.e. synchrotron) origin arising from the operating jet.

Another striking feature of RX J2314.9+2243 is its very broad O[III] 5007 Å emission line component. This shows an unusually high kinematical blueshift of 1260 km s^{-1} that could be due to the presence of an outflow powered by a radio jet feedback, in a face-on orientation with respect to our line of sight (Komossa *et al.* 2015).

4. Direct imaging of 1H 0323+342 at 15 GHz

1H 0323+342 is the most nearby radio-loud (R ~ 50) γ-ray emitting NLS1 at z = 0.0629 (Zhou *et al.* 2007). It exhibits a high Eddington luminosity ratio ($L/L_{\rm Edd} \sim 0.1$) with a low BH mass of $\sim10^7 M_\odot$ confirmed by several independent estimates (Zhou *et al.* 2007; Yao *et al.* 2015; Wang *et al.* 2016; Landt *et al.* 2016). 1H 0323+342 is highly variable at radio bands (cm to mm; see Fig. 1) and is also a special case because of the morphology of its host galaxy which appears ring-like or as a one-armed spiral (Zhou *et al.* 2007; Antón *et al.* 2008; León Tavares *et al.* 2014).

We studied the parsec-scale structure and kinematics of the source employing high-resolution VLBI images at 15 GHz obtained in eight epochs from 2010 October until 2013 July†. Additionally, with our single-dish observations we infer the T_b and δ for the source. This combination provides a good estimate for the viewing angle towards 1H 0323+342 (see Karamanavis 2015, for first results).

1H 0323+342 shows a core–jet morphology characterized by the prominent core, a straight jet, and a total absence of emission for the receding side of the jet; i.e. no counter-jet is detected. A stationary feature is seen close to the core while six components move

† Data from the MOJAVE monitoring program (Lister *et al.* 2009).

along the relativistic jet. Five jet components show superluminal motion with velocities from 1 c up to 7 c (Karamanavis 2015; Fuhrmann *et al.* 2016).

The source shows fast variability seen with single-dish and VLBI monitoring alike. The flaring events are of low to mild amplitude, but more rapid compared to the typical long-term behavior of blazars. 1H 0323+342 shows high brightness temperatures. While Angelakis *et al.* (2015) report a Doppler factor of 3.6 at 14.6 GHz, and a higher one ($\delta = 4.3$) at 2.64 GHz, we deduce a more extreme value of $T_b = 5.7 \times 10^{12}$ K and an associated Doppler factor of 5.2, from the most rapid and largest flux density variation. Combining the knowledge on the source kinematics (i.e. the speeds of jet components) with the Doppler factors inferred from the observed variability, allowed us, under the assumption of causality, to constrain the viewing angle towards the source. The range of plausible values is $\theta \leqslant 4°–13°$ (Karamanavis 2015; Fuhrmann *et al.* 2016). The smaller value of this range is consistent with viewing angles typical of blazars, but nevertheless a larger θ with the respect to the observer's line of sight cannot be excluded.

5. Concluding remarks

The remarkable characteristics of NLS1 galaxies separate them as prominent class of AGN and made them, in recent years, the focal point of intensive research. These systems accrete matter at a rate near the Eddington limit and rapidly grow the low-mass black holes driving them. The study of the small population of radio-loud and gamma-ray detected NLS1s probes a parameter space not constrained by classical blazars and can provide a deeper understanding on the formation and evolution of powerful radio jets and the interplay with their complex environments.

References

Abdo, A. A., Ackermann, M., Ajello, M., *et al.* 2009a, *ApJ*, 707, 727
Abdo, A. A., Ackermann, M., Ajello, M., *et al.* 2009b, *ApJ*, 707, L142
Angelakis, E., Fuhrmann, L., Marchili, N., *et al.* 2015, *A&A*, 575, A55
Angelakis, E., Hovatta, T., Blinov, D., *et al.* 2016, *MNRAS*, 463, 3365
Antón, S., Browne, I. W. A., & Marchã, M. J. 2008, *A&A*, 490, 583
Boller, T., Brandt, W. N., & Fink, H. 1996, *A&A*, 305, 53
Boroson, T. A. & Green, R. F. 1992, *ApJS*, 80, 109
D'Ammando, F., Orienti, M., Finke, J., *et al.* 2013, *MNRAS*, 436, 191
Fuhrmann, L., Karamanavis, V., Komossa, S., *et al.* 2016, arXiv:1608.03232
Grupe, D. 2004, *AJ*, 127, 1799
Karamanavis, V. 2015, PhD thesis, University of Cologne, 2015
Komossa, S. 2008, in , Vol. 32, Revista Mexicana de Astronomia y Astrofisica Conference Series, 86
Komossa, S., Voges, W., Xu, D., *et al.* 2006, *AJ*, 132, 531
Komossa, S., Xu, D., Zhou, H., Storchi-Bergmann, T., & Binette, L. 2008, *ApJ*, 680, 926
Komossa, S., Xu, D., Fuhrmann, L., *et al.* 2015, *A&A*, 574, A121
Landt, H., Ward, M. J., Balokovic, M., *et al.* 2016, arXiv:1609.08002
León Tavares, J., Kotilainen, J., Chavushyan, V., *et al.* 2014, *ApJ*, 795, 58
Lister, M. L., Aller, H. D., Aller, M. F., *et al.* 2009, *AJ*, 137, 3718-3729
Myserlis, I. 2015, PhD thesis, University of Cologne, 2015
Osterbrock, D. E. & Pogge, R. 1985, *ApJ*, 297, 166
Wang, F., Du, P., Hu, C., *et al.* 2016, *ApJ*, 824, 149
Yao, S., Yuan, W., Komossa, S., *et al.* 2015, *AJ*, 150, 23
Zhou, H., Wang, T., Yuan, W., *et al.* 2006, *ApJS*, 166, 128
Zhou, H., Wang, T., Yuan, W., *et al.* 2007, *ApJ*, 658, L13

New Frontiers in Black Hole Astrophysics
Proceedings IAU Symposium No. 324, 2016
Andreja Gomboc, ed.

doi:10.1017/S1743921317001740

The jet detection in radio-loud narrow-line Seyfert 1 galaxies

Minfeng Gu

Key Laboratory for Research in Galaxies and Cosmology, Shanghai Astronomical Observatory, Chinese Academy of Sciences, 80 Nandan Road, Shanghai, 200030, China
email: gumf@shao.ac.cn

Abstract. With relatively small black hole masses and high accretion rates, narrow-line Seyfert 1 galaxies are thought to be young AGNs. About 7 % of them are radio-loud narrow-line Seyfert 1 galaxies (RLNLS1s). RLNLS1s allow us to re-address some of the key questions regarding the physics of jet formation. As the first step of the systematic study on the jet properties of RLNLS1s, we present the radio structure of fourteen RLNLS1s from VLBA observations at 5 GHz in 2013. Although all these sources are very radio-loud with $R > 100$, their jet properties are diverse, in terms of their pc-scale morphology and overall radio spectral shape. The core brightness temperatures of our sources are significantly lower than those of blazars, therefore, the beaming effect is generally not significant, compared to blazars. This implies that the bulk jet speed may likely be low in our sources.

Keywords. galaxies: active, galaxies: jets, galaxies: Seyfert, radio continuum: galaxies.

1. Introduction

Narrow-line Seyfert 1 galaxies (NLS1s) are a subclass of active galactic nuclei (AGNs), with strong permitted optical/UV Fe II emission lines, relatively weak forbidden-line emission (i.e., [OIII]5007/Hβ < 3), and their broad lines are narrower than those of normal broad-line Seyfert 1 galaxies with FWHM(Hβ) less than 2000 km s^{-1} (Osterbrock & Pogge, 1987; Goodrich *et al.* 1989). NLS1s are often thought to be young and still evolving AGNs with relatively small black hole masses and high accretion rates (see review by Komossa 2008). NLS1s as a class show a low probability to be radio loud at $\sim$ 7%, compared to normal broad line AGNs (e.g., Komossa *et al.*, 2006; Zhou *et al.* 2006). Interestingly, radio-loud NLS1s (RLNLS1s) are inhomogeneous in their radio properties. As shown in Komossa *et al.* (2006), most RLNLS1s in their sample are compact, steep spectrum sources, and hence likely associated with compact steep-spectrum (CSS) radio sources, while only a few NLS1s showed blazar-like behaviour. On the other hand, observational evidence has recently shown that a significant fraction of RLNLS1s at the highest radio-loudnesses, does display the characteristics of blazars, including large-amplitude radio flux and spectral variability, compact radio cores, very high variability brightness temperatures, enhanced optical continuum emission, flat X-ray spectra, and blazar-like spectral energy distributions (e.g., Yuan *et al.* 2008). RLNLS1s are at an opposite end of AGN correlation space than classical radio-loud AGNs (e.g., Sulentic *et al.* 2008), in terms of small black hole masses, high accretion rate, and very strong Fe II emission (Komossa *et al.* 2006). Therefore, RLNLS1s allow us to re-address some of the key questions regarding the physics of jet formation, for example, the physical conditions under which a jet can be launched.

2. The jet properties of RLNLS1s

Relativistic jets were shown to exist in a few RLNLS1s using Very Long Baseline Interferometry (VLBI) observations, based on their high brightness temperatures, and their inverted radio spectra (Doi *et al.* 2007), which was later confirmed by the detection of flaring γ-ray emission (Abdo *et al.* 2009a,b). The previous work based on VLBA data for a few sources argued that RLNLS1s can be either intrinsically radio loud, or apparently radio loud due to jet beaming effects (Gu & Chen 2010; Doi *et al.* 2011). These observational results may be contrary to the well-known paradigm that jets are generally associated with elliptical host galaxies in typical radio-loud AGNs, while there is tentative evidence suggesting that at least a few low-redshift RLNLS1s are hosted by spiral galaxies (e.g., Zhou *et al.* 2007; Antón *et al.* 2008; León Tavares *et al.* 2014). Although a number of RLNLS1s have been imaged with VLBI (Gu & Chen 2010; Doi *et al.* 2011; Giroletti *et al.* 2011; D'Ammando *et al.* 2013b; Wajima *et al.* 2014; Richards & Lister 2015), comprehensive large-sample studies of their jet properties on parsec scales have just started (Gu *et al.* 2015; Gu *et al.* 2016).

2.1. *γ-ray detected sources*

Nine NLS1s have been detected in the γ-ray band by the *Fermi*/LAT with high significance hitherto† (PMN J0948+0022, Abdo *et al.* 2009a; PKS 1502+036, 1H 0323+342 and PKS 2004−447, Abdo *et al.* 2009b; SBS 0846+513, D'Ammando *et al.* 2012; FBQS J1644+2619, D'Ammando *et al.* 2015; B3 1441+476, Liao *et al.* 2015; SDSS J122222.55+041315.7, Yao *et al.* 2015; SDSS J124634.65+023809.0, D'Ammando *et al.* 2016). All these γ-ray NLS1s possess a one-sided jet extending from a dominant central core, reminiscent of blazars (e.g., D'Ammando *et al.*, 2013a; Wajima *et al.*, 2014), except for SDSS J124634.65+023809.0, in which only a compact core is detected (see Fig. 1, and also Gu *et al.* 2015). The brightness temperature is usually high, well in excess of the equipartition limit and in several cases above the inverse-Compton limit (Schulz *et al.* 2016). The values are comparable to those of blazars, strongly favoring the beaming effect in these gamma-ray objects. So far, the apparent superluminal motion has been detected in three objects, SBS 0846+513 ($\beta_{\rm app} = (9.3 \pm 0.6)c$, D'Ammando *et al.* 2013b), PMN J0948+0022 and 1H 0323+342 (Lister *et al.* 2016). A maximum apparent speed of $0.94c$ is reported for SDSS J122222.55+041315.7‡, the γ-ray NLS1 with largest redshift ($z = 0.966$) (Yao *et al.*, 2015). B3 1441+476, the only NLS1 galaxy both detected in γ-ray band with high significance and with radio properties similar to CSS sources¶ (Liao *et al.* 2015), shows a core-jet structure with jet extending to about 30 milliarcsecond (Gu *et al.* 2015). The core brightness temperature is $\sim 10^{10.3}$ K at 5 GHz‖. The value is at least one order of magnitude lower than most flat-spectrum γ-ray NLS1 galaxies, indicating intrinsic differences in Doppler boosting of the jet emission.

2.2. *The systematic studies*

In order to understand the jet formation in RLNLS1s, the first systematic investigation on the jet properties based on VLBA observations of 14 sources has been carried out (Gu *et al.* 2015). All fourteen NLS1 galaxies were detected as single unresolved sources

† Two more tentative candidates, FBQS J1102+2239 and RX J2314.9+2243 were mentioned by Foschini(2011) and Komossa *et al.* (2015).

‡ http://www.physics.purdue.edu/astro/MOJAVE/sourcepages/1219+044.shtml

¶ The CSS-like NLS1 galaxy, PKS 2004−447, was also detected by *Fermi*/LAT, however, the optical spectral classification as a NLS1 galaxy continues to remain uncertain.

‖ However, the core identification is uncertain (Gu *et al.* 2015).

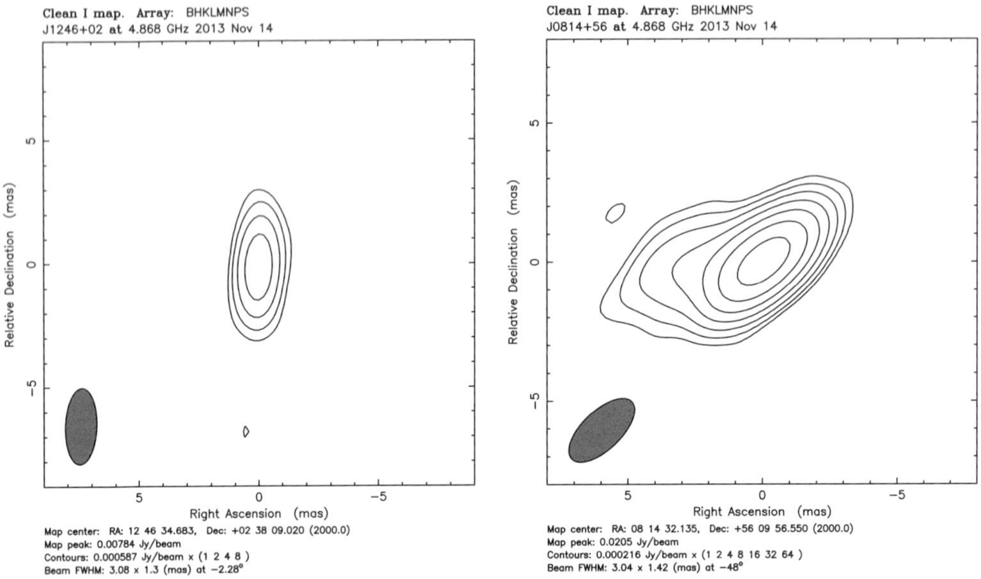

Figure 1. Example VLBA 5 GHz images. *Left*: SDSS J124634.65+023809.0 with a compact core only. *Right*: SDSS J081432.11+560956.6 with a core-jet structure.

within 1 arcmin of the SDSS position in both FIRST and NVSS, indicating compact radio structure at FIRST resolution (5 arcsec). Although all these sources are very radio-loud with $R > 100$, their jet properties are diverse, in terms of their milli-arcsecond (mas) scale (pc scale) morphology and their overall radio spectral shape. From VLBA 5 GHz images, seven sources show a compact core only, while the remaining seven objects have core-jet structure (see example images in Fig. 1). The majority of these sources is compact at VLBI mas scales with all components being within ~ 5 mas (~ 24 - 38 pc depending on source redshift). In combination with archival multi-frequency data, seven sources show flat or even inverted radio spectra ($\alpha \leqslant 0.5$, $f_\nu \propto \nu^{-\alpha}$), while steep spectra ($\alpha > 0.5$) are revealed in the remaining seven objects.

From the high-resolution VLBA images, the brightness temperature of the radio core in the rest frame can be estimated (e.g., Ghisellini *et al.* 1993), which is often used to constrain the jet Doppler factor in combination with the assumed intrinsic brightness temperature, either as the equipartition brightness temperature of 5×10^{10} K (Readhead 1994), or the inverse Compton catastrophe brightness temperature of $\sim 10^{12}$ K (Kellermann & Pauliny-Toth 1969). The core brightness temperature ranges from $10^{8.4}$ to $10^{11.4}$ K with a median value of $10^{10.1}$ K. This indicates that the radio emission is from non-thermal jets, confirming that powerful jets can be formed in accretion systems with relatively small black hole masses and high accretion rates (Gu & Chen 2010, Doi *et al.* 2011). However, the core brightness temperatures of these sources are significantly lower than those of blazars, as the VLBA core brightness temperatures of blazars typically range between 10^{11} and 10^{13} K with a median value near 10^{12} K, and can even extend up to 5×10^{13} K (Kovalev *et al.* 2005, 2009). The values are also well below those of flat-spectrum γ-ray NLS1 galaxies. Therefore, the beaming effect is generally not significant in these sources, implying that the bulk jet speed may likely be low. The evidence for slow jet speeds (i.e., less relativistic jets), in combination with the low kinetic/radio power, may offer an explanation for the compact VLBA radio structure in most sources. The mildly relativistic jets in these high accretion rate systems are consistent with a scenario, where jets are accelerated from the hot corona above the disk by the magnetic field and the radiation force of the accretion disk (Cao 2014). Alternatively, a low jet bulk velocity can be explained by low spin in the Blandford-Znajek mechanism (Blandford & Znajek 1977).

Acknowledgements

This work is supported by the National Science Foundation of China (grants 11473054 and U1531245) and by the Science and Technology Commission of Shanghai Municipality (grant 14ZR1447100).

References

Abdo, A. A., *et al.* 2009a, *ApJ*, 699, 976
Abdo, A. A., *et al.* 2009b, *ApJ*, 707, L142
Antón, S., Browne, I. W. A., & Marchã, M. J. 2008, *A&A*, 490, 583
Blandford, R. D. & Znajek, R. L. 1977, *MNRAS*, 179, 433
Cao, X. 2014, *ApJ*, 783, 51
D'Ammando, F., Orienti, M., Finke, J., *et al.* 2012, *MNRAS*, 426, 317
D'Ammando, F., Orienti, M., Doi, A., *et al.* 2013a, *MNRAS*, 433, 952
D'Ammando, F., Orienti, M., Finke, J., *et al.* 2013b, *MNRAS*, 436, 191
D'Ammando, F., Orienti, M., Larsson, J., & Giroletti, M. 2015, *MNRAS*, 452, 520
D'Ammando, F., Orienti, M., Finke, J., *et al.* 2016, arXiv:1609.04434
Doi, A., *et al.* 2007, PASJ, 59, 703
Doi, A., Asada, K., & Nagai, H. 2011, *ApJ*, 738, 126
Foschini, L. 2011, Narrow-Line Seyfert 1 Galaxies and their Place in the Universe, 24
Ghisellini, G., Padovani, P., Celotti, A., & Maraschi, L. 1993, *ApJ*, 407, 65
Giroletti, M., Paragi, Z., Bignall, H., *et al.* 2011, *A&A*, 528, L11
Goodrich, R. W. 1989, *ApJ*, 342, 22
Gu, M. & Chen, Y. 2010, *AJ*, 139, 2612
Gu, M., Chen, Y., Komossa, S., *et al.* 2015, *ApJS*, 221, 3
Gu, M., Chen, Y., Komossa, S., Yuan, W., & Shen, Z. 2016, Astronomische Nachrichten, 337, 125
Kellermann, K. I., Pauliny-Toth I. I. K. 1969, *ApJ*, 155, L71
Komossa, S., Voges, W., Xu, D., Mathur, S., Adorf, H. M., Lemson, G., Duschl, W., & Grupe D. 2006, *AJ*, 132, 531
Komossa, S. 2008, Rev. Mex. AA Ser. Conf., 32, 86
Komossa, S., Xu, D., Fuhrmann, L., *et al.* 2015, *A&A*, 574, A121
Kovalev, Y. Y., Kellermann, K. I., Lister, M. L., *et al.* 2005, *AJ*, 130, 2473
Kovalev, Y. Y., Aller, H. D., Aller, M. F., *et al.* 2009, *ApJL*, 696, L17
León Tavares, J., Kotilainen, J., Chavushyan, V., *et al.* 2014, *ApJ*, 795, 58
Liao, N.-H., Liang, Y.-F., Weng, S.-S., Gu, M.-F., & Fan, Y.-Z. 2015, arXiv:1510.05584
Lister, M. L., Aller, M. F., Aller, H. D., *et al.* 2016, *AJ*, 152, 12
Osterbrock D. E., Pogge R. W. 1985, *ApJ*, 297, 166
Readhead, A. C. S.. 1994, *ApJ*, 426, 51
Richards, J. L. & Lister, M. L. 2015, *ApJL*, 800, L8
Schulz, R., Kreikenbohm, A., Kadler, M., *et al.* 2016, *A&A*, 588, A146
Sulentic, J. W., Zamfir, S., Marziani, P., & Dultzin, D. 2008, Revista Mexicana de Astronomia y Astrofisica Conference Series, 32, 51
Wajima, K., Fujisawa, K., Hayashida, M., *et al.* 2014, *ApJ*, 781, 75
Yao, S., Yuan, W., Komossa, S., *et al.* 2015, *AJ*, 150, 23
Yuan, W., Zhou, H. Y., Komossa, S., Dong, X. B., Wang T. G., Lu, H. L., & Bai J. M. 2008, *ApJ*, 685, 801
Zhou, H., Wang, T., Yuan, W., Lu, H., Dong, X., Wang, J., & Lu Y. 2006, *ApJS*, 166, 128
Zhou, H., *et al.*, 2007, *ApJ*, 658, L13

New Frontiers in Black Hole Astrophysics
Proceedings IAU Symposium No. 324, 2016
Andreja Gomboc, ed.

doi:10.1017/S1743921317002265

Elusive Accretion Discs in Low Luminosity AGN

J.A. Fernández-Ontiveros[1,2], M.A. Prieto Escudero[1,2], S. Markoff[3], L. Reb[4], D. Espada[5,6] and O. González-Martín[7]

[1]Instituto de Astrofísica de Canarias (IAC), E–38205 La Laguna, Tenerife, Spain
[2]Universidad de La Laguna, Dept. Astrofísica, E–38206 La Laguna, Tenerife, Spain
[3]Anton Pannekoek Institute of Astronomy, University of Amsterdam, Science Park 904, 1098 XH Amsterdam, The Netherlands
[4]Ludwig-Maximilians-Universität München, Geschwister-Scholl-Platz 1, 80539 Munich, Germany
[5]Joint ALMA Observatory, Alonso de Cordova 3107, Vitacura, Santiago 763-0355, Chile
[6]Instituto de Radioastronomía y Astrofísica (UNAM), 3–72 (Xangari), 8701, Morelia, México

Abstract. Low luminosity AGN (LLAGN) represent the vast majority of the AGN population in the near universe. However, they show intrinsic differences when compared to the Unified Model scenario, i.e. a lack of both the big blue bump and the IR bump in their spectral energy distribution (SED), which are the signatures of an accretion disk and a torus, respectively. In this work we present SED for a sample of six LLAGN in the nearby Universe obtained from subarcsec resolution observations. The nature of the IR continuum emission is discussed in the context of non-thermal radiation powered by compact jets.

---

1. Introduction

The vast majority of active galaxies in the Local Universe are Low Luminosity Active Galactic Nuclei (LLAGN), characterised by low accretion rates and modest luminosities (Ho 2008). However, LLAGN are not just scale-down versions of their bright counterparts, showing significant differences with Seyfert galaxies and quasars: *i)* absence of the big blue bump (Ho 1996), signature of the accretion disk in the Spectral Energy Distribution (SED); *ii)* LLAGN tend to be "radio-loud", showing compact cores and parsec-scale radio jets (Nagar *et al.* 2005); and *iii)* they are frequently associated with low-ionisation nuclear emission-line regions (LINERs, Heckman 1980).

Structural changes predicted at low luminosities might explain these differences: the broad-line region and the "torus", keystones of the Unified Model (Antonucci 1993), are expected to vanish at $L_{bol} \lesssim 10^{42}$ erg/s (Laor 2003; Hönig & Beckert 2007). The absence of the blue bump emission led to the wide use of the advection-dominated accretion solution (ADAF) which has been applied successfully from X-ray binaries to Sgr A* (Narayan *et al.* 1995), to explain the spectral shape of LLAGN and their low luminosity (Fabian & Rees 1995; Di Matteo *et al.* 2003; Yuan *et al.* 2009; Eracleous *et al.* 2010). However, ADAF models have been mainly applied to SEDs with scarce wavelength coverage and/or based on flux measurements using large apertures. This is especially critical in the optical and IR ranges, where the host galaxy usually outshines these faint nuclei.

2. High-angular resolution data

This work is based on a multiwavelength, high-angular resolution dataset for the inner few kiloparsecs of a sample of six LLAGN. The use of Adaptive Optics (AO) in the near-IR (NACO) and diffraction-limited imaging in the mid-IR (VISIR) at the Very Large

Telescope (VLT) allowed us to sample the 1–20 μm range with nuclear flux measurements using subarcsec apertures. The dataset was completed with optical/UV imaging with the *Hubble Space Telescope* (*HST*) scientific archive. Nuclear fluxes were measured using aperture photometry of the unresolved component in the centre ($\lesssim 0.''1$), subtracting the local background around ($0.''2$–$0.''3$) in order to remove the contribution of stars and dust in the host galaxy. In the mid-IR, PSF photometry was taken from Reunanen *et al.* (2010); Asmus *et al.* (2014). Measurements in the radio and X-ray ranges have been collected after an extensive and careful search in the literature. These correspond mainly to Very Large Array (VLA) and Very Long Baseline Interferometry (VLBI) in radio and *Chandra*, *XMM*-Newton and *Integral* at X-rays. Despite of their lower spatial resolution, the flux at high energies is dominated by the AGN and not by the host galaxy, and thus can be consistently compared with subarcsec measurements at other wavelength ranges. The characteristics of this dataset allow us to build, for each LLAGN in the sample, a consistent SED of the same physical region, very well sampled over a wide range in wavelength. Additionally, a low-spatial resolution SED based on apertures larger than few arcsec was also built using the NED database† and measurements published in the literature. The low-spatial resolution data is complementary to the subarcsec SED, as they permit to identify those spectral ranges in which the host galaxy shows a strong contribution, that might contaminate the nuclear spectrum.

This project is part of *The central parsecs of the nearest galaxies*‡ (Reunanen *et al.* 2010; Prieto *et al.* 2010). The objects included in the present work correspond to the nearest LLAGN accessible from the Southern Hemisphere. Three of the objects are the canonical reference for definition of the LLAGN class (NGC 1052 in Heckman 1980; NGC 1097 in Keel 1983; M87 in Fabian & Rees 1995).

2.1. *The nature of the IR emission*

The multiwavelength high-angular resolution SED allow us to: *i)* probe the genuine SED of the LLAGN; and *ii)* identify the physical mechanisms that dominate the energy output in different wavelength ranges. Fig. 1 shows the subarcsec SEDs for the sample of LLAGN (black dots), sorted by increasing distance. Low-spatial resolution data (grey spikes) trace the contribution of the host galaxy, which dominates the overall emission in most of the cases. For comparison, the radio-loud quasar template from Elvis *et al.* (1994) is also shown (grey-solid line). The self-absorbed synchrotron emission is represented as a broken power law (see below Marscher & Gear 1985), which is fitted to the radio-to-UV range for each subarcsec aperture SED (dashed line).

2.2. *LLAGN at parsec scales*

The SEDs of LLAGN in our sample show: *i)* a flat radio continuum, more similar to radio-loud quasars rather than Seyfert nuclei; *ii)* mid-IR and X-ray ranges drive the bolometric luminosity of these nuclei; and *iii)* subarcsec SEDs show a flat-to-inverted shape from radio to IR/optical wavelengths, which is characteristic of non-thermal emission in a jet. Note the steep slope in the IR/optical continuum ($1 < \alpha < 3$; $S_\nu \propto \nu^{-\alpha}$), far from the classical synchrotron cooling slope of $\alpha \approx 0.7$. The Sombrero galaxy (NGC 4594) shows a soft thermal bump at $\sim 1\,\mu$m that might be caused by a receding accretion disc.

At high-angular resolution, LLAGN show significantly different SEDs when compared with quasars and Seyfert galaxies. The torus, if present, is not evidenced by the continuum emission, which seems to follow a non-thermal distribution (Fig. 1). Radio-loud quasars are more similar to LLAGN in the radio and X-ray ranges, but the absence of a big blue

† `http://ned.ipac.caltech.edu/`
‡ `http://www.iac.es/project/parsec/`

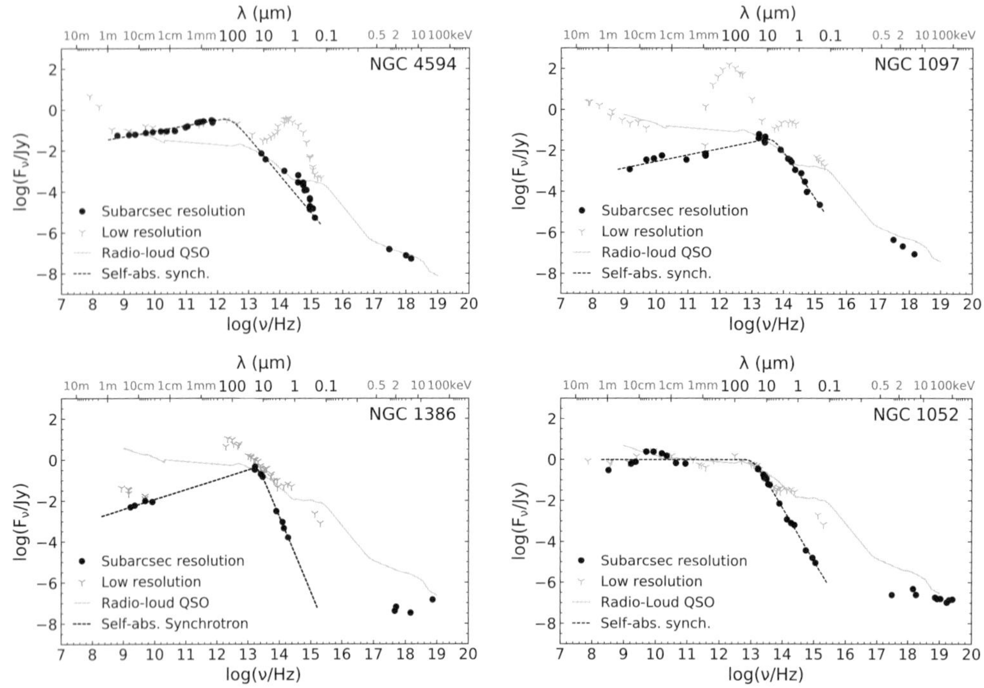

Figure 1. Subarcsec SEDs (black dots) and low-spatial resolution data (grey spikes) for four of the LLAGN in our sample. The grey line corresponds to the radio-loud quasar template from Elvis *et al.* (1994). A broken power law, i.e. representative of self-absorbed synchrotron emission from a jet, has been fitted to the radio, IR and optical/UV ranges (dashed line).

bump in the latter suggest important structural differences between both classes. The LLAGN in our sample are in agreement with the optical to X-ray colours and the IR to X-ray correlation from Maoz (2007); Asmus *et al.* (2011). However, this behaviour is not produced by thermal features in the SED. In contrast, both spectral slopes are produced by a steep power-law continuum from the mid-IR to optical/UV wavelengths. The non-thermal behaviour can be well described, at first order, by a self-absorbed synchrotron model (dashed line in Fig. 1 Marscher & Gear 1985), suggesting a jet as the component that drives the overall continuum emission in these objects. In Section 3 we further explore this possibility using a more detailed model of a jet. Nonetheless, one should keep in mind that spectral ratios are not able to discriminate bright AGN with strong thermal contributions in the mid-IR (torus) and the optical/UV (accretion disk) from LLAGN dominated by a steep non-thermal continuum as those included in our work.

3. Jet model

Data for M87 are carefully selected from quiescent periods in which the core was in a quiet state of activity (see Fig. 2, left panel), plus an active period between 2005–2006 during which the nucleus and the jet knot HST-1 were in a loud state, see Prieto *et al.* (2016) for a detailed discussion on the M87 data. The right panel in Fig. 2 shows a fit of the jet+disk model developed by S. Markoff (Markoff *et al.* 2005, 2008) to the SED of M87. The jet emission can explain the overall shape of the continuum emission for this LLAGN.

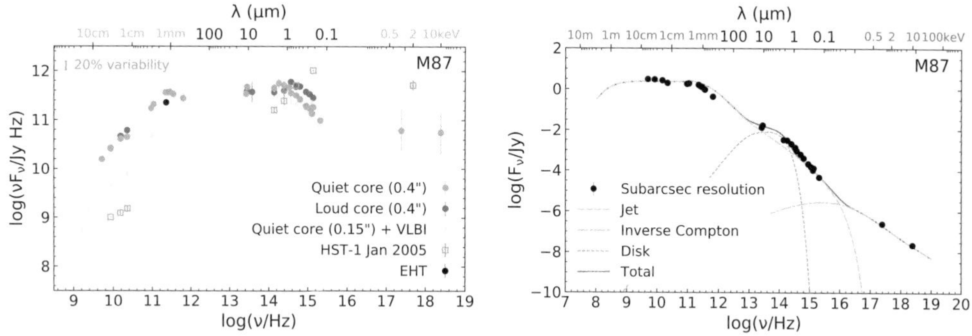

Figure 2. *Left:* Subarcsec resolution SED for the quiet and loud states of M87. *Right:* jet+disk model from Markoff *et al.* (2005, 2008), applied to the subarcsec SED of M87. Individual components of the model are also plotted. A colour version of this figure is available in the electronic version of this journal.

4. Summary

The LLAGN in our sample show remarkably homogeneous SEDs at subarcsec resolution. The torus and the big blue bump, if present, are not detected. Most of the nuclei show a featureless shape well described by a flat-to-inverted continuum from radio to the optical range, favouring a compact jet as the dominant source of the continuum emission. However, the IR/optical continuum does not follow the classical synchrotron cooling slope, but shows instead a much steeper spectrum ($1 < \alpha < 3$). Very high radiative losses or a thermal lepton distribution at the base of the jet can explain this behaviour.

References

Antonucci, R. 1993, ARA&A, 31, 473
Asmus, D., Gandhi, P., Smette, A., Hönig, S. F., & Duschl, W. J. 2011, A&A, 536, A36
Asmus, D., Hönig, S. F., Gandhi, P., Smette, A., & Duschl, W. J. 2014, MNRAS, 439, 1648
Di Matteo, T., Allen, S. W., Fabian, A. C., Wilson, A. S., & Young, A. J. 2003, ApJ, 582, 133
Elvis, M., Wilkes, B. J., McDowell, J. C., *et al.* 1994, ApJS, 95, 1
Eracleous, M., Hwang, J. A., & Flohic, H. M. L. G. 2010, ApJS, 187, 135
Fabian, A. C. & Rees, M. J. 1995, MNRAS, 277, L55
Heckman, T. M. 1980, A&A, 87, 152
Ho, L. C. 1996, PASP, 108, 637
Ho, L. C. 2008, ARA&A, 46, 475
Hönig, S. F. & Beckert, T. 2007, MNRAS, 380, 1172
Keel, W. C. 1983, ApJ, 269, 466
Laor, A. 2003, ApJ, 590, 86
Maoz, D. 2007, MNRAS, 377, 1696
Markoff, S., Nowak, M., Young, A., *et al.* 2008, ApJ, 681, 905
Markoff, S., Nowak, M. A., & Wilms, J. 2005, ApJ, 635, 1203
Marscher, A. P. & Gear, W. K. 1985, ApJ, 298, 114
Nagar, N. M., Falcke, H., & Wilson, A. S. 2005, A&A, 435, 521
Narayan, R., Yi, I., & Mahadevan, R. 1995, Nature, 374, 623
Prieto, M. A., Fernández-Ontiveros, J. A., Markoff, S., Espada, D., & González-Martín, O. 2016, MNRAS, 457, 3801
Prieto, M. A., Reunanen, J., Tristram, K. R. W., *et al.* 2010, MNRAS, 402, 724
Reunanen, J., Prieto, M. A., & Siebenmorgen, R. 2010, MNRAS, 402, 879
Yuan, F., Yu, Z., & Ho, L. C. 2009, ApJ, 703, 1034

New Frontiers in Black Hole Astrophysics
Proceedings IAU Symposium No. 324, 2016
Andreja Gomboc, ed.

doi:10.1017/S1743921317001375

A thin disk model for the high efficiency jet in powerful lobe-dominated FRII radio galaxies

Shuang-Liang Li

Key Laboratory for Research in Galaxies and Cosmology, Shanghai Astronomical Observatory, Chinese Academy of Sciences, 80 Nandan Road, Shanghai 200030, China;
email: lisl@shao.ac.cn

Abstract. We propose a thin accretion disk with magnetically driven winds model to investigate the high jet efficiency phenomenon in lobe-dominated luminous FRII radio galaxies, which was reported by Fernandes *et al.* (2011) and Punsly (2011). It was found that the temperature of a thin disk with winds was much lower comparing with that of a standard thin disk because the winds take the most of energy released in the disk away, resulting in a much smaller raditive efficiency. Therefore, the jet efficiency can be very high even for conventional jet power. Our model can explain the observations quite well.

Keywords. accretion, accretion disks, galaxies: active, black hole physics, galaxies: magnetic fields.

1. Introduction

Radio galaxies can be divided into two classed (FRI and FRII) according to their lobe morphology. It was usually believed that FRI and FRII radio galaxies should have different accretion model due to their different accretion rates (Ghisellini & Celotti 2001). A radiatively efficient accretion disk is suggested to be present in FRII radio galaxies. Punsly (2011) and Fernandes *et al.* (2011) reported that a maximum jet efficiency line $R = 25$ $(R = L_{\rm jet}/L_{\rm bol})$ exists in some lobe-dominated luminous FRII galaxies.

The bolometric luminosity of a thin disk can be expressed as $L_{\rm bol} = \eta_{\rm th}\dot{M}c^2$, where $\eta_{\rm th}$ is the radiative efficiency. Together with the jet power $(L_{\rm jet} = \eta_{\rm Q}\dot{M}c^2$, here $\eta_{\rm Q}$ is the jet production efficiency), the jet efficiency $R(= L_{\rm jet}/L_{\rm bol} = \eta_{\rm Q}/\eta_{\rm th})$ will be decided by both $\eta_{\rm Q}$ and $\eta_{\rm th}$. The jet production efficiency for a magnetically-arrested-disk (MAD) can reach $\sim 30\%$ and 140% for $a = 0.5$ and 0.99, respectively (Tchekhovskoy *et al.* 2011). Therefore, the jet efficiency can reach about 14 for a thin disk surrounding Kerr black hole ($\eta_{\rm th}$=0.1 is adopted). However, if the radiative efficiency $\eta_{\rm th}$ of a thin disk were much smaller than 0.1, we can get a high jet efficiency R even for conventional jet production efficiency $\eta_{\rm Q}$. As presented in Li (2014) and Li & Begelman (2014), they do find the temperature of a thin disk with magnetically driven winds/jets can decrease more than one order of magnitude, which directly results in the decrease of $\eta_{\rm th}$ and the increase of jet efficiency R.

2. Models

We consider a thin disk with magnetically driven winds surrounding a spinning black hole. The continuity, radial momentum, angular momentum and energy equations of the disk are as follows:

$$\frac{d}{dr}(2\pi\Delta^{1/2}\Sigma v_{\rm r}) + 4\pi r \dot{m}_{\rm w} = 0, \tag{2.1}$$

$$\frac{\gamma_\phi AM}{r^4\Delta}\frac{(\Omega-\Omega_{\rm k}^{+})(\Omega-\Omega_{\rm k}^{-})}{\Omega_{\rm k}^{+}\Omega_{\rm k}^{-}} + g_{\rm m} = 0, \tag{2.2}$$

$$-\frac{\dot{M}}{2\pi}\frac{dL}{dr} + \frac{d}{dr}(rW_\phi^r) + T_{\rm m}r = 0, \tag{2.3}$$

$$\nu\Sigma\frac{\gamma_\phi^4 A^2}{r^6}\left(\frac{d\Omega}{dr}\right)^2 = \frac{16acT^4}{3\bar{\kappa}\Sigma}, \tag{2.4}$$

where the equations and the meaning of parameters are all the same as in Li (2014).

The bolometric luminosity of accretion disk can be calculated as

$$L_{\rm bol} = \int_{r_{\rm in}}^{r_{\rm out}} Q_{\rm rad} 2\pi r dr = \eta_{\rm th}\dot{M}c^2. \tag{2.5}$$

Jets can be powered by both BZ (Blandford & Znajek 1977) and BP (Blandford & Payne 1982) processes. In the general form of the BZ process, the jet power $L_{\rm BZ}$ can be calculated with

$$L_{\rm BZ} = \frac{1}{32}\omega_{\rm F}^2 B_\perp^2 r_{\rm H}^2 (J/J_{\rm max})^2 c. \tag{2.6}$$

The jet power from an accretion disk (BP process) can be given as in Cao (2002)

$$L_{\rm BP} = \int_{r_{\rm in}}^{r_{\rm out}} \frac{B_{\rm p}B_\phi}{4\pi} r\Omega 2\pi r dr, \tag{2.7}$$

where B_ϕ and $B_{\rm p}$ are the toroidal and poloidal components of magnetic field, respectively.

Therefore the total jet power $L_{\rm jet}$ is:

$$L_{\rm jet} = L_{\rm BZ} + L_{\rm BP} = \eta_{\rm Q}\dot{M}c^2. \tag{2.8}$$

With $\eta_{\rm th}$ and $\eta_{\rm Q}$, the jet efficiency is given by:

$$R = \frac{\eta_{\rm Q}}{\eta_{\rm th}}. \tag{2.9}$$

3. Results And Discussion

Previous studies mainly focused on the improvement of jet power. However, the strong large-scale magnetic field threading on the disk can not only improve jet power, but also decrease the radiative efficiency (Tchekhovskoy *et al.* 2011). We suggest that in order to explain the high jet efficiency in luminous FRII radio galaxies, the jet power is not necessary very high if the radiative efficiency is low. This can be realized because the winds/jets can take away most of gravitational energy released in the disk (see figure 1). A theoretical maximum jet efficiency $R \sim 1000$ is found, which is large enough even taking the episodic activity of jets into account.

The key point of this work is the formation of a large-scale magnetic field in a thin disk. One popular mechanism is assumed that there is initial weak large-scale magnetic field at the outer boundary of a thin disk, which can be magnified with the accretion of gas. However, the advection isn't effective for a standard thin disk because of the fast diffusion therein (Lubow *et al.* 1994). This problem can be solved by including the feedback of

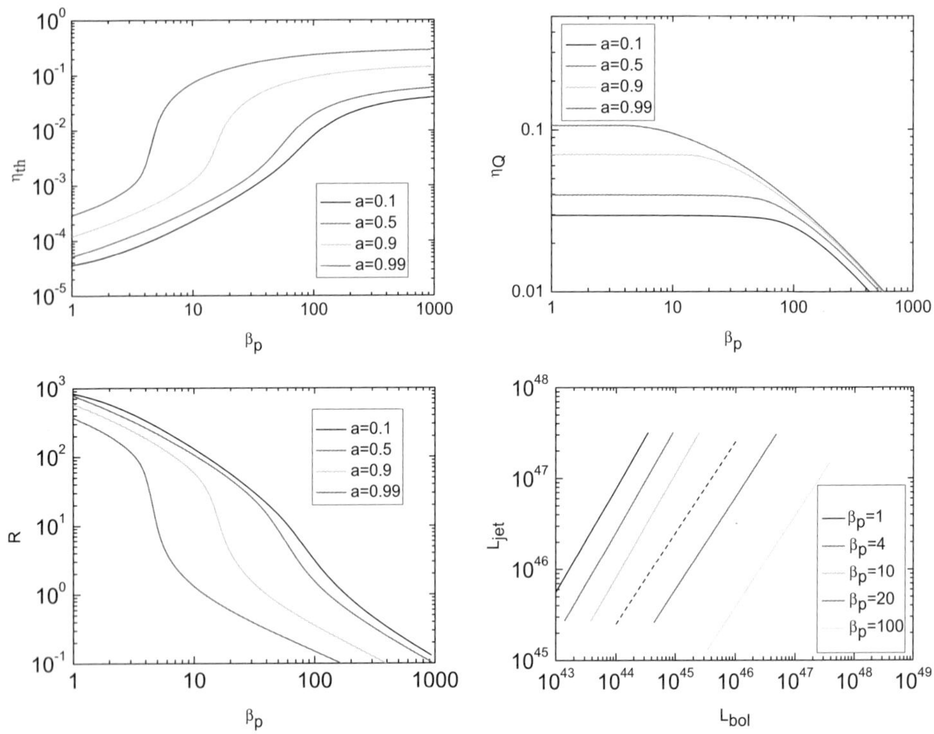

Figure 1. Panels (a), (b) and (c) are for the radiative efficiency $\eta_{\rm th}$, the jet production efficiency $\eta_{\rm Q}$ and the jet efficiency R of a relativistic thin accretion disk with magnetically driven jets as functions of $\beta_{\rm p}$, respectively. And panel (d) represent the jet power as functions of bolometric luminosity for different $\beta_{\rm p}$ with $a = 0.9$, where the black hole mass varies from $10^8 M_{\odot}$ to $10^{10} M_{\odot}$. The dashed line corresponds to the observational jet efficiency lines $R = 25$.

winds on accretion disk (Cao & Spruit 2013; Li & Begelman 2014). The formation of large scale magnetic field also depends on the initial strength and morphology of magnetic field in the MHD simulations (e.g., McKinney *et al.* 2012), which is still an open issue so far.

Acknowledgements

This work is supported by the NSFC (grants 11373056, 11233006).

References

Blandford, R. D. & Payne, D. G. 1982, *MNRAS*, 199, 883
Blandford, R. D. & Znajek, R. 1977, *MNRAS*, 179, 433
Cao, X., 2002, *MNRAS*, 332, 999
Cao, X., & Spruit, H. C., 2013, *ApJ*, 765, 149
Fernandes C. A. C., Jarvis, M. J., Rawlings, S., *et al.*, 2011, *MNRAS*, 411, 1909
Ghisellini, G. & Celotti, A. 2001, *A&A*, 379, L1
Li, S.-L., 2014, *ApJ*, 788, 71
Li, S.-L. & Begelman, M. C., 2014, *ApJ*, 786, 6
Lubow, S. H., Papaloizou, J. C. B., & Pringle, J. E., 1994, *MNRAS*, 267, 235
McKinney, J. C., Tchekhovskoy, A., & Blandford, R. D., 2012, *MNRAS*, 423, 3083
Punsly, B., 2011, *ApJ*, 728, 17
Tchekhovskoy, A., Narayan, R., & McKinney, J. C., 2011, *MNRAS*, 418, L79

New Frontiers in Black Hole Astrophysics
Proceedings IAU Symposium No. 324, 2016
Andreja Gomboc, ed.

doi:10.1017/S1743921316012722

Particle-in-cell Simulations of Global Relativistic Jets with Helical Magnetic Fields

Ioana Duţan[1], Ken-Ichi Nishikawa[2], Yosuke Mizuno[3], Jacek Niemiec[4], Oleh Kobzar[4], Martin Pohl[5,6], Jose L. Gómez[7], Asaf Pe'er[8], Jacob T. Frederiksen[9], Åke Nordlund[9], Athina Meli[10], Helene Sol[11], Philip E. Hardee[12] and Dieter H. Hartmann[13]

[1]Institute of Space Science, Atomiştilor 409, Bucharest-Magurele RO-077125, Romania
email: idutan@spacescience.ro

[2]Department of Physics, University of Alabama, Huntsville, AL 35899, USA
email: ken-ichi.nishikawa@nasa.gov

[3]Institute for Theoretical Physics, Goethe University, Frankfurt am Main D-60438, Germany
email: mizuno@th.physik.uni-frankfurt.de

[4]Institute of Nuclear Physics PAN, ul. Radzikowskiego 152, Kraków 31-342, Poland
email: Jacek.Niemiec@ifj.edu.pl (J.N.); oleh.kobzar@ifj.edu.pl (O.K.)

[5]Institut fur Physik und Astronomie, Universität Potsdam, Potsdam-Golm D-14476, Germany
[6]DESY, Platanenallee 6, Zeuthen 15738, Germany
email: marpohl@uni-potsdam.de

[7]Instituto de Astrofísica de Andalucía, CSIC, Apartado 3004, Granada 18080, Spain
email: jlgomez@iaa.csic.es

[8]Physics Department, University College Cork, Cork T12 YN60, Ireland
email: a.peer@ucc.ie

[9]Niels Bohr Institute, University of Copenhagen, Blegdamsvej 17, Copenhagen DK-2100, Denmark; email: trier@nbi.ku.dk (J.T.F.), aake@nbi.dk (Å.N.)

[10]Department of Physics and Astronomy, University of Gent, Proeftuinstraat 86, Gent B-9000, Belgium; email: ameli@ulg.ac.be

[11]LUTH, Observatore de Paris-Meudon, 5 place Jules Jansen, Meudon Cedex 92195, France
email: helene.sol@obspm.fr

[12]Department of Physics and Astronomy, University of Alabama, Tuscaloosa, AL 35487, USA
email: pehardee@gmail.com

[13]Department of Physics and Astronomy, Clemson University, Clemson, SC 29634, USA
email: hdieter@g.clemson.edu

Abstract. We study the interaction of relativistic jets with their environment, using 3-dimensional relativistic particle-in-cell simulations for two cases of jet composition: (i) electron-proton ($e^- - p^+$) and (ii) electron-positron ($e^{\pm}$) plasmas containing helical magnetic fields. We have performed simulations of "global" jets containing helical magnetic fields in order to examine how helical magnetic fields affect kinetic instabilities such as the Weibel instability, the kinetic Kelvin-Helmholtz instability and the Mushroom instability. We have found that these kinetic instabilities are suppressed and new types of instabilities can grow. For the $e^- - p^+$ jet, a recollimation-like instability occurs and jet electrons are strongly perturbed, whereas for the $e^{\pm}$ jet, a recollimation-like instability occurs at early times followed by kinetic instability and the general structure is similar to a simulation without a helical magnetic field. We plan to perform further simulations using much larger systems to confirm these new findings.

Keywords. plasmas, instabilities, magnetic fields, shock waves, relativity, galaxies: jets, active

1. Introduction

Particle-in-cell (PIC) simulations of collisionless shock formation and instability growth, such as the Weibel instability, the kinetic Kelvin-Helmholtz instability (kKHI), and the Mushroom instability (MI), have been performed to study magnetic field generation, particle acceleration, and emission of radiation with applications to astrophysical plasma jets (e.g., Alves *et al.* 2012, Nishikawa *et al.* 2003, 2005, 2013, 2014, 2016a,b). Furthermore, Nishikawa *et al.* (2016a) have performed global simulations involving injection of a cylindrical jet into an ambient plasma in order to investigate shock (Weibel instability) and velocity shear instabilities (kKHI and MI) simultaneously. Previously, these two processes have been investigated separately.

To our knowledge, Nishikawa *et al.* (2016b) were the first to present results of numerical PIC simulations of global relativistic jets containing helical magnetic fields. The kinetic features of secondary magnetic reconnection in a single flux rope undergoing internal kink instability are studied by means of three-dimensional particle-in-cell simulations where the single flux rope is modeled with a simple screw-pinch configuration as in Markidis *et al.* (2014).

The presence of helical magnetic fields is suggested by twisted structures that have been observed in many active galactic nuclei (AGN) jets, from sub- to kiloparsec scales (e.g., Lobanov & Zensus (2001), Perucho *et al.* (2012), Gómez *et al.* (2016)). These jet structures were explained through relativistic magnetohydrodynamic (MHD) modeling, where simulations of current-driven (kink) instabilities were performed (e.g., Mizuno *et al.* (2015), Singh *et al.* (2016)). The main result obtained by Nishikawa *et al.* (2016b) has revealed that new types of shocks, similar to the recollimation shocks attained in relativistic MHD simulations, occur when a relativistic plasma jet contains a helical magnetic field. The simulations presented in this paper were designed to further test these new findings.

2. Numerical methods and simulation setup

We use a fully kinetic approach to model the formation of shocks in relativistic plasma jets containing helical magnetic fields. We apply PIC methods to numerically simulate the injection of a cylindrical relativistic jet with a Lorentz factor $\gamma = 15$ into an ambient plasma at rest, using a modified version (e.g., Nishikawa *et al.* (2003), Nishikawa *et al.* (2014), Nishikawa *et al.* (2016a)) of the TRISTAN code (Buneman (1993)).

For plasma composition, we use (i) an electron-proton ($e^- - p^+$) plasma with a realistic proton-electron mass ratio ($m_{\rm p}/m_{\rm e} = 1836$) and (ii) an electron-positron ($e^{\pm}$) plasma. The simulations were performed with a numerical grid of $(L_{\rm x}, L_{\rm y}, L_{\rm z}) = (645\Delta, 131\Delta, 131\Delta)$, where $\Delta = 1$ is the cell size, and periodic boundary conditions in traverse directions. The plasma jet, with a radius of $r_{\rm jt} = 20\Delta$, is injected in the middle of the $y - z$ plane ($(y_{\rm jc}, z_{\rm jc}) = (63\Delta, 63\Delta)$) at $x = 100\Delta$. For the complete set of parameters, see Nishikawa *et al.* (2016b).

The helical magnetic field structure is implemented using the equations in Mizuno *et al.* (2015), with an exponential damping function for the magnetic fields external to the jet in order for plasma instabilities to grow (Nishikawa *et al.* (2016b)). However, our simulations use Cartesian coordinates. We set $\alpha = 1$, thus eqs. (9), (10) and (11) from Mizuno *et al.* (2015) are reduced to eq. (2.1) and the magnetic field takes the form:

$$B_x = \frac{B_0}{[1 + (r/a)^2]}, \quad B_\phi = \frac{(r/a)B_0}{[1 + (r/a)^2]} \tag{2.1}$$

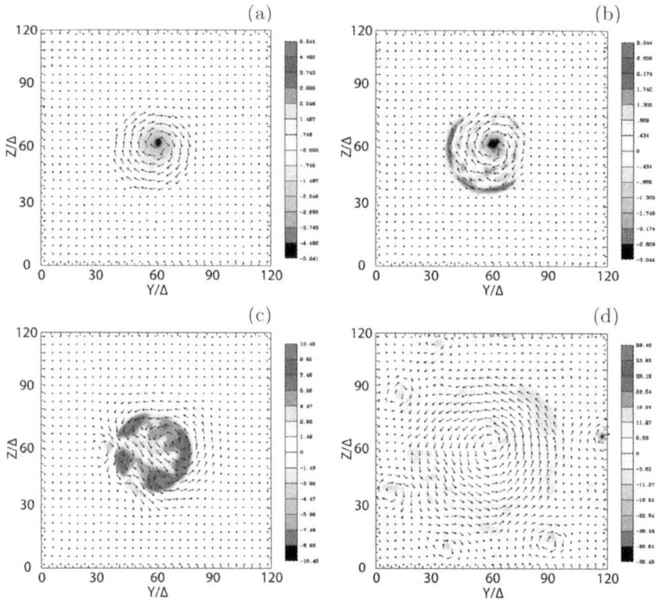

Figure 1. Panels show 2D plots of the isocontour of J_x in the $y-z$ plane: (a) and (c) for $e^- - p^+$ and (b) and (d) for $e^\pm$ plasma jets, at time $t = 500\omega_{\rm pe}^{-1}$. Arrows show the magnetic fields ($B_{y,z}$). Panels (a) and (b) show J_x at $x = 180\Delta$, whereas panels (c) and (d) show J_x at $x = 480\Delta$. For both cases, the original left--handed polarity of $B_{y,z}$ found at $x = 180\Delta$ has switched to right-handed polarity at $x = 480\Delta$.

The toroidal magnetic field is created by a current $+J_x(y,z)$ in the positive x-direction, so that defined in Cartesian coordinates:

$$B_y(y,z) = \frac{((z - z_{\rm jc})/a)B_0}{[1 + (r/a)^2]}, \quad B_z(y,z) = -\frac{((y - y_{\rm jc})/a)B_0}{[1 + (r/a)^2]}. \tag{2.2}$$

Here a is the characteristic length-scale of the helical magnetic field, $(y_{\rm jc}, z_{\rm jc})$ is the jet center, and $r = \sqrt{(y - y_{\rm jc})^2 + (z - z_{\rm jc})^2}$. The choice of helicity is defined by eq. (2.2), and has left-hand polarity with positive B_0.

In the simulations, the initial magnetic field amplitude parameter $B_0 = 0.1c$, $(c = 1)$, $(\sigma = B^2/n_{\rm e}m_{\rm e}\gamma_{\rm jet}c^2 = 2.8 \times 10^{-3})$, and $a = 5.0\Delta = 0.25 * r_{\rm jt}$, with $r_{\rm jt} = 20\Delta$.

3. Simulation results

Results of global jet PIC simulations containing helical magnetic fields were presented in Nishikawa *et al.* (2016b). In the $e^- - p^+$ jet case recollimation-like shocks are developed. In the $e^\pm$ jet case small recollimation structures are initially formed, and after instabilities have grown, currents extend outside the jet and the current density becomes turbulent. Nishikawa *et al.* (2016b) also compared their results to two different relativistic MHD simulations of (a) recollimation (Mizuno *et al.* (2015)) and (b) current-driven kink instability (Singh *et al.* (2016)).

In Figure 1 we show 2D isocontour plots of the x-component of the current density $J_{\rm x}$ in the $y-z$ plane for the electron-proton ($e^- - p^+$) case and the electron-positron ($e^\pm$) case at time $t = 500\omega_{\rm pe}^{-1}$. We have found in both cases that, due to the growth of jet instabilities, the original left-handed (clockwise viewed from the jet front) polarity of the magnetic field ($B_{y,z}$) is switched to right-handed polarity, as shown in Fig. 1.

At $x = 180\Delta$ instabilities start to grow and, for the $e^- - p^+$ jet at $x = 480\Delta$, the magnetic field structure is distorted and $B_{y,z}$ shows linear polarity inside the jet, as the magnitude of J_x decreases towards a null value. This is what we might expect from recollimation-like shocks.

4. Conclusions

Using a fully kinetic approach, the simulation results presented in Nishikawa *et al.* (2016b) demonstrate the formation, at the microphysics level, of new types of shock structures in relativistic plasma jets due to the presence of helical magnetic fields, whereas kinetic instabilities, such as the Weibel instability, kKHI, and MI are suppressed. These new types of shocks have structures similar to those obtained using relativistic MHD methods (Mizuno *et al.* (2015), Singh *et al.* (2016)). Here, we have also shown that due to the presence of instabilities, the polarity of the magnetic field $B_{y,z}$ switches sign and, in the case of the $e^- - p^+$ jet, $B_{y,z}$ shows linear polarity inside the jet.

Further systematic study of relativistic jets containing helical magnetic fields along with associated processes such as reconnection, turbulence, and generation of ultra-high energy particles will provide more advanced interpretation of observed phenomena such as gamma-ray burst (GRB) emission and polarized emission in Blazars and GRB jets.

Acknowledgements

This work is supported by NSF AST-0908010, AST-0908040, NASA-NNX09AD16G, NNX12AH06G, NNX13AP-21G, and NNX13AP14G grants. I.D. was partially supported by the NUCLEU Program (Contr. No. 4N/2016/PN16470201). The work of J.N. and O.K. has been supported by Narodowe Centrum Nauki through research project DEC-2013/10/ E/ST9/00662. Y.M. is supported by the ERC Synergy Grant “BlackHoleCam - Imaging the Event Horizon of Black Holes” (Grant No. 610058). M.P. acknowledges support through grant PO 1508/1-2 of the Deutsche Forschungsgemeinschaft. Simulations were performed using Pleiades and Endeavor facilities at NASA Advanced Supercomputing (NAS), and using Gordon and Comet at The San Diego Supercomputer Center (SDSC), and Stampede at The Texas Advanced Computing Center, which are supported by the NSF. This research was started during the program “Chirps, Mergers and Explosions: The Final Moments of Coalescing Compact Binaries” at the Kavli Institute for Theoretical Physics, which is supported by the National Science Foundation under grant No. PHY05-51164. The first velocity shear results using an electron–positron plasma were obtained during the Summer Aspen workshop “Astrophysical Mechanisms of Particle Acceleration and Escape from the Accelerators” held at the Aspen Center for Physics (1–15 September 2013).

References

Alves, E. P., Grismayer, T., Martin, S. F., Fiúza, F., *et al.* 2012, *ApJ* (Lett.), 746, L14

Buneman, O. 1993, in Computer Space Plasma Physics: Simulation Techniques and Software; Eds: Matsumoto, H., Omura, Y., & Terra Scientific Publishing Company: Tokio, Japan, 1993, pp. 67-79

Gómez, J. L., Lobanov, A. P., Bruni, G., Kovalev, Y. Y., *et al.* 2016, *ApJ*, 817, 2

Lobanov, A. P. & Zensus, J. A. 2001, *Science*, 294, 128

Markidis, S., Lapenta, G., Delzanno, G. L., Henri, P., *et al.* 2014, *PPCF*, 56, 6, article id. 064010

Mizuno, Y., Gómez, J. L., Nishikawa, K.-I., Meli, A., *et al.* 2015, *ApJ*, 809, 38

Nishikawa, K.-I., Frederiksen, J. T., Nordlund, Å., Mizuno, Y., *et al.* 2016a, *ApJ*, 820, 94

Nishikawa, K.-I., Mizuno, Y., Niemiec, J., Kobzar, O., *et al.* 2016b, *Galaxies*, 4, 38

Nishikawa, K.-I., Hardee, P. E., Duţan, I., Niemiec, J., *et al.* 2014, *ApJ*, 793, 60

Nishikawa, K.-I., Hardee, P. E., Zhang, B., Duţan, I., *et al.* 2013, *Ann. Geophys.*, 31, 1535

Nishikawa, K.-I., Hardee, P. E., Richardson, G., Preece, R., *et al.* 2005, *ApJ*, 622, 927

Nishikawa, K.-I., Hardee, P. E., Richardson, G., Preece, R., *et al.* 2003, *ApJ*, 595, 555

Perucho, M., Martí-Vidal, I., Lobanov, A. P., & Hardee, P. E. 2012, *A&A*, 545, 65

Singh, C. B., Mizuno, Y., & de Gouveia Dal Pino, E. M. 2016, *ApJ*, 824, 48

New Frontiers in Black Hole Astrophysics
Proceedings IAU Symposium No. 324, 2016
Andreja Gomboc, ed.

doi:10.1017/S174392131700014X

Multi-wavelength flares and magnetic field in blazars: a case study of IBL S5 0716+714

S. Chandra[1], K. P. Singh[1] and K. S. Baliyan[2]

[1]Tata Institute of Fundamental Research, Mumbai-400005, India
email: sunil.chandra@tifr.res.in

[2]Physical Research Laboratory, Ahmedabad, Gujarat - 380009

Abstract. We present a study performed to understand the role of magnetic field during a historical outburst of IBL S5 0716+714 witnessed in early 2015. The two month long profile of the outburst reveals several episodes of sub-flares superimposed over the diminishing trend of the peak-flux. The broadband X-ray spectrum from *Swift* + *Nustar* exhibits a break at $4.9^{+0.4}_{-0.5}$ keV which is consistent with the valley in the SED predicted by model. The spectral index variations, closely correlated to the flux variations during the first prominent sub-flare, indicate the alterations in the particle energy distribution coinciding the onset of the flare. The PA rotations approximately at the same epoch are used to constrain the field and its geometry.

Keywords. AGN, Blazar, Blazar-jet-magnetic fields, etc.

1. Introduction

The blazars are a subclass of Active Galactic Nuclei (AGN) seen at very small angles (<15°) to the relativistic jet emanated from very close to the central engine. The constraints on the role and morphologies of the magnetic fields in blazar jets are crucial ingredients for a complete understanding of the radiative processes. The presence of the magnetic field in blazars is supported by the detection of highly polarized emission in the optical and radio, parametrized by the degree of polarization (DP) and the position angle (PA). The DP has imprints of both, the particle energy distribution (PED) and the local magnetic field, whereas, the PA presents a non-contaminated representation of the morphology of the magnetic field. Therefore, PA and DP provide the most effective proxies for the jet magnetic field. The blazar emission is characterized by the strong variability comprising episodes of the multi-wavelength (MW) outbursts which may or may not be associated with the PA rotations. The formation of the shock in downstream jet and re-accelerations of the relativistic plasma is the most preferable scenario to explain the MW variability. The spectral energy distribution (SED) and the flux/polarization variability are used as a tool for the detailed understanding of the outbursts. The two kind of widely accepted models, leptonic and hadronic, corroborate the shock-in-jet framework, are employed to recreate the SEDs and lightcurves (see Böttcher *et al.* 2013 for the comparison of these two classes of models). The basic framework of both the classes rely on the synchrotron emission from ultra-relativistic leptons (e^-/e^+) to explain the low energy emission (radio- to X-rays). However, the ultra-fast variability is better explained by the leptonic approaches. The leptonic models induct the two emission scenarios for the high energy radiation (X-rays to TeV), namely SSC (Self Synchrotron Comptonization), and EC (External Comptonization), where they assume that the seed population for the Compton up-scattering comes from synchrotron photons or the photons external to the emission zone (disk, torus, BLR, CMBR etc), respectively. The complete understanding of the outbursts requires the simultaneous modeling of the the lightcurves, the DP & PA variability and SEDs. This requirement is primarily facilitated by two models, namely the Turbulent Extreme Multi-zone Model (TEMZ; Marscher 2014) and Helical Magnetic Field Model (HMFM; Zhang *et al.* 2015). Most importantly, these models use two different assumptions for the magnetic fields (small-scale turbulent-chaotic v/s large-scale ordered) and its morphology (straight v/s helical). The HMFM model is more favorable for the correlated PA rotations($> 180°$) during MW-flares and incorporates a self consistent tracing of the particle dynamics, radiation transfer and the polarization.

In the following sections we summarize the study of an outburst for the intermediate energy peaked BL Lac object (IBL) S5 0716+714 (z$\sim$0.31) during early 2015.

2. Data analysis

S5 0716+714 is one of the blazars being regularly monitored using Mt Abu Infrared observatory (MIRO), India, since late 2005. This source underwent a major outburst from 11 January 2015 to 28 February 2015. Being a very prominent flare (rising by 2.5 mag to reach 11.27 mag in NIR J band, between 01-11 January 2015; Carrasco *et al.* 2015), this outburst is followed by the facilities across the globe. We have performed the optical photometric and polarimetric observations using the facilities at MIRO during 16-30 January 2015.

The present work extends the studies by Chandra *et al.*(2015b), hereafter C15. The MW data from *Fermi*, *Swift* and MIRO for the durations of 01 January - 28 February 2015 are analyzed using the methods adopted by C15. The only available long *Nustar* observations of the source (ObsID: 90002003002; Date-Obs: 24 January 2015 & Exp. 18.5 ks) are analysed to understand the broader X-rays spectrum (0.3 - 80 keV), using standard prescriptions by *Nustar* team. The *Swift*+*Nustar* spectra are simultaneously fitted using absorbed broken-powerlaw model + a constant multiplicative factor to the *Nustar* spectrum to take care of variability introduced by $\approx$1-day difference in the observations. The intra-night R-band photometry data for 4 nights (23-26 January, 2015) are also analyzed using the methodologies adopted by Chandra *et al.* (2012) (see Fig. 1b for micro-variability on 23 January 2015).

3. Results

The multi-wavelength lightcurves shown in Fig. 1a & 1b clearly establish the MW-nature of 2015 outburst. The optical flux (V= 12.06 mag) communicated by Chandra *et al.* (2015a) corresponds to the historically highest flux value till date for this source. A significant micro-variability is witnessed on 23 January 2016, characterized by the timescale of $\sim$ 1.2 hrs. Rapid variations in the DP un-correlated with total flux and rotations in the PA by $>180°$ are observed during the onset of the first sub-flare. Fig. 1a depicts the lightcurves lasting for almost two months, covering the full outburst. As evident from Fig. 1a, the 2015 outburst consists of a number of small sub-flares superimposed over a longer trend. The peak of the first two X-ray flares are of similar level, whereas all the six sub-flares in UV/optical show a tendency of diminishing peak flux with time. A frequency dependent inter-night optical variations are also observed. The X-ray spectral indices, derived from XRT-data, show the flux dependent variations. The first sub-flare is accompanied with a correlated changes in the spectral index. Afterwards, it remained at same level ($\alpha \sim 2.6$) throughout the second sub-flare and ahead. The X-rays spectra, derived from the nearest possible observations by Swift & Nustar shows a satisfactory fit ($\chi_\nu \sim 1.2$, $\nu = 328$) with a break at $4.9^{+0.4}_{-0.5}$ keV. The spectral index before the break, derived from this combined fitting ($\alpha = 2.6\pm0.04$, $\beta = 1.8^{+0.06}_{-07}$) is in agreement with the estimates obtained from the XRT spectrum alone. A detailed qualitative analysis of the constraints on the spectral shape of S5 0716+714 using same dataset is presented by Wierzcholska & Siejkowski (2016). The Fig. 2a and Fig. 2b shows the simultaneous modeling of the polarization variations and the SED with HMFM model for the duration of MJD 57046 - 57048, the span of the first polarization swing (borrowed from C15). The SED is superimposed with *Swift+Nustar* spectrum showing the spectral break at 4.9 keV to be consistent with the model-predicted turn-over or the "valley" which is an estimate of the frequency corresponding to the equal cooling rates for the synchrotron and SSC processes.

4. Discussion

The 2015 outburst comprises a series of sub-flares associated with the major event and is seen at very broad spectrum. The flux variations in the optical bands are frequency dependent, making the shock driven processes more favourable over the geometry dependent models. The micro-variability seen during 23 January 2015, exhibiting sub-hour optical variations, may possibly be due to small fluctuations introduced in the shock front. The variations in the spectral indices closely following the flux variations in 0.3-10.0 keV band, during the first X-ray flare (MJD 57040-57054), can be explained in the following manner: The PED is forced to evolve during the interaction of the plasma with the shock which again tries to retain its initial configuration afterwards. The absence of similar correlations during the second X-ray sub-flare and ahead (MJD 57055-57075), indicate that the particle accelerations, here, somehow try to preserve the PED of the flaring state. The detailed modelling of the simultaneous MW-lightcurves, polarization and the SED, for a part of the data (MJD 57046-57053), shows the magnetic field alterations (poloidal$-$ $>$toroidal$-$ $>$poloidal) during the passage of the shock through the emission region (see Table 2 of C15 for the derived parameters). The second PA rotations and the corresponding flux variations indicate the possibilities of the magnetic-reconnection or the shock driven reconnection, inferred from the lowering in the synchrotron flux (optical) coinciding with the rise in the EC (γ-rays) flux, however, a detailed modelling lacks because of the large data gap. The similar argument can also be associated to the spectral index and flux variations, specially for the span of MJD 57046-57053. The multiple MW sub-flares associated with the major outbursts in blazars are becoming more evident because of the good cadence monitoring provided by the coordinated campaigns. Such patterns reflect the several episodes of particle energization before the major one diminishes. However, the proper constraints on the radiative processes are still limited because of the unavailability of strictly simultaneous data.

5. Prospects of the blazars' study with AstroSat

AstroSat, the first Indian multi-wavelength observatory in space, launched on 28 October 2015, provides a strictly simultaneous coverage over a broad spectrum (550 nm 130 nm in optical/UV and 0.3-($>$)120 keV in X-rays, with unprecedented timing and spectral accuracies (ref. http://astrosat.iucaa.in). We refer to the Science Support Center† for in-orbit performances for various instruments and data policies. The blazars are one of the key science objects for the AstroSat. The consortium of HCT + MIRO + SAAO + AstroSat is prepared to organize the MW campaigns during ongoing and upcoming cycles, aiming to uncover the role of magnetic fields in the wake of the more sophisticated blazars' emission models, e.g. Zhang *et al.* (2016) and references therein.

† http://astrosat-ssc.iucaa.in/

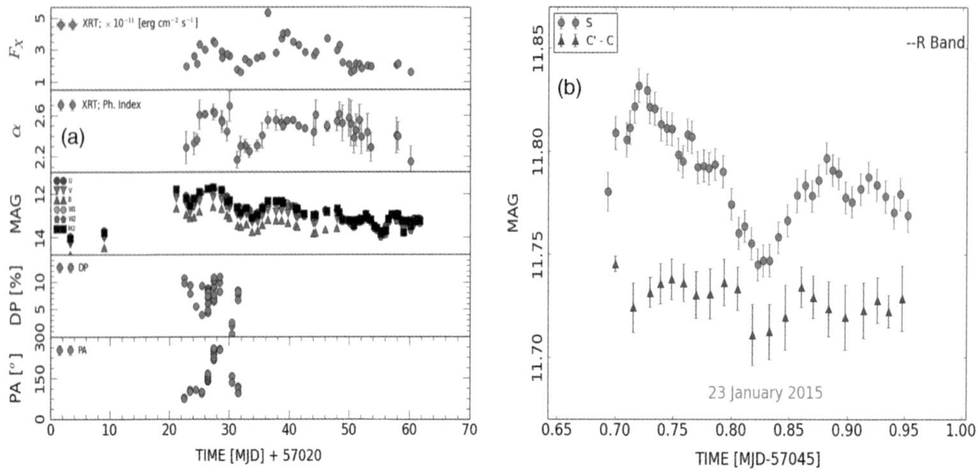

Figure 1. a: The MW lightcurves observed during 01 January-28 February 2015 with the top panel as X-ray lightcurve derived from spectral fitting. The next two panels represent the spectral indices & UV lightcurves, respectively. The last two panels highlight the DP variations and PA rotations during first sub-flare. **b:)** The intra-night R-band lightcurve from MIRO dated on 23 January 2015 with the source magnitudes (red) and differential magnitudes of the comparison and control star (blue).

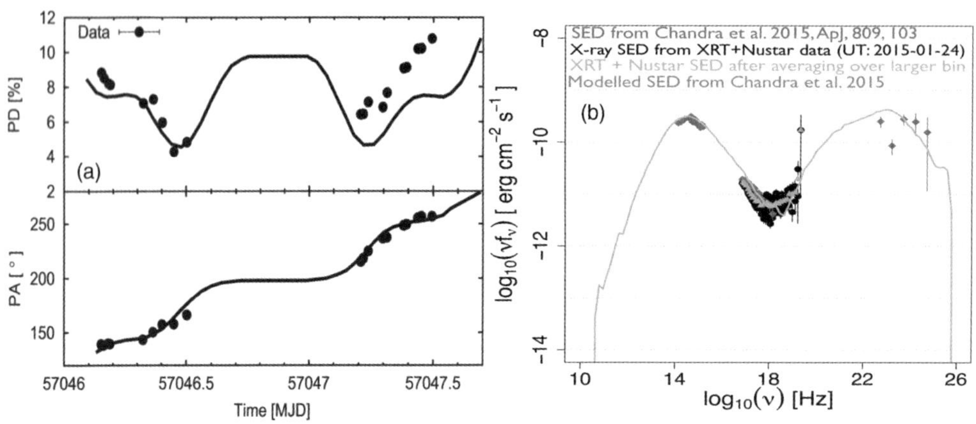

Figure 2. a): The HMFM model and the data for the DP and the PA variations (C15). **b):** The modeled broad-band SED of first sub-flare, borrowed from C15 with the *Swift+Nustar* SED over-plotted. Note that the SED shown in C15 is strictly simultaneous.

References

Böttcher M., Reimer, A., Sweeney, K., & Prakash A., **2013**, *ApJ*, 768, 54

Carrasco, L., Porras, A., Recillas, E., *et al.***2015** *ATel* #6902, 1

Chandra, S., Baliyan, K. S., Ganesh, S., & Joshi, U. C., **2015**, *ApJ*, 731, 118

Chandra, S., Kushwaha, P., Ganesh, S., *et al.* **2015**, *ATel*, #6962, 1

Chandra, S., Zhang, H., Kushwaha, P., *et al.* **2015**, *ApJ*, 809, 130

Marscher, A. P. **2014**, *ApJ*, 780, 87

Wierzcholska, A., & Siejkowski, H. **2016**, *MNRAS*, 458, 2350

Zhang, H., Chen, X., Böttcher M., Guo, F., & Li, H. **2015** *ApJ* 804 58

Zhang, H., Deng, W., Li, H., & Böttcher, M. **2016**, *ApJ*, 817, 63

New Frontiers in Black Hole Astrophysics
Proceedings IAU Symposium No. 324, 2016
Andreja Gomboc, ed.

doi:10.1017/S1743921317000023

Producing ultra high energy cosmic rays from AGN magnetic luminosity

Carlos H. Coimbra-Araújo and Rita C. Anjos

Departamento de Engenharias e Exatas, Universidade Federal do Paraná, Rua Pioneiro 2153, 85950-000, Palotina, Brazil.
email: carlos.coimbra@ufpr.br

Abstract. The present work proposes a method to calculate the AGN magnetic luminosity fraction to be converted into ultra high energy cosmic rays (UHECRs) luminosities for nine UHECR AGN Seyfert sources based on the respective observation of gamma ray upper limits. The motivation for such calculation comes attached to the fact that a fraction of the magnetic luminosity (L_B) produced by Kerr black holes in some AGNs can produce the necessary energy to accelerate UHECRs beyond the GZK limit, observed, e.g., by the Auger experiment. Nevertheless, the direct detection of those UHECRs has a lack of information about the direction of the source from where those cosmic rays are coming, since charged particles are deflected by the intergalactic magnetic field. Such problem arises alternative methods to evaluate the luminosity of UHECRs from the observation of upper limits during the propagation.

1. Introduction

The nature and the origin of the ultra high energy cosmic rays (UHECRs) with energies higher than 10^{19} eV are subject of many investigations. It is a common sense in the literature about the topic that the main extragalactic candidates for accelerating particles to those highest energies are active galactic nuclei (AGNs) (Henri *et al.* 1999), the most powerful radiogalaxies (Takahara 1990, Rachen & Biermann 1993), and also gamma ray bursts, fast spinning newborn pulsars, interacting galaxies, large-scale structure formation shocks and a plenty of other phenomena (Kotera & Olinto 2011). All the potential of a source to accelerate particles up to the exposed extreme energies is summarized in the well-known Hillas plot (Hillas 1984).

From the point of view of Earth experiments, the deduction of cosmic ray luminosities can possibly shed some light on radiative bounds of UHECR potential sources, as can be seen in Supanitsky & de Souza (2013) and in Anjos *et al.* (2014), where it is shown that the methods of UHECR propagation from the source to Earth and the measured upper limit on the integral flux of GeV-TeV gamma-rays can infer upper limits of proton and total UHECR (iron) luminosities. This comes from the fact that gamma-rays can be produced as a result of the cosmic ray propagation and contribute to the total flux measured from the source. Cascading background to produce GeV-TeV gammas comes from the interaction of UHECRs with the Extragalactic Background Light (EBL) ($\approx 10-10^5$ GeV gammas), with the Cosmic Microwave Background (CMB) ($\approx 10^5 - 10^{10}$ GeV) and with the radio background ($10^{10} - 10^{13}$ GeV), see more details in (Blumenthal 1970, Puget *et al.* 1976, Kelner & Aharonian 2008, De Angelis *et al.* 2013). In this way, gamma-ray observation and respective upper limits are advantageous in comparison to direct UHECR observations, mainly in the aspect of the reconstruction of trajectories to see the main probable candidate sources and respective luminosities in UHECRs.

Following the method summarized above, the present work investigates nine AGNs as UHECR sources (see Table 1). It is calculated upper limits of UHECRs luminosities to be compared to the theoretical magnetic/jet luminosity of those AGNs to therefore calculate the η_{CR} conversion fraction.

2. Upper Limits of UHECRs

The reconstructed direction of a charged cosmic-ray particle might not point back exactly to its source due to deviation in the magnetic fields. Given this limitation, the study of the origin of the cosmic rays is restricted to limited classes of sources and any important information involves the convolution of measured quantities with propagation models and catalogs of objects. On the other hand, gamma-rays are not deviated by magnetic fields. At TeV energies a few hundred gamma-ray sources have been discovered by the past and current Imaging Atmospheric Cherenkov Telescopes (IACTs). Also, high energy gamma rays (10 GeV$< E_\gamma <$ 300 GeV) have been detected by space and ground instruments, as FERMI-LAT (Adbo *et al.* 2010), VERITAS (Galante *et al.* 2009), H.E.S.S (Aharonian *et al.* 2008) and MAGIC (Antonelli *et al.* 2010). The method described in Supanitsky & de Souza (2013) and in Anjos *et al.* (2014) connects those measured upper limits with the source UHECR cosmic ray luminosity (L_{CR}^{UL}) by

$$L_{CR}^{UL} = \frac{4\pi D_s^2 (1+z_s)\langle E\rangle_0}{K_\gamma \int_{E_{th}}^{\infty} dE_\gamma \; P_\gamma(E_\gamma)} \; I_\gamma^{UL}(> E_\gamma^{th}), \tag{2.1}$$

where $I_\gamma^{UL}(> E_\gamma^{th})$ is the upper limit on the integral gamma-ray flux for a given confidence level and energy threshold, K_γ is the number of gamma rays generated from the cosmic ray particles, $P_\gamma(E_\gamma)$ is the energy distribution of the gamma-rays arriving on Earth, E_γ is the energy of gamma-rays, $\langle E\rangle_0$ is the mean energy, D_s is the comoving distance and z_s is the redshift of the source. This method allows one to calculate upper limits on the proton and total luminosities for energies above 10^{18} eV. Also, it illustrates techniques to study the origin of UHECRs from multi-messenger GeV-TeV gamma-rays and it has been used to calculate at least upper limits for thirty sources (AGNs), with redshift smaller than 0.048 and UHECR spectra measured by the Pierre Auger (Schulz *et al.* 2013) and Telescope Array (TA) (Bergman 2013) observatories. For this same method, CTA Observatory (Actis *et al.* 2011) will also allow, in the future, a range of new UHECR luminosity calculations.

3. AGN Magnetic luminosities

In what follows it is described the production of UHECR luminosities based on the possible relation between magnetic flux accumulation and jet production efficiency. To mimic AGN powering fluxes, let us consider a system with a rotating central BH that is producing electromagnetic Poynting flux, i.e., it drives magnetic field to accelerate particles via Blandford-Znajek mechanism. The necessary condition to use Blandford-Znajek mechanism (Blandford & Znajek 1977) is that $\Phi_d > \Phi_{BH,\max}(\dot{M})$, i. e., the net poloidal magnetic flux Φ_d trapped in the disk is larger than the maximum that can be confined on the BH caused by pressure of the accreting plasma. Satisfied this condition, the rate of energy extraction from the rotating BH via the Blandford-Znajek (BZ) mechanism yields the magnetic luminosity (see Coimbra-Araújo & Anjos (2015)). As AGN jets can be the main extragalactic sources of UHECRs, one can write the UHECR luminosity L_{CR}^{Theory} as a fraction η of the magnetic luminosity:

$$L_{CR}^{Theory} = \eta L_B. \tag{3.1}$$

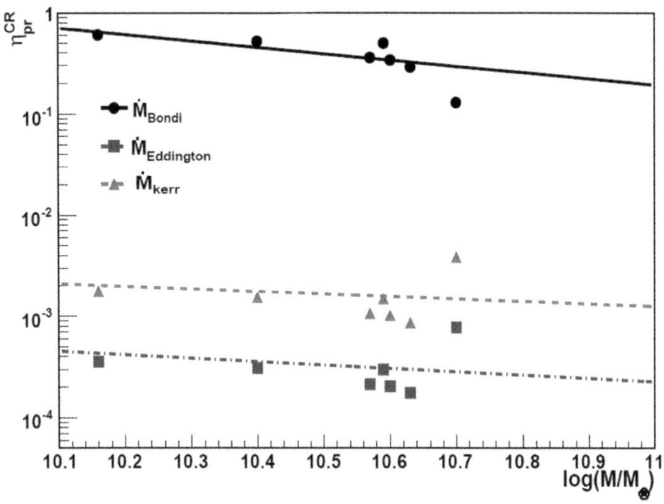

Figure 1. Production of UHECRs in AGNs with Kerr black holes (axisymmetric accretion by Kerr black holes). In comparison we show the limits using the Eddington and Bondi accretion rates.

Table 1. Comparison between cosmic ray luminosity L_{pr}^{UL} (protons) from the method derived from eq. (2.1) and the theoretically calculated range $L_{CRmin}^{Theory} - L_{CRmax}^{Theory}$ of cosmic ray luminosities, from nine sources. Here it is considered the case for $a = 0.7$.

Source name	$\mathbf{L_{pr}^{UL}}$ (Proton) [erg s^{-1} $\times 10^{45}$]	$\mathbf{L_{CRmin}^{Theory} - L_{CRmax}^{Theory}}$ [erg s^{-1} $\times 10^{45}$]
NGC 985	1.03	0.77 - 2.79
NGC 1142	0.49	0.97 - 3.52
2MASX J07595347+2323241	1.01	0.72 - 2.61
CGCG 420-015	0.95	0.83 - 2.99
MCG-01-24-012	0.65	0.28 - 1.02
2MASX J11454045-1827149	1.30	0.19 - 0.70
LEDA 170194	1.48	0.75 - 2.73
NGC 5995	0.90	0.15 - 0.54
Mrk 520	0.98	0.49 - 1.77

For example, bounds on the fraction η_{pr} of L_B to be converted in relativistic protons, from upper limits as a function of BH spin a is written as

$$\eta(a)_{pr} = \frac{4\pi D_s^2(1+z_s)}{10(\phi/50)x_a^2 f_a(x_a)\dot{M}c^2} \frac{\langle E\rangle_0 I_\gamma^{UL}(>E_\gamma^{th})}{K_\gamma \int_{E_{th}}^{\infty} dE_\gamma P_\gamma(E_\gamma)}, \tag{3.2}$$

where ϕ and $f(x_a)$ are BZ functions described in McKinney & Tchekhovskoy (2012). For the spin range $0.44 < a < 0.8$ where most black holes are expected to lie, the fraction η_{pr} varies from $\sim 5\%$ to 50% for $a = 0.45$ and from 2% to 10% for $a = 0.8$ Anjos *et al.* (2016), Coimbra-Araújo & Anjos (2015). Such limits are calculated with the aid of Table

1, which shows nine sources with their respective calculated upper limits in UHECR luminosity for protons.

4. Results and discussion

Fig. 1 shows how Kerr black holes, using a axisymmetric accretion with a central spinning hole, according to AGN masses, will produce UHECRs. A $M^{-5.8}$ power law is obtained, according to the AGN mass. The interplay between L_{CR}^{Theory} upper limits and the magnetic luminosity L_B is explicitly investigated here by a conversion rate η_{CR} calculated here and developed in Coimbra-Araújo & Anjos (2015) and Anjos *et al.* (2016). The comparison among L_{CR}^{Theory}, the upper limit L_{pr} and L_B offer limits to the conversion fraction η_{pr} for each source. Here, the accretion mechanism has spherical symmetry (Bondi mechanism) and also axial symmetry (Fig. 1), where it is assumed the general case of a Randall-Sundrum-Kerr central black hole, see Coimbra-Araújo & Anjos (2015).

References

Abdo, A. A., Ackermann, M., Ajello, M., Allafort, A., Antolini, E., Atwood, W., Axelsson, M., Baldini, L., Ballet, J., Barbiellini, G., *et al.*, *The Astrophysical Journal* 715, p. 429 (2010).

Actis, M., Agnetta, G., Aharonian, F., Akhperjanian, A., Aleksic, J., Aliu, E., Allan, D., Allekotte, I., Antico, F., Antonelli, L., *et al.*, *Experimental Astronomy* 32, 193-316 (2011).

Aharonian, F., Akhperjanian, A., De Almeida, U. B., Bazer-Bachi, A., Behera, B., Beilicke, M., Benbow, W., Bernlohr, K., Boisson, C., Bolz, O., *et al.*, *Astronomy & Astrophysics* 478, 387-393 (2008).

Anjos, R., De Souza, V., & Supanitsky, A., *Journal of Cosmology and Astroparticle Physics* 2014, p. 049 (2014).

Anjos, R. C., Coimbra-Araújo, C. H., da Rocha, R., & de Souza, V., *Journal of Cosmology and Astroparticle Physics* 2016, p. 014 (2016).

Antonelli, L., Antoranz, P., Backes, M., Barrio, J., Bastieri, D., Bednarek, W., Berdyugin, A., Berger, K., Bernardini, E., Biland, A., *et al.*, *The Astrophysical Journal Letters* 723, p. L207 (2010).

Bergman, D., "Ta spectrum summary," Proceedings of the 33rd International Cosmic Ray Conference, Rio de Janeiro, Brazil (2013).

Blandford, R. D. & Znajek, R. L., *Monthly Notices of the Royal Astronomical Society* 179, 433-456 (1977).

Blumenthal, G. R., *Physical review* D 1, p. 1596 (1970).

Coimbra-Araújo, C. & Anjos, R., *Physical Review* D92, p. 103001 (2015).

De Angelis, A., Galanti, G., & Roncadelli, M., *Monthly Notices of the Royal Astronomical Society* 432, 3245-3249 (2013).

Dutan, I. & Caramete, L. I., *Astroparticle Physics* 62, 206-216 (2015).

Galante, N., *et al.*, arXiv preprint arXiv:0912.3850 (2009).

Henri, G., Pelletier, G., Petrucci, P.-O., & Renaud, N., *Astroparticle Physics* 11(3) , 347-356 (1999).

Hillas, A., *Annual review of astronomy and astrophysics* 22, 425-444 (1984).

Kelner, S. & Aharonian, F., *Physical Review* D78, p. 034013 (2008).

Kotera, K. & Olinto, A. V., *Annual Review of Astronomy and Astrophysics* 49, 119-153 (2011).

McKinney, J. C., Tchekhovskoy, A., & Blandford, R. D., *Monthly Notices of the Royal Astronomical Society* 423, 3083-3117 (2012).

Puget, J., Stecker, F., & Bredekamp, J., *The Astrophysical Journal* 205, 638-654 (1976).

Rachen, J. P. & Biermann, P. L., *Astronomy and Astrophysics* 272, p. 161 (1993).

Schulz, A., *et al.*, *Proceedings of the 33rd ICRC*, Rio de Janeiro, Brasil (2013).

Supanitsky, A. & de Souza, V., *Journal of Cosmology and Astroparticle Physics* 2013, p. 023 (2013).

Takahara, F., *Progress of Theoretical Physics* 83(6), 1071-1075 (1990).

New Frontiers in Black Hole Astrophysics
Proceedings IAU Symposium No. 324, 2016
Andreja Gomboc, ed.

doi:10.1017/S1743921317002290

The Energetic Particle Population in Centaurus A

Paula Chadwick[1], Anthony M. Brown[1], Celine Boehm[1], Jamie Graham[1], Thomas Lacroix[2] and Joseph Silk[3,4]

[1]Dept. of Physics, University of Durham, South Road, Durham DH1 3LE, UK
[2]LAPTh U. de Savoie, CNRS, BP 110, 74941 Annecy-le-Vieux
[3]Institut d'Astrophysique de Paris, UMR 7095, CNRS, UPMC Université Paris 6, Sorbonne Université, 98 bis boulevard Arago, 75014 Paris, France
[4]Johns Hopkins University, Baltimore, MD 21218, United States

Abstract. We report a significant hardening of the *Fermi*-LAT gamma-ray spectrum from the core of Cen A at E > 2.4 GeV, suggesting there is a source of high energy particles in the core of Cen A which is in addition to the jet component. We show that the observed gamma-ray spectrum is compatible with either a spike in the dark matter halo profile or a population of millisecond pulsars. This work gives a strong indication of new gamma-ray production mechanisms in active galactic nuclei and could even provide evidence for the clustering of heavy dark matter particles around black holes.

Keywords. gamma rays: observations, dark matter, galaxies: active, pulsars: general

1. Introduction

Cen A is the closest active galaxy to Earth (Rejkuba, 2004) and emits radiation at all wavelengths, including γ-rays (Sreekumar *et al.*, 1999; H.E.S.S. Collaboration, 2009, *Fermi*-LAT Collaboration, 2010a). Observations made with the Large Area Telescope (LAT) of the *Fermi Gamma-ray Space Telescope* confirmed the existence of a bright γ-ray core (*Fermi*-LAT Collaboration, 2010a) and extended emission spatially coincident with Cen A's giant radio lobes (*Fermi*-LAT Collaboration, 2010b). A possible hardening of the spectrum above 4 GeV was noted by Sahakyan *et al.* (2013), along with some evidence for variability on 45-day timescales, much less than other γ-ray bright radio galaxies such as M87 and NGC 1275 (H.E.S.S. Collaboration, 2006; Brown & Adams, 2011). This paper focuses on the core emission from Cen A, using the new Pass 8 data release from *Fermi*-LAT. After establishing the presence of a statistically significant hardening of the spectrum and assessing the evidence for variability, we consider possible interpretations of this result.

2. Analysis

All photons from within 10° of Centaurus A were extracted from 8 years of *Fermi*-LAT data since August 2008 and a summed likelihood analysis was performed using the *Fermi*-LAT PSF classes. These apply instrument response functions which depend on the quality of the PSF reconstruction and enable better resolution of Cen A's radio lobes and γ-ray core at lower energies. A zenith cut of 90° was applied to each PSF quartile, using only `SOURCE` photons. The corresponding emission models and the Pass 8 corrected Galactic diffuse emission template `gll_iem_v06.fit` were used, with the *Fermi* 4-year point source catalog, 3FGL, being used to produce a preliminary sky model (*Fermi*-LAT collaboration, 2010c).

Model	Norm [MeV cm^{-2} s^{-1}]	Shape Parameters		Scale E [MeV]	$2\Delta\log(\mathcal{L})$[a]	DoF
PL	$(4.36 \pm 0.06) \times 10^{-9}$	$\gamma = 2.690 \pm 0.014$		$E_0 = 481.17$	0.0	2
BPL	$(4.21 \pm 1.38) \times 10^{-11}$	$\gamma_1 = 2.73 \pm 0.02$	$\gamma_2 = 2.29 \pm 0.07$	$E_b = 2593 \pm 314$	28.600	4[b]
LP	$(4.15 \pm 0.08) \times 10^{-11}$	$\alpha = 2.70 \pm 0.01$	$\beta = -0.04 \pm 0.01$	$E_0 = 481.17$	16.637	3

Table 1. Spectral fits using several models, the definitions of which can be found in Equations 3.1 and 3.2. [a] This is the maximum likelihood probability compared to that of the best-fit power-law model. [b] The broken power-law model does not satisfy Wilks' theorem as it is not nested with a power-law model.

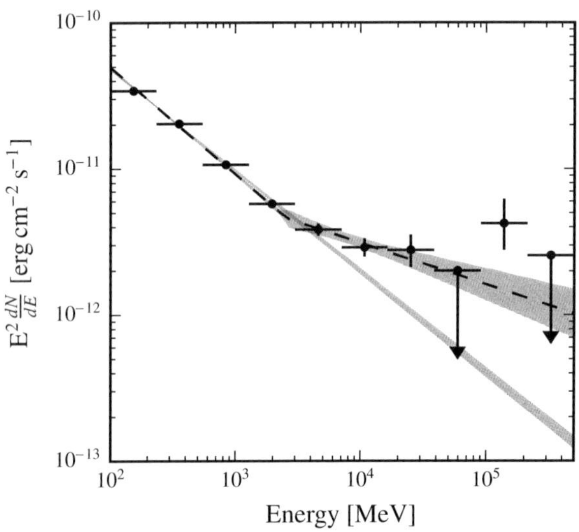

Figure 1. The spectrum of Centaurus A fitted with broken power-law and power-law models. The data are binned into 10 bins between 100 MeV and 500 GeV, with 95% upper limits plotted if the bin has a TS< 25.

3. The Spectrum

The spectrum of Cen A between 0.1 and 300 GeV was fitted with a power-law (PL), a broken power-law (BPL) spectrum:

$$\frac{dN(E)}{dE} = N \times \begin{cases} \left(\frac{E}{E_b}\right)^{-\gamma_1} & \text{if } E < E_b, \\ \left(\frac{E}{E_b}\right)^{-\gamma_2} & \text{otherwise,} \end{cases} \tag{3.1}$$

and a log-parabola (LP) model:

$$\frac{dN(E)}{dE} = N\left(\frac{E}{E_0}\right)^{-(\alpha+\beta\log(E/E_0))}, \tag{3.2}$$

The spectrum of Centaurus A, fitted with Equation 3.1 and a PL model, is shown in Figure 1; the best-fit parameters are in Table 1, including the fit to Equation 3.2. The BPL model is preferred over the PL fit, with a test statistic of over 28, a significance slightly greater than 5σ. This compares well with the HESS spectrum observed at higher energies; extrapolating the power law found at energies below 2.4 GeV to the HESS energies would under predict the flux above 250 GeV by a factor of 10 (see Figure 1).

4. Variability

Binning the data on 15, 30 and 45-day timescales, as in Sahakyan *et al.* (2013), shows no significant variability using either a χ^2 comparison to a constant value or the TS_{var} (Nolan *et al.*, 2012). As our analysis suggests that there may be two separate emission components, we split the data at the best-fit break energy of 2.4 GeV and searched for variability above and below the break, binning the data on a 6-month timescale to retain sufficient statistics at high-energies. We find minimal evidence of variability (at the 95% level) in the data either above or below the break. When treating bins with a test statistic less than 16 as upper limits, which therefore do not contribute to the χ^2 value, no variability at all could be found in the high-energy component.

5. Implications

Petropoulou *et al.* (2014) noted a one-zone synchrotron self-Compton model cannot account for the emission of Cen A above a few GeV, and attributed the high-energy radiation to a relativistic proton component. Such a model predicts strong variability in the high energy component, which we have not found, although this lack of evidence could be due to the relatively low statistics in the high-energy regime. Nonetheless, we explore emission mechanisms in which the high-energy emission would be non-variable.

5.1. *Dark Matter*

To investigate whether the emission between 2.4 GeV and $\sim$ 5 TeV could be produced by dark matter (DM), we allowed the DM mass and annihilation cross-section to be free parameters, as well as the normalisation and slope of the spectral emission below 2.4 GeV. We considered DM self-annihilations into leptons and quarks, and tested two DM density profiles, the NFW profile $\rho(r) \propto r^{-1}$ (Navarro, Frenk and White, 1996) and a 'spiky' profile, $\rho(r) \propto r^{-\frac{7}{3}}$ (Gondolo and Silk, 1999). Our best fit favours a DM candidate of mass 3 TeV, annihilating into a pair of top and anti-top quarks with cross-section $\sigma\mathrm{v}\approx 1.6 \times 10^{-32}\mathrm{cm}^3\mathrm{s}^{-1}$, and a 'spiky' density profile. The fit has a χ^2 of 1.7, or a χ^2/d.o.f. of 0.24, suggesting that the fit is dominated by statistical errors. The best-fit annihilation cross-section of $\sigma\mathrm{v} \approx 1.6 \times 10^{-32}\mathrm{cm}^3\mathrm{s}^{-1}$ is too small to explain the observed DM fraction in the Universe, but might suggest a rich DM sector with several (non thermal) DM particles (Boehm, Fayet and Silk, 2004; Zurek, 2009) or both velocity-independent and dependent terms in the annihilation cross-section.

5.2. *Millisecond Pulsars*

A further explanation for the high energy emission is the presence of millisecond pulsars (MSPs), which are a possible explanation for the low energy γ-ray excess observed in the central region of the Milky Way (e.g. Goodenough and Hooper, 2009). The important assumptions for modelling the expected emission from MSPs are the electron injection spectrum (which must extend to a few 10s of TeV) and a large enough ambient photon field for the inverse Compton losses to dominate over the synchrotron losses. For the magnetic field, we assume a constant value of 10 μG, and use a density profile for the MSPs consistent with the Galactic centre γ-ray excess of $\rho_{MSP} \propto r^{-1.2}$. The normalisation and slope of the jet contribution are kept as free parameters. Assuming MSPs are entirely responsible for the emission, this model leads to a fit with a $\chi^2 \sim 10$ or $\chi^2/dof \sim 1.4$.

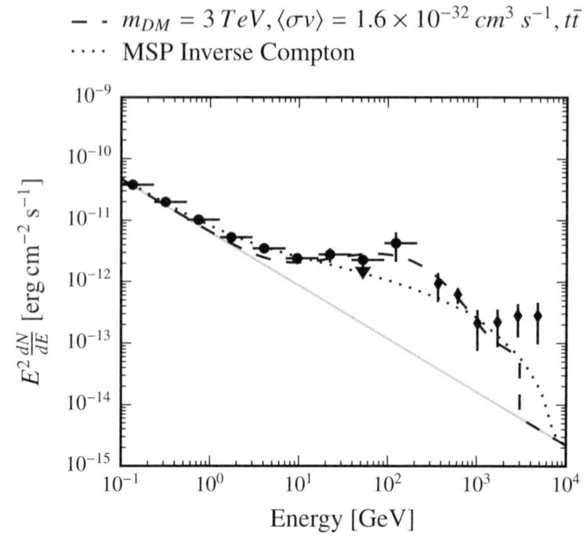

Figure 2. Dark matter and MSP fits to both Fermi and H.E.S.S. data. The dark matter fit finds $\chi^2/dof = 0.27$, and the MSP fit finds $\chi^2/dof = 1.4$. The sharp cutoff in the predictions from the dark matter model is a result of the mass of the dark matter particle, as no DM particle acceleration is considered.

6. Conclusions

We have shown evidence at the 5σ level for the existence of a hardening in the *Fermi*-LAT spectrum from Cen A, and that either heavy DM particles or a population of MSPs could explain this spectral feature. We cannot exclude the possibility that the jet is the origin of the spectral hardening, but the absence of variability above 2.4 GeV argues against jet-induced leptonic models. To understand better the nature of the high-energy component requires more data. In particular, the forthcoming Cherenkov Telescope Array, together with more *Fermi* data will be important in this respect. The detection of variability in either component, for example, would significantly constrain the models.

References

Boehm, C., Fayet, P., & Silk, J. 2004, *Phys. Rev. D.*, 69, 101302
Brown, A. M. & Adams, J. 2011, *Monthly Notices of the Royal Astronomical Society* 413, 2785
Fermi-LAT, Falcone, A., *et al.* 2010, *Astrophysical Journal*, 719, 1433
Fermi-LAT, Abdo, A. A., *et al.* 2010, *Science*, 328, 725
Fermi-LAT, Abdo, A. A., *et al.* 2010, *Astrophysical Journal Supplement*, 218, 23
Gondolo, P. & Silk, J. 1999, *Physical Review Letters*, 83, 1719
Goodenough, L. & Hooper, D. 2009, ArXiv e-prints 0910.2998
H. E. S.S, Aharonian, F., *et al.* 2009, *Astrophysical Journal Letters*, 695, L40
Navarro, J. F., Frenk, C. S., & White, S. D. M.. 1996, *Astrophysical Journal*, 462, 563
Nolan, P. L., *et al.* 2012, *Astrophysical Journal Supplement*, 199, 31
Petropoulou, M., *et al.* 2014, *Astronomy & Astrophysics*, 562, A12
Rejkuba, M. 2004, *Astronomy & Astrophysics*, 413, 903
Sahakyan *et al.* 2013, *Astrophysical Journal Letters*, 770, L6
Sreekumar, P., *et al.* 1999, *Astroparticle Physics*, 11, 221
Zurek, K. M. 2009, *Phys. Rev. D.*, 115002, 0811.4429

New Frontiers in Black Hole Astrophysics
Proceedings IAU Symposium No. 324, 2016
Andreja Gomboc, ed.

doi:10.1017/S1743921316012680

Space Telescope and Optical Reverberation Mapping Project: A Leap Forward in Reverberation Mapping

Bradley M. Peterson[1,2,3]

[1]Space Telescope Science Institute, 3700 San Martin Drive, Baltimore, MD 21218, USA

[2]Department of Astronomy, The Ohio State University, 140 West 18th Ave., Columbus, OH 43210, USA

[3]Center for Cosmology and AstroParticle Physics, The Ohio State University, 191 West Woodruff Ave., Columbus, OH 43210, USA
email: peterson.12@osu.edu

Abstract. In 2014, a 179-orbit allocation of *Hubble Space Telescope* time anchored a massive reverberation-mapping program on the well-studied Seyfert 1 galaxy NGC 5548. Supporting imaging and spectrophotometric observations were provided by *Swift*, *Chandra*, *Spitzer*, and a world-wide network of ground-based telescopes. Understanding the data remains a significant challenge, partly because the level of detail is far beyond what has been seen before and partly because the behavior of the AGN was not typical of its past behavior. Based on analysis to date, the following conclusions can be reached: (1) the AGN accretion disk has a temperature profile that is consistent with that predicted by the Shakura–Sunyaev model, but is about three times larger than expected; (2) at least part of the broad-line region appears to be a Keplerian disk seen at intermediate inclination, and (3) the broad-line emission response from the far side of the disk is weaker than expected.

Keywords. black hole physics, galaxies: active, galaxies: nuclei, quasars: emission lines, ultraviolet: galaxies

1. Introduction

Reverberation mapping is a technique for studying the structure and kinematics of the broad-line region (BLR) in AGNs by making use of the intrinsic flux variability of the central continuum source (Blandford & McKee 1982, Peterson 1993, Peterson 2014). The relatively dense BLR gas responds to continuum variations with a time delay due to the light-travel time across the BLR. The amplitude of continuum and line flux variations is generally small on the relevant timescales, so a successful reverberation-mapping campaign requires high S/N, high precision spectrophotometry, a high cadence, and a long duration (Horne *et al.* 2004) in order to recover velocity–delay maps, which are the projection of the BLR into the two observables of line-of-sight (Doppler) velocity and time delay (relative to the continuum variations). However, even with more modest data, it is possible to infer the mean BLR size simply by measuring the average response time for the entire emission line to the continuum variations — this is in fact what has been done in most reverberation studies for the past 25 years. Such measurements have been made for over 50 AGNs (see Bentz & Katz 2015 for a compilation), and in some cases for multiple lines in a single AGN, but more often only for Hβ.

The particular value of these measurements are that they can be combined with the width of the emission line to obtain a mass for the central black hole (see Peterson 2014 for details). In the case of AGNs for which multiple emission-line lag measurements have been made, the relationship between time lag τ and line width ΔV is found to be $\tau \propto \Delta V^{-1/2}$, which is as expected in system dominated by gravity. The mass of the central black hole is thus taken to be

$$M_{\rm BH} = f\left(\frac{c\tau\,\Delta V^2}{G}\right), \tag{1.1}$$

where c is the speed of light, G is the gravitational constant, and f is a dimensionless scale factor of order unity that subsumes all of the details we do not know, such as the inclination of the system and its structure and kinematics. The factor f is different for each AGN, but a statistical average value for AGNs can be inferred by assuming that AGNs follow the same relationship between black hole mass and host-galaxy bulge velocity dispersion as quiescent galaxies (Onken *et al.* 2004). Woo *et al.* (2015) provide the most recent update to this parameter, $\langle f \rangle = 4.47 \pm 1.25$.

A key result from reverberation studies is the relationship between the BLR radius and the AGN luminosity, the $R - L$ relationship (Wandel, Peterson, & Malkan 1999, Kaspi *et al.* 2000, Kaspi *et al.* 2005, Bentz *et al.* 2006, Bentz *et al.* 2009, Bentz *et al.* 2013). The great value of this relationship is that it allows us to bypass observationally expensive reverberation mapping by using a single spectrum to determine the AGN luminosity and the emission-line width to estimate the mass of the central black hole. *All quasar mass estimates trace back to this relationship.*

2. Velocity-Resolved Reverberation Mapping

While black-hole mass measurement has turned out to be an important byproduct of reverberation mapping, the actual goal remains characterization of the structure and kinematics of the BLR. After around a decade of experience with reverberation mapping, the continuum and emission line behaviors were sufficiently well characterized that we could address through simulations the question of what would it take to recover high-fidelity velocity–delay maps (Horne *et al.* 2004). Even at the present time, velocity–dependent lags have been identified for a limited number of objects (e.g., Bentz *et al.* 2009, Denney *et al.* 2009, Barth *et al.* 2011a, Barth *et al.* 2011b, Grier *et al.* 2013), and the number of AGNs for which velocity–delay maps have been recovered remains limited (Bentz *et al.* 2010, Grier *et al.* 2013). It must also be noted that a complementary approach of direct Bayesian modeling of the spectra has also shown to be very promising (Pancoast, Brewer, & Treu 2011, Brewer *et al.* 2011, Li *et al.* 2013, Pancoast *et al.* 2014).

There is a perhaps interesting bit of sociology to note here: most reverberation results are based primarily on observations made on ground-based proprietary telescopes that are usually not the largest telescopes available to the observers. While reverberation mapping has been practiced in some form for over 25 years, it is a time-intensive risky† technique that is generally regarded as under development. As a consequence, most reverberation programs have been undertaken on smaller telescopes. The "reverberation sample" thus has several well-known biases compared to the AGN population as a whole: the reverberation-mapped AGNs *tend* to be lower-luminosity, apparently bright, high Eddington rate local AGNs that are primarily in the northern hemisphere. And space-based reverberation-mapping in the ultraviolet (UV) essentially ended with the decommissioning of the *International Ultraviolet Explorer* in 1996.

It is only when velocity–delay maps were successfully recovered from ground-based data (Bentz *et al.* 2010, Grier *et al.* 2013) that confidence in the reverberation-mapping

† Our experience with multiobject campaigns lasting months is that somewhat more than half of the targets will yield useful emission-line lags.

technique was high enough that a proposal to carry out a UV reverberation program with *Hubble Space Telescope (HST)* was finally approved. The target selected for this study was NGC 5548, which is by far the best-studied of all the reverberation-mapped AGNs ("reverberated to death," in the words of one critic...) for a number of reasons, including that (1) its reverberation characteristics were of course well known, (2) historically it has little self-absorption in the UV resonance lines, and (3) it is bright enough to obtain a high S/N spectrum in one *HST* orbit, and (4) its luminosity and line lags are almost optimal for efficient use of *HST* — daily observations for somewhat less than an entire observing season. The approved program consisted of 179 daily observations with the Cosmic Origins Spectrograph‡. The *HST* program was complemented by high-cadence observations with *Swift* and with ground-based imagining and spectroscopy, and we refer to the combined program as the AGN Space Telescope and Optical Reverberation Mapping (AGN STORM) program. The initial *HST* results are reported by De Rosa *et al.* (2015) and the *Swift* observations are described by Edelson *et al.* (2015). Fausnaugh *et al.* (2016) present the ground-based optical photometry and an analysis of the continuum behavior from the shortest *HST* band (1158 Å) to the Sloan z band (9157 Å). Pei *et al.* (2016) present the results of the ground-based spectroscopy program, and Goad *et al.* (2016) and Starkey *et al.* (2016) describe some of the early analysis of the data.

The principal results to date include the following:

(1) The continuum variations at shorter wavelengths precede those at longer wavelengths. Throughout the UV to near-IR, the light curves at longer wavelengths look like shifted and smoothed versions of those at shorter wavelengths, as expected in simple lamp-post models (Edelson *et al.* 2015, Fausnaugh *et al.* 2016). The X-ray variations precede those in the UV-optical, but the X-ray light curves have more structure than those in the UV-optical — in other words, the UV-optical light curves are *not* shifted and smoothed versions of the X-ray light curves (Edelson *et al.* 2015, Gardner & Done 2016).

(2) If the continuum variations are caused by irradiation of the accretion disk by a variable driving continuum, the variations can be used to map the accretion-disk structure. The disk temperature profile is consistent with the predictions of a standard Shakura–Sunyaev thin disk, but the size of the disk is larger than expected by about a factor of three.

(3) The lag between the shortest-wavelength UV continuum and the optical continuum is approximately the same as the lag between the UV continuum and the most rapidly responding emission lines, He II λ1640 and He II λ4686 (De Rosa *et al.* 2015, Pei *et al.* 2016). This has important implications for optical-only reverberation programs.

(4) The emission-line behavior is not as simple as seen in previous reverberation observations. The responsivity of the emission lines varies during the campaign, which is unusual (Goad *et al.* 2016, Pei *et al.* 2016).

(5) Compared to previous reverberation campaigns on NGC 5548, the emission-line lags are unexpectedly small compared to expectations based on its luminosity. Based on luminosity, the Hβ lag, for example, should have been ~ 20 days, and the observed value is ~ 6 days (Pei *et al.* 2016). The equivalent width of Hβ is also surprisingly small, near its historically low value.

(6) The velocity–delay maps show that at least part of the BLR is a Keplerian disk viewed at intermediate inclination. However, the response of the far side of the BLR is weaker than expected (Horne *et al.*, in preparation).

‡ Due to safing events, only 171 of the observations were successfully executed, but the largest gap in the coverage is only two days.

(7) The small lag, low equivalent width, and weak response of the far side of the BLR suggest that much of the ionizing radiation from the central source is not reaching the far side of the BLR, either because of absorption near the central source or anisotropic continuum emission.

(8) The resonance lines show both strong absorption in both broad and narrow components, which complicates the analysis. Historically, the absorption features have been weak in NGC 5548, but the situation has changed dramatically recently (Kaastra *et al.* 2014). We have carefully modeled the absorption in all the UV spectra in order to remove the absorption features. Further analysis based on these models is in progress (Kriss *et al.*, in preparation).

We are grateful for support of this program by NASA through grant HST-GO-13330 from the Space Telescope Science Institute.

References

Barth, A. J., *et al.* 2011a, *ApJ*, 732:121
Barth, A. J., *et al.* 2011b, *ApJ*, 742, L4
Bentz, M. C. & Katz, S. 2015, *PASP*, 127, 67
Bentz, M. C., *et al.* 2006, *ApJ*, 644, 133
—. 2009, *ApJ*, 697, 160
—. 2010, *ApJ*, 720, L46
—. 2013, *ApJ*, 767:149
Blandford, R. D. & McKee, C. F. 1982, *ApJ*, 255, 419
Brewer, B. J., Treu, T., Pancoast, A., Barth, A. J., Bennert, V. N., Bentz, M. C., Filippenko, A. V., Greene, J. E., Malkan, M. A., & Woo, J.-H. 2011, *ApJ*, 733, L33
De Rosa, G., *et al.* 2015, *ApJ*, 806:128
Denney, K. D., *et al.* 2009, *ApJ*, 704, L80
Edelson, R., *et al.* 2015, *ApJ*, 806:129
Fausnaugh, M. M., *et al.*, *ApJ*, 821:56
Gardner, E. & Done, C. 2016, *MNRAS*, in press
Goad, M. R., *et al.* 2016, *ApJ*, 824:11
Grier, C. J., *et al.* 2013, *ApJ*, 764:47
Horne, K., Peterson, B. M., Collier, S. J., & Netzer, H. 2004, *PASP*, 116, 465
Kaastra, J., *et al.* 2014, *Science*, 345, 64
Kaspi, S., Maoz, D., Netzer, H., Peterson, B. M., Vestergaard, M., & Jannuzi, B. T. 2005, *ApJ*, 629, 61
Kaspi, S., Smith. P. S., Netzer, H., Maoz, D., Jannuzi, B. T., & Giveon, U. 2000, *ApJ*, 533, 631
Li, Y.-R., Wang, J.-M., Ho, L. C., Du, P., & Bai, J.-M. 2013, *ApJ*, 779:110
Onken, C. A., Ferrarese, L., Merritt, D., Peterson, B. M., Pogge, R. W., Vestergaard, M., & Wandel, A. 2004, *ApJ*, 615, 645
Pancoast, A., Brewer, B. J., & Treu, T. 2011, *ApJ*, 730: 139
Pancoast, A., Brewer, B. J., Treu, T., Park, D., Barth, A. J., Bentz, M. C., & Woo, J.-H. 2014, *MNRAS*, 445, 3073
Pei, L., *et al.* 2016, submitted to *ApJ*
Peterson, B. M. 1993, *PASP*, 105, 247
—. 2014, Space Sci. Rev., 183, 253
Starkey, D., *et al.* 2016, submitted to *ApJ*
Wandel, A., Peterson, B. M., & Malkan, M. A. 1999, *ApJ*, 526, 591
Woo, J.-H., *et al.* 2015, *ApJ*, 801:38

New Frontiers in Black Hole Astrophysics
Proceedings IAU Symposium No. 324, 2016
Andreja Gomboc, ed.

doi:10.1017/S1743921316013107

Studying the outskirts of reverberation mapped AGNs

Shai Kaspi

School of Physics & Astronomy and the Wise Observatory
The Raymond and Beverly Sackler Faculty of Exact Sciences
Tel-Aviv University, Tel-Aviv 69978, Israel
email: shai@wise.tau.ac.il

Abstract. About 100 AGNs have their black hole mass measured directly using the reverberation mapping technique over the past few decades. By now we have high enough numbers to explore unique subsamples within these objects and to study phenomena across variety of AGNs. I will review recent reverberation mapping studies which focus on high-redshift high-luminosity AGNs and on AGNs with super-Eddington accreting massive black holes. These studies enable to investigate the BLR size, mass, and luminosity relations in different subsamples of AGNs and to check whether there are differences in these relations in different types of AGNs. In particular I will discuss the following questions: Is the BLR size - luminosity relation the same over the whole AGNs luminosity range? Are there different relations for different types of AGNs? What are these studies teaching us about theory of accretion into black holes in AGNs?

Keywords. accretion, accretion disks, galaxies: active, galaxies: nuclei

---

1. Introduction

The reverberation mapping (RM) method uses time lags between the emission of broad lines and the continuum in Active Galactic Nuclei (AGNs) in order to determine the distance of the Broad Line Region (BLR) from the central emission source (accretion to the central black hole), and from this to infer the mass of the black hole (BH) in the Center of the AGN (Peterson 1993; Kaspi *et al.* 2000). RM can also be used to determine the geometry and velocity field of the BLR given sufficiently high resolution spectroscopy over long enough period of time in which the AGN varied and with high cadence of observations.

Over the past three decades almost 100 AGNs with sufficient data to determine their BLR size were studied (see a partial compilation in Bentz & Katz 2015). This has resulted with a firm determination of a relation between the BLR size (R_{BLR}) and the AGN luminosity in various wavelength bands (e.g., Kaspi et al. 2005, Bentz *et al.* 2013). The R_{BLR}—Luminosity relation is used together with the width of the broad lines as a proxy to the BLR velocity and some assumptions about the geometry to determine the BH mass in the center of the AGNs. This yields also a Mass—Luminosity relation for the AGNs.

The R_{BLR}—Luminosity relation is broadly used to estimate the BH mass of thousands of AGNs using a single epoch spectrum of each AGN from which the luminosity and the width of the broad lines are measured. These samples are used for high- and low-luminosity AGNs in order to study the cosmological evolution of AGNs (mass function, accretion history, etc. ; e.g., Vestergaard *et al.* 2008). However, the measured R_{BLR}—Luminosity relation span only on a limited luminosity range, from $\lambda L_\lambda(5100\mathring{A})$ of 10^{42} to 10^{46} erg s^{-1}, and there are hardly no RM measurements outside of that range. In the

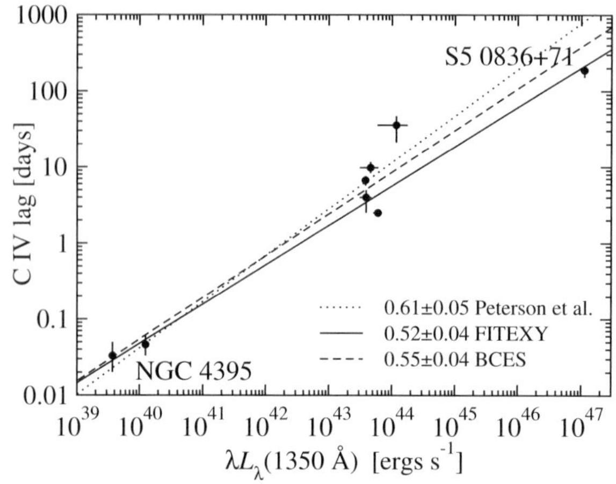

Figure 1. Size–Luminosity relation based on C IV(1550Å) emission line and the UV continuum. Data are reproduced from Peterson *et al.* (2005) to which the result for S5 0836+71 from Kaspi *et al.* (2007) was added. See Kaspi *et al.* (2007) for more details. Two more objects are now added to that result and are consistent with the point of S5 0836+71.

first part of the this contribution results from a RM campaign of high-luminosity RM campaign will be presented.

With tens of AGNs with measured R_{BLR}, studies of subsamples within the AGN phenomenon can now be carried out. In the second part of this contribution results from a RM campaign of Supper Eddington Accreting Massive Black Holes will be described with a comparison to normal accreting AGNs.

2. Reverberation Mapping of High-Luminosity AGNs

RM of high-luminosity AGNs (quasars) have not been much successful so far due to several difficulties: RM requires a lot of telescope time for monitoring the AGNs. High-luminosity quasars have much larger R_{BLR} and much longer variability time scales. Thus, a monitoring period of about a decade or more is needed. High-luminosity AGNs are fainter in general and thus a larger telescope is needed. The combination of a large telescope to be dedicated for more than a decade is very hard to to find. The light curves are also smeared due to time dilation and by the large BLR, and thus the response of the line to the continuum variation is very hard to detect.

In 1995 we started a photometric monitoring of 11 high-redshift high-luminosity quasars at the Wise Observatory 1 m telescope in Israel which is operated by Tel-Aviv University. In 1999 we started to spectroscopically monitor 6 of these quasars at the Hobby-Eberly Telescope (HET) 9 m telescope which is partly owned by Pennsylvania State University. This yield a RM campaign which lasted for 14 years. The sample ranges on redshift of $2.1 < z < 3.2$, luminosity of $10^{46} < \lambda L_\lambda(5100\AA) < 10^{47.5}$, and the lines that were monitored for each object are two of Lyα, C IV(1550Å), and C III(1909Å), depending on the object's redshift. First preliminary results from the first five years were published by Kaspi *et al.* (2007; and see Fig. 1).

After 14 years of spectroscopic monitoring and 18 years of photometric monitoring we found that the continuum variation for all object is around the level of 20–50% and we found time lags of the C IV(1550Å) line in three objects, in one of which we found a time lag for Lyα and for another object we found a time lag of C III(1909Å). The time lags are all around 200 light days in the rest frame and fall around the preliminary result which

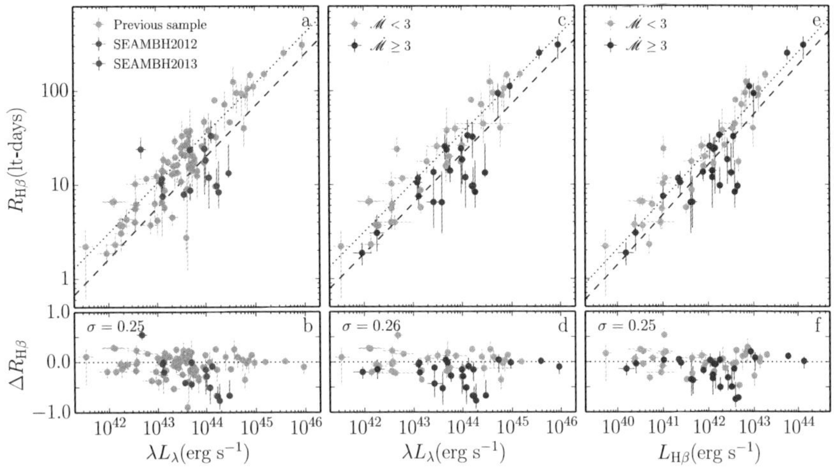

Figure 2. The R_{BLR}–Luminosity relations (panels *a, c, e*) and the deviations of R_{BLR} from the regression relationships (panel *b, d, f*). The dotted line is the fit for the normal AGNs sample whereas the dashed line is the fit for the SEAMBHs sample in all panels (see Du *et al.* 2015).

is shown in Fig. 1. The relation of C IV(1550Å) size and the UV luminosity seems to be the same over 9 orders of magnitude in luminosity, and high-luminosity AGNs keep the R_{BLR} and Luminosity relation.

3. Reverberation Mapping of Supper Eddington Accreting Massive Black holes

Supper Eddington Accreting Massive Black holes (SEAMBHs) are AGNs which accrete into their BHs above their Eddington rate. We carried out a RM campaign with photometric observations from the Wise Observatory 1 m telescope in Israel and spectroscopic observations from the 2.5 m telescope in Lijiang observatory in China. Ten AGNs were monitored during 2012 and for nine of then we determined the time lag of the Hβ line. During 2013 eight AGNs were monitored and for five of them a time lag was determined. The results are presented in the four papers by Wang et al. (2014), Hu *et al.* (2015), and Du *et al.* (2014; 2015).

One of the main results of this RM campaign is shown in Fig,. 2. The R_{BLR}–Luminosity relation for the SEAMBHs lays below this relation for AGNs which are accreting at sub-Eddington rate ("Normal AGNs"). Beyond accretion rate which is about three times the Eddington rate the deviation of individual objects from the R_{BLR}–Luminosity relation increases with the accretion rate (higher accretion rate objects have higher deviation from the "Normal" relation). This implies that the Hβ time lags (BLR size) become shorter (smaller) with increasing accretion rate relative to normal AGNs with the same luminosity.

4. Slim Accretion Disks Around Supper-Massive BHs

While thin accretion disk has been suggested to power the BH in the center of sub-Eddington accreting BHs, Slim accretion disks (which are thicker than thin disks) were suggested to power supper-Eddington accretion disks (e.g. Sodowsky 2011). However, numerical simulations have contradict this paradigm and suggest that radial advection is not important and much of the accretion energy can be released via fast winds. Thus, The

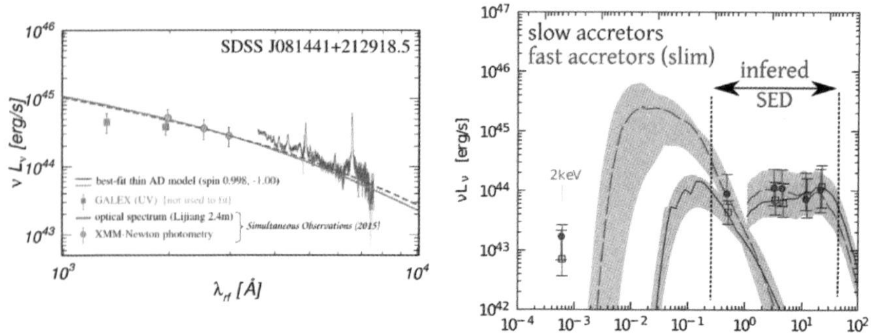

Figure 3. Left: An example for thin accretion disk model fitted to the data of the intrinsic 0.2–1 μm SEAMBH SED. Right: The median SED for sub-Eddington and supper-Eddington sources show no statistical evidence for different optical/UV spectrum or torus emission. See Castello-Mor, Netzer & Kaspi (2016) for more details.

spectral energy distribution of slim discs is rather uncertain and raise several important questions for high accretion rate AGN (e.g. Sadowsky & Narayan 2015; Wang *et al.* 2015).

To test whether thin accretion disk can fit the optical/UV spectral energy distribution (SED) of SEAMBHs we used a RM sample of both SEAMBHs and normal AGNs and fitted them with a model of thin accretion disk. The fitting takes into account the measured BH mass, accretion rate, BH spin, and intrinsic reddening of the sources. This study was published in Castello-Mor, Netzer & Kaspi (2016) and Castello-Mor *et al.* (2017). We found that thin accretion disk models can be used to obtain a satisfactory fits to the intrinsic 0.2–1 μm SEAMBH SED (Fig. 3 left). The median SED for sub-Eddington and supper-Eddington sources show no statistical evidence for different optical/UV spectrum or torus emission.

We found Various indications for transition in X-ray and UV properties at AGNs with very high Eddington ratio. Additional physics, occurring in the nuclear regions of SEAMBHs and responsible for making the slim disk SED consistent with that of thin disk, must be at work. One option can be that powerful winds can carry away a significant amount of the accreted energy from the innermost regions, decreasing its UV radiation and perhaps quenching the disk corona.

References

Bentz, M., *et al.* 2013, *ApJ*, 127, 67
Bentz, M. & Katz, S. 2015, *PASP*, 767, 149
Castell-Mor, N., Netzer, H., & Kaspi, S. 2016, *MNRAS*, 458, 1839
Castell-Mor, N., *et al.* 2017, *MNRAS*, submitted
Kaspi, S., Brandt, W. H., Maoz, D., Netzer, H., Schneider, D.P., & Shemmer O. 2007, *ApJ*, 659, 997
Kaspi, S., Smith, P. S., Netzer, H., Maoz, D., Jannuzi, B., & Giveon, U. 2000, *ApJ*, 533, 631
Du, P., *et al.* 2014, *ApJ*, 782, 45
Du, P., *et al.* 2015, *ApJ*, 806, 22
Hu, C., *et al.* 2015, *ApJ*, 804, 138
Peterson, B. M. 1993, *PASP*, 105, 247
Peterson, B. M., *et al.* 2005, *ApJ*, 632, 799 (erratum 641, 638 [2006])
Sadowski, A. 2011, *Ph.D. thesis*, arXiv:1108.0396
Sadowski, A, Narayan, R. 2015, *MNRAS*, 454, 2372
Vestergaard, M., Fan, X., Tremonti, C. A., Osmer, P. S., & Rechards, G. T. 2008, *ApJL*, 674, L1
Wang, J.-M., *et al.* 2014, *ApJ*, 793, 108

New Frontiers in Black Hole Astrophysics
Proceedings IAU Symposium No. 324, 2016
Andreja Gomboc, ed.

doi:10.1017/S1743921317000722

Searching for a pair of accreting supermassive black holes in J1425+3231

K. É. Gabányi[1,2], S. Frey[1,2], Z. Paragi[3], T. An[4] and S. Komossa[5]

[1]FÖMI, Satellite Geodetic Observatory, P.O. Box 585, 1592 Budapest, Hungary
email: krisztina.g@gmail.com
[2]Konkoly Observatory, MTA CsFK, P.O. Box 67, 1525 Budapest, Hungary
[3]Joint Institute for VLBI ERIC, Postbus 2, 7990 AA Dwingeloo, the Netherlands
[4]Shanghai Astronomical Observatory, CAS, 80 Nandan Road, 200030 Shanghai, P.R. China
[5]QianNan Normal University for Nationalities, Longshan Street, Duyun City of Guizhou Province, P.R. China

Abstract. In hierarchical structure formation scenarios, merging galaxies are expected to be seen in different phases of their coalescence. Simulations suggest that simultaneous activity of the supermassive black holes (SMBHs) in the centres of the merging galaxies may be expected at kpc-scale separations. Currently, there are no direct observational methods which allow the selection of a large number of such dual active galactic nuclei (AGN) candidates. SDSS J142507.32+323137.4 was reported as a promising candidate source based on its optical spectrum. Here we report on our sensitive e-MERLIN observations performed at 1.6 and at 5 GHz, which show that the optical spectrum of the source can be more straightforwardly explained with jet–cloud interactions instead of the dual AGN scenario.

Keywords. galaxies: active, quasars: individual: SDSS J142507.32+323137.4, radio continuum: galaxies, techniques: interferometric

1. Introduction

Hierarchical structure formation models predict that galaxies and their central supermassive black holes (SMBHs) form and grow via merging events. These events may increase activity of the central SMBHs and/or star formation in the host galaxies. Simulations suggest that simultaneous activity of the black holes may be expected at kpc-scale separations (Van Wassenhove *et al.*, 2012). Additionally, according to Steinborn *et al.* (2015) and Volonteri *et al.* (2016), the number of luminous dual AGN increases with decreasing separation of the supermassive black holes. Directly resolving the tightest AGN pairs is only feasible in the most nearby Universe with the current optical and X-ray instrumentation, while cosmological simulations of Volonteri *et al.* (2016) show that less dual AGN are expected at low redshift.

At radio wavelengths, the technique of very long baseline interferometry (VLBI) provides the highest achievable angular resolution, however only $\approx 10\,\%$ of AGN are radio-loud. Since a large fraction of low-luminosity AGN are very compact on milliarcsecond resolution (Panessa & Giroletti, 2013), VLBI is a powerful tool to confirm such AGN candidates. But low-luminosity AGN may have very resolved structures on parsec scales as well (e.g., Nagar *et al.*, 2002), which can only be imaged by arrays with shorter baselines. Also VLBI instruments are not ideal for large surveys due to their limited field of view. Therefore, it is essential to assemble reliable samples of candidate dual AGN. Currently, however, no method is known which can be used to select such sources. Searching for double-peaked narrow optical emission lines emerging from the two separate narrow-line regions (NLR) of dual AGN was proposed as such a method (e.g. Wang *et al.*, 2009).

Table 1. Details of the fitted circular Gaussian components to the e-MERLIN observations of J1425+3231.

ν (GHz)	ID	Position (mas)	Flux density (μJy)	FWHM (mas)
1.6	C	–	712 ± 63	33 ± 2
	S	428	1122 ± 81	104 ± 6
	N	202	1835 ± 116	204 ± 12
5	C	–	319 ± 57	38 ± 6

However, the double-peaked spectral lines can also arise due to other effects occurring in a single AGN, e.g. peculiar kinematics or jet–cloud interactions (Heckman *et al.*, 1984), a rotating, disk-like NLR (Xu & Komossa, 2009), and the combination of a blobby NLR and extinction effects (Crenshaw *et al.*, 2010). Therefore it is important to follow-up those radio-emitting sources claimed to be promising dual AGN candidates using VLBI technique. The SMBHs in dual AGN, once they finally merge and coalesce, are the loudest sources of gravitational wave emission in the universe. Observing active SMBHs in all stages of merging from wide to small separation is therefore of great interest.

SDSS J142507.32+323137.4 (hereafter J1425+3231, $z = 0.478$) was reported as a promising dual AGN candidate by Peng *et al.* (2011) based upon its double-peaked narrow optical emission lines. Frey *et al.* (2012) found two compact radio emitting features at a projected separation of ~ 2.6 kpc at 1.7 GHz in a European VLBI Network (EVN) observation. At 5 GHz, only the brighter feature was detected with EVN. Below we report on our e-Multiple Element Remotely Linked Interferometer Network (e-MERLIN) dual-frequency observations of J1425+3231 (project: CY2205), which were aimed at finding out whether the two detected radio features are the cores of two radio-emitting AGN, or they belong to one AGN, and represent core and extended features.

2. Observations and Data Reduction

The 1.6 GHz e-MERLIN observations took place between 2015 October 2 and 4, the interferometric array consisted of the following antennas: Jodrell Bank Mk2, Defford, Pickmere, Darnhall, and Cambridge. The on-source integration time was 27 h. The 5 GHz e-MERLIN observations took place on 2014 November 1 and 11, the array consisted of: Jodrell Bank Mk2, Cambridge, Defford, Pickmere, and Knockin. The on-source integration time was 19 h. In both cases the phase reference source was J1422+3223; this source was used as phase reference source also in the EVN observations.

Data reduction was done using the National Radio Astronomy Observatory (NRAO) Astronomical Image Processing System (AIPS, Greisen, 2003), following the e-MERLIN cookbook†, and with help from the e-MERLIN science team (R. Beswick).

We used the modelfit task of the Difmap package (Shepherd *et al.*, 1994) to model the brightness distribution with circular Gaussian components. The full-width at half maximum (FWHM) sizes, flux densities, and projected separations from component C of the fitted Gaussian components are given in Table 1.

3. Results

At 1.6 GHz, J1425+3231 shows a rich structure in the e-MERLIN observation (see left panel of Fig. 1). We detected component C and S also seen in the 1.7 GHz EVN image,

† `http://www.e-merlin.ac.uk/data_red/tools/e-merlin-cookbook_V3.0_Feb2015.pdf`

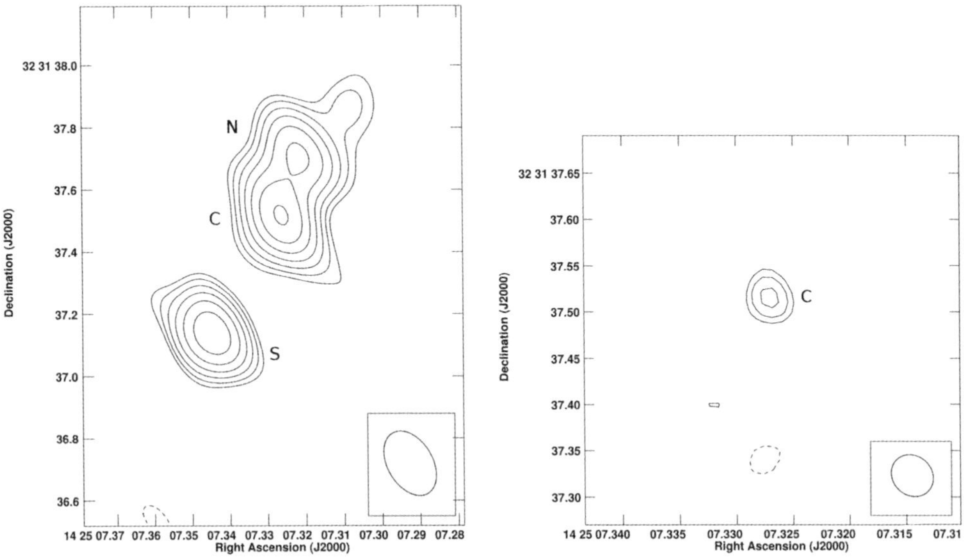

Figure 1. *Left panel:* 1.6-GHz e-MERLIN image of J1425+3231. The peak brightness is 837μJy beam^{-1}. The lowest positive contour level is at 100μJy beam^{-1} (4σ image noise level), further contour levels increase by a factor of $\sqrt{2}$. The beamsize is 226×149 mas at a position angle of 33° and shown in the lower right corner of the image. *Right panel:* 5-GHz e-MERLIN image of J1425+3231. The peak brightness is 154μJy beam^{-1}. The lowest positive contour level is at 69μJy beam^{-1} (4σ image noise level), further contour levels increase by a factor of $\sqrt{2}$. The beamsize is 49×43 mas at a position angle of 52° and shown in the lower right corner of the image.

and additionally a bright northern feature, component N. At 5 GHz, similar to the EVN results, we detected only component C (see right panel of Fig. 1). There is no additional radio emission in the 5-GHz image down to $\sim 100\mu$Jy/beam (6σ noise level).

The sum of the flux densities of the components detected at 1.6 GHz (3.67 ± 0.16 mJy) is in good agreement with the flux density measured for the source at the close frequency of 1.4 GHz in the lower resolution observation conducted with the Very Large Array (3.28 ± 0.14 mJy) within the framework of the Faint Images of the Radio Sky at Twenty-centimeters (FIRST, Becker *et al.*, 1995) survey. The e-MERLIN observations thus recovered all arcsec-scale radio emission in the source measured in the FIRST survey. The slightly larger e-MERLIN flux density can be caused by source variability.

Compared with the 1.7-GHz EVN data, both detected components (C and S) show larger flux densities in the e-MERLIN observation. This can indicate that at this frequency, e-MERLIN picked up the large-scale structure resolved out in the EVN observation and/or it may be caused by flux density variability. However, the large FWHM size of component S (104 ± 6 mas) indicates that the resolution effect is more likely in that case. At 5 GHz, the flux densities of the only detected component (component C) agree within the uncertainties in the e-MERLIN and EVN observations.

Components N and S have significantly larger FWHM sizes (104 ± 6 mas, and 204 ± 12 mas, respectively) in the e-MERLIN data than component C (33 ± 2 mas). This suggests that component C is a compact core, while components N and S are more extended features probably associated with the jet or lobe of the AGN. On the other hand, the detection of emission at the position of component S with the EVN shows that it contains a compact radio-emitting feature.

There are two scenarios that can explain the radio structure of J1425+3231. (i) There is a single low-power radio-emitting nucleus in the source at the position of component C, the synchrotron-emitting jet base. A jet-like feature (component N) to the north, and a lobe (component S) to the south could be imaged with e-MERLIN at 1.6 GHz. The latter contains a compact hotspot, which was detected with the EVN. However, component N was resolved out with EVN completely. In this scenario, the double-peaked optical emission lines are related to the jet-driven outflows and not to dual AGN. A similar observational character has been found in other radio AGN with double-peaked narrow optical lines (e.g. 3C 316 An *et al.*, 2013). (ii) There are two low-power radio emitting nuclei in the source at the position of components C and S. Their compact cores were detected in the EVN observation at 1.7 GHz. One of them (C) has a jet-like extension to the north (N). The large-scale structure of the secondary nucleus (S) was detected with e-MERLIN. The second option is less likely since S was not detected at any of the higher-frequency radio observations, indicating a very steep radio spectrum ($\alpha < -2$, the spectral index is defined as $S \sim \nu^{\alpha}$), which is at odds with its identification as an AGN core.

4. Summary

J1425+3231 was proposed to be a dual AGN candidate because of its double-peaked optical spectrum (Peng *et al.*, 2011). High-resolution radio observations with the EVN showed that it has two compact radio emitting feature. Our follow-up e-MERLIN observations revealed a rich morphology in the source at 1.6 GHz, but detected only one radio-emitting feature at 5 GHz. While the scenario involving dual radio-emitting AGN cannot be completely ruled out, our results are more indicative of one low-luminosity AGN with two-sided jet and/or lobe-like features in J1425+3231.

The e-MERLIN is a National Facility operated by the University of Manchester at Jodrell Bank Observatory on behalf of the UK Science and Technology Facilities Council (STFC). This research was supported by the Hungarian National Research, Development and Innovation Office (OTKA NN110333) and by the China–Hungary Collaboration and Exchange Programme by the International Cooperation Bureau of the Chinese Academy of Sciences.

References

An, T., Paragi, Z., Frey, S., *et al.* 2013 *MNRAS*, 433, 1161
Becker, R. H., White, R. L., & Helfand, D. J. 1995, *ApJ*, 450, 559
Crenshaw, D. M., Schmitt, H. R., Kraemer, S. B., *et al.* 2010, *ApJ*, 708, 419
Frey, S., Paragi, Z., An, T., & Gabányi, K. É. 2012, *MNRAS*, 425, 1185
Greisen, E. W. 2003, *Information Handling in Astronomy - Historical Vistas*, 285, 109
Heckman, T. M., Miley, G. K., & Green R. F. 1984, *ApJ*, 281, 525
Helfand, D. J., White, R. L., & Becker, R. H. 2015, *ApJ*, 801, 26
Nagar, N. M., Falcke, H., Wilson, A. S., *et al.* 2002, *A&A*, 392, 53
Panessa, F. & Giroletti, M. 2013, *MNRAS*, 432, 1138
Peng, Z.-X., Chen, Y., Gu, Q., *et al.* 2011, *Research in Astronomy and Astrophysics*, 11, 411
Shepherd, M. C., Pearson, T. J., & Taylor, G. B. 1994, *BAAS*, 26, 987
Steinborn, L. K., Dolag, K., Hirschmann, M., *et al.* 2015, *MNRAS*, 458, 1013
Van Wassenhove, S., Volonteri, M., Mayer, L., *et al.* 2012, *ApJ* (Letters) 748, L7
Volonteri, M., Dubois, Y., Pichon, C., & Devriendt, J. 2016, *MNRAS*, 460, 2979
Wang, J.-M., Chen, Y.-M., Hu, C., *et al.* 2009, *ApJ* (Letters) 705, L76
Xu, D. & Komossa, S. 2009, *ApJ* (Letters) 705, L20

New Frontiers in Black Hole Astrophysics
Proceedings IAU Symposium No. 324, 2016
Andreja Gomboc, ed.

doi:10.1017/S1743921317000734

A Statistical Method for Detecting Gravitational Recoils of Supermassive Black Holes in Active Galactic Nuclei

P. Raffai[1,2]†, B. Bécsy[1,2], Z. Haiman[3] and Z. Frei[1,2]

[1]Institute of Physics, Eötvös Loránd University, 1117 Budapest, Hungary
[2]MTA-ELTE EIRSA "Lendület" Astrophysics Research Group, 1117 Budapest, Hungary
[3]Department of Astronomy, Columbia University, New York, NY 10027, USA

Abstract. We propose an observational test for gravitationally recoiling supermassive black holes in active galactic nuclei, based on a positive correlation between the velocities of black holes relative to their host galaxies, $|\Delta v|$, and their obscuring dust column densities, $\Sigma_{\rm dust}$, both measured along the line of sight. Our findings using a set of toy models implemented to a Monte Carlo simulation imply that models of the galactic centre and of recoil dynamics can be tested by future observations of the potential $\Sigma_{\rm dust}$–$|\Delta v|$ correlation. We have also found that the fraction of obscured quasars decreases with $|\Delta v|$, for which the predicted trend can be compared to the observed fraction of type II quasars, and can further test combinations of models we may implement.

Keywords. black hole physics, galaxies: active, galaxies: nuclei, methods: statistical

1. Introduction

Numerical simulations of black hole binary mergers (see, e.g. Healy *et al.* 2014 and references therein) suggest that merger remnant supermassive black holes (SMBHs) in galactic centres receive a recoil velocity of typically several hundred of km/s due to highly anisotropic gravitational-wave emission in the final merger phase. Taking into account the spatial distribution of mass in the galactic centre region, and corresponding dynamical friction, this means that the SMBH can engage in a damped oscillating motion (Madau & Quataert 2004, Tanaka & Haiman 2009) with an amplitude comparable to, or exceeding the $\mathcal{O}$(10-100 pc) size of the optically thick dusty molecular torus ("dust torus") believed to be surrounding galactic centres (for an overview, see Hönig 2008). As accreting material can stay gravitationally bound to the moving SMBH, and thus the SMBH can remain active for $10^7 - 10^8$ yrs after the merger event (e.g. Loeb 2007), kinematic and spatial signatures of the recoil could be found in the spectra of quasars (QSOs; see, e.g. Bonning *et al.* 2007, Komossa 2012, Blecha *et al.* 2016). These signatures could confirm the gravitational recoil of merger remnant SMBHs, and could probe models of the process and of the galactic centre region.

We propose that in the presence of dust tori obscuring galactic nuclei, the gravitational recoil of active merger remnant SMBHs should introduce a positive correlation between dust column mass densities along the line-of-sight ($\Sigma_{\rm dust}$) and magnitudes of line-of-sight peculiar velocities of SMBHs relative to their host galaxies ($|\Delta v|$). Using a selected combination of models of gravitational recoil, SMBH trajectories, and obscuring dust tori, in Raffai *et al.* (2016) we have demonstrated the feasibility of detecting this $\Sigma_{\rm dust}$–$|\Delta v|$ correlation with simulating random observations of recoiled QSOs. We have also

† E-mail: praffai@caesar.elte.hu

calculated the fraction of QSOs obscured by their dust tori, $\mathcal{F}_{\rm obs}(|\Delta v|)$, which, compared to the observed fraction of type II (i.e. obscured) QSOs could provide an independent test for a chosen combination of models. In this paper we highlight our findings presented in details in Raffai *et al.* (2016), and propose an observational method to find the $\Sigma_{\rm dust}$–$|\Delta v|$ correlation using the SDSS-DR12 QSO Catalog (Pâris *et al.* 2016).

2. Black Hole Dynamics and Torus Model

We adopted the analytical formulae given in eq. (4) of Baker *et al.* (2008) to construct the distribution of recoil velocity magnitudes, $v_{\rm recoil}$. We calculated $v_{\rm recoil}$ for a given pair of masses ($m_{1,2}$) and dimensionless spin vectors of the merging SMBHs ($\vec{\alpha}_{1,2}$). For simplicity, we first assigned each merging SMBH one of two observed $|\vec{\alpha}_{1,2}|$ populations (see Reynolds 2013) with equal probability, and then drew a random $|\vec{\alpha}_{1,2}|$ value from the corresponding $\vec{\alpha}_{1,2}$ interval. The directions of spin vectors were chosen independently from a uniform distribution covering a whole sphere. We randomised pairs of masses from the SMBH mass function given in Aller & Richstone (2002), downscaled both mass values by a factor of two, and used them as values for m_1 and m_2. We chose the mass of the merger remnant as $M = m_1 + m_2$.

Using the calculated set of $v_{\rm recoil}$'s with uniformly distributed random directions, we simulated the resulting SMBH trajectories based on the model presented in Madau & Quataert (2004). This model assumes that the SMBH is embedded in a spherical stellar bulge, and is decelerated by dynamical friction. We simulated the trajectories with a $\Delta t = 10^3$ yr time resolution up to a randomly picked final time $T \in [\Delta t, 3 \times 10^4 \Delta t]$. The position, velocity, and Δv of the SMBH in its final state were calculated and stored, as if they were results of a QSO observation made at a random time during the QSO activity after the recoil.

We implemented a hydrostatic model of smooth dust tori presented by Schartmann *et al.* (2005). The dust torus is assumed to form from gas released through stellar winds and ejection of planetary nebulae in the nuclear star cluster. Schartmann *et al.* treat the torus as a continuous medium characterised by the density distribution $\rho_{\rm d}$ (see their eq. 8). The effective potential created by the central SMBH and the angular momentum distribution in the stellar core makes $\rho_{\rm d}$ axisymmetric around the galactic centre.

The density distribution of a torus in this model is fully determined by two parameters: the mass of the nuclear star cluster, M_*, and the mass of the central SMBH, M. Consistently with observations presented in Leigh *et al.* (2012), we chose masses of the nuclear star clusters $M_* = 10^{5+\beta} M_\odot$, where β was drawn from a uniform distribution in the range $\beta \in [0, 4]$. For details about how other parameters of the tori were derived from M_* and M, check Raffai *et al.* (2016).

Using the random values of M_* and M, we calculated $\rho_{\rm d}$ for each torus. In our simulation, only the dust column mass density, $\Sigma_{\rm dust}$, was recorded as the final output, which was calculated by numerically integrating $\rho_{\rm d}$ along the line-of-sight from each SMBH position to the observer for a randomly oriented torus.

3. Results

We simulated a total number of 2.5×10^5 random observations of recoiling QSOs. We have calculated Pearson's r between $\Sigma_{\rm dust}$ and $\log_{10}(|\Delta v|)$ for these QSOs, restricted to various ranges of $|\Delta v|_{\rm min} \leqslant |\Delta v| \leqslant |\Delta v|_{\rm max}$. We have found that obscured QSOs with $|\Delta v| < 5$ km/s show no correlation between their $\Sigma_{\rm dust}$ and $\log_{10}(|\Delta v|)$ values ($r_{<5} \simeq 0$, and the corresponding p–value for the hypothesis of no correlation is $p \simeq 0.95$), while

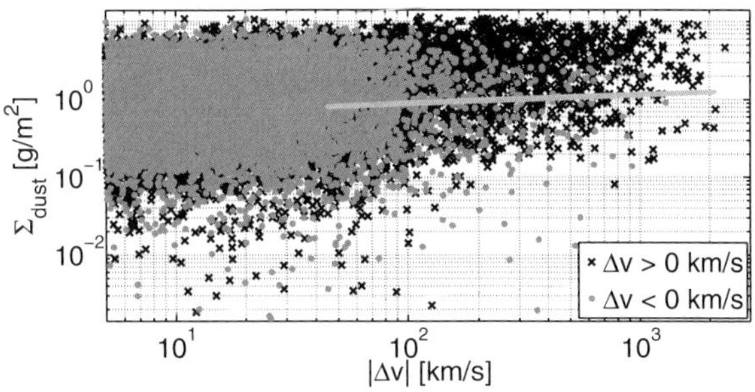

Figure 1. $\Sigma_{\rm dust}$ *vs.* $|\Delta v|$ for the 17,157 obscured QSOs with $|\Delta v| \geqslant 5$ km/s. Receding QSOs are shown by blue crosses, while approaching QSOs are shown by red solid circles. We also show the $\Sigma_{\rm dust}(\log_{10}(|\Delta v|)) = (1.7 \pm 0.1)$ g/m^2 $\times$ $(\log_{10}(|\Delta v|/[1 \text{ km/s}])) - (1.3 \pm 0.2)$ g/m^2 curve (green solid line) we obtained by fitting $\Sigma_{\rm dust}$ versus $\log_{10}(|\Delta v|)$ values for obscured QSOs with $|\Delta v| \geqslant 45$ km/s.

QSOs with $|\Delta v| \geqslant 5$ km/s show a significant correlation, with $r_5 = 0.22 \pm 0.01$ and $p \simeq 0$. r increases with $|\Delta v|_{\rm min}$ until it reaches its maximum at $|\Delta v|_{\rm min} = 45$ km/s with $r_{45} = 0.28 \pm 0.02$. Allowing $|\Delta v|_{\rm min}$ and $|\Delta v|_{\rm max}$ to vary simultaneously does not change the result that r is the highest for the 3,824 obscured QSOs with $|\Delta v| \geqslant 45$ km/s.

We show $\Sigma_{\rm dust}$ *vs.* $|\Delta v|$ in the subsample of 17,157 obscured QSOs with $|\Delta v| \geqslant 5$ km/s in Figure 1, and indicate whether they are moving away from ($\Delta v > 0$, shown as blue crosses) or towards ($\Delta v < 0$, shown as red full circles) the observer. Receding QSOs are slightly more common than the approaching cases, and $\Sigma_{\rm dust}$ values of the receding QSOs also tend to be higher, especially when the velocity offset is large ($|\Delta v| \geqslant 100$ km/s), This is because receding QSOs are more likely to be found behind, rather than in front of the tori.

Allowing uncorrelated uncertainties on the observational measurements of $\Sigma_{\rm dust}$ and $|\Delta v|$ reduces the statistical significance of a correlation between them, and thus increase the number of QSOs required for a detection. We repeated our analysis after adding a Δv component drawn from a Gaussian distribution with a standard deviation of $\sigma_{\Delta v}$ to the velocity offset of each simulated QSO. We measured the expectation value of Pearson's r ($\langle r \rangle$) between $\Sigma_{\rm dust} > 0$ and $|\Delta v|$ in 1,000 randomly drawn sets of N QSOs from the $|\Delta v| \geqslant 45$ km/s subset, and determined the minimum number $N_{\rm min}$ of QSOs required to infer $r > 0$ with 3σ significance 95% of the time. As expected, the number of QSOs required to detect a correlation increases monotonically with $\sigma_{\Delta v}$, rising to $N_{\rm min} = 3,560$ for the $\Sigma_{\rm dust}$–$|\Delta v|$ correlation and for $\sigma_{\Delta v} = 100$ km/s; this is more than an order of magnitude increase over the $N_{\rm min} = 260$ in the idealised case for the $\Sigma_{\rm dust}$–$\log_{10}(|\Delta v|)$ correlation without any velocity-offset error. $\langle r \rangle$ decreases monotonically with $\sigma_{\Delta v}$, reaching $\langle r \rangle = 0.091 \pm 0.016$ for the $\Sigma_{\rm dust}$–$|\Delta v|$ correlation and for $\sigma_{\Delta v} = 100$ km/s. This is a decrease from the $\langle r \rangle = 0.28 \pm 0.02$ in the idealised case for the $\Sigma_{\rm dust}$–$\log_{10}(|\Delta v|)$ correlation by a factor of three. We also evaluated the impact of random Gaussian errors in $\Sigma_{\rm dust} > 0$, and we have found that realistic values of such errors have a very small effect on the $\Sigma_{\rm dust}$–$\log_{10}(|\Delta v|)$ correlation and thus can be neglected.

Finally, we have studied the fraction of obscured QSOs within the whole sample of QSOs, $\mathcal{F}_{\rm obs}(|\Delta v|)$. We have found that this fraction decreases monotonically with $|\Delta v|$ from $\mathcal{F}_{\rm obs}(< 10 \text{ km/s}) \gtrsim 0.8$ to $\mathcal{F}_{\rm obs}(> 10^3 \text{ km/s}) \lesssim 0.4$ (for details, see Figure 11. in Raffai *et al.* 2016). This result can be compared in the future to the observed fraction of obscured (e.g. type II) QSOs, which can potentially provide an independent test for a chosen combination of recoil, trajectory, and dust tori models.

4. Future Work

We propose to use the SDSS-DR12 QSO (DR12Q) catalog (Pâris *et al.* 2016) to perform a search for the correlation between $\Sigma_{\rm dust}$ and $|\Delta v|$. The DR12Q catalog contains data on nearly 300,000 QSOs, for which the BOSS optical spectra are also publicly available. We propose to use the $(g-i)$ color excess of QSOs ($\Delta(g-i)$) as a quantity that is believed to strongly correlate with $\Sigma_{\rm dust}$ (see e.g. Ledoux *et al.* 2015). $\Delta(g-i)$ is defined as the difference in the Galactic extinction corrected $(g-i)$ color for the QSO and that of the mean of the QSOs at that redshift (see Pâris *et al.* 2016). $\Delta(g-i)$ values for different QSOs are given in the DR12Q catalog, and thus can be used in the analysis as a proxy for $\Sigma_{\rm dust}$.

Broad lines in QSO spectra are thought to arise from high-velocity gas remaining bound to the recoiled SMBH, while narrow lines from the gaseous region left unaffected by the recoil in the galactic centre. The line-of-sight velocity offset of a QSO relative to its host can therefore be estimated as $\Delta v = c(z_{\rm B} - z_{\rm N})(1+z_{\rm N})^{-1}$ using the redshifts corresponding to centroids of broad spectral lines ($z_{\rm B}$), relative to those of narrow lines ($z_{\rm N}$) (see e.g. Bonning *et al.* 2007). We propose to use the BOSS spectra of DR12 QSOs and the method described in Greig *et al.* (2016) to fit both narrow and broad line profiles to the following spectral lines: CIV (λ1549Å), Hβ (λ4863Å), and OIII (λ5007Å). We will then use the most optimal combination of these line profiles to obtain $z_{\rm B}$ and $z_{\rm N}$ values and estimate Δv for the highest possible number of DR12 QSOs.

We will report on the findings of our observational search in a follow-up paper to be published in the near future.

Acknowledgments

Péter Raffai is grateful for the support of the Hungarian Academy of Sciences through the "Bolyai János" Research Scholarship programme. Bence Bécsy was supported by the Hungarian Templeton Program that was made possible through the support of a grant from Templeton World Charity Foundation, Inc. We also acknowledge financial support from the NASA Astrophysics Theory Program under grant #NNX11AE05G (to ZH) and from OTKA under grant no. 101666 (to ZF).

References

Aller, M. C. & Richstone, D. 2002, *AJ*, 124, 3035
Baker, J. G., *et al.* 2008, *ApJ*, 682, L29
Blecha, L., *et al.* 2016, *MNRAS*, 456, 961
Bonning, E. W., Shields, G. A., & Salviander, S. 2007, *ApJ*, 666, L13
Greig, B., *et al.* 2016, eprint arXiv:1605.09388
Healy, J., Lousto, C. O., & Zlochower, Y. 2014, *PRD*, 90, 104004
Hönig, S. F. 2008, Dr. rer. nat. dissertation, Rheinische Friedrich-Wilhelms-Universität, Bonn
Komossa, S. 2012, Adv. in Astr., 2012, 364973
Ledoux, C., *et al.* 2015, *A&A*, 580, A8
Leigh, N., Böker, T., & Knigge, C. 2012, *MNRAS*, 424, 2130
Loeb, A. 2007, *Phys. Rev. Lett.*, 99, id. 041103
Madau, P. & Quataert, E. 2004, *ApJ*, 606, L17
Pâris, *et al.* 2016, eprint arXiv:1608.06483
Raffai, P., Haiman, Z., & Frei, Z. 2016, *MNRAS*, 455, 484
Reynolds, C. S. 2013, *CQG*, 30, 244004
Schartmann, M., *et al.* 2005, *A&A*, 437, 861
Tanaka, T. & Haiman, Z. 2009, *ApJ*, 696, 1798

New Frontiers in Black Hole Astrophysics
Proceedings IAU Symposium No. 324, 2016
Andreja Gomboc, ed.

doi:10.1017/S1743921317001508

Simulation of AGN feedback and its impact on galaxies

Martin A. Bourne

Institute of Astronomy and Kavli Institute for Cosmology, University of Cambridge, Madingley Road, Cambridge, CB3 0HA, UK
email: mabourne@ast.cam.ac.uk

Abstract. Feedback released during the growth of supermassive black holes is expected to play a key role in shaping black hole-host galaxy co-evolution. Powerful, accretion disc driven winds have been invoked to explain both observed scaling relations (e.g., $M - \sigma$) and large-scale outflows with mass outflow rates of $\sim 100 - 1000$ $M_\odot$ yr^{-1} and momentum rates of up to ~ 30 L_{AGN}/c. Critically, how these winds couple to the host galaxy depends on if they are momentum or energy conserving. I outline observational signatures that could distinguish between these regimes and discuss their roles in establishing galaxy properties. Furthermore, I discuss high-resolution simulations exploring feedback in a multi-phase medium, highlighting how structural properties of galaxies can impact feedback efficiency. Finally, feedback, in the form of collimated jets, is expected to regulate cooling in galaxy clusters. I discuss new simulations of jet feedback using the moving-mesh code AREPO and outline the scope of our new study.

Keywords. galaxies: active, galaxies: ISM, (galaxies:) quasars: general, galaxies: clusters: general, (galaxies:) cooling flows, galaxies: evolution, galaxies: jets, methods: numerical

1. Introduction

It is widely accepted that at the centres of most massive galaxies resides a supermassive black hole (SMBH). Active galactic nuclei (AGN) represent phases of accretion onto the SMBH that results in the release of vast amounts of energy or feedback. This feedback is expected to play an important role in shaping the host galaxy and has become a key ingredient in many galaxy formation models and simulations (Di Matteo *et al.*, 2005, Bower *et al.*, 2006, Croton *et al.*, 2006, Sijacki *et al.*, 2015). Scaling relations between the properties of the central SMBH and the host galaxy, such as $M - \sigma$ (Kormendy & Ho, 2013), are usually interpreted as evidence for feedback mediated co-evolution. Comparing the energy released growing a SMBH to it's $M - \sigma$ mass, $E_{\rm BH} \sim \eta M_{\rm BH} c^2$, with the binding energy of the gas within a bulge $E_{\rm b} \sim f_{\rm g} M_{\rm b} \sigma^2 \sim 200 f_{\rm g} M_{\rm BH} \sigma^2$, one finds a ratio of $\sim 10^3 - 10^4$. This suggests firstly that more than enough energy is released to have a serious impact on the gas within a galaxy and secondly that the energy released must couple very weakly (see, King & Pounds, 2015, for a review).

Broadly, AGN feedback can be split into two categories (see e.g., Fabian, 2012); *quasar mode* feedback, which represents phases of rapid SMBH growth, at high Eddington ratios and *radio mode* feedback, which is expected during more sedate episodes of accretion with lower Eddington ratios. Feedback during the quasar mode is expected to take the form of winds and outflows which push gas out of the galaxy, quench star formation and facilitate the production of the $M - \sigma$ relation. On the other hand, the radio mode is expected to proceed through the production of highly collimated jets, which act to regulate heating and cooling within the hot haloes of galaxies, for example in galaxy clusters, preventing gas cooling onto the galaxy and hence inhibiting further star formation.

2. AGN winds and outflows

Large scale *outflows*, likely driven by small scale AGN *winds*, have been observed in a number of objects (e.g.,Rupke & Veilleux, 2011, Sturm *et al.*, 2011, Cicone *et al.*, 2014, Cicone *et al.*, 2015, Feruglio *et al.*, 2015, Tombesi *et al.*, 2015) out to kpc scales and across a range of redshifts ($z \gtrsim 6$). They are seen in ionised, neutral atomic and molecular gas with large mass outflow rates of $\dot{M}_{\rm out} \sim 100 - 1000$ $M_\odot$ yr^{-1}, high velocities $v_{\rm out} \sim 1000$ km s^{-1}, high energy content $\dot{E}_{\rm out} \sim 0.05 L_{\rm AGN}$ and large momentum boosts of up to $\sim 30 \times L_{\rm AGN}/c$ (Faucher-Giguère *et al.*, 2012, King & Pounds, 2015). The kinetic energy content of these winds therefore requires that a few percent of the total AGN luminosity couples to the interstellar medium (ISM) to drive these outflows. Yet the $M - \sigma$ relation, if driven by AGN feedback, suggests much weaker coupling.

Fast accretion disc winds, such as ultra-fast outflows (UFOs) observed in blue shifted X-ray absorption features (Tombesi *et al.*, 2010, Tombesi *et al.*, 2010), have provided a particularly successful feedback channel to explain both the $M-\sigma$ relation and large-scale outflows. The UFOs, which have fairly simple properties - $v_{\rm w} \sim 0.1$ c, $\dot{E}_{\rm w} \sim 0.05 L_{\rm Edd}$ and $\dot{p}_{\rm w} \sim L_{\rm Edd}/c$ (King & Pounds, 2015), collide with and shock against the ISM, forming a reverse *wind shock* and forward *ISM shock*. The wind shock will reach temperatures of $T_{\rm ws} \sim 10^{10}$ K and subsequent evolution of the outflow depends on whether or not the wind shock retains the thermalised kinetic energy. If the wind shock can cool efficiently, then it is in the momentum conserving regime and only the pre-shock ram pressure drives the large-scale outflow. King (2003 , 2005) has shown that a momentum conserving outflow results in a limiting SMBH mass similar to the $M-\sigma$ relation. However, if the wind shock is unable to cool, i.e., it is energy conserving, then the thermal pressure of the shocked wind bubble drives the outflow. This regime could potentially have a much greater impact on the host galaxy ISM and can explain both the energy content and momentum boosts observed in large-scale outflows (Faucher-Giguère & Quataert, 2012, Zubovas & King, 2012).

2.1. *Inverse-Compton cooling of the wind shock*

The dominant cooling mechanism of the wind shock would likely be inverse-Compton (IC) cooling of the wind shock electrons by the AGN radiation field photons (King, 2003). Assuming a single temperature plasma for the wind shock, IC cooling is effective within the central $\sim$ 500 pc of a galaxy (King, 2003) resulting in an outflow that is momentum driven within this radius and energy driven beyond it (King, 2005). In this model, the $M - \sigma$ relation can be established at small radii while at larger radii powerful outflows can be produced. Given that IC cooling only acts on the electrons, there needs to be efficient coupling between the protons, which carry most of the energy, and the electrons in the wind shock. However, Faucher-Giguère & Quataert (2012) have shown that the low-density, high-temperature conditions mean that energy exchange between protons and electrons through Coulomb collisions is inefficient and hence IC cooling is ineffective. We, therefore, proposed an observational test of cooling.

In Bourne & Nayakshin (2013), we calculated the IC signature of the AGN photons up-scattered by the hot wind electrons. Assuming that all of the wind kinetic energy is radiated, the components have luminosities of $L_{\rm IC} = (1/2)\dot{M}_{\rm w} v_{\rm w}^2$ and so higher wind velocities produce higher luminosity components. Further, increased wind velocities result in higher shock temperatures and hence higher energy photons can be produced. This is illustrated in Fig. 4 of Bourne & Nayakshin (2013); as the wind velocity increases the IC component moves from being predominantly in softer X-rays with low luminosity to being in harder X-rays with high luminosity. For velocities of $v_{\rm w} = 0.1$ c and above the component becomes comparable to a typical AGN spectrum and we expect such a component to be readily observable if it exists. However, the only observed component we are aware of is in NGC 4051 (Pounds & Vaughan, 2011), which is at a radius of less than 1 pc. The lack of observed components on larger scales, beyond the central few parsecs, and the theoretical modelling of Faucher-Giguère & Quataert (2012), points towards outflows essentially always being energy driven. Therefore we pose the question, how can SMBHs grow so massive?

2.2. *AGN feedback in a multi-phase medium*

In an attempt to answer this question, we performed hydrodynamical simulations, presented in Bourne *et al.* (2014). We compared the evolution of a hot gas bubble, mimicking the energy conserving hot wind shock, in a homogeneous versus inhomogeneous medium. The homogeneous scenario behaves as expected, with the hot bubble expanding and sweeping up all material in its path. On the other hand, in a turbulent and inhomogeneous medium, the hot bubble escapes via low-density channels, while high-density gas is affected far less. High-density clumps can be compressed (potentially triggering star formation), ablated or even completely destroyed depending on their location and physical properties.

Interestingly, while in the homogeneous case mass and energy flows are dominated by outflowing material, this is not the case for a turbulent and inhomogeneous medium. In the latter, we find that the mass and energy flows are able to decouple, with the energy flow being dominated by outflowing material and the mass flow tending to consist more of in-flowing gas. This provides a situation in which, despite feedback occurring, most of the energy can escape the system relatively freely while gas clumps can continue to flow inwards, potentially providing fuel for SMBH growth. Similar to previous work (e.g., Wagner *et al.*, 2013), we find that it is the ram pressure of the outflow that acts on the high-density clumps and thus it is the clump column density that determines whether or not a clump can be swept outwards. Assuming that the column density of dense clumps is limited by star formation, Nayakshin (2014) has shown that the $M - \sigma$ relation can be recovered.

3. Jet mode feedback

As well as the quasar mode of feedback discussed in the previous section, a second major mode of AGN feedback exists as the jet mode. To investigate this regime we have implemented a jet feedback routine into the moving mesh code AREPO (Springel, 2010), making use of the super-Lagrangian refinement scheme of Curtis & Sijacki (2015). This scheme allows improved resolution close to the SMBH, which in turn gives the ability to better capture gas flows close to the SMBH and in the case of this work, inject a jet on small scales. Similar to other works (e.g., Omma *et al.*, 2004, Cattaneo & Teyssier, 2007), we inject mass, momentum and energy into a cylinder centred on the SMBH in order to produce the jet. We have compared various jet injection techniques including pure momentum driven jets as well as purely kinetic jets and jets with a momentum and thermal component, finding marked differences in their evolution and content of different energy components. Further, we have tested precessing kinetic jets and compared these to those mentioned previously. We have been able to study the evolution of the jets, the inflation of cocoons and their subsequent interaction with the intracluster medium (ICM), including measurements of jet-induced turbulence. A detailed account of this research will be presented in future publications that are currently in preparation.

4. Summary

SMBH scaling relations suggest that coupling of AGN feedback to the ISM is weak and potentially it is only the momentum that is important in establishing scaling relations, although the plethora of physical processes involved means this is still rather uncertain. Yet, the powerful outflows observed on kpc scales require that approximately a few percent of the AGN energy couples to the host ISM. If IC cooling of the wind shock occurs, we predict that it should produce an X-ray feature, which has not been observed at large radii. This lack of an observation could hint at inefficient IC cooling, potentially due to weak electron-proton coupling (Faucher-Giguère & Quataert, 2012). Modelling structure in the ISM in simulations makes it more resilient to AGN feedback, even in the energy driven regime. It is the ram pressure of the outflow that is communicated to the bulk of the gas and this may, therefore, negate the need for IC cooling of the wind shock in order to explain both the $M - \sigma$ relation and powerful large scale outflows.

Finally, we have implemented a jet feedback scheme into AREPO in combination with a super-Lagrangian refinement technique that allows the jet to be injected on small scales. We have compared various jet implementation schemes to investigate their impact on both jet and halo properties. Preliminarily it appears that the jet is unable to drive significant bulk turbulence within the ICM, which is consistent with Hitomi observations of the Perseus Cluster (Hitomi Collaboration *et al.*, 2016).

Acknowledgements

I would like to acknowledge support by the ERC starting grant 638707 "BHs and their host galaxies: co-evolution across cosmic time."

References

Bourne, M. A., Nayakshin, S., & Hobbs, A. 2014, *MNRAS*, 441, 3055
Bourne, M. A. & Nayakshin, S. 2013, *MNRAS*, 436, 2346
Bower, R. G., Benson, A. J., Malbon, R., *et al.* 2006, *MNRAS*, 370, 645
Cattaneo, A. & Teyssier, R. 2007, *MNRAS*, 376, 1547
Cicone, C., Maiolino, R., Sturm, E., *et al.* 2014, *A&A*, 562, A21
Cicone, C., Maiolino, R., Gallerani, S., *et al.* 2015, *A&A*, 574, A14
Croton, D. J., Springel, V., White, S. D. M., *et al.* 2006, *MNRAS*, 365, 11
Curtis, M. & Sijacki, D. 2015, *MNRAS*, 454, 3445
Di Matteo, T., Springel, V., & Hernquist, L. 2005, *Nature*, 433, 604
Fabian, A. C. 2012, *ARAA*, 50, 455
Faucher-Giguère, C.-A., Quataert, E., & Murray, N. 2012, *MNRAS*, 420, 1347
Faucher-Giguère, C.-A. & Quataert, E. 2012, *MNRAS*, 425, 605
Feruglio, C., Fiore, F., Carniani, S., *et al.* 2015, *A&A*, 583, A99
Hitomi Collaboration, Aharonian, F., Akamatsu, H., *et al.* 2016, *Nature*, 535, 117
King, A. 2003, *ApJ*, 596, L27
King, A. 2005, *ApJ*, 635, L121
King, A. & Pounds, K. 2015, *ARAA*, 53, 115
Kormendy, J. & Ho, L. C. 2013, *ARAA*, 51, 511
Nayakshin, S. 2014, *MNRAS*, 437, 2404
Omma, H., Binney, J., Bryan, G., & Slyz, A. 2004, *MNRAS*, 348, 1105
Pounds, K. A. & Vaughan, S. 2011, *MNRAS*, 413, 1251
Rupke, D. S. N. & Veilleux, S. 2011, *ApJ*, 729, L27
Sijacki, D., Vogelsberger, M., Genel, S., *et al.* 2015, *MNRAS*, 452, 575
Springel, V. 2010, *MNRAS*, 401, 791
Sturm, E., González-Alfonso, E., Veilleux, S., *et al.* 2011, *ApJ*, 733, L16
Tombesi, F., Cappi, M., Reeves, J. N., *et al.* 2010, *A&A*, 521, A57
Tombesi, F., Sambruna, R. M., Reeves, J. N., *et al.* 2010, *ApJ*, 719, 700
Tombesi, F., Meléndez, M., Veilleux, S., *et al.* 2015, *Nature*, 519, 436
Wagner, A. Y., Umemura, M., & Bicknell, G. V. 2013, *ApJ*, 763, L18
Zubovas, K. & King, A. 2012, *ApJ*, 745, L34

New Frontiers in Black Hole Astrophysics
Proceedings IAU Symposium No. 324, 2016
Andreja Gomboc, ed.

doi:10.1017/S1743921317002344

MAGIC detection of sub-TEV emission from gravitationally lensed blazar QSO B0218+357

D. Dominis Prester[1], J. Sitarek[2], J. Becerra[3,4], S. Buson[5,4], E. Lindfors[6], M. Manganaro[7], D. Mazin[8], M. Nievas Rosillo[9], K. Nilsson[6], A. Stamerra[10], F. Tavecchio[11], Ie. Vovk[12] for the MAGIC Coll.[13] and the *Fermi*-LAT Coll.[14]

[1]University of Rijeka, Department of Physics, 51000 Rijeka, Croatia, email: `dijana@phy.uniri.hr`; [2]University of Łódź, PL-90236 Lodz, Poland; [3]NASA Goddard Space Flight Center, Greenbelt, MD 20771, USA; [4]Department of Physics and Department of Astronomy, University of Maryland, MD 20742, USA; [5]Università di Padova and INFN, I-35131 Padova, Italy; [6]Tuorla Observatory, Department of Physics and Astronomy, University of Turku, Finland; [7]IAC, E-38200 La Laguna; Universidad de La Laguna, Dpto. Astrofísica, E-38206 La Laguna, Tenerife, Spain; [8]ICRR, The University of Tokyo, 277-8582 Chiba, Japan; [9]Universidad Complutense, E-28040 Madrid, Spain; [10]INAF Osservatorio Astrofisico di Torino, I-10025 Pino Torinese, Italy; [11]INAF National Institute for Astrophysics, I-00136 Rome, Italy; [12]Max-Planck-Institut für Physik, D-80805 München, Germany; [13]http://wwwmagic.mppmu.mpg.de; [14]http://fermi.gsfc.nasa.gov/

Keywords. Gamma rays: galaxies – Gravitational lensing – Galaxies: jets – Quasars: individual: QSO B0218+357

The blazar QSO B0218+357 is the first gravitationally lensed blazar detected in the very high energy (VHE, $E > 100\,\mathrm{GeV}$) gamma-ray spectral range (Ahnen *et al.* 2016). It is gravitationally lensed by the intervening galaxy B0218+357G ($z_l = 0.68466 \pm 0.00004$, Carilli *et al.*1993), which splits the blazar emission into two components, spatially indistinguishable by gamma-ray instruments, but separated by a 10-12 days delay. In July 2014 a flare from QSO B0218+357 was observed by the *Fermi*-LAT (Large Area Telescope, Atwood *et al.* 2009, Ackermann *et al.* 2012), and followed-up by the MAGIC (Major Atmospheric Gamma Imaging Cherenkov) telescopes, a stereoscopic system of two 17m Imaging Atmospheric Cherenkov Telescopes located on La Palma, Canary Islands (Aleksić *et al.* 2016a, 2016b), during the expected time of arrival of the delayed component of the emission. MAGIC could not observe the leading image due to the Full Moon. The MAGIC and *Fermi*-LAT observations were accompanied by optical data from KVA telescope at La Palma, and X-ray observations by *Swift*-XRT (Fig. 1 left). Variability in gamma-rays was of the order of one day, while no variability correlated with gamma-rays was observed at lower energies. The flux ratio of the leading to trailing image in HE gamma-rays was larger than in the flare of QSO B0218+357 observed by *Fermi*-LAT in 2012 (Cheung *et al.* 2014). Changes in the observed flux ratio can be caused by gravitational microlensing on individual stars in the host galaxy (Neronov *et al.* 2015), or by other compact objects like for ex. clumps in giant molecular clouds (Sitarek & Bednarek 2016).

We derive gamma-ray spectral energy distribution (Fig. 1, right top) and model it in multi-wavelength context (Fig. 1, right bottom). Large separation of the two peaks and strong Compton dominance, together with the MWL light curve features described above, lead us to the conclusion that a two zone external Compton model can be suitable for describing this source. According to this model, gamma-rays (GeV and sub-TeV) are produced beyond the Broad Line Region (BLR) as a sum od the SSC and EC components on the radiation field from BLR and dusty torus (Jet out), while optical emission and X-rays are produced within the BLR (Jet in).

QSO B0218+357 is one of the two most distant AGN (z=0.944) detected so far in the VHE gamma-ray range, together with PKS 1441+25 (z=0.940) which was also detected by

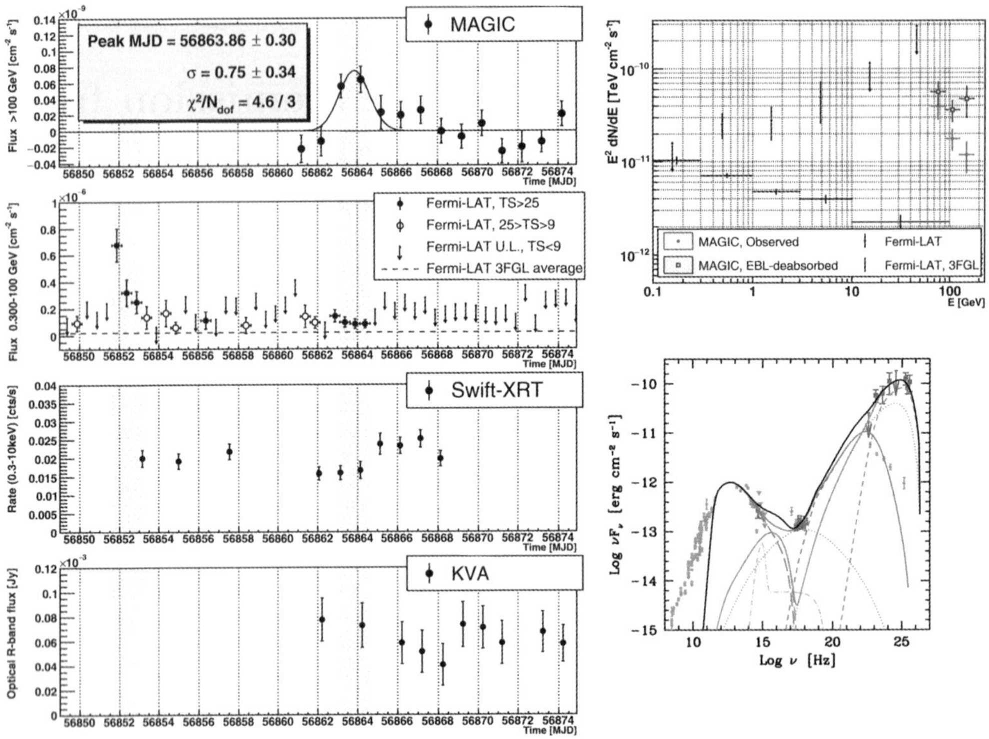

Figure 1. July 2014 QSO B0218+357 flare (Ahnen *et al.* 2016). **Left**: MWL light curve. The two shaded regions are separated by 11.46 days. **Right top**: Gamma-ray SED. MAGIC observations during MJD = 56863-56864 before and after deabsorption due to EBL compared to the *Fermi*-LAT spectrum from the same time and the average emission of QSO B0218+357 in the 3FGL catalog. **Right bottom**: Broadband SED. The reconstructed fluxes (squares) are corrected for different magnifications in different energy ranges. Filled circles: historical measurements (flux upper limit). Emission from the region located within (left thin solid line) and beyond (right thin solid line) the BLR. Long dashed curves: synchrotron radiation; Dotted: SSC emission; Short dashed: the external Compton emission. Dash-dot line: accretion disk emission and its X-ray corona. Thick black solid line: the sum of the non-thermal emission from both regions.

the MAGIC telescopes (Ahnen *et al.* 2015). The GeV and sub-TeV data, obtained by Fermi-LAT and MAGIC, are used to set some constraints on the extragalactic background light (EBL) that are compatible with current models.

References

Ackermann, M., *et al.* 2012, *ApJS*, 203, 4
Ahnen, M. L., *et al.* 2015, *ApJL*, 815, 2, L23
Ahnen, M. L., *et al.* 2016, http://adsabs.harvard.edu/abs/2016arXiv160901095M
Aleksić, J., *et al.* 2016a, *Astroparticle Physics*, 72, 61
Aleksić, J., *et al.* 2016b, *Astroparticle Physics*, 72, 76
Atwood W. B., *et al.* 2009, *ApJ*, 697, 1071
Cheung, C. C., Larsson, S., Scargle, J. D., *et al.* 2014, *ApJL*, 782, L14
Carilli, C. L., Rupen, M. P., & Yanny, B. 1993, *ApJL*, 412, L59
Domínguez, A., Primack, J. R., Rosario, D. J., *et al.* 2011, *MNRAS*, 410, 2556
Neronov, A., Vovk, I., & Malyshev, D. 2015, *Nature Physics*, 11, 664
Sitarek, J. & Bednarek, W. 2016, *MNRAS*, 459, 1959

New Frontiers in Black Hole Astrophysics
Proceedings IAU Symposium No. 324, 2016
Andreja Gomboc, ed.

doi:10.1017/S1743921317000746

The effect of AGN feedback on Sunyaev-Zeldovich properties of simulated galaxy clusters

Dunja Fabjan[1,2]†, S. Planelles[3,4]‡, S. Borgani[2,4,5], G. Murante[2], E. Rasia[2,12], V. Biffi[2,4], N. Truong[6], C. Ragone-Figueroa[9,2], G. L. Granato[2], K. Dolag[7,8], E. Pierpaoli[10], A. M. Beck[7], Lisa K. Steinborn[7] and M. Gaspari[11]

[1]Faculty of Mathematics and Physics, Univ. of Ljubljana, Jadranska 19, 1000 Ljubljana, Slovenia
[2]INAF, Osservatorio Astronomico di Trieste, via Tiepolo 11, I-34131 Trieste, Italy
[3]Departamento de Astronomía y Astrofísica, Universidad de Valencia, c/ Dr. Moliner, 50, 46100 - Burjassot (Valencia), Spain
[4]Astronomy Unit, Department of Physics, University of Trieste, via Tiepolo 11, I-34131 Trieste, Italy
[5]INFN – National Institute for Nuclear Physics, Via Valerio 2, I-34127 Trieste, Italy
[6]Dipartimento di Fisica, Università di Roma Tor Vergata, via della Ricerca Scientifica, I-00133, Roma, Italy
[7]University Observatory Munich, Scheinerstr. 1, 81679 Munich, Germany
[8]Max-Planck-Institut für Astrophysik, Karl-Schwarzschild Strasse 1, 85748 Garching bei München, Germany
[9]Instituto de Astronomía Teórica y Experimental (IATE), Consejo Nacional de Investigaciones Científicas y Técnicas de la República Argentina (CONICET), Observatorio Astronómico, Universidad Nacional de Córdoba, Laprida 854, X5000BGR, Córdoba, Argentina
[10]University of Southern California, Los Angeles, CA 90089
[11]Department of Astrophysical Sciences, Princeton University, Princeton, NJ 08544, USA; Einstein and Spitzer Fellow
[12]Department of Physics, University of Michigan, 450 Church St., Ann Arbor, MI 48109

Abstract. We studied the imprints that feedback from Active Galactic Nuclei (AGN) leaves on the intracluster plasma during the assembly history of galaxy clusters. To this purpose we used state-of-the-art cosmological hydrodynamical simulations based on an updated version of the Tree-PM SPH GADGET-3 code, comparing three sets of simulations with different prescriptions for the physics of baryons (including AGN and/or stellar feedback). We explore the effect of these different physics, in particular AGN feedback, on IntraCluster medium (ICM) properties observed via Sunyaev-Zel'dovich (SZ) effect using an extended set of galaxy clusters (~ 100 clusters with M_{500} masses above $5 \times 10^{13} M_{\odot}/\mathrm{h}$). Some of the main findings are that the scaling relation between the integrated SZ flux and the galaxy cluster total mass is in good accordance with several observed samples, especially for massive clusters, and does not show any clear redshift evolution, with the slope of the relation close to the theoretical one in the AGN feedback case. As for the scatter of this relation, we obtain a mild dependence on the cluster dynamical state.

Keywords. galaxies: clusters: general, (galaxies:) intergalactic medium, galaxies: active, methods: numerical, cosmology: observations

† e-mail: dunja.fabjan@fmf.uni-lj.si
‡ e-mail: susana.planelles@uv.es

The accurate measure of the masses of galaxy clusters is one of the most important goals in order to use them as tracers of cosmic evolution (Borgani&Kravtsov 2011). To calibrate cluster total mass a number of observable quantities can be used. However, physical processes in galaxy clusters (e.g. galactic winds powered by supernovae and AGN feedback) can affect the observed properties of the IntraCluster Medium (ICM), having a non negligible impact on the determined masses (Planelles *et al.* 2016, Biffi *et al.* 2016).

We used a simulated cluster sample that consists of 29 Lagrangian regions extracted from a parent Dark Matter (DM) only simulation (described in Bonafede *et al.* 2011), resimulated at higher resolution including the baryonic component. The complete sample of $\sim$ 100 clusters with $M_{500} > 3\times 10^{13} M_{\odot}/h$ was resimulated including different prescriptions for the physics of baryons: NR - non-radiative physics; CSF - cooling, star formation, feedback by supernovae, metal enrichment and galactic winds, and AGN - as CSF but adding AGN feedback (see Planelles *et al.* 2016 and Steinborn *et al.* 2015 for details). Clusters in the AGN model were further classified into (i) regular/disturbed, according to their global dynamical state and in (ii) CC(cool-core)/ NCC(non-cool-core)-like halos according to their core thermodynamical properties (see Rasia *et al.* 2015).

We were interested in the impact that different physical processes as AGN feedback have on the observed Sunyaev-Zeldovich (SZ) effect and the related scaling relations (see Giodini *et al.* 2013). The integrated SZ effect $Y_{\rm SZ}$ is proportional to the integrated thermal pressure of the intracluster medium along the line of sight and therefore it is an ideal proxy for the cluster total mass M. We found that the $Y_{\rm SZ} - M$ relation is close to the theoretical self-similar prediction for simulated clusters, with a scatter of $0.06 - 0.07$ that is lower than the one reported in observations. Consistently with previous numerical analysis, results at R_{500} for the complete sample do not show any significant dependence on the included physics. When analyzing the pressure profiles of the ICM and its evolution we found that small changes in this profiles, depending on either the dynamical state or the cool-coreness of the considered systems, have different impact on the corresponding $Y_{SZ} - M$ relation. In particular, cluster dynamical state measured at R_{500} is not strongly correlated with its cool-coreness.

We further explore the evolution with redshift of the $Y_{\rm SZ} - M$ relation. For all the three models we observe that the normalization and slope for this relation do not show any evolution with redshift, while only the scatter increases from 0.06 (at $z = 0$) to 0.10 (at $z = 1$). The scatter is reduced to $0.03 - 0.05$ when we restrict to massive clusters, in agreement with previous numerical studies. We also observe that residual variation of cluster properties in radiative simulations (CSF, AGN) with respect to non-radiative ones (NR) are a consequence of overcooling that removes gas from the hot phase. Including AGN feedback partially prevents cooling and brings the slope of the $Y_{SZ} - M$ relation in closer agreement with the theoretical value.

References

Biffi, V., Borgani, S., Murante, G., *et al.* 2016, *ApJ*, 827, 112
Bonafede, A., Dolag, K., Stasyszyn, F., Murante, G., & Borgani, S. 2011, *MNRAS*, 418, 2234
Borgani, S. & Kravtsov, A. 2011, *Advanced Science Letters*, 4, 204
Giodini, S., Lovisari, L., Pointecouteau, E., *et al.* 2013, *Space Science Reviews*, 177, 247
Planelles, S. & Fabjan, D., 2016, *to appear in MNRAS*
Rasia, E., Borgani, S., Murante, G., *et al.* 2015, *ApJ* (Letters), 813, L17
Steinborn, L. K., Dolag, K., Hirschmann, M., *et al.* 2015, *MNRAS*, 448, 1504

New Frontiers in Black Hole Astrophysics
Proceedings IAU Symposium No. 324, 2016
Andreja Gomboc, ed.

doi:10.1017/S1743921317001284

Empirical multi-wavelength prediction method for Very High Energy Gamma-ray emitting BL Lacs

V. Fallah Ramazani[1], **E. Lindfors**[1] **and K. Nilsson**[2]

[1]Tuorla Observatory, Department of Physics and Astronomy, University of Turku, Finland
[2]FINCA (Finnish Center for Astronomy with ESO), University of Turku, Finland

Abstract. We present the most up-to-date and complete multi-wavelength correlation analysis on luminosity properties of TeV BL Lacs. Correlation function (power law or linear) parameters are calculated based on linear regression method. Using the lower energy luminosities of a sample of 182 non-TeV BL Lacs and the generated functions, minimum level of VHE gamma-ray emission was calculated for each non-TeV BL Lacs. This multi wavelength prediction method gives us a list of best candidates to be observed with current generation of Imaging Air Cherenkov Telescopes.

Keywords. BL Lacs, VHE gamma-ray

1. Introduction

The majority of extragalactic objects detected in Very High Energy (VHE) γ-ray band ($>$ 100 GeV) are from rare type of radio-loud Active Galactic Nuclei (AGN) called BL Lacs. As a subclass of blazars, BL Lacs' jets are nearly aligned to the observer's line of sight. As bright distant objects, VHE γ-ray emitting (TeV) BL Lacs can be used to study the universe between us and BL Lacs. The observed properties of BL Lacs in VHE γ-rays band are measured using targeted observation by Imaging Air Cherenkov Telescopes (IACTs). Only 57 BL Lacs are detected in VHE γ-rays band. Therefore, IACTs are expected to discover many BL Lacs in VHE gamma-rays band. But due to the restrictions on current technology a prediction method based on the lower energy data is demanding. We constructed a prediction method based on the cross bands luminosity of TeV BL Lacs.

2. Samples, Data and methodology

Radio (4.85 GHz), Optical (R-band), X-ray (2-10 keV) and Gamma-ray (1-100 GeV) data are collected from literature and databases for TeV (#47) and non-TeV (#182) BL Lacs. VHE γ-ray flux ($>$ 200GeV) are collected for TeV sample by dividing this sample to multiple (Group A) and single detected (Group B). Five VHE γ-ray data sets are formed based on different states of group A (high state and low state), group B and combination of group B with each state of group A. Figure 1 illustrate the distribution of TeV sample in different Luminosity-Luminosity planes in logarithmic scale. Figure 2 illustrates the approach to check the existence of correlation and calculating the best fitted model parameters. The lower energy band luminosities of non-TeV BL Lacs are used as variable input for the prediction functions. We calculate at least 4 levels of VHE gamma-ray luminosity for each non-TeV BL Lac. The non-TeV BL Lacs with highest predicted VHE gamma-ray flux are selected as the promising TeV candidates.

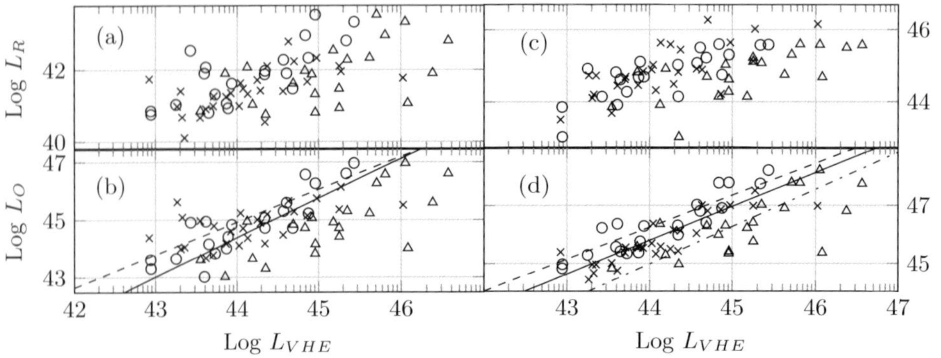

Figure 1. Luminosity in the VHE γ-ray band versus the luminosity in radio (a), optical (b), X-ray (c) and γ-ray (d) bands in logarithmic scale. All the axis are in [erg/s] unit. The different symbols represent the data of different groups (see text): Low state group A (circle), High state group A (triangle) and group B (cross). The correlation functions of the low state of group A is shown with dashed line. The solid/dot-dashed lines represent the correlation functions of the combined data sets of group B and low/high state of group A.

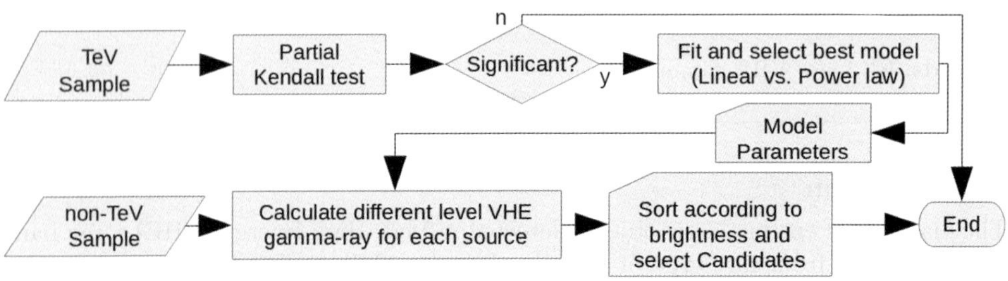

Figure 2. Statistical approach for selecting TeV BL Lacs Candidates.

Table 1. Promising TeV BL Lacs candidates. The fluxes are in (erg/cm^2/s) unit.

Source name	z	$S_{VHE,min}$	$S_{VHE,med}$	$S_{VHE,max}$
PMN J0444-6014	0.097	5.40×10^{-12}	1.71×10^{-11}	2.31×10^{-11}
3C 371	0.046	5.29×10^{-12}	3.37×10^{-11}	1.79×10^{-10}
B2 1811+31	0.117	4.96×10^{-12}	9.41×10^{-12}	2.11×10^{-11}
1RXS J195815.6-301119	0.119	3.33×10^{-12}	1.06×10^{-11}	1.71×10^{-11}
1H 1914-194	0.137	3.04×10^{-12}	9.37×10^{-12}	2.95×10^{-11}
PKS 0829+046	0.174	2.98×10^{-12}	1.60×10^{-11}	3.86×10^{-11}
RBS 0351	0.265	2.53×10^{-12}	1.47×10^{-11}	2.57×10^{-11}
B2 0806+35	0.083	2.49×10^{-12}	5.17×10^{-12}	1.32×10^{-11}
PMN J1936-4719	0.265	2.34×10^{-12}	5.47×10^{-12}	4.96×10^{-11}

3. Results and Conclusions

There are 7 significant correlations between lower energy bands and TeV band. All the significant correlations can be described better with power law model. We build our prediction functions based on these correlations. The lower energy band luminosities of non-TeV BL Lacs are used as variable input for the prediction functions. At least 4 levels of VHE gamma-ray luminosity are calculated for each non-TeV BL Lac. The non-TeV BL Lacs with highest predicted VHE gamma-ray flux are selected as the promising TeV candidates and listed in table 1.

New Frontiers in Black Hole Astrophysics
Proceedings IAU Symposium No. 324, 2016
Andreja Gomboc, ed.

doi:10.1017/S1743921317002241

Observations of possible jet formation in the binary blazar OJ287

Helen Jermak[1,2], Iain A. Steele[1], Gavin P. Lamb[1], Mauri Valtonen[3] and Staszek Zola[4]

[1]Astrophysics Research Institute, Liverpool John Moores University,
146 Brownlow Hill, Liverpool, L3 5RF, UK.
email: i.a.steele@ljmu.ac.uk, g.p.lamb@2010.ljmu.ac.uk

[2]Dept. of Physics, Lancaster University, Bailrigg Campus,
Lancaster, LA1 4YW, UK.
email: h.e.jermak@ljmu.ac.uk

[3] FINCA, University of Turku, Väisäläntie 20, 21500 Piikkiö, Finland.
email:mvaltonen2001@yahoo.com

[4] Jagiellonian University, Gołębia 24, 31-007 Kraków, Poland.
email: szola@oa.uj.edu.pl

Abstract. The blazar, OJ287, underwent a predicted period of flaring in late 2015, triggering multiple observing campaigns to conduct intensive monitoring. We present the polarisation data taken during this event with the Ringo3 polarimeter on the Liverpool Telescope and discuss two possible causes of the flares.

Keywords. polarization, galaxies: active, galaxies: jets, accretion, accretion disks

1. Introduction

In November- December 2015, OJ287, the binary supermassive black holes system, went into predicted outburst; interpreted as the interaction of a supermassive black hole with the accretion disk of the primary supermassive black hole (Valtonen *et al.* 2016). 20 days after the initial flare, a second flare was observed, with a simultaneous high optical degree of polarisation flare.

2. Overview

The first flare at MJD$\sim$57360 (see Valtonen *et al.* (2016) for full light curve) is predominantly thermal and shows low degree of polarisation. Therefore, any of the typical non-thermal emission from the primary's jet is diluted by the increase in thermal emission from the impact of the secondary supermassive black hole (SMBH2) with the primary's accretion disk.

To explain the second, strongly polarised flare (at MJD$\sim$57380, see Figure 1) there must be an ordered magnetic field or magnetic reconnection. While it is possible that the polarisation flare is coincidentally from the primary jet in OJ287, it is not possible that this is caused by the impact due to light crossing times across the accretion disk. It may be coincidental that the primary jet flares shortly after the impact, however, the second strongly polarised peak is also present during the 1984 outburst which would suggest the events are connected (Smith *et al.* 1987). This may suggest the formation of a jet on the SMBH2 or magnetic reconnection occurring as the secondary approaches the primary.

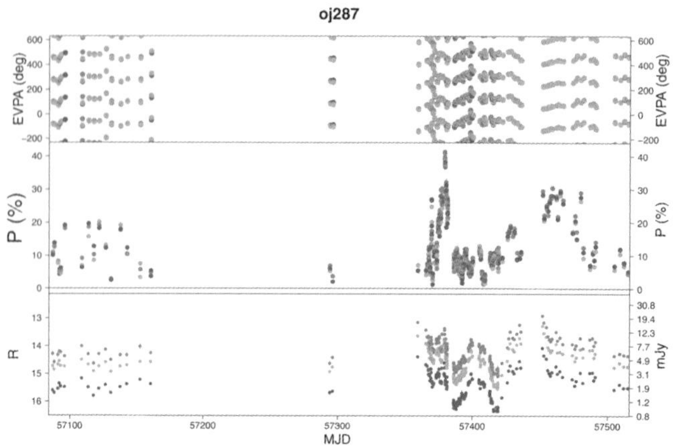

Figure 1. The multi-colour Ringo3 polarisation plots and light curve from MJD= 57100-57500. The blue, green and red points represent data from the respective Ringo3 cameras.

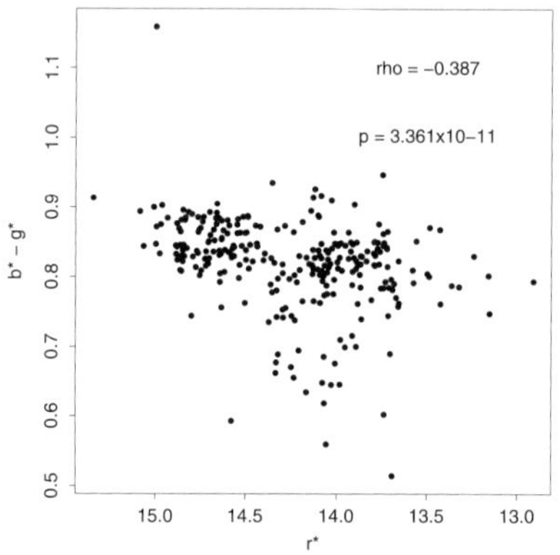

Figure 2. The colour magnitude plot for OJ287 using the Ringo3 bands b*, g* and r*. The results from the Spearman Rank Correlation analysis are shown on the plot.

2.1. *Colour analysis*

The Ringo3 polarimeter measures photometry and polarimetry simultaneously at three wavelengths (Arnold *et al.* 2012). The three Ringo3 wavelength bands are: r* ('red') = 770 - 1000 nm, g* ('green') = 650 - 740 nm and b* ('blue') = 350 - 640 nm.

Figure 2 shows the b* - g* colour plotted against the r* magnitude. The Spearman rank correlation coefficient, rho and the p value are shown on the plot and suggest a significant negative correlation of rho = -0.387 for OJ287 over the Ringo3 observing period. This correlation suggests that the source is 'bluer' when brighter, like many BL Lac-type blazars. The spectral properties of the source during the flaring period show significant negative correlations of a similar value. This suggests that the contribution from red light (i.e. jet emission) and blue light (accretion disk and/or high energy jet emission) does not significantly change during the impact.

References

Arnold, D. M., Steele, I. A., Bates, S. D., Mottram, C. J., & Smith, R. J. 2012, in SPIE Conference Series, Vol. 8446, SPIE Conference Series

Smith, P. S., Balonek, T. J., Elston, R., & Heckert, P. A. 1987, *ApJS*, 64, 459

Valtonen, M. J., Zola, S., Ciprini, S., *et al.* 2016, ApJL, 819, L37

New Frontiers in Black Hole Astrophysics
Proceedings IAU Symposium No. 324, 2016
Andreja Gomboc, ed.

doi:10.1017/S1743921316013065

Optical variability patterns of radio-quiet and radio-loud quasars

P. Marziani[1], E. Bon[2], A. Grieco[3], N. Bon[2], D. Dultzin[4], A. Del Olmo[5] and M. D'Onofrio[6]

[1]INAF, Osservatorio Astronomico di Padova, Italy
email: paola.marziani@oapd.inaf.it

[2]Belgrade Observatory, Serbia

[3]Via Carpane 78, Vigonza, Italy

[4]Instituto de Astronomía, UNAM, Mexico

[5] IAA (CSIC), Granada, Spain

[6] Università di Padova, Italy

Abstract. We analyzed the light curves (LCs) of several radio-quiet and radio-loud quasars belonging to the same parameter space volume in the 4D Eigenvector 1 (4DE1) quasar sequence, using data from the Catalina Real-time Transient Survey (CRTS). We report preliminary results on detected variability pattern, and discuss possible cases of periodic variability.

Keywords. quasars: general, black hole physics, accretion, accretion disks, galaxies: jets

1. Introduction

The origin of the radio-quiet/radio-loud (RQ/RL) dichotomy is still not understood. In the present contribution, the optical variability patterns of RQ and RL low-z ($\lesssim$ 0.6) quasars are compared in the region of RL source maximum occupation along the eigenvector 1 sequence of Sulentic *et al.* (2000) in the plane FWHM of broad Hβ vs. FeII prominence ($R_{\rm FeII} = I(\text{FeII}\lambda4570)/I(\text{H}\beta)$). Periodical variability may ultimately indicate the presence of a sub-parsec supermassive binary black hole (SMBBH). As a preliminary test to this hypothesis, we analyze whether RL and RQ quasars show systematic differences in variability measures, and identify some apparent cases of periodic variability.

2. Quasar V-band variability patterns from the CRTS

RL quasars are not distributed uniformly along the eigenvector 1 sequence, but are predominantly found in the area B1 with $R_{\rm FeII} \lesssim 0.5$ and 4000 km s^{-1} $\lesssim$ FWHM(Hβ) $\lesssim$ 8000 km s^{-1} where also RQ quasars are found with similar accretion parameters ($\log L \sim 45.5$ [erg s^{-1}], $\log M_{\rm BH} \sim 8.5$, $L/L_{\rm Edd} \sim 0.1$). The CRTS (Drake *et al.* 2009) provided V-band magnitudes at 200 – 400 epochs for our sample of 77 B1 quasars (31 RL), over a time span of $\sim$ 3000 d. The data allow for the computation of variability statistics as well as of the auto-correlation (ACF) and structure function (SF) of individual sources.

A two-sample comparison between RL and RQ sources in spectral bin B1 shows that (1) the excess variance distributions are not statistically different, according to Kolmogorov-Smirnov tests; (2) the rest-frame ACF 0 peak half-width (ACF-HW) distributions are marginally different (at slightly less than 2σ confidence level; at 2σ if the ACF-HW is normalized by the black hole mass). Two extreme cases are represented by III Zw 2 that shows a narrow ACF and is associated with no variability, and by PKS 2128-12 that shows

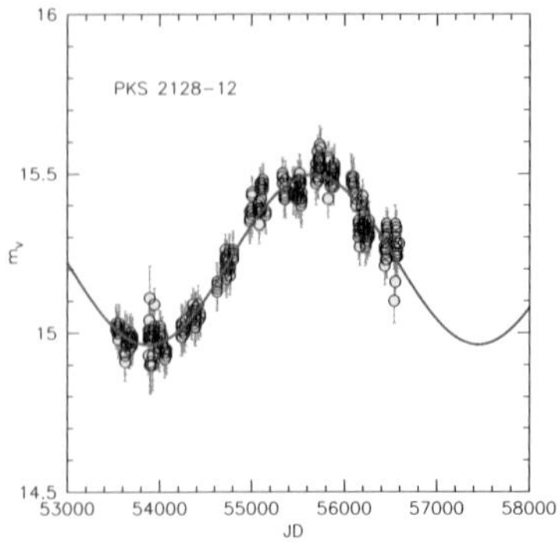

Figure 1. V-band light curve for PKS 2128-12 from CRST data. The blue curve represents a sine fit that corresponds to a period $P \approx 3576$ d. The excellent fit by a periodic function may indicate (but is absolutely not yet a proof of) periodic behavior.

a high degree of correlation between distant observations. Large ACF-HW values ($\gtrsim 500$ d) are anyway possible for RL and RQ alike, which may indicate a stochastic process with a "red" frequency spectrum, or relatively long time-scale deterministic physical processes for both RQ and RL sources. In addition, Kolmogorov-Smirnov tests show that (3) the SF slopes are also not statistically different. The times and amplitudes of the SF maximum are slightly larger for RL than for RQ sources, although the two-sample difference is again not statistically significant. Nonetheless, RQ sources more often show an irregular multi-peaked SF than RL ones. A larger sample of sources in the same domain the eigenvector 1 sequence is needed to confirm the results hinting at systematic RL/RQ differences in optical variability.

The inspection of several LCs suggests periodical behavior. The most impressive case in our sample is PKS 2128-12 (Fig. 1), with a light curve similar to the one of PKS 1302-102 (Graham *et al.* 2015), and of the compact-steep spectrum radio source 3C 57 (Sulentic *et al.* 2015). The Lomb-Scargle periodograms indicate that a large fraction of sources exhibit periodicity to some extent. However, quasars apparently show an optical LC consistent with damped random walks (e.g., MacLeod *et al.* 2012) where the stochastic variability amplitude grows larger at lower frequencies, thus creating the illusion of periodic curves (Vaughan *et al.* 2016). For instance, the peak corresponding to a period $P \approx 3600$ d for PKS 2128-12 is highly significant against conventional Lomb-Scargle significance tests (associated with measurement errors) but not against red noise.

The CRTS is now covering the domain of deterministic processes dynamical timescales that can give rise to a periodic signal in luminous quasars. However, longer time coverages are needed to confirm periodicity in the candidates we identify, as for NGC 4151 and NGC 5548 (Bon *et al.* 2012, 2016). In both cases, periodicity was also detected in the radial velocity curves measured from the Hβ and Hα Balmer lines. Additional spectroscopic monitoring is needed for the CRTS candidates as well to track periodic behavior that may ultimately provide evidence in favor of sub-pc SMBBHs.

References

Bon, E., Jovanović, P., Marziani, P., *et al.* 2012, *ApJ*, 759, 118
Bon, E., Zucker, S., Netzer, H., *et al.* 2016, *ApJS*, 225, 29
Drake, A. J., Djorgovski, S. G., Mahabal, A., *et al.* 2009, *ApJ*, 696, 870
Graham, M. J., Djorgovski, S. G., Stern, D., *et al.* 2015, *Nature*, 518, 74
MacLeod, C. L., Ivezić, Ž., Sesar, B., *et al.* 2012, *ApJ*, 753, 106
Sulentic, J. W., Martínez-Carballo, M. A., Marziani, P., *et al.* 2015,*MNRAS*, 450, 1916
Sulentic, J. W., Zwitter, T., Marziani, P., & Dultzin-Hacyan, D. 2000, *ApJL*, 536, L5
Vaughan, S., Uttley, P., Markowitz, A. G., *et al.* 2016, *MNRAS*, 461, 3145

New Frontiers in Black Hole Astrophysics
Proceedings IAU Symposium No. 324, 2016
Andreja Gomboc, ed.

doi:10.1017/S1743921316012655

Highly accreting quasars: a tool for cosmology?

P. Marziani[1], C. A. Negrete[2], D. Dultzin[2], M. L. Martinez-Aldama[3], A. Del Olmo[3], D. Esparza[2], J. W. Sulentic[3], M. D'Onofrio[4], G. M. Stirpe[5], E. Bon[6] and N. Bon[6]

[1]INAF, Osservatorio Astronomico di Padova, Italy
email: paola.marziani@oapd.inaf.it
[2]Instituto de Astronomía, UNAM, Mexico
[3]IAA (CSIC), Granada, Spain
[4]Università di Padova, Italy
[5]INAF, Osservatorio Astronomico di Bologna, Italy
[6]Belgrade Observatory, Serbia

Abstract. Highly accreting quasars are possible cosmological probes, as their Eddington ratio is expected to saturate toward values of order unity. We present preliminary estimates of redshift-independent source luminosities and the Hubble diagram for quasars in the redshift range $0.1 \lesssim z \lesssim 2.6$.

Keywords. quasars: general, quasars: emission lines, line: profiles, cosmological parameters, distance scale, dark matter

1. Introduction

Several new techniques have been proposed in recent years to use quasars as redshift-independent distance indicators with the goal of deriving Ω_M and Ω_Λ. One of the most promising methods involves highly accreting quasars (xA for extreme accretors or extreme Population A) which may be considered as "Eddington standard candles:" their Eddington ratio is expected to converge toward values of order unity, with low scatter (Marziani & Sulentic 2014). Using optical/UV spectral line widths as virial broadening estimators, we are then able to derive the black hole mass and a luminosity estimate avoiding the use of scaling laws. Highly accreting quasars are recognized from diagnostic ratios in their rest-frame UV and optical spectra. They yield a preliminary Hubble diagram with SDSS and dedicated Gran Telescopio Canarias (GTC) data.

2. Highly accreting quasars: selection and cosmological exploitation

The optical, UV, and X properties of quasars can be organized along a sequence defined by FeII prominence $R_{\rm FeII} = I(\rm FeII\lambda4570)/\mathit{I}(H\beta)$ and FWHM of broad Hβ (the optical plane of the "eigenvector 1"). The optical criterion $R_{\rm FeII} \gtrsim 1$ is found equivalent to a UV-based criterion: (a) I(AlIIIλ1860)/I(SiIII] λ1892)$\gtrsim$0.5, (b) I(SiIII] λ 1892)/I(CIII]λ1909)$\gtrsim$1 which can be applied to high-z quasars (Marziani & Sulentic 2014). We are currently carrying out an analysis of survey data and of data collected from dedicated observations at GTC to obtain high-S/N spectra of xA sources.

It is possible to link accretion luminosity L and line width ($\delta v \equiv$ FWHM of Hβ or of the intermediate ionization lines AlIIIλ1860 and SiIII]λ1892) through multiplicative terms that can be assumed to have a small dispersion around an average value. Their product $\mathcal{L}_0$ can be assumed equal for all xA sources as a first approximation. The equation

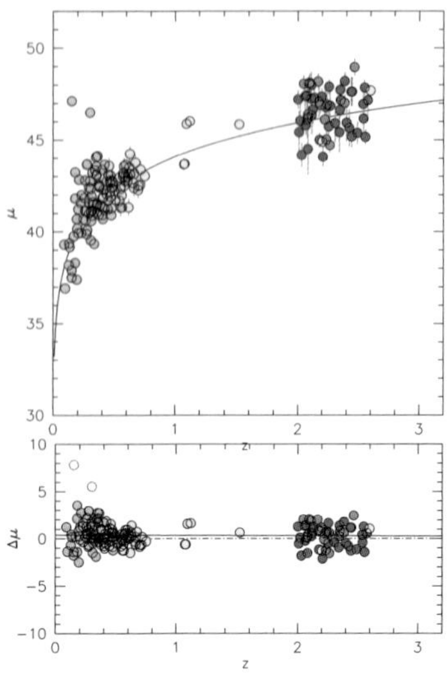

Figure 1. Preliminary Hubble diagram obtained from the analysis of the Marziani & Sulentic (2014) data (yellow: Hβ, navy blue: AlIIIλ1860 and SiIII]λ1892) supplemented by new Hβ measurements from the SDSS (green) and from GTC observations (magenta). Red circles indicate outliers whose profile appears composite at a second inspection, suggesting the presence of a strong broadening component not associated with virial motions. The value of log $\mathcal{L}_0 \approx 45.06$ was used in the computation of the distance modulus μ for this work. The lower panel shows the distance modulus residuals with respect to concordance cosmology. The filled lines represents a lsq fit to the residuals as a function of z. Its slope ($b \approx -0.035 \pm 0.130$) is not significantly different from 0, indicating good statistical agreement between luminosities derived from concordance cosmological parameters and from the virial equation.

$L = \mathcal{L}_0(\delta v)^4$ was derived assuming virial line broadening, broad line region size scaling with luminosity as $\propto L^{\frac{1}{2}}$, as well as a well-defined L/M ratio. If $L/M = const.$, then L can be retrieved once the mass is known. The "virial luminosity" equation is the equivalent for a black hole system of the original Faber & Jackson (1976) relation linking the luminosity and velocity dispersion of galaxies assuming a fixed mass-to-luminosity ratio.

The Hubble diagram of Fig. 1 includes 92 sources from Marziani & Sulentic (2014), and is supplemented by 77 new SDSS low-luminosity xA sources at low z covering the Hβ range, and by 10 new high-z quasar observations with GTC of the fifty sources at redshift $2 \lesssim z <\lesssim 3$ that were observed with the OSIRIS camera at the GTC, covering the λ1900 Å blend. The sample rms is $\sigma_{\Delta\mu} \approx 1.1$ (≈ 0.44 dex in luminosity), not improved with respect to the previous work, due to the slightly higher scatter found among the low-luminosity sources. The reanalysis of some quasars with $\Delta\mu \approx 3-4$ indicates that they are heavily affected by excess emission on the blue side of Hβ profile (represented by an unshifted Lorentz function). Outlying quasars show composite profile whose interpretation is currently not clear, and contamination by their host galaxy spectrum. However, the new data confirm the feasibility of the method, yielding a behavior consistent with concordance cosmology $\Omega_{\rm M} \approx 0.31$ and $\Omega_\Lambda \approx 0.69$ (Ade *et al.* 2016). Significant results require an increase in sample size, reduction of intrinsic scatter and skew as a function of z (believed to be associated with orientation effects, and non-virial motions due to outflows), and precise knowledge of the flux ratio at 5100 and 1800 Å if Hβ and the 1900 Å blend data are used together. Mock samples show that the method can in principle provide significant constraints especially on $\Omega_{\rm M}$, if applied to a quasar sample uniformly distributed over a redshift range $0.2 \lesssim z \lesssim 3$ (Marziani & Sulentic 2014a).

References

Ade, P. A. R., Planck collaboration 2016, *A&Ap*, 594, 13
Faber, S. M. & Jackson, R. E. 1976, *ApJ*, 204, 668
Marziani, P. & Sulentic, J. W. 2014, *MNRAS*, 442, 1211
Marziani, P. & Sulentic, J. W. 2014a, *Adv. Spa. Res.*, 54, 1331
Wang, J.-M., Du, P., Valls-Gabaud, D., Hu, C., & Netzer, H. 2013, *Phys. Rev. Lett.*, 110, 081301

New Frontiers in Black Hole Astrophysics
Proceedings IAU Symposium No. 324, 2016
Andreja Gomboc, ed.

doi:10.1017/S1743921317000680

Open our eyes to wider fields in VLBI surveys

Kristóf Rozgonyi[1] **and Sándor Frey**[2,3]

[1]Department of Physics of Complex Systems, Eötvös Loránd University, Pázmány Péter sétány 1/A, H-1117 Budapest, Hungary
e-mail: rstofi@gmail.com

[2]FÖMI Satellite Geodetic Observatory, PO Box 585, H-1592 Budapest, Hungary

[3]Konkoly Observatory, MTA CSFK, PO Box 67, H-1525 Budapest, Hungary
e-mail: sandor.frey@gmail.com

Abstract. The observation and imaging of hundreds or thousands of radio sources with the technique of very long baseline interferometry (VLBI) is a computationally intensive task. However, these surveys allow us to conduct statistical investigations of large source samples, and also to discover new phenomena or types of objects. The field of view of these high-resolution VLBI imaging observations is typically a few arcseconds at cm wavelengths. For practical reasons, often a much smaller fraction of the field, the central region is imaged only. With an automated process we imaged the ~1.5-arcsec radius fields around more than 1000 radio sources, and found a variety of extended radio structures. Some of them are yet unknown in the literature.

Keywords. surveys, techniques: interferometric, radio continuum: galaxies

1. Treasure-house of data: VIPS

We examined the VLBA Imaging and Polarimetry Survey (VIPS, Helmboldt *et al.* 2007). It contains 1127 AGN observations taken at 5 GHz with the Very Long Baseline Array (VLBA). The calibrated VLBI visibility data sets can be downloaded from the VIPS data collection index page†. The original field of view of the interferometer was ~1.5 arcsec but the imaging was typically preformed only in the central area of 128 mas × 128 mas. Even though large structures were reported in the original publication and subsequent studies (Helmboldt *et al.* 2007, Tremblay *et al.* 2016), we could find further objects in the full-field images with structures extending beyond the central area.

2. Imaging of the extended field

The large sample size and the extended fields motivated us to image the sources and analyse the data in an automated way. The imaging and model fitting were performed through standard procedures using Difmap (Shepherd *et al.* 1994). We modeled the visibility data with a central elliptical Gaussian component. In addition, extended structures were fitted with circular Gaussians. The number of these components was determined by the signal-to-noise ratio (SNR) in the residual map. We applied a 6-σ threshold for fitting a new component, as done in Helmboldt *et al.* (2007). Thereafter, we calculated the separation of the circular components from the central elliptical Gaussian. We marked the sources with at least one fitted component farther away than 50 mas from the centre as candidates for having large-scale structure within the full VLBA field of view.

† http://www.phys.unm.edu/~gbtaylor/VIPS/vipscat/vipsncapindx.shtml

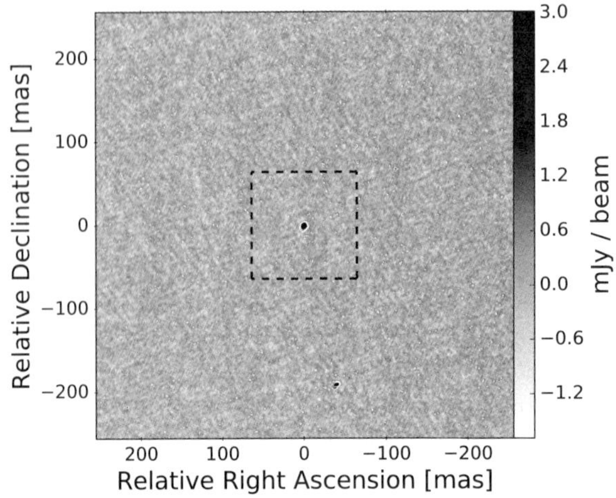

Figure 1. 5-GHz VLBI image of J15048+5438. A compact component is seen ∼200 milli-arcseconds (mas) south of the central component. Follow-up multi-frequency observations could be useful to decide if this is a kpc-separation dual radio AGN, a gravitationally lensed background source with two images, or an AGN core plus a hot spot embedded in an extended lobe that is resolved out with the VLBA. The area originally imaged in VIPS is enclosed by the dashed line.

3. Results and plans for the future

We found 60 sources as candidates with extended structure among the 1127 VIPS sources. To check the quality of the candidate sample, we visually inspected all the wide-field images. We found that in a few cases the distant components are unreliable because of the low flux density of the fitted components. An increased SNR threshold in the automated model-fitting procedure would eliminate these questionable cases. Further quantitative classification of the sample is needed (and planned). Published information from the literature, analysis of archival data, and eventually new targeted observations would be necessary to reveal the true nature of these objects, on a case-by-case basis. These sources show great diversity in morphology and brightness. The large, typically ∼0.1–1 kpc structures could most probably be extended jet structures. However, the possibility of dual radio AGN or gravitationally lensed radio sources should also be investigated. As an illustration, we present an example from the candidate sample (Fig. 1).

We plan to perform the same automated procedure with other publicly available survey data, in a hope to find further objects with VLBI structure extended beyond what is already known.

Acknowledgements

This work was supported by the Hungarian National Research, Development and Innovation Office (OTKA NN110333). KR thanks the New National Excellence Program of the Ministry of Human Capacities for support.

References

Helmboldt, J. F., *et al.* 2007, *ApJ*, 658, 203
Shepherd, M. C., Pearson, T. J., & Taylor, G. B. 1994, *BAAS*, 26, 987
Tremblay, S. E., *et al.* 2016, *MNRAS*, 459, 820

New Frontiers in Black Hole Astrophysics
Proceedings IAU Symposium No. 324, 2016
Andreja Gomboc, ed.

doi:10.1017/S1743921317000783

Ensemble quasar spectral variability from the XMM-Newton Serendipitous Source Catalogue

Roberto Serafinelli[1], **Fausto Vagnetti**[1] **and Riccardo Middei**[2]

[1]Dipartimento di Fisica, Università di Roma "Tor Vergata"
Via della Ricerca Scientifica 1, 00133, Rome, Italy

[2]Dipartimento di Matematica e Fisica, Università Roma Tre
Via della Vasca Navale 84, 00146, Rome, Italy

Abstract. Variations of the X-ray spectral slope have been found in many Active Galactic Nuclei (AGN) at moderate luminosities and redshifts, typically showing a "softer when brighter" behaviour. However, similar studies are not usually performed for high-luminosity AGNs. We present an analysis of the spectral variability based on a large sample of quasars in wide intervals of luminosity and redshift, measured at several different epochs, extracted from the fifth release of the XMM Newton Serendipitous Source Catalogue. Our analysis confirms a "softer when brighter" trend also for our sample, extending to high luminosity and redshift the general behaviour previously found. These results can be understood in light of current spectral models, such as intrinsic variations of the X-ray primary radiation, or superposition with a constant reflection component.

Keywords. Quasars: general, Galaxies: active, X-rays: galaxies, Surveys

1. Introduction

The spectral slope optical/UV variations in AGNs have been quantified by Trevese and Vagnetti (2002), by means of the *spectral variability parameter* $\beta = \Delta\alpha/\Delta \log F$, α being the slope of the spectrum and F its flux in the given band. A positive value was found, which means that the spectrum is harder when the flux is higher.
However, in the X-ray band, the opposite behaviour has been found for individual sources, and very few systematic studies have been performed, e.g. Sobolewsa and Papadakis (2009), who found this trend for a sample of 10 nearby Seyfert galaxies.
In our study we investigated the spectral variability of quasars, using the MEXSAS catalogue (Vagnetti *et al.* 2016), created cross-matching the multi-epoch observations of the XMM-Newton Serendipitous Source Catalogue, Data Release 5 (Rosen *et al.* 2016) with two partially overlapping Sloan Digital Sky Survey catalogues, SDSS-DR7Q (Schneider *et al.* 2010) and SDSS-DR12Q (Pâris *et al.* 2016), obtaining a catalogue of 7,837 X-ray observations of 2,700 quasar sources.

2. Ensemble and single source analyses

In order to study the spectral variability in the X-ray band, we redefine the spectral variability parameter in terms of the photon index Γ, defined after $N(E) \propto E^{-\Gamma}$. Then, the spectral variability parameter becomes $\beta = -\Delta\Gamma/\Delta \log F$, since $\Gamma = 1 - \alpha$. In order to better compare sources we computed the linear fit between the variations of Γ and $\log F_S$ (F_S in band $0.5 - 2$keV) from the source mean values. The computed spectral variability parameter is $\beta = -0.69 \pm 0.03$ (see Fig. 1). A negative β implies that the

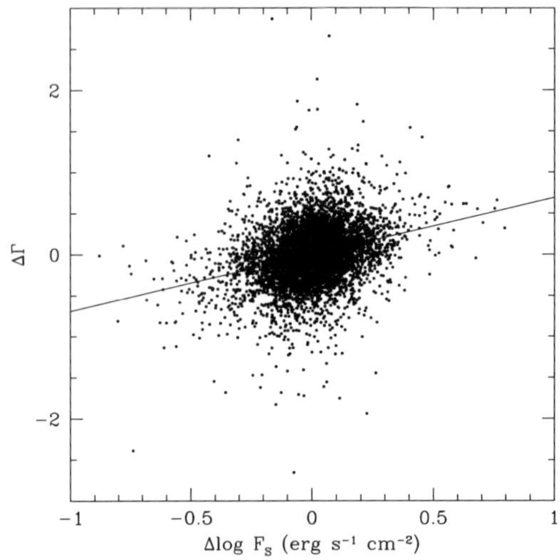

Figure 1. $\Delta\Gamma - \Delta \log F_S$ (band $0.5 - 2$keV) correlation of the whole sample.

spectral slope becomes higher for increasing flux, making the spectrum steeper. This means that the spectrum is softer when brighter.

We tried to investigate the dependence of β from some source parameters, such as black hole mass, Eddington ratio, redshift and X-ray luminosity, finding no evidence of such dependence. The value of β in some bins deviates significantly from the ensemble value, though, suggesting that different sources may have different values of β between each other. Therefore we studied single sources, selecting the ones with low probability of finding a $\Gamma - \log F_S$ correlation by chance ($p \leqslant 10^{-3}$). We find all negative βs in a wide range of values, from $\beta = -3.54 \pm 0.54$ to $\beta = -0.62 \pm 0.08$.

3. Discussion

The softer when brighter trend, found in both the ensemble and single-source analysis, may be caused by the superposition of the primary X-ray emission with an additional reflected component. According to some models (e.g., Shih *et al.* 2002) the primary emission could be variable in flux, but not in spectrum, while the constant reflected component is not variable at all, producing the observed spectra. Other models (e.g., Liang 1979), suggest that the primary component may be variable in spectrum as well. As for the single source analysis, the wide range of β values may be dependent on several unconsidered features of the source, such as black hole spin, angle of view or radio-loudness, or it may be influenced by stochastic processes.

References

Liang, E. P. T., 1979, *ApJ*, **231**, L111.
Pâris, I., Petitjean, P., Ross, N. P., *et al.*, 2016, `arXiv:1608.06483`
Rosen, S. R., Webb, N. A., Watson, M. G., *et al.*, 2016, *A&A*, **590**, A1.
Schneider, D. P., Richards, G. T., & Hall, P. B., 2010, *AJ*, **139**, 2360.
Shih, D. C., Iwasawa, K., & Fabian, A. C., 2002, *MNRAS*, **333**, 687.
Sobolewska, M. A. & Papadakis, I. E., 2009, *MNRAS*, **399**, 1597.
Trevese, D. & Vagnetti, F., 2002, *ApJ*, **564**, 624.
Vagnetti, F., Middei, R., Antonucci, M., *et al.* 2016, *A&A*, **593**, A55.

New Frontiers in Black Hole Astrophysics
Proceedings IAU Symposium No. 324, 2016
Andreja Gomboc, ed.

doi:10.1017/S1743921317001387

Peculiar emission from the new VHE gamma-ray source H1722+119

T. Terzić[1], A. Stamerra[2], the MAGIC Coll.[3], F. D'Ammando[4,5] (for the *Fermi*-LAT Coll.), C. M. Raiteri[6], M. Villata[6], F. Verrecchia[7,8] and O. Kurtanidze[9]

[1]University of Rijeka, 51000 Rijeka, Croatia, email: tterzic@phy.uniri.hr; [2]INAF National Institute for Astrophysics, I-00136 Rome, Italy; [3]The full list of collaborators can be found at: wwwmagic.mppmu.mpg.de; [4]Dip. Di Fisica e Astronomia, Universit degli Studi di Bologna, I-40127 Bologna, Italy; [5]INAF Istituto di Radioastronomia, I-40129 Bologna, Italy; [6]INAF Osservatorio Astrofisico di Torino, I-10025 Pino Torinese (TO), Italy; [7]INAF Osservatorio Astronomico di Roma, I-00040 Monteporzio Catone, Italy; [8]ASI Science Data Center (ASDC), I-00133 Roma, Italy; [9]Abastumani Observatory, Mt. Kanobili, 0301 Abastumani, Georgia

Keywords. BL Lacertae objects: individual (H1722+119), gamma-rays, multiwavelength

The BL Lac object H1722+119 was observed in the very high energy band (VHE, $E >$ 100 GeV) by the MAGIC (Major Atmospheric Gamma-ray Imaging Cherenkov) telescopes (Aleksić *et al.* 2016a, b) between 2013 May 17 and 22, following a state of high activity in the optical band measured by the KVA (Kungliga Vetenskapsakademien†) telescope. Optical high states are often used to trigger MAGIC observations, which result in the VHE γ-ray signal detection (see e.g. Aleksić *et al.* 2015, Ahnen *et al.* 2016 and references therein).

Integrating 12.5 h of observation, the source was detected at VHE with a statistical significance of 5.9 sigma. The flux appears to be constant at the level of $f = (6.3 \pm 1.6) \times 10^{-12}$ ph cm^{-2} s^{-1} above 150 GeV with $\chi^2/N_{\mathrm{d.o.f.}} = 3.5/5$ (fig. 1 right top). This corresponds to (2.0 ± 0.5) % of the Crab Nebula flux in the same energy range. Contemporaneous observations were performed by the *Fermi*-LAT (Large Area Telescope, Atwood *et al.* 2009, Ackermann *et al.* 2012) in the high energy range (HE, 100 MeV$< E <$ 100 GeV); by the *Swift*-XRT: 200 eV$< E <$ 10 keV (Burrows *et al.* 2005) and the *Swift*-UVOT: 170 nm $< \lambda <$ 600 nm (Roming *et al.* 2005); by the KVA and the Abastumani telescope‡ in the R-band, and by the OVRO (Owens Valley Radio Observatory) telescope in the 15 GHz radio band (Richards *et al.* 2011).

The multiwavelength light curve (fig. 1 left) shows no significant temporal variability in the HE band either. The flux was $f = (3.4 \pm 0.6) \times 10^{-8}$ ph cm^{-2} s^{-1}, with $\chi^2/N_{\mathrm{d.o.f.}} = 3.9/5$, while the spectrum appears to be constant as well, with the spectral index of 1.96 ± 0.07, with $\chi^2/N_{\mathrm{d.o.f.}} = 1.7/5$. The optical flux varied with no apparent regularity, and the radio data show a linear trend ($F[\mathrm{Jy}] = p_0 + p_1 \times (t - 56300)[\mathrm{MJD}]$) of increasing flux on a time-scale of one year, with the fit parameters $p_0 = (5.2 \pm 0.2) \times 10^{-2}$ Jy, $p_1 = (1.07 \pm 0.10) \times 10^{-4}$ Jy/day and $\chi^2/N_{\mathrm{d.o.f.}} = 39.68/36$.

Using the HE and VHE gamma-ray observations, and the expected imprint of the extragalactic background light (EBL, Franceschini *et al.* (2008)) absorption, we estimated the redshift of the source to be $z = 0.34 \pm 0.15_{\mathrm{stat}} \pm 0.05_{\mathrm{meth}}$ with upper limit (UL) of $z < 1.06$ (95% C.L.), using the method from Prandini *et al.* (2010). Using the method of Mazin & Goebel (2007) we obtain an UL of $z < 0.95$. Our reconstructed redshift value is in agreement with the latest Landoni *et al.* (2014) and Farina *et al.* (2013, priv. comm.) results. A quasi simultaneous multi-wavelength SED (spectral energy distribution, fig. 1 right bottom) shows an unexpected feature in the $3 \times 10^{14} - 10^{18}$ Hz frequency range. A possible explanation using an inhomogeneous helical jet synchrotron self-Compton model is suggested (Villata & Raiteri 1999, Ahnen *et al.* 2016).

† http://users.utu.fi/kani/1m
‡ http://www.oato.inaf.it/blazars/webt/aba_info.html

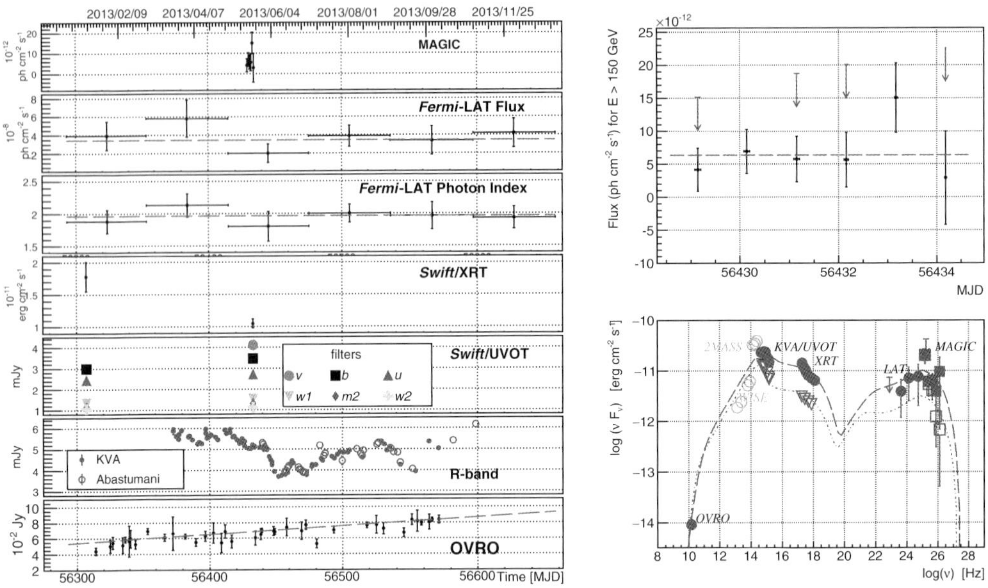

Figure 1. Light curve and the SED of H1722+119 for data collected during 2013. **Left**: Multi-wavelength light curve during 2013. The data were collected (from top to bottom) by MAGIC, *Fermi*-LAT, *Swift*/XRT, *Swift*/UVOT, KVA and Abastumani and OVRO. The HE γ-ray light curve and the evolution of the spectral index, as measured by *Fermi*-LAT, are shown in 2-month time bins, while daily observations time-scale is used in the light curves in the other energy bands. **Right top**: MAGIC nightly light curve for energies above 150 GeV. Horizontal error bars represent the duration of each observation. Vertical arrows represent ULs for points whose relative error of the excess is > 0.5. **Right bottom**: The H1722+119 SED. Full circles represent data contemporaneous to MAGIC observations. The MAGIC measured data are shown by empty squares, while full squares indicate the de-absorbed points. *Fermi*-LAT data are shown by full circles, while arrows represent *Fermi*-LAT ULs. The *Swift* data indicated by full circles were taken on 2013 May 20 (MJD 56432). The KVA *R*-band point represented by a full circle was taken on the same night. The OVRO measurement represented by a full circle was taken on 2013 May 22 (MJD 56434). Empty triangles show the *Swift* data from 2008 May 31 (MJD 54617). The solid line in the $10^{14} - 10^{15}$ Hz range indicating data from Landoni *et al.* (2014) was not considered for modelling, as well as archival data from 2MASS and WISE shown by empty circles. The long-dashed line indicates the fit of the helical jet model to the 2013 data, while the fit to the 2008 data is indicated by the dash-dot line. Both models represent the intrinsic VHE emission.

References

Ackermann, M., *et al.* 2012, *ApJS*, 203, 4

Ahnen M. L., *et al.* 2016, *MNRAS*, 459, 3271

Aleksić J., *et al.*, 2015, *MNRAS*, 451, 739

Aleksić J., *et al.* 2016a, *Astroparticle Physics*, 72, 61

Aleksić J., *et al.* 2016b, *Astroparticle Physics*, 72, 76

Atwood W. B., *et al.* 2009, *ApJ*, 697, 1071

Burrows D. N., *et al.* 2005, *Space Sci. Rev.*, 120, 165

Franceschini, A., Rodighiero, G., & Vaccari M. 2008, *A&A*, 487, 837

Landoni, M., *et al.* 2014, *A&A*, 570, A126

Mazin, D., Goebel F. 2007, *ApJ*, 655, L13

Prandini, E., *et al.* 2010, *MNRAS*, 405, L76

Richards J. L., *et al.* 2011, *ApJS*, 194, 29

Roming, P., *et al.* 2005, *Space Sci. Rev.*, 120, 95

Villata, M. & Raiteri, C. M. 1999, *A&A*, 347, 30

New Frontiers in Black Hole Astrophysics
Proceedings IAU Symposium No. 324, 2016
Andreja Gomboc, ed.

doi:10.1017/S174392131700148X

Influence of self-gravity on the equilibrium structures of magnetized tori

Audrey Trova[1], **Vladimír Karas**[1], **Petr Slaný**[2] **and Jiří Kovář**[2]

[1]Astronomical Institute, Czech Academy of Sciences,
Boční II 1401, Prague, CZ-141 31, Czech Republic

[2]Institute of Physics, Faculty of Philosophy and Science, Silesian University in Opava
Bezručovo nám. 13, CZ-746 01 Opava, Czech Republic

Abstract. We present some results obtained with a toy model developed in Trova *et al.* 2016 used to study the influence of the self-gravity on the equilibrium configurations of magnetized rotating self-gravitating gaseous tori, in the context of gaseous/dusty tori surrounding supermassive black holes in galactic nuclei. While the central black hole dominates the gravitational field and it remains electrically neutral, the surrounding material has a non-negligible self-gravitational effect on the torus structure. The vertical and radial structures of the torus are influenced by the balance between the gravitational and the magnetic force. By comparison with a previous work without self-gravity (Slany *et al.* 2016), we show that the conditions of existence of these configurations can change.

Keywords. gravitation – magnetic fields – methods: numerical.

---

1. Introduction

The equilibrium structures of dusty tori present in active galactiv nuclei are influenced by the self-gravitational field of the torus. Due to the large distance from the central source, frequently a massive black hole, the self-gravitational field of these tori has to be taken into account. Equilibrium figures of rotating self-gravitating fluids have been described in great detail since decades under various assumptions (Masuda *et al.* 1998, Otani *et al.* 2009). Here, the rotating torus, modelled by a perfect fluid with barotropic equation of state and with some net electric charge spread through the fluid, is embedded in a spherical gravitational and dipolar magnetic field. The existence of these structures is possible for certain values of the model parameters, such as the rotation law, the polytropic index, the magnetic field and the self-gravity intensity.

2. Conditions of existence of equilibrium stuctures

The torus equilibrium is governed by the Euler equation (Landau & Lifshitz 1987, equation (2.3)), which can be integrated and rewritten as the Bernouilli's equation:

$$aH + d_t \Psi_{\rm Sg} + \Psi_c + b\phi + e\mathcal{M} = c, \tag{2.1}$$

where H, Ψ_{Sg}, Ψ_c, Φ, $\mathcal{M}$ are, respectively, the enthalpy, the self-gravitational potential of the torus, the gravitational potential of the central mass, the centrifugal and the "magnetic" potential (linked to the dipolar field due to the central mass). The constants a, b, c, d_t, and e, are linked to physical parameters (the angular momentum, strength of the magnetic field...) and to the geometry of the torus. The conditions of integrability give insight on the rotation of the fluid and on the charge density. By setting these two parameters, we can find the conditions of existence of the equilibrium solutions. Working

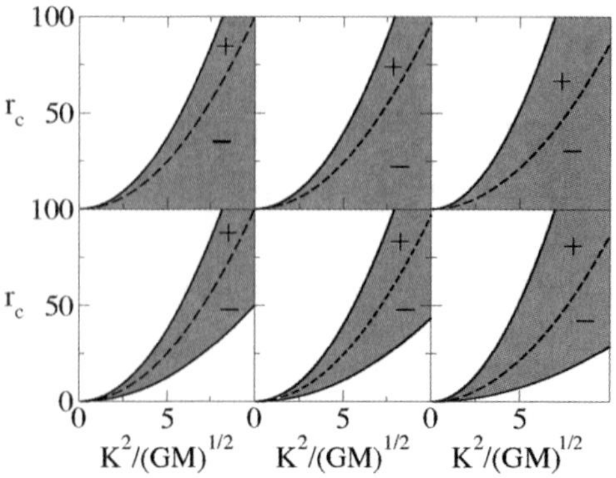

Figure 1. Representation of conditions of existence: from the left to the right, $d_t = 0, 0.1, 0.5$. The x-axis represents the constant b and the y-axis the maximum of pressure r_c.

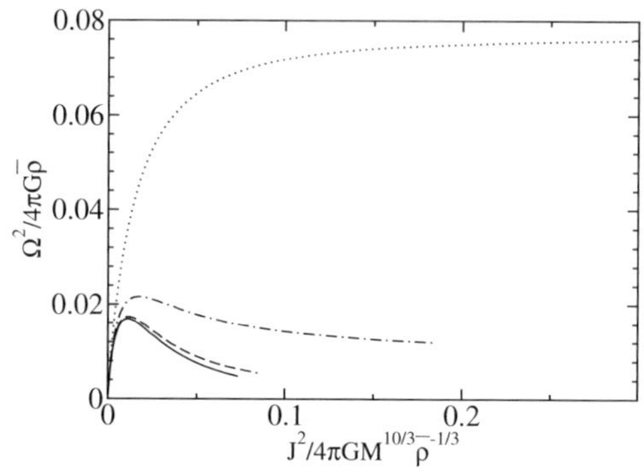

Figure 2. One-ring sequences for various values of $d_t = 0$ (full line), 0.01 (dashed line), 0.1 (dot-dashed line), 0.5 (point line).

with a constant angular momentum rotation and two different charge distributions (given by Slany *et al.* 2016), we obtain, from Equation 2.1, the conditions for three mass ratios (Figure 1). The colored region represents values for which we have a solution. The $+(-)$ region corresponds to positively(negatively) charged tori. In Figure 2, we show the impact of the value of d_t on the one ring sequence (as shown in Hachisu 1986, for self-gravitating tori only).

3. Conclusion

We have found that the self gravity has a direct impact on the conditions of existence of a magnetized self-gravitating torus. The range of possible solutions increases with the value of d_t. This result can also been found on the one-ring sequence presented in Figure 2. Moreover, for the same parameters, the tori can be positively or negatively charged in function of d_t. In Trova *et al.* 2016, we also showed that the morphologies of the solution, meaning the shape and the density profile, can be impacted by the self-gravity.

Acknowledgments

We thank the Czech Science Foundation Center (GAČR No. 14-37086G) – "Albert Einstein Center for Gravitation and Astrophysics" and the IAU for the grant allocation.

References

Hachisu, I. 1986, *ApJS*, 479-507, 6
Landau, L. D. & Lifshitz, E. M. 1987, *Fluid Mechanics: Vol 6 (Course of Theoretical Physics)*
Otani, J., Takahashi, R., & Eriguchi, Y. 2009, *MNRAS*, 2152-2166, 396
Masuda, N., Nishida, S., & Eriguchi, Y. 2006, *MNRAS*, 1139-1144, 297
Trova, A., Karas, V., Slaný, P., & Kovář, J. 2016, *ApJS*, 12, 226
Slaný, P., Kovář, J., Stuchlík, Z., & Karas, V. 2013, *ApJS*, 3, 205

Topic 5: Tests of fundamental theories of physics using black hole systems

New Frontiers in Black Hole Astrophysics
Proceedings IAU Symposium No. 324, 2016
Andreja Gomboc, ed.

doi:10.1017/S174392131700076X

Unveiling early black holes with *JWST*

Priyamvada Natarajan

Department of Astronomy, Yale University, 52 Hillhouse Avenue, New Haven, CT 06520, U.S.A.
email: priyamvada.natarajan@yale.edu

Abstract. The formation of direct collapse black hole seeds with masses $\sim 10^4 - 10^5$ $M_\odot$ could help explain the assembly of supermassive black holes powering high redshift quasars. Conditions conducive to the formation of these massive initial seeds exist at high redshift. Halos hosting these massive seeds merge promptly with a nearby galaxy. These early stage mergers at high redshift produce a new class of transient galaxies that contain an accreting black hole that is over-massive compared to the newly acquired stellar component - Obese Black hole Galaxies (OBGs). During this phase, the accretion luminosity of the direct collapse black hole seed exceeds that of the acquired stellar component. Here we calculate the multi-wavelength spectrum of this short-lived OBG stage, and show that there exist unique observational signatures in long wavelengths spanning near, mid to far-infrared that should be detectable by instruments aboard the upcoming James Webb Space Telescope (JWST).

Keywords. black hole physics, cosmology: early universe, infra-red: galaxies

---

1. Introduction

Discovery of populations of luminous, high redshift quasars at $z \gtrsim 6$, suggests that supermassive black holes with masses of the order 10^{8-10} $M_\odot$ are already in place at these early epochs in the universe as reported in observations by Fan *et al.* (2001), Mortlock *et al.* (2011) and Wu *et al.* (2015). This requires early and extremely rapid assembly of these black holes, within the first billion years after the Big Bang, which poses a challenge to our current models of supermassive black hole (SMBH) formation and growth. One possible solution that has been proposed to alleviate this timing crunch is the formation of massive initial black hole seeds from the direct collapse of gas in pristine haloes in the cosmological context (Lodato & Natarajan (2006)). This new channel likely operates in addition to the standard one that leads to the formation of seeds from the remnants of the first stars. The first generation of stars referred to as Population III (Pop III hereafter) form from H_2-cooled, pristine gas at $z \sim 25$ onward. The initial mass function for Pop III stars is expected to be skewed high, resulting in the production of remnant black hole seeds with $M_{\bullet,III} \sim 100$ $M_\odot$. Under optimal conditions, these seeds could potentially grow via Eddington limited accretion to produce the SMBHs that power the detected high redshift quasars. This mechanism however requires extremely optimistic conditions post seed-formation: steady accretion from a large gas reservoir in the vicinity, and simultaneously, inefficient radiation feedback from the rapidly growing seed (see e.g. Tanaka *et al.* 2012, Pacucci *et al.* 2015 & Park *et al.* 2016). Additionally, the initial mass function of Pop III stars is also poorly constrained at present, resulting in uncertainty in the masses of the Pop III remnants. Recent reviews that present details of these models can be found in Volonteri (2012), Natarajan (2014) and Latif & Ferrara (2016).

The tuning required to produce SMBHs from Pop III remnants led to the proposal of alternate models for the formation of more massive black hole seeds from the direct collapse of pristine gas. In the early universe conditions feasible for the production of these direct collapse black holes (DCBHs), with initial masses of 10^{3-5} $M_\odot$ (see e.g. Eisenstein & Loeb (1995), Omukai (2001), Oh & Haiman (2002), Bromm & Loeb (2003), Koushiappas *et al.* (2004), Lodato & Natarajan (2006)) exist. This additional DCBH seed formation channel has been actively explored in the past decade. What is required for this channel to operate are physical processes that prevent fragmentation and hence star-formation in early gas, rich haloes. This suppression of cooling requires the dissociation of molecular hydrogen which is an efficient coolant. Photons in the Lyman-Werner band ($11.2 \leqslant h\nu \leqslant 13.6$ eV) from star-forming sources nearby can effectively destroy molecular hydrogen. Photons in this band can photo–dissociate H_2 via $H_2 + \gamma_{LW} \rightarrow H + H$ (see e.g. Omukai *et al.* 2001). Atomic H is an inefficient coolant that can only cool gas down to 8000 K; H_2 instead can cool the gas down to $\sim$ 200 K. Therefore, DCBH collapse can occur feasibly in satellite halos in the vicinity of star-forming galaxies that are the source of external Lyman-Werner photons. The presence of an external radiation field from a neighboring early star-forming galaxy can prevent a pristine halo from forming further Pop III stars. This suppression could instead lead to the isothermal collapse of the gas at 8000 K to extremely high densities, causing a runaway process that leads to the formation of a DCBH seed with $M_{\bullet,DC} \sim 10^{4-5}$ $M_\odot$ (see e.g. Bromm & Loeb (2003), Lodato & Natarajan (2006), Begelman *et al.* (2006), Volonteri *et al.* (2008), Shang *et al.* (2010), Johnson *et al.* (2012)Johnson, Whalen, Fryer, & Li, Ferrara *et al.* (2014)). Given their large initial seed masses, these DCBHs can grow and produce SMBHs by $z \sim 7$ even accreting at sub–Eddington rates.

There has been much recent progress in exploring the formation of these DCBH seeds, their environment, and their early growth history (Volonteri *et al.* (2008), Natarajan (2014), Agarwal *et al.* (2012), Agarwal *et al.* (2016), Pacucci & Ferrara (2015), Pacucci *et al.* (2015)). Current work suggests that the lighter Pop III remnant seeds and massive seeding models (DCBHs) can in principle be distinguished with multi-wavelength data at high redshift. This discrimination is possible due to the fact that DCBH seeds that assemble in satellite halos transit through a unique, short-lived stage post-formation upon merging with their parent haloes. During this stage, which we refer to as Obese Black hole Galaxies (OBGs), this first merger produces a source in which the black hole mass exceeds that of the acquired stellar component (Agarwal *et al.* 2013). This is in sharp contrast to what is seen locally, where the stellar masses of the bulges of nearby galaxies vastly exceeds that of the black hole harbored in the galactic nucleus (Tremaine *et al.* (2002), Ferrarese & Merritt (2000)). In addition at high redshifts, what is key is that the accretion luminosity of the growing black hole exceeds that of the stellar component. Therefore, calculating the shape of the emergent spectrum from accretion at these early epochs, we find that OBGs are potentially detectable with instruments aboard the upcoming *JWST* mission due to their distinct signatures in near and mid-infra-red wavelengths spanning 1 - 28 microns (Natarajan *et al.* 2016) .

2. The formation of direct collapse black holes

Physical conditions that permit the formation of DCBHs are available amply in the early universe (Agarwal *et al.* 2012). The host halos where DCBHs form are mini-halos with a virial temperature of $2000 < T < 10^4$ K. As outlined above, star formation in these haloes could be delayed or even halted by a modest level of external LW radiation field (Machacek *et al.* (2001), O'Shea & Norman (2008)), keeping them pristine till they grow to the atomic cooling limit $T \sim 10^4$ K. At this juncture, if the external LW radiation field prohibits molecular hydrogen formation by dissociation of H_2 (into H), DCBH formation would ensue due to the suppression of H_2 cooling and inhibition of fragmentation into

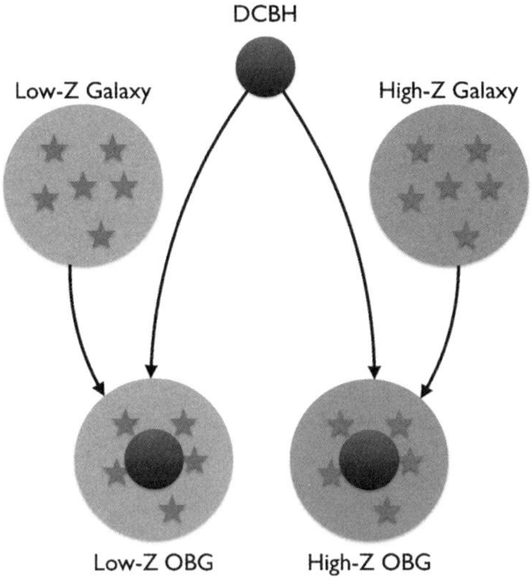

Figure 1. Schematic figure showing the DCBH seed formation scenario explored here. DCBH seeds form at the center of pristine, satellite halos that are illuminated by a nearby SF galaxy. Star formation is quenched in these DCBH host halos, although a stellar component is accrued promptly via merging with nearby SF galaxies. In the models studied here, we examine the growth following the merger of the DCBH (initial seed mass of 10^5 $M_\odot$) with a high and low metallicity stellar population. The gas in the post-merger halo is assumed to have the same metallicity as the stellar population. We also explore subsequent growth via accretion from a standard thin disk accretion as well as a radiatively inefficient slim disk.

stars. Viable sites are satellite halos in the vicinity of the first generation of star forming (SF) galaxies in the early universe (Agarwal *et al.* (2013), Agarwal *et al.* (2014)). At these satellite sites, pristine gas-rich proto-galactic disks could form that do not fragment or cool but rather go dynamically unstable (Q-unstable, as suggested by Toomre (1964), leading to rapid runaway accretion of gas to the center and to the formation of a massive DCBH seed. The physics and evolution of this process has been calculated analytically by Lodato & Natarajan (2006), and conducive conditions (assembly of massive proto-galactic disks) are increasingly seen to occur in state of the art high-resolution cosmological simulations of early structure formation as noted by Regan & Haehnelt(2009), Choi *et al.* (2013) and Latif *et al.* (2013). It is estimated that the typical masses of these DCBH seeds would lie in the range 10^{4-5} $M_\odot$ and the collapse process would be limited entirely by the mass available in the gas reservoir (e.g. Johnson *et al.* (2012)Johnson, Whalen, Fryer, & Li & Park *et al.* (2016)). Radiative feedback from these assembling DCBH seeds would halt SF in these halos (Aykutalp *et al.* 2014). Recently, Pacucci *et al.* (2015) have shown that $> 90\%$ of the gas in these proto-galactic core regions would be accreted and the rest blown away in $T_{\rm vir} \sim 10^4$ K DCBH host halos, under radiatively inefficient conditions.

Tracking the merging histories of these early dark matter halos, we find that these DCBHs' host satellite halos rapidly merge with the parent SF halo in their vicinity. This leads to the swift acquisition of a stellar component by the DCBH halo. This new object corresponds to the formation of a special transitory class of galaxies that we have referred to as OBGs (Agarwal *et al.* 2013). This is under the assumption that star formation is halted or highly suppressed in the DCBH halo itself due to radiative feedback from the assembling seed. Meanwhile, post-merger, stars and the DCBH continue to grow self consistently in an OBG in the sense that the same gas reservoir likely feeds the BH and

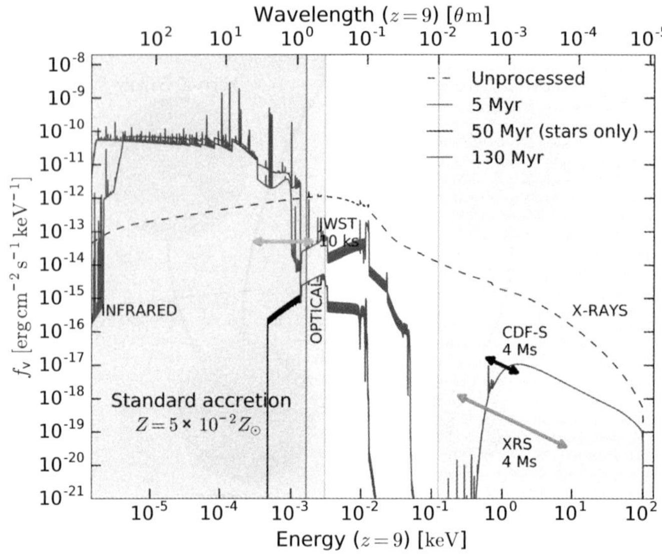

Figure 2. The computed multi-wavelength OBG spectrum for growth via standard accretion for an initial 10^5 $M_\odot$ DCBH seed, the higher metallicity case (0.05 $Z_\odot$) is plotted at two time slices $t = 5$ Myr and $t = 130$ Myr. The sensitivity limits for *JWST* for a 10 ks observation, the CDF-S (4 Ms) and the *X-ray Surveyor* (4 Ms) are highlighted. For contrast, we over-plot the spectrum of just the stellar component at a time slice $t = 60$ Myr (the high metallicity case).

forms stars in the merged remnant during this transient stage (see Fig. 1 for a schematic representation of the scenario explored here). For galaxies that host a rapidly growing DCBH seed, the energy output from BH accretion vastly exceeds the emission from the stellar component. In what follows, we outline the computation of the multi-wavelength SED of an OBG under these specific circumstances.

3. Calculating JWST signatures of DCBHs

We compute the multi-wavelength spectral energy distribution (SED) template for a growing DCBH seed in the OBG stage using a synthetic model to calculate the combined contribution of fluxes from the stellar and the accreting black hole components. We assemble these two components in the context of the merging hierarchy of LCDM halos and a background LW radiation field using the self-consistent prescription developed in Agarwal *et al.* (2012). Once a DCBH forms, its subsequent growth and energy output from accretion are simulated using a state–of–the–art 1D radiation-hydrodynamic code *GEMS*. This code simulates the spherical accretion onto a high-redshift black hole seed, computing the emitted luminosity self-consistently from the accretion rate. The full spectral analysis is then performed in a post-processing step with the code *Cloudy* Ferland *et al.* (2013). A full description of *GEMS* is available in Pacucci & Ferrara (2015). The growth and resulting energy output for the DCBH seed are modeled in two limiting cases. In the first, that we refer to as the *standard* case the growth involves accretion via a classic α-disk (Shakura & Sunyaev 1976), which is geometrically thin and optically thick. This accretion disk is radiatively efficient and accretion occurs, on average, at the Eddington rate. As described in Pacucci *et al.* (2015), this regime is entirely feedback-limited and the growth of the DCBH seed gets stunted once the feedback from the accretion process heats the gas, as it disrupts the inflow. In addition to this BH growth mode, we also model the case of radiatively inefficient accretion via a *slim disk* as proposed

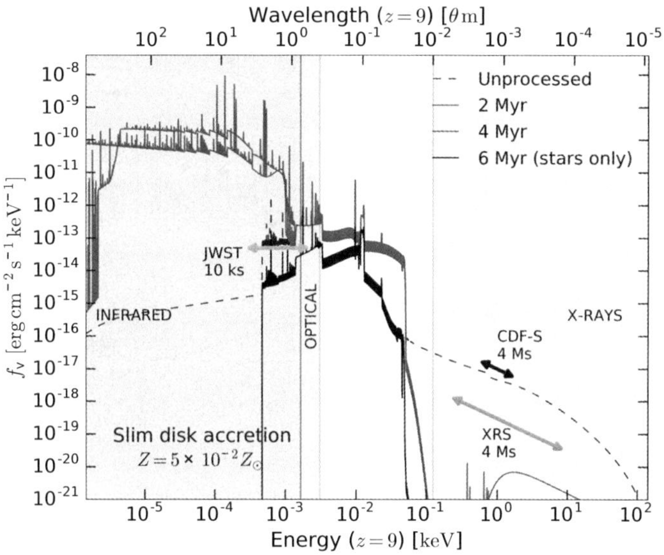

Figure 3. Here we show the spectrum with initial DCBH seed mass of 10^5 M$_\odot$, but now for growth via slim disk accretion for the higher metallicity case (0.05 Z$_\odot$) at two time slices $t = 2$ Myr and $t = 4$ Myr. Once again for contrast, we over-plot the spectrum of just the stellar component at a time slice of $t = 3$ Myr (high metallicity). The reference sensitivity limits for for *JWST*, the CDF-S and the *X-ray Surveyor* are highlighted here as well.

by Paczynski & Abramowicz (1982) and Abramowicz *et al.* (1988), and more recently explored in Sadowski (2009), and McKinney *et al.* (2014). In this case, radiation pressure is less efficient in quenching the gas inflow, due to photon trapping. Therefore, highly super-Eddington accretion rates may be reached and the mass growth is now entirely gas supply limited. We study the accretion of metal-enriched gas onto the DCBH during the OBG stage, where the metallicity is indexed to that of the merged stellar population. At this point, growth of the DCBH is jump-started by the merger. For the two modes of accretion outlined above, we explore the case of the higher metallicity for the OBG stage with $Z = 5\times10^{-2}$ $Z_\odot$. The stellar population that merges to form the OBG is aged accordingly, with the metallicity of 5×10^{-2} Z$_\odot$ and an age of 800 Myr attributed to older Pop II stars. To model the stellar SED component of the OBG stage, we used *Yggdrasil*, a stellar population synthesis code that employs Raiter *et al.* (2010) for the lower metallicity, and *Starburst99* Leitherer *et al.* (1999) with Geneva high mass-loss tracks for the higher metallicity case. The nebular emission component was computed using *Cloudy* adopting the parameters described in Zackrisson *et al.* (2011). The SEDs were then rescaled to the Kroupa (2001) stellar initial mass function. The nebular emission is also computed assuming the same ambient gas metallicity as that of the stars. We model the post–merger star formation in the OBG host halo with a constant star formation rate, producing a final stellar mass of $M_* = 10^7$ M$_\odot$ at the end of the OBG phase. The total duration of the OBG phase varies from ~ 10 Myr in the slim disk cases to ~ 100 Myr in standard accretion cases. Detailed treatment and additional results are presented in Natarajan *et al.* (2016). Here we plot the output OBG spectra for the high-metallicity case in Figs. 2 and 3.

The emergent spectrum from our fiducial OBGs shows distinct observational signatures wherein UV photons are reprocessed by the gas as infra-red emission primarily between 1 - 30 microns. Therefore, we expect *JWST* to be particularly efficient in finding OBGs. We show the template infra-red spectra, observable in the *JWST* NIRCAM and MIRI bands

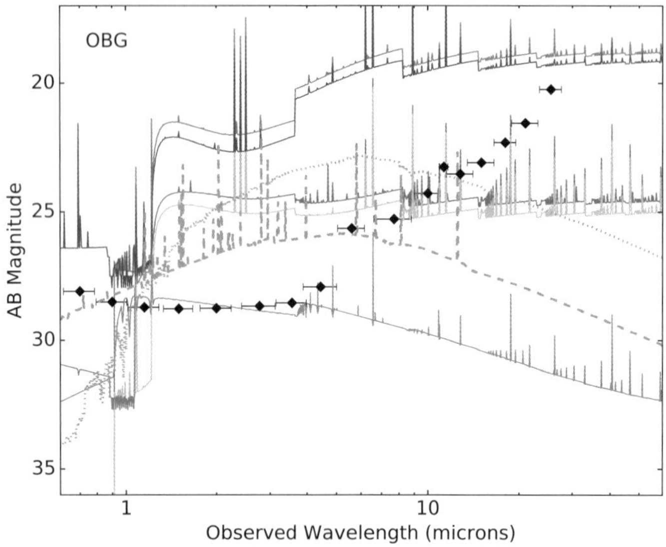

Figure 4. Predicted *JWST* template spectra for the DCBH seeding model and other contaminating high and low redshift sources. For this seeding model, the SED snapshots with the maximum-IR flux have been selected for plotting here. An OBG with an active DCBH accreting in the standard and slim disk cases for high and low metallicity at $z = 9$ are plotted. Here, for standard accretion the high metallicity case is shown in green; and the low metallicity case in red. In the slim disk accretion case, the lower metallicity template is plotted in orange and the higher one in purple. For comparison with potential contaminants, the spectrum of a typical actively star-forming high redshift galaxy (blue); and the measured spectra for low-redshift interlopers - an old and dusty star forming galaxy (grey dotted) at $z = 2.5$ and an Extreme Emission Line Galaxy (EELG) at $z = 2.1$ (grey, dashed) taken from Oesch *et al.* (2016) are over-plotted. The black symbols correspond to the MIRI and NIRCAM filter band passes on *JWST* and the error bars denote the width of the filters.

for both accretion modes in Fig. 4. We note the following trends: increasing the metal content of the host halo modifies the emerging spectrum in two ways: (i) it increases the absorption of high-energy ($\gtrsim 1\,\mathrm{keV}$) photons, and (ii) increases the power irradiated in the infra-red. From the calculated overall normalization of the spectrum, OBGs have estimated bolometric luminosities that are a factor of 100 or so higher than that of their growing Pop III seed counter-parts. We grow DCBH seeds to the final mass $\sim 10^7\ \mathrm{M}_\odot$, which is the total gas mass in the original halo. For a given BH seed mass, any change in the accretion model and in the metallicity essentially amounts to a re-scaling of the total growth time needed to arrive at the final mass. The effects of changing metallicity and accretion model are, in fact, degenerate. The reason is that we obtain the same high energy absorption with a small column of high metallicity gas or with a large, obscuring column of lower metallicity gas. The metallicity of the halo gas therefore changes the overall energy distribution due to photon scattering. A higher metallicity corresponds to an enhancement in the number of low-energy photons as higher energy photons with $E > 1$ keV are preferentially absorbed. There is also an enhancement in the re-radiated photons in the infra-red bands due to higher order Auger-like processes. Regardless of the metallicity of the stellar population which is also the adopted metallicity for the halo gas, the luminosity of the stellar population is always sub-dominant for the DCBH seeds. Aside from the discrimination offered by OBG candidate spectra in infra-red wavelengths, there are also clear signatures in X-ray wavelengths that we discuss further in Natarajan *et al.* (2016). OBG candidates as we show in Fig. 4 are easily detected in the MIRI bands.

These candidates would be characterized by a red slope between 1 and 10 microns and a flat slope beyond. This is in contrast to interloping low redshift sources that are extremely blue in the far IR as seen in Fig. 4 (Oesch *et al.* 2016).

4. Conclusions

Summarizing our results, for a growing DCBH seed with initial mass $M_{\rm seed} \sim 10^5$ $M_\odot$, we find that for the standard (Eddington–limited) growth, the signatures of the OBG stage are particularly distinct for the high metallicity case, as the overall infra-red flux is elevated relative to the lower metallicity case by about two orders of magnitude. We have shown the spectrum for the higher metallicity case here in Fig. 2. However, the X-ray flux is significantly higher for the lower metallicity case (discussed in detail in Natarajan *et al.* 2016). Despite their lower X-ray fluxes, the higher metallicity OBGs will be clearly detected in the NIRCAM and MIRI bands as seen in Fig. 4. For accretion via a slim–disk (super-Eddington growth), which is relatively short lived (5-10 Myr), OBGs would have diminished X-ray emission, and enhanced flux in the infra-red for the higher metallicity case shown here in Fig. 3. Due to the overall lowered X-ray flux though, detection will require deeper X-ray exposures than the ones currently available in the *Chandra* archive. Once again, despite their lower X-ray fluxes, the higher metallicity OBGs accreting via slim disks will also be detected unambiguously in the MIRI bands. In the context of the fiducial models studied here, high and low metallicity OBGs that are growing either via standard accretion or slim disk accretion will be unambiguously detected photometrically by *JWST* in the MIRI and NIRCAM bands. Additionally due to their high luminosities in the infra-red bands, OBGs could be feasibly followed up spectroscopically by NIRSPEC. Therefore, the upcoming *JWST* mission promises to provide new insights into the formation of the first black holes in the early universe and help discriminate between initial seeding models.

Acknowledgments

PN acknowledges support from a Theoretical and Computational Astrophysics Network grant with award number 1332858 from the National Science Foundation. She is grateful to her collaborators on this work: Fabio Pacucci, Andrea Ferrara, Bhaskar Agarwal, Angelo Ricarte, Erik Zackrisson and Nico Cappelluti.

References

Abramowicz, M. A., Czerny, B., Lasota, J. P., & Szuszkiewicz, E. 1988, *ApJ*, 332, 646
Agarwal, B., Dalla Vecchia, C., Johnson, J. L., Khochfar, S., & Paardekooper, J.-P. 2014, *MNRAS*, 443, 648
Agarwal, B., Davis, A. J., Khochfar, S., Natarajan, P., & Dunlop, J. S. 2013, *MNRAS*, 432, 3438
Agarwal, B., Johnson, J. L., Zackrisson, E., *et al.* 2016, *MNRAS*, 460, 4003
Agarwal, B., Khochfar, S., Johnson, J. L., *et al.* 2012, *MNRAS*, 425, 2854
Agarwal, B., Smith, B., Glover, S., Natarajan, P., & Khochfar, S., 2016, *MNRAS*, 459, 4209
Aykutalp, A., Wise, J. H., Spaans, M., & Meijerink, R. 2014, *ApJ*, 797, 139
Basu-Zych, A. R., Lehmer, B. D., Hornschemeier, A. E., *et al.* 2013, *ApJ*, 762, 45
Begelman, M. C., Volonteri, M., & Rees, M. J. 2006, *MNRAS*, 370, 289
Bromm, V. & Loeb, A. 2003, *ApJ*, 596, 34
Cappelluti, N., Comastri, A., Fontana, A., *et al.* 2016, *ApJ*, 823, 95
Choi, J.-H., Shlosman, I., & Begelman, M. C. 2013, *ApJ*, 774, 149
Decarli, R., Walter, F., Yang, Y., *et al.* 2012, *ApJ*, 756, 150

Devecchi, B. & Volonteri, M. 2009, *ApJ*, 694, 302
Dijkstra, M., Ferrara, A., & Mesinger, A. 2014, *MNRAS*, 442, 2036
Eisenstein, D. J. & Loeb, A. 1995, *ApJ*, 443, 11
Fan, X., *et al.* 2001, *AJ*, 122, 2833
Ferland, G. J., Porter, R. L., van Hoof, P. A. M., *et al.* 2013, Revista Mexicana de Astronomia y Astrofisica, 49, 137
Ferrara, A., Salvadori, S., Yue, B., & Schleicher, D. 2014, *MNRAS*, 443, 2410
Ferrarese, L. & Merritt, D. 2000, *ApJ Letters* 539, L9
Guo, Y., Ferguson, H. C., Giavalisco, M., *et al.* 2013, *ApJ Supplement*, 207, 24
Häring, N. & Rix, H.-W. 2004, *ApJl*, 604, L89
Johnson, J. L., Whalen, D. J., Fryer, C. L., & Li, H. 2012, *ApJ*, 750, 66
Kormendy, J. & Ho, L. C. 2013, *ARAA*, 51, 511
Koushiappas, S. M., Bullock, J. S., & Dekel, A. 2004, *MNRAS*, 354, 292
Kroupa, P. 2001, *MNRAS*, 322, 231
Latif, M. A. & Ferrara, A. 2016, *PASA*, 33, 51
Latif, M. A., Schleicher, D. R. G., Schmidt, W., & Niemeyer, J. 2013, *MNRAS*, 433, 1607
Leitherer, C., Schaerer, D., Goldader, J. D., *et al.* 1999, *ApJs*, 123, 3
Lodato, G. & Natarajan, P. 2006, *MNRAS*, 371, 1813
—. 2007, *MNRAS*, 377, L64
Machacek, M. E., Bryan, G. L., & Abel, T. 2001, *ApJ*, 548, 509
McKinney, J. C., Tchekhovskoy, A., Sadowski, A., & Narayan, R. 2014, *MNRAS*, 441, 3177
Mortlock, D. J., Warren, S. J., Venemans, B. P., *et al.* 2011, *Nature*, 474, 616
Natarajan, P., *et al.* 2016, *preprint, arXiv:1610.05312*
Natarajan, P., 2014, General Relativity and Gravitation, 46, 1702
Natarajan, P. & Volonteri, M. 2012, *MNRAS*, 422, 2051
Oesch, P. A., Brammer, G., van Dokkum, P. G., *et al.* 2016, *ApJ*, 819, 129
Oh, S. P. & Haiman, Z. 2002, *ApJ*, 569, 558
Omukai, K. 2001, *ApJ*, 546, 635
O'Shea, B. W. & Norman, M. L. 2008, *ApJ*, 673, 14
Pacucci, F. & Ferrara, A. 2015, *MNRAS*, 448, 104
Pacucci, F., Ferrara, A., Grazian, A., *et al.* 2016, *MNRAS*, 459, 1432
Pacucci, F., Ferrara, A., Volonteri, M., & Dubus, G. 2015, *MNRAS*, 454, 3771
Pacucci, F., Volonteri, M., & Ferrara, A. 2015, *MNRAS*, 452, 1922
Paczynski, B. & Abramowicz, M. A. 1982, *ApJ*, 253, 897
Park, K., Ricotti, M., Natarajan, P., Bogdanović, T., & Wise, J. H. 2016, *ApJ*, 818, 184
Raiter, A., Schaerer, D., & Fosbury, R. A. E. 2010, *Astronomy & Astrophysics*, 523, A64
Regan, J. A. & Haehnelt, M. G. 2009, *MNRAS*, 396, 343
Sadowski, A. 2009, *ApJ Supplement*, 183, 171
Shakura, N. I. & Sunyaev, R. A. 1976, *MNRAS*, 175, 613
Shang, C., Bryan, G. L., & Haiman, Z. 2010, *MNRAS*, 402, 1249
Tanaka, T., Perna, R., & Haiman, Z. 2012, *MNRAS*, 425, 2974
Toomre, A. 1964, *ApJ*, 139, 1217
Tremaine, S., Gebhardt, K., Bender, R., *et al.* 2002, *ApJ*, 574, 740
Volonteri, M. 2012, *Science*, 337, 544
Volonteri, M., Lodato, G., & Natarajan, P. 2008, *MNRAS*, 383, 1079
Wu, X.-B., Wang, F., Fan, X., *et al.* 2015, *Nature*, 518, 512
Zackrisson, E., Rydberg, C.-E., Schaerer, D., Östlin, G., & Tuli, M. 2011, *ApJ*, 740, 13

New Frontiers in Black Hole Astrophysics
Proceedings IAU Symposium No. 324, 2016
Andreja Gomboc, ed.

doi:10.1017/S1743921316012965

Black holes in Einstein-Gauß-Bonnet-dilaton theory

Jose Luis Blázquez-Salcedo[1], Vitor Cardoso[2], Valeria Ferrari[3], Leonardo Gualtieri[3], Panagiota Kanti[4], Fech Scen Khoo[5], Burkhard Kleihaus[1], Jutta Kunz[1], Caio F. B. Macedo[6], Sindy Mojica[7], Paolo Pani[3] and Eugen Radu[8]

[1]Institut für Physik, Universität Oldenburg, Postfach 2503, D-26111 Oldenburg, Germany
[2]CENTRA, Departamento de Física, Instituto Superior Técnico, Universidade de Lisboa Avenida Rovisco Pais 1, 1049 Lisboa, Portugal
[3]Dipartimento di Fisica, "Sapienza" Università di Roma & Sezione INFN Roma1 Piazzale Aldo Moro 5, 00185 Roma, Italy
[4]Division of Theoretical Physics, Department of Physics, University of Ioannina Ioannina GR-45110, Greece
[5]Department of Physics and Earth Sciences, Jacobs University, D-28759 Bremen, Germany
[6]Faculdade de Física, Universidade Federal do Pará 66075-110 Belém, Pará, Brazil
[7]Escuela de Física, Universidad Industrial de Santander A. A. 678, Bucaramanga 680002, Colombia
[8]Departamento de Física da Universidade de Aveiro and CIDMA Campus de Santiago, 3810-183 Aveiro, Portugal

Abstract. Generalizations of the Schwarzschild and Kerr black holes are discussed in an astrophysically viable generalized theory of gravity, which includes higher curvature corrections in the form of the Gauss-Bonnet term, coupled to a dilaton. The angular momentum of these black holes can slightly exceed the Kerr bound. The location and the orbital frequency of particles in their innermost stable circular orbits can deviate significantly from the respective Kerr values. Study of the quasinormal modes of the static black holes gives strong evidence that they are mode stable against polar and axial perturbations. Future gravitational wave observations should improve the current bound on the Gauss-Bonnet coupling constant, based on observations of the low-mass x-ray binary A 0620-00.

Keywords. Relativity, gravitation, black hole physics, gravitational waves

1. Introduction

So far General Relativity (GR) has passed all tests in the Solar System and beyond. Still we expect that GR will be superseded by a generalized theory of gravity, which encompasses GR as a limit. Reasons for this expectation reside on the one hand on the theoretical side, where the incompatibility of GR with quantum mechanics has sparked the search for a theory of quantum gravity. On the other hand, when GR is applied to the evolution of the Universe as a whole, the necessity for the presence of dark matter and dark energy arises, the nature of which is currently unknown.

By now a large number of alternative theories of gravity have been proposed and some of their implications for astrophysical objects have been studied. The review *Testing General Relativity with Present and Future Astrophysical Observations* by Berti *et al.* (2015) gives a broad account of recent activities in this area.

Here we consider Einstein–Gauss-Bonnet-dilaton (EGBd) gravity, which represents a very interesting and well motivated extension of GR. The EGBd action is obtained by adding a real scalar field, a dilaton, to the GR action which is non-minimally coupled to the Gauss-Bonnet (GB) term. The resulting theory of gravity has quadratic curvature terms, but it has only second order equations of motion. Moreover, the EGBd action arises naturally in the framework of the low-energy effective string theories (see *e.g.* Moura & Schiappa (2007)).

To represent a viable alternative theory of gravity, EGBd theory must satisfy theoretical and observational constraints. These yield bounds on the GB coupling constant α. Solar system measurements, as e.g., measurements of the Shapiro time delay, yield the rather weak bound, $\sqrt{\alpha} \lesssim 10^{13}$cm (Bertotti, Iess & Tortora (2003)). In contrast, black holes in low-mass x-ray binaries can give a much stronger bound, since black holes in EGBd theory carry a scalar charge. Observations of the black hole low-mass x-ray binary A0620-00 (Cantrell *et al.* (2010)) then lead to a constraint based on the orbital decay rate (Yagi (2012))

$$\sqrt{\alpha} \lesssim 10^6 \text{cm}. \tag{1.1}$$

From the theoretical side the mere existence of black hole solutions implies an upper bound (Kanti *et al.* (1996), Pani & Cardoso (2009))

$$\frac{\alpha}{M^2} \lesssim 0.691. \tag{1.2}$$

Future observations of quasi-periodic oscillations (QPOs) from accreting black holes could be used to further constrain the theory (Maselli *et al.* (2015)).

Here we explore the physical properties of static and rotating EGDd black holes (Mignemi & Stewart (1993), Kanti *et al.* (1996), Torii, Yajima & Maeda (1996), Pani & Cardoso (2009), Kleihaus, Kunz & Radu (2011), Pani *et al.* (2011), Ayzenberg & Yunes (2014), Kleihaus, Kunz & Mojica (2014), Maselli *et al.* (2015), Kleihaus *et al.* (2016), Blazquez-Salcedo *et al.* (2016)). In particular, we study their domain of existence and point out their differences to the Schwarzschild and Kerr black holes of GR. By computing the quasinormal modes of the static EGBd black holes, we infer their mode stability. Finally we address how future gravitational wave observations may improve the current bound on the GB coupling constant.

2. Einstein-Gauss-Bonnet-dilaton gravity

The action of EGBd gravity is given by

$$S = \frac{1}{16\pi} \int d^4x \sqrt{-g} \left[R - \frac{1}{2}(\partial_\mu \phi)^2 + \frac{\alpha}{4} e^{\phi} R^2_{\rm GB} \right], \tag{2.1}$$

where R is the curvature scalar, ϕ is a real scalar field, and α is the GB coupling constant. The dilaton coupling constant has been given its string theory value throughout, and $R^2_{\rm GB}$ is the Gauss-Bonnet term, which is quadratic in curvature

$$R^2_{\rm GB} = R_{\mu\nu\rho\sigma} R^{\mu\nu\rho\sigma} - 4 R_{\mu\nu} R^{\mu\nu} + R^2. \tag{2.2}$$

The resulting set of equations of motion are of second order. The "modified" Einstein equations read

$$G_{\mu\nu} = \frac{1}{2} T_{\mu\nu} \tag{2.3}$$

with the effective stress-energy tensor

$$T_{\mu\nu} = T^{(\phi)}_{\mu\nu} + \frac{\alpha}{2} e^{\phi} T^{(\rm GBd)}_{\mu\nu}, \tag{2.4}$$

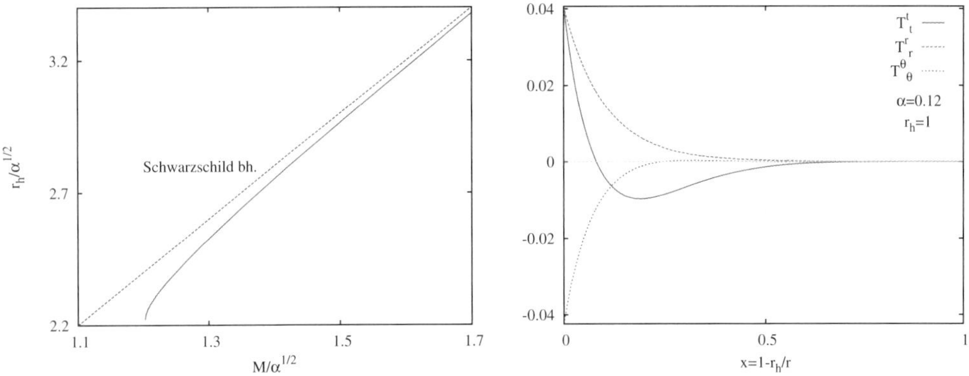

Figure 1. Scaled horizon radius r_h versus scaled mass M of static EGBd black holes with comparison to Schwarzschild black holes (left). Profile of the components of the effective stress-energy tensor, T_t^t, T_r^r and T_θ^θ, for a typical static EGBd black hole (right).

where

$$T_{\mu\nu}^{(\phi)} = \nabla_\mu \phi \nabla_\nu \phi - \frac{1}{2} g_{\mu\nu} \nabla_\lambda \phi \nabla^\lambda \phi \tag{2.5}$$

$$T_{\mu\nu}^{(\mathrm{GBd})} = H_{\mu\nu} + 4 \left(\nabla^\rho \phi \nabla^\sigma \phi + \nabla^\rho \nabla^\sigma \phi \right) P_{\mu\rho\nu\sigma} \ , \tag{2.6}$$

where $H_{\mu\nu}$ is quadratic and $P_{\mu\rho\nu\sigma}$ is linear in the curvature.

The dilaton equation is given by

$$\nabla^2 \phi = \frac{\alpha}{4} e^\phi R^2_{\mathrm{GB}} \ . \tag{2.7}$$

Because of the contributions from the GB term on the right hand side of the Einstein equations, the theory allows for negative "effective" energy densities giving rise to black holes with scalar "hair", although secondary. Moreover, this theory allows for wormholes without the need for exotic matter (Kanti, Kleihaus & Kunz (2012)).

3. Black hole properties

The vacuum black holes in GR consist of the family of static Schwarzschild and rotating Kerr black holes, which in general form the basis of current analyses of astrophysical observations. The GR black holes possess very special properties, in particular, all Kerr black holes are uniquely described by only two parameters, their mass M and their angular momentum J. For Kerr black holes the ratio $j = J/M^2$ is bounded, $|j| \leqslant 1$, with the extremal black holes saturating the limit.

Static EGBd black holes.

The full set of static EGBd black holes was first constructed by Kanti *et al.* (1996). In this analysis it was realized, that unlike GR black holes the EGBd black holes possess a lower bound for the size and for the mass, when the GB coupling is kept fixed, i.e., for a given theory. This bound arises when the metric and dilaton functions are expanded at the horizon r_h, since the expansion for the dilaton involves a coefficient with a square root

$$\sqrt{1 - 6 \frac{\alpha^2}{r_h^4} e^{2\phi_h}} \ , \tag{3.1}$$

whose radicand should not be negative in order to retain a real value for the scalar field.

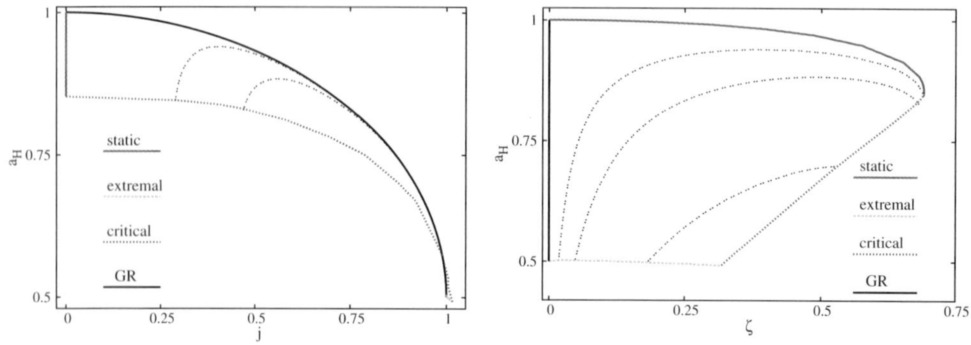

Figure 2. The scaled area versus the scaled angular momentum (left). The scaled area versus the scaled GB coupling constant (right). Also indicated are the static, the extremal, the critical and the GR black holes. Kleihaus, Kunz & Radu (2011) and Kleihaus *et al.* (2016).

We demonstrate the dependence of the horizon radius r_h on the mass M for the static EGBd black holes in Fig. 1 in the vicinity of the minimal mass and the minimal radius. It is the presence of the minimal mass, which leads to the bound (1.2). We also note from the figure, that whereas for a given α the deviations from Schwarzschild are quite pronounced for small masses, i.e., the EGBd black holes have a much smaller area for the same mass, the EGBd solutions tend toward the Schwarzschild solutions for larger masses.

Fig. 1 also exhibits the relevant components of the stress-energy tensor of a typical static black hole. It reveals that the energy density $\rho = -T_t^t$ is negative in the vicinity of the horizon. This allows these static black holes to circumvent the no-hair theorem and carry scalar hair.

Rotating EGBd black holes.

The inclusion of rotation is essential for astrophysical applications, but also from a theoretical point of view. The domain of existence of rotating EGBd black holes has been explored in Kleihaus, Kunz & Radu (2011), Kleihaus *et al.* (2016) and is exhibited in Fig. 2. This domain of existence is limited by the static EGBd black holes, the Kerr black holes, the extremal rotating EGBd black holes and the set of critical EGBd black holes. The latter arise analogous to the static case, when the radicand of a square root in the horizon expansion vanishes.

On the left of Fig. 2 the scaled area is exhibited versus the scaled angular momentum. It comes as a surprise, that EGBd theory allows for black holes violating the Kerr bound, i.e., possessing $|j| > 1$, although this violation is small. On the right of Fig. 2 for comparison the scaled area is shown versus the scaled GB parameter $\zeta = \alpha/M^2$.

As seen on the left of Fig. 3, the scaled angular momentum reaches its maximum roughly at half the maximal value of the scaled GB coupling, where the critical and the extremal rotating EGBd black holes merge. We note that the extremal rotating EGBd black holes possess a regular metric on the horizon, whereas the dilaton field diverges on the horizon at the poles.

The quadrupole moment of the Kerr black holes is completely fixed by the global charges, $Q = -J^2/M$. Fig. 3 (right), however, shows, that the quadrupole moment of EGBd black holes can considerably differ from the Kerr case. The Kerr values for the scaled moment of inertia are given by $J/(\Omega_{\rm H} M^3) = 2(1 + \sqrt{1 - j^2})$, thus they are fixed by the value of j. For the EGBd black holes this is no longer the case, where they decrease monotonically from the Kerr value for a fixed j.

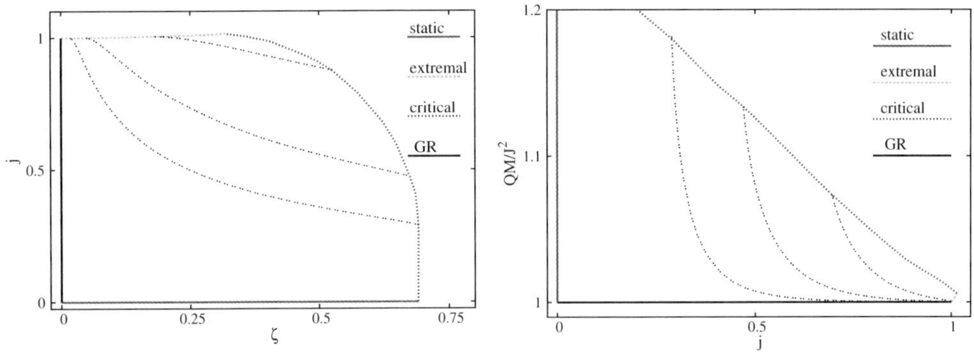

Figure 3. The scaled angular momentum versus the scaled GB coupling constant (left). The scaled quadrupole moment versus the scaled angular momentum (right). Also indicated are the static, the extremal, the critical and the GR black holes. Kleihaus *et al.* (2016) and Kleihaus, Kunz & Mojica (2014).

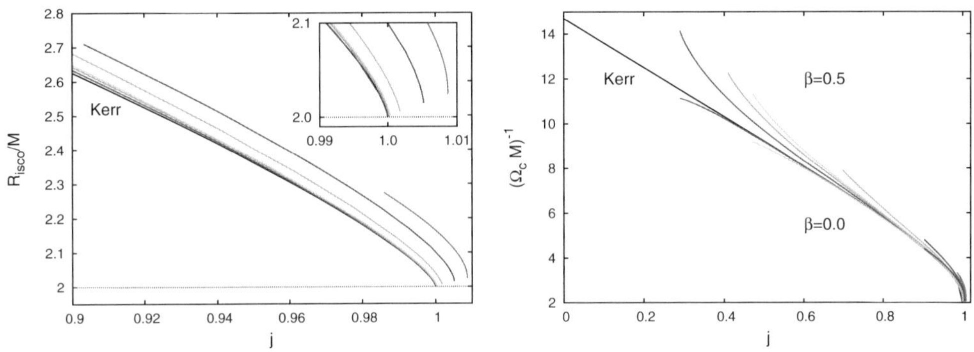

Figure 4. The scaled ISCO radius versus the scaled angular momentum (left). The inverse of the scaled ISCO frequency versus the scaled angular momentum for matter coupling constant $\beta = 0$ and $\beta = 1/2$ (right). The curves correspond to different horizon velocities of the black hole. Kleihaus, Kunz & Radu (2011) and Kleihaus *et al.* (2016).

Geodesics.

Also the study of the geodesics of these black holes is both of intrinsic interest and of astrophysical relevance. The analysis of the geodesics is based on the Lagrangian

$$\mathcal{L} = \frac{1}{2} e^{-2\beta\phi} g_{\mu\nu} \dot{x}^\mu \dot{x}^\nu \ , \tag{3.2}$$

where β is a coupling constant ($\beta = 1/2$ for string theory). Since the general geodesic equations are not separable (see the discussion in Kleihaus *et al.* (2016), showing the EGBd black holes are of Petrov type I), it is simpler to consider equatorial motion. In particular, we now restrict the analysis to the innermost stable circular orbits (ISCOs) of these black holes.

A perturbative analysis for slow rotation was performed in Pani & Cardoso (2009), while the analysis of the general rotating case was performed in Kleihaus, Kunz & Radu (2011), Kleihaus *et al.* (2016). We exhibit the scaled ISCO radius versus the scaled angular momentum in Fig. 4 (left) for coupling constant $\beta = 1/2$. (The ISCO radius here represents the circumferential radius, not the Boyer-Lindquist coordinate radius.) We note from the figure that the ISCO radius of EGBd black holes exceeds the one of Kerr black holes. Also shown in the figure is the inverse of the scaled ISCO frequency versus

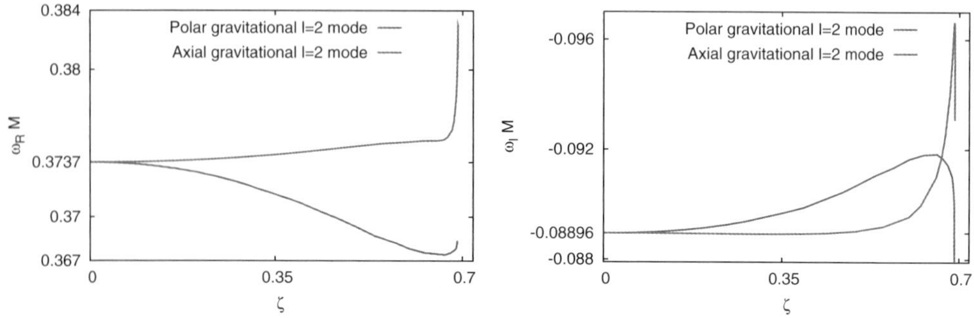

Figure 5. Real (left) and imaginary (right) parts of the gravitational axial and gravitational-led polar $l = 2$ fundamental modes. The Schwarzschild values are recovered at $\zeta = 0$. Blazquez-Salcedo *et al.* (2016).

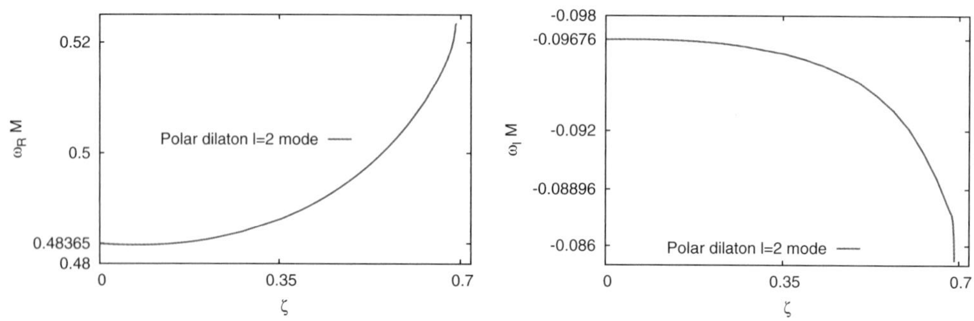

Figure 6. Real (left) and imaginary (right) parts of the scalar-led polar $l = 2$ fundamental modes. The Schwarzschild values are recovered at $\zeta = 0$. Blazquez-Salcedo *et al.* (2016).

the scaled angular momentum (right). As compared to the Kerr case, the frequencies are smaller when $\beta = 1/2$. For $\beta = 0$, however, they are larger.

Quasinormal modes and gravitational waves.

When a black hole is perturbed it reacts with the emission of gravitational waves, to settle down again in a stationary state. In an astrophysical setting such perturbations can for instance arise through the infall of a smaller mass, or through the coalescence of black holes. Observation of the latter process was announced earlier this year by the LIGO collaboration Abbott *et al.* (2016). The process consists of three major phases, the inspiral, the merger and the ringdown. We will now briefly address the quasinormal modes of EGBd black holes, relevant for the ringdown phase.

In the following we restrict to the quasinormal modes of static EGBd black holes. The perturbations of the metric and the scalar field are given by

$$g_{ab} = g_{ab}^{(0)} + \varepsilon\, h_{ab}\,, \quad \phi = \phi_0(r) + \varepsilon\, \delta\phi\,, \tag{3.3}$$

where the subscript zero denotes the unperturbed fields. They can be expanded in terms of spherical harmonics and Fourier transformed, where ω is the Fourier frequency.

To determine the quasinormal modes, we first note, that the modes decouple according to their behavior under parity transformations. The axial modes involve only the metric, while the polar modes involve the dilaton field as well. Imposing physical boundary conditions on the modes, such that the solutions are purely ingoing at the horizon and purely outgoing at infinity, the coupled equations can be solved for the axial and the polar modes (Blazquez-Salcedo *et al.* (2016)).

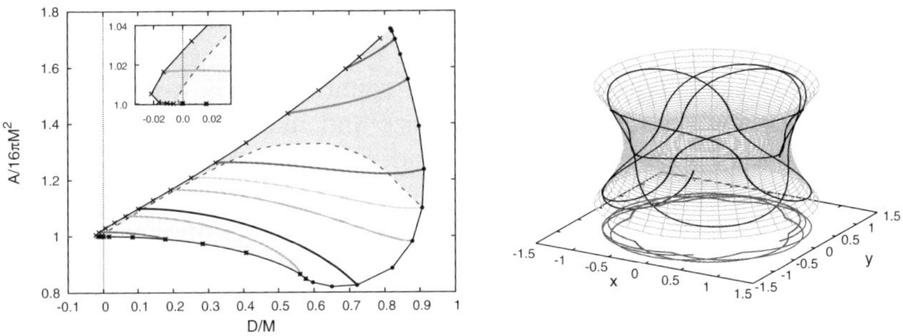

Figure 7. Domain of existence of wormholes in terms of the scaled throat area versus the scaled scalar charge. Wormholes in the lower part of the diagram are expected to be stable (left). Bound orbit of a test particle traveling between the universes (right). Kanti, Kleihaus & Kunz (2012).

We exhibit the frequencies of the lowest fundamental axial modes in Fig. 5, where the real part is shown on the left and the imaginary part on the right, versus the scaled GB coupling constant ζ. The Schwarzschild values are recovered at $\zeta = 0$. The deviations from the Schwarzschild values are small for $\zeta \leqslant 0.3$, but increase significantly in the vicinity of the critical solution.

For the polar modes we have to distinguish between gravitational-led modes and scalar-led modes. In the limit $\zeta \to 0$ the gravitational-led EGBd modes reduce to the gravitational modes of the Schwarzschild metric, while the scalar-led EGBd modes reduce to the modes of a test scalar field in the Schwarzschild background. Fig. 5 also shows the frequencies of the lowest gravitational-led EGBd modes, while the scalar-led modes are shown in Fig. 6. As seen in the figures, the deviations from the Schwarzschild modes are somewhat larger for the polar modes.

4. Wormholes

As mentioned in Sec. 2, EGBd theory allows for traversable wormholes without the need for exotic matter, since the quadratic gravitational terms contribute to the effective stress-energy tensor, enabling a violation of the energy conditions (Kanti, Kleihaus & Kunz (2012)). The domain of existence of the resulting static wormholes is exhibited in Fig. 7 (left), where the scaled area of the throat is shown versus the scaled scalar charge. At the left boundary wormholes with a double throat arise, at the lower boundary a transition to black holes is encountered, while at the right boundary solutions with a singularity arise. The lower part of this domain of existence is likely to contain (mode) stable wormholes.

EGBd wormholes possess bound states of test particles, as illustrated in Fig. 7 (right). These may reside in either asymptotically flat universe, but also move between universes. To obtain a low surface gravity at the throat, which is on the order of the gravitational acceleration on the surface of the earth, these wormholes need an astronomical size with a throat radius on the order of 10 - 100 lightyears. Still, smaller EGBd wormholes with higher surface gravity may be considered as potential astrophysical objects, whose observational signatures can be studied.

5. Conclusions and outlook

We have studied the physical properties of black holes in EGBd theory, an astrophysically viable alternate theory of gravity. The coupling constant of the theory has a theoretical bound, and can be constrained by current and future astrophysical observations.

The EGBd black holes differ from their GR counterparts in a number of significant ways. For instance, they possess a minimal size and mass for a given coupling, their angular momentum can exceed the Kerr bound, and their quadrupole moment can be far bigger. Also, depending on the coupling to ordinary matter, the radii and frequencies of their ISCOs can change considerably. A general analysis of their geodesics is still missing, but their shadow is currently being analyzed. Here further surprises may arise.

We have also analyzed the quasinormal modes of the static EGBd black holes, leaving the analysis of the rotating black holes as a challenge for the future. This analysis leads to the conclusion, that the static EGBd black holes should be mode stable, since all frequencies (calculated) have a negative imaginary part and thus decay exponentially in time. This is in accordance to the results of an early study Kanti *et al.* (1998) where a semi-analytic method demonstrated the stability of these black holes under linear perturbation. On the other hand, the quasinormal modes with the smallest imaginary part control the late-time dynamics of perturbed black holes.

The constraints on the GB coupling are expected to be improved by future observations with third generation gravitational wave detectors. A simple estimate performed in Blazquez-Salcedo *et al.* (2016) leads to the following upper bound on the GB coupling constant

$$\sqrt{\alpha} \lesssim 11 \left(\frac{50}{\rho}\right)^{1/4} \left(\frac{M}{10M_\odot}\right) \text{km}, \tag{5.1}$$

where ρ is the signal-to-noise ratio in the ringdown waveform. The Einstein telescope may achieve a signal-to-noise ratio $\rho \approx 100$ for an event like GW150914. This would then translate into the bound

$$\sqrt{\alpha} \lesssim 8 \left(\frac{M}{10M_\odot}\right) \text{km} \tag{5.2}$$

and $\zeta \lesssim 0.4$.

References

Berti, E. *et al.* 2015, *Class. Quant. Grav.* 32, 243001
Moura, F. & Schiappa, R. 2007, *Class. Quant. Grav.* 24, 361
Bertotti, B., Iess, L., & Tortora, P. 2003, *Nature* 425, 374
Cantrell, A. G. *et al.* 2010, *Astrophys. J.* 710, 1127
Yagi, K. 2012, *Phys. Rev. D* 86, 081504
Mignemi, S. & Stewart, N. R. 1993, *Phys. Rev. D* 47, 5259
Kanti, P. *et al.* 1996, 1996, *Phys. Rev. D* 54, 5049
Torii, T., Yajima, H., & Maeda, K. i. 1996, *Phys. Rev. D* 55, 739
Pani, P. & Cardoso, V. 2009, *Phys. Rev. D* 79, 084031
Kleihaus, B., Kunz, J., & Radu, E. 2011, *Phys. Rev. Lett.* 106, 151104
Pani, P., Macedo, C. F. B., Crispino, L. C. B., & Cardoso, V. 2011, *Phys. Rev. D* 84, 087501
Ayzenberg, D. & Yunes, N. 2014, *Phys. Rev. D* 90, 044066, E: *Phys. Rev. D* 91, 069905
Kleihaus, B., Kunz, J., & Mojica, S. 2014, *Phys. Rev. D* 90, 061501 (2014)
Maselli, A., Pani, P., Gualtieri, L., & Ferrari, V. 2015, *Phys. Rev. D* 92, 083014 (2015)
Kleihaus, B., Kunz, J., Mojica, S., & Radu, E. 2016, *Phys. Rev. D* 93, 044047
Blázquez-Salcedo, J. L. *et al.* 2016, arXiv:1609.01286 [gr-qc].
Kanti, P., Kleihaus, B., & Kunz, J. 2012 *Phys. Rev. Lett.* 107, 271101; *Phys. Rev. D* 85, 044007
Abbott, B. P. *et al.* 2016, *Phys. Rev. Lett.* 116, 061102
Maselli, A. *et al.* 2015, *Astrophys. J.* 801, 115
Kanti, P. *et al.* 1998, *Phys. Rev. D* **57**, 6255

New Frontiers in Black Hole Astrophysics
Proceedings IAU Symposium No. 324, 2016
Andreja Gomboc, ed.

doi:10.1017/S1743921317002319

Black hole superradiance as a probe of ultra-light new particles

Robert Lasenby
Perimeter Institute for Theoretical Physics, 31 Caroline Street N, Waterloo, Ontario N2L 2Y5, Canada
email: rlasenby@perimeterinstitute.ca

Abstract. Bosonic fields around a spinning black hole can be amplified via 'superradiance', a wave analogue of the Penrose process, which extracts energy and momentum from the black hole. For hypothetical ultra-light bosons, with Compton wavelengths on $\gtrsim$ km scales, such a process can lead to the exponential growth of gravitationally bound states around astrophysical Kerr black holes. If such particles exist, as predicted in many theories of beyond Standard Model physics, then these bosonic clouds give rise to a number of potentially-observable signals. Among the most promising are monochromatic gravitational radiation signals which could be detected at Advanced LIGO and future gravitational wave observatories.

1. Overview

Astrophysical and cosmological evidence, along with the theoretical problems of Planck-scale physics, point towards new physics beyond the Standard Model (SM). However, direct evidence, such as signals in laboratory experiments, has so far been elusive. This may be because any new particles have masses too heavy to be produced by existing experiments such as the LHC. Hypothesised examples include the supersymmetric partners of SM particles (though in many models these should have been observed at the LHC), or the extra particles predicted by Grand Unified Theories. Alternatively, it is possible that new light states exist, but couple very weakly to SM matter, so that their effects are weak enough to have been undetected, similarly to how the neutrino was not detected until 60 years ago. Such new particles arise naturally in many models of Beyond Standard Model physics — examples include the QCD axion, string theory axions, 'fifth forces', and many others (see e.g. Essig *et al.* 2013).

The one way in which any new particles *must* interact with SM states is through gravity. However, the spacetime curvature scales achieved in the experimentally-accessible universe are $\gtrsim$ km, which corresponds to a very low effective source density in particle physics terms. For example, a 10 Solar mass black hole has a Hawking temperature of $T_H = 1/(8\pi G M_{\rm BH}) \sim 10^{-8}\,{\rm K}(10 M_\odot/M_{\rm BH})$ (using natural units). Gravitational production in particle physics process is suppressed by multiple powers of the energy of the process over the Planck mass, again giving tiny effects.

One way to get around the very small spacetime curvatures available is to take advantage of coherence enhancement — as in a laser, this can give rise to an exponential amplification of a very small initial population. The 'black hole bomb' thought experiment of Press & Teukolsky (1972) used the wave analogue of the Penrose process, 'superradiant scattering', through which energy and angular momentum can be extracted from a spinning black hole. Along with a mirror to obtain multiple passes of amplification, this gives an exponentially growing wave around the hole. The key idea behind using black hole superradiance to probe new light particles is that, if they have a non-zero mass, such particles will form bound states around the black hole, acting as 'their own

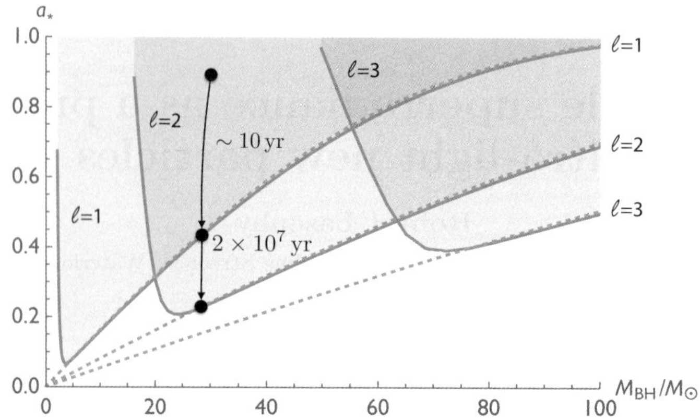

Figure 1. Illustration of how a black hole of initial mass $30M_{\odot}$ and initial dimensionless spin 0.9 evolves, through superradiance of a spin-0 field with mass $\mu = 6 \times 10^{-13}$ eV. The first arrow shows how the angular momentum $l = 1$ level grows over a timescale of ~ 10 yr, extracting a significant fraction of the black hole's spin, up to the critical spin at which this bound state is no longer superradiant. The $l = 2$ level takes much longer to grow, $\sim 2 \times 10^7$ yr, but will still extract all of the available angular momentum on shorter-than-cosmological timescales. The $l \geqslant 3$ levels take longer than the age of the universe to grow, so the black hole will remain on the critical $l = 2$ line. Repeating the same story for other initial masses and spins, if a black hole begins life within the shaded region, then it will end up at the boundary of the shaded region over the lifetime of the universe. Since, for the case shown, the boson's Compton wavelength is significantly larger than the size of the black hole, the bound state wavefunctions are mostly localised far away from the hole. Thus, they can extract a significant proportion of the black hole's angular momentum while only extracting a small fraction of its mass, as illustrated by the almost vertical evolution arrows.

mirror' (Detweiler 1980, Arvanitaki *et al.* 2010). For a spinning black hole, some of these bounds states can be superradiant — rather than decaying into the black hole, they will be amplified with time. Though such growth rates will be much less than the black hole crossing time, and thus extremely small on particles physics scales, if the new particle is bosonic then its occupation number can grow exponentially. Thus, a significant proportion of the black hole's energy and angular momentum can be extracted on a timescale that is short compared to astrophysically-relevant periods.

If such an exponential growth occurred, it could have a number of astrophysical signatures. The most obvious one is that initially fast-spinning black holes would be spun down over time by the formation of a bosonic cloud, as illustrated in Figure 1 (Section 2 goes over this process in more detail). The most accurate spin determinations come from black holes in X-ray binaries. Since, as per Figure 2, a few of these are observed to have near-extremal spins incompatible with significant spin-down, we can rule out the presence of a new light, weakly-coupled boson across a certain range in masses and couplings.

Recently, the LIGO gravitational wave observatory detected its first binary black hole merger, and through analysis of the waveform, was able to put some constraints on the masses and spins of the pre-merger black holes. In the near future, upgraded gravitational wave observatories will likely detect hundreds more such mergers. Though the errors on each individual spin measurement will be fairly large, the statistical information from many such events may provide indications of a spin-down effects, if one exists (since we do not know the pre-merger history of such systems, robust detections or exclusions would be difficult).

Gravitational wave observatories provide another, more promising probe of superradiance — the gravitational waves from the bosonic cloud itself. As explained in Section 2,

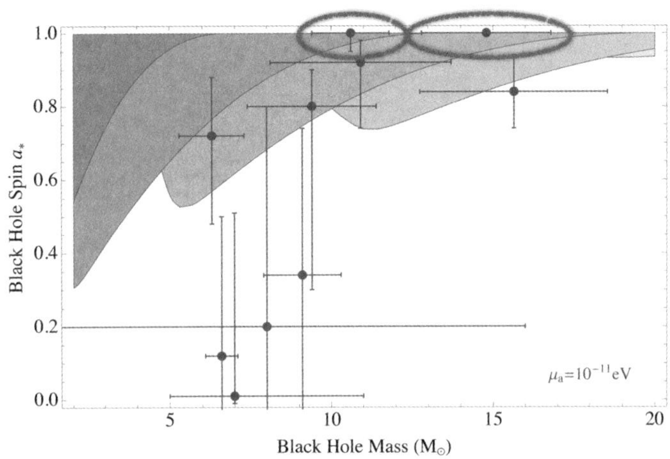

Figure 2. Illustration of how black hole spin measurements can place bounds on the existence of a light spin-0 particle (from Arvanitaki *et al.* 2015). The shaded areas, as per Figure 1, show the region within which an initially-fast-spinning black hole gets spun down to the boundary of the shaded region (in this case, for a particle of mass $\mu_a = 10^{-11}$ eV). The dots with error bars show the masses and spins of black holes, as measured via X-ray observations of accretion disks (Arvanitaki *et al.* 2015). The two circled measurements are well within the shaded region, so under the assumption that nothing would have disturbed the growth of a bosonic cloud (which is supported by the measured properties of these systems (Arvanitaki *et al.* 2015)), a weakly-coupled scalar boson of this mass is ruled out. The spin measurements errors for fast-spinning black holes are very small, since the properties of the accretion disk depend non-linearly on a_* at near-extremal spins.

the superradiant growth rate is largest when the boson's Compton wavelength is around the size of the black hole. This means that oscillations of the bosonic field, having a frequency of roughly the boson's mass, are naturally in the LIGO frequency range for roughly-stellar-mass black holes. These oscillations source monochromatic gravitational radiation, at a frequency set (to within a few percent (Arvanitaki *et al.* 2016)) by the boson's mass.

If there is a black hole with a large bosonic cloud sufficiently close to Earth — which, for most of the parameter space, means being inside our galaxy — then it could be detected as a monchromatic signal at LIGO, similar to that expected from a slightly asymmetric neutron star. However, signals from superradiant clouds would have the added feature that separate signals would be expected to cluster at frequencies within a a few percent of each other (as well as displaying subtly different frequency drift properties from a neutron star). As shown in Figure 3, depending on the mass and spin distribution of black holes, it is possible that such a signal could be seen within the next few years, if there is a light boson in the correct mass range.

The effects discussed so far have assumed that the new boson interacts sufficiently weakly with SM matter, and with itself, that gravity dominates its evolution around the black hole. This is expected to be true for some beyond Standard Model particle candidates — for example, a very light QCD axion also has an extremally small quartic self-coupling — but not for others. When self-interactions become important, very complicated behaviour can occur, though for large enough self-interactions, we would expect superradiant growth to saturate well before extracting a significant fraction of the black hole's energy and angular momentum.

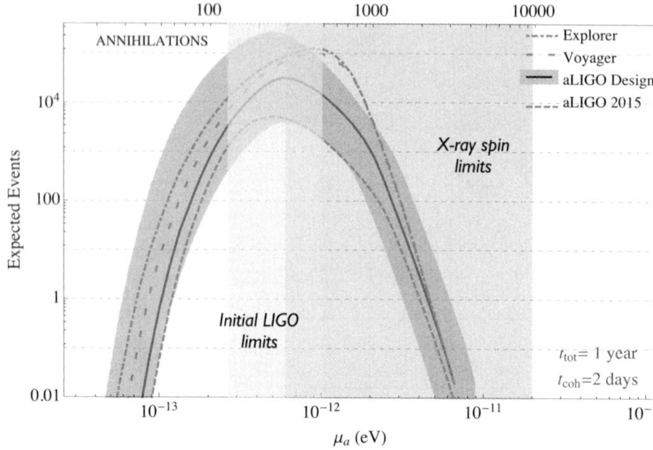

Figure 3. Expected number of monochromatic gravitational wave signals from black hole superradiance of a light scalar, detectable with one year of gravitational wave observations, for realistic black hole mass and spin distributions (from Arvanitaki *et al.* 2016). The shaded bands correspond to the range between pessimistic and optimistic black hole distribution with Advanced LIGO at design sensitivity. The various dotted lines correspond to gravitational wave detectors with different sensitivies. Boson masses μ_a in the greyed-out region are disfavoured by black hole spin measurements (Figure 2). In the light-gray region, null results from initial LIGO searches mean there is less chance of observing a new signal.

2. Black hole spin-down by superradiance

In this section, we give a slightly more detailed account of how the superradiant growth of bosonic bound states around a Kerr black hole can spin it down.

Rotational superradiance — amplification of scattered waves by rotating objects — is not a phenomenon that is unique to black holes. For example, as shown by Zeldovich (1971), a rotating conductor (with some resistivity) can amplify incident electromagnetic waves. The effect is a direct kinematic counterpart to the ability of a non-rotating object to absorb radiation. In the case of a black hole, we can see directly from the area theorem that, in absorbing a particle with $E/L > \Omega_H$, where E is the particle's energy, L its angular momentum, and Ω_H the angular velocity of the horizon, the black hole's horizon area would decrease. This is classically disallowed, but the converse is that emitting such a particle leads to an increase in horizon area, so is entropically favoured.

In many situations, we can also estimate the amplification factor (for bound states, the growth rate) from superradiance. Parametrically, the energy flux across the horizon due to a scalar field configuration $\varphi = \varphi_0 e^{i(\omega t - m\phi)}$ near the horizon is

$$P \sim A_H |\varphi_0|^2 \omega(\omega - m\Omega_H) \tag{2.1}$$

We see the occurrence of the $\omega - m\Omega_H$ factor expected from the area theorem — when $\omega/m < \Omega_H$, i.e. when the angular phase velocity of the field is less than the angular velocity of the horizon, the energy flux into the hole is negative. Bound states necessarily have $\omega < \mu$, where μ is the mass of the scalar field, so the bound states with the highest growth rate will generally be those with the largest field values near the horizon, among those with $m > \omega/\Omega_H$.

The states most localised near the black hole are those with the lowest energy and angular momentum. If $\mu < \Omega_H$ (for a fast-spinning black hole, if the Compton wavelength is large compared to the size of the hole), then all states with $m \geqslant 1$ are superradiant. If $\mu \gg \Omega_H$, then only states with $m > \mu/\Omega_H$ are superradiant; due to the large angular

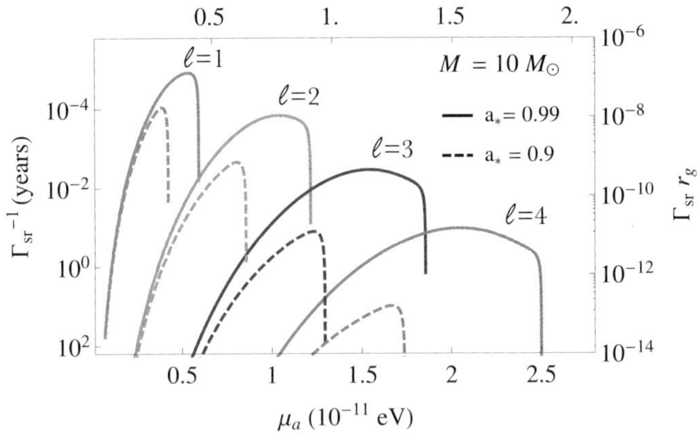

Figure 4. Superradiant growth rates Γ_{sr} for a scalar field around a Kerr black hole of 10 Solar masses, with dimensionless spin 0.99 (solid lines, 0.9 dotted lines), from Arvanitaki *et al.* 2015. The different curves show the growth rate of the fastest-growing level for each angular momentum value l. As discussed in the text, higher-l levels have smaller growth rates, and for each l level, the fastest growth rate occurs for μ near the maximum allowed for superradiance (the upper horizontal scale shows $GM_{\mathrm{BH}}\mu = \mu r_S/2$).

momentum barrier, their value near the horizon is very small. Consequently, as shown in Figure 4, the maximum growth rate is achieved for $\mu \sim 0.4/(GM_{\mathrm{BH}})$ — for lower values, the bound states are localised futher away from the black hole, while for larger values, only higher angular momentum states are superradiant, and their value near the horizon is suppressed by the angular momentum barrier.

Following the above discussion, Figure 1 illustrates how the existence of a light weakly-coupled boson would affect the mass and spin evolution of an initially-fast-spinning black hole. The dotted lines show the critical spin values below which superradiance does not occur, for each angular momentum value m. If a black hole begins life above one of these, and the superradiance rate of the fastest-growing level is fast enough, then it will be spun down by the formation of a bosonic cloud, until it is almost on the critical line. This process will then repeat with the next m level, again if the growth rates are fast enough, with the result that we expect all sufficiently old (and undisturbed) black holes to have spins below a mass-dependent value, with initially-fast-spinning holes ending up at these critical spins. As discussed above, such a pattern can be tested or ruled out by black hole spin measurements, in addition to gravitational wave signatures from the bosonic clouds themselves.

3. Summary

Black hole superradiance provides a unique probe of light bosons beyond the Standard Model of particle physics, with the ability to exclude or discover arbitrarily weakly-coupled particles in suitable mass ranges. The advent of gravitational wave astronomy opens multiple windows through which such an effect may be discovered, with the potential for observable signals in near-future experiments (Arvanitaki *et al.* 2015, Arvanitaki *et al.* 2016).

Acknowledgements: Most of the material in the text is derived from Arvanitaki *et al.* 2015 and Arvanitaki *et al.* 2016. We have not attempted to provide comprehensive references to the extensive literature on black hole superradiance, links to which can be found in some of the cited papers. Research at Perimeter Institute is supported by the Government of Canada through Industry Canada and by the Province of Ontario through the Ministry of Economic Development & Innovation.

References

Essig *et al.* 2013, *Working Group Report: New Light Weakly Coupled Particles*, arXiv:1311.0029
Detweiler 1980, *Phys. Rev.* D22, 2323
Arvanitaki *et al.* 2010, *Phys. Rev.* D81, 123530
Arvanitaki *et al.* 2011, *Phys. Rev.* D83, 044026
Arvanitaki *et al.* 2015, *Phys. Rev.* D91, 084011
Arvanitaki *et al.* 2016, arXiv:1604.03958
Press & Teukolsky 1972, *Nature*, 238, 211-212
Y. B. Zeldovich 1971, *JETP Lett.*, 14, 180

New Frontiers in Black Hole Astrophysics
Proceedings IAU Symposium No. 324, 2016
Andreja Gomboc, ed.

doi:10.1017/S1743921316012941

Quantum electromagnetic phenomena far from small evaporating black holes

Slava Emelyanov

Institute of Theoretical Physics, Karlsruhe Institute of Technology (KIT),
76128 Karlsruhe, Germany
email: viacheslav.emelyanov@kit.edu

Abstract. One might expect far away from physical black holes that quantum field quantisation performed in Minkowski space is a good approximation. Indeed, all experimental tests in particle colliders reveal no deviations so far. Nevertheless, the black holes should leave certain imprints of their presence in quantum processes. In this paper, we shall discuss several local imprints of small, primordial evaporating black holes in quantum electrodynamics in the weak gravity regime. Physically this can be interpreted as being macroscopic manifestations of vacuum fluctuations.

Keywords. Black hole physics, gravitation, elementary particles

1. Introduction

In quantum field theory observables are self-adjoint, local operators. These local operators could be of various nature. In what follows we consider observables corresponding to the electromagnetic field. Specifically, these could be an electromagnetic wave propagating from an emitter to a receiver and an electrostatic field sourced by a point-like charge, both being studied in a box of about one-cubic-meter size.

It is a consequence of the requirements needed to formulate a physically acceptable quantum field theory (e.g., Haag 1996) that quantum operators localised in a certain space-time region probe the vacuum as if it is a non-empty state. The physical interpretation of this effect can be readily given in terms of the quantum or vacuum fluctuations. These fluctuations or quantum noise can show up at microscopic as well as macroscopic scales. The formation of inhomogeneities like galaxies and clusters of galaxies in our universe is their most remarkable macroscopic manifestation (Mukhanov & Chibisov 1981).

According to the Tomita-Takesaki theorem (Haag 1996), the quantum vacuum satisfies a Kubo-Martin-Schwinger aka thermal condition with respect to any localised set of the field operators and a certain one-parametric group of its automorphism. This group could be of a geometrical as well as non-geometric origin, i.e. generated by a Killing or non-Killing vector, respectively. The lesson of this is that the quantum vacuum can respond as a many-particle state which is mainly a property of quantum field operators, rather than of the vacuum state.

2. Quantum vacuum as plasma-like medium

Vacuum as dielectric. The quantum vacuum possesses a *dielectric*-like property in the absence of any external field with one of its observable/measurable manifestation being a part of the Lamb shift. The "dipoles" of the vacuum are virtual e^+e^- pairs. This effect follows from taking into account the one-loop correction to the photon self-energy which leads to the running electric charge (e.g., Peskin & Schroeder 1995).

Specifically, the full photon propagator in quantum field theory is diagrammatically

given by

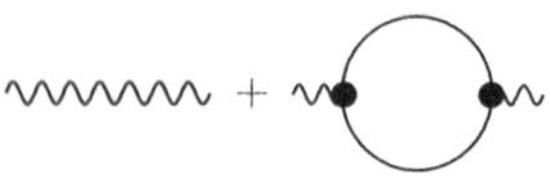

+ higher-loop diagrams. (2.1)

The loop diagrams lead to the modification of the classical field equation. This becomes in momentum space

$$\left(k^2\eta^{\mu\nu} - k^\mu k^\nu - \Pi^{\mu\nu}(k)\right)a_\nu(k) \quad = \quad 0\,, \tag{2.2}$$

where $\Pi^{\mu\nu}(k)$ is a polarisation tensor vanishing in the limit of the vanishing fine structure constant α. The gauge symmetry entails $\Pi^{\mu\nu}(k) = (k^2\eta^{\mu\nu} - k^\mu k^\nu)\Pi(k^2)$. Thus, one obtains $k^2(1-\Pi(k^2))a^\mu = 0$ for the photon ($k^\mu a_\mu = 0$). The extra factor $1-\Pi(k^2)$ leads to the charge renormalisation and, hence, to the modification of the classical Coulomb interaction at scales of the order of the Compton length of the electron. The resulting effect is the screening of the charge by the virtual e^+e^- pairs like in a dielectric.

Vacuum as magnetised plasma. Probing the quantum vacuum by studying how an electromagnetic wave propagates through a spacetime region with a super-strong magnetic field ($B \gg (\pi/\alpha)B_c$, where $B_c = m_e^2/e$), the vacuum shows up a property inherent to a magnetised plasma, i.e. the plasma held at a static external magnetic field (Melrose & Stoneham 1976 and Dittrich & Gies 2000).

Due to a collective response of the plasma particles being tied to the external magnetic field lines, the propagation of an electromagnetic wave is significantly dependent on its polarisation (e.g., Bittencourt 2004). Specifically, an electromagnetic wave polarised in the direction coplanar to the plane defined by its spreading direction $\mathbf{k}$ and the external magnetic field $\mathbf{B}$ propagates along the magnetic field lines. This is known as the Alfvén wave.

The external magnetic field B in the vacuum entails the modification of the vacuum polarisation tensor $\Pi^{\mu\nu}(k, B)$ which becomes non-diagonal. This is due to $B \neq 0$ which enters the field equations and, hence, modifies the electron and photon propagator. At one-loop approximation it turns out that a light wave polarised in the direction perpendicular to the plane defined by $\mathbf{k}$ and $\mathbf{B}$ is oblivious to the external magnetic field, i.e. $\omega^2 = |\mathbf{k}|^2$. This resembles a behaviour of the fast magneto-acoustic wave in case of the Alfvén velocity $v_A = |\mathbf{B}|/\sqrt{4\pi\rho} \gg 1$, where ρ is a plasma-like density of the vacuum. If the polarisation vector a_μ is coplanar to that plane, then, as it was found by Melrose & Stoneham (1976), $\omega^2 = |\mathbf{k}|^2\cos^2\theta$, where θ is an angle between $\mathbf{k}$ and $\mathbf{B}$. This corresponds to the Alfvén wave in case the condition $v_A \gg 1$ is fulfilled.

Vacuum as neutral plasma. The quantum vacuum responds non-trivially in the background of evaporating black holes, namely it acquires plasma-like properties (Emelyanov 2016a).

To demonstrate the effect, one needs to compute the vacuum polarisation tensor. It must depend *a priori* on a mass M of the hole and a distance R to its centre. The former dependence arises since when $M = 0$ the problem reduces to that in flat space-time. The latter one is, however, due to our expectation that *local* physics in the asymptotically flat region ($R \to \infty$) should reduce to the familiar Minkowskian one.

A part of the tensor $\Pi^{\mu\nu}(k, M, R)$ was computed at one-loop level by Drummond & Hathrell (1980). However, it gets an extra term due to the quantum effect entailing the black-hole evaporation. Specifically for the semi-classical theory be reliable, a field operator in the background of any black hole should be expanded over modes in which the vacuum expectation value of stress tensor does not diverge on the (future) horizon.

This yields the electron propagator in momentum space of the following form

$$S(k,k') \approx (\not{k} + m_e)\left(\frac{i}{k^2 - m_e^2 + i\varepsilon} - 2\pi\left(\frac{27r_H^2}{16R^2}\right)\frac{\delta(k^2 - m_e^2)}{e^{k_0/T_H} + 1}\right)\delta(k - k')\,, \quad (2.3)$$

where $T_H = M_{\rm Pl}^2/8\pi M$ is Hawking's parameter, $M_{\rm Pl} = (\hbar/G)^{\frac{1}{2}}$ the Planck mass, $r_H \equiv 2MG$ a size of the black-hole horizon. We have omitted in (2.3) terms of sub-leading order vanishing as $1/R^3$ at the spatial infinity. The propagator (2.3) is valid only for $|r - r'| \ll R$ and $r_H \ll R$ (Emelyanov 2016a & 2016b). We should mention that it is legitimate in this regime to employ flat coordinates as space-time is *locally* Minkowski.

The geometrical part of $\Pi^{\mu\nu}(k, M, R)$ drops out at least as fast as $1/R^3$ at spatial infinity, whereas its vacuum part vanishes as $1/R^2$ as can be seen in (2.3). Therefore, the main contribution to $\Pi^{\mu\nu}(k, M, R)$ is due to the second term in (2.3).

The modification of the electron propagator has a thermal-like structure. In case of a large black hole ($m_e \gg T_H$) the one-loop self-energy of the photon is exponentially suppressed by the Boltzmann factor $\exp(-m_e/T_H)$. Therefore, we consider a small black hole with $T_H \gg m_e$ in the following. This corresponds to the black-hole mass $M \ll 10^{16}$ g and, thus, justifies the hard thermal loop approximation (e.g., Le Bellac 1996) in which one can omit the electron mass m_e as being negligible with respect to the temperature parameter T_H.

At one-loop level the photon propagator acquires two poles. One of these corresponds to a transverse mode, the photon, with $\omega^2 = |\mathbf{k}|^2 + \pi_T(\omega, |\mathbf{k}|)$, another one to a longitudinal mode with $\omega^2 = |\mathbf{k}|^2 + \pi_L(\omega, |\mathbf{k}|)$ which is known as a plasmon. The functions $\pi_T(\omega, |\mathbf{k}|)$ and $\pi_L(\omega, |\mathbf{k}|)$ are the same as found by Weldon (1982), but with the temperature

$$T_L \approx \frac{3\sqrt{3}}{16\pi}M_{\rm Pl}\frac{L_{\rm Pl}}{R}\,, \quad (2.4)$$

where $L_{\rm Pl} = (\hbar G)^{\frac{1}{2}}$ is the Planck length (Emelyanov 2016a). In the leading-order approximation T_L does not directly depend on the mass M of the black hole. The quantum vacuum unlike a physical hot plasma described by α and T (but not, e.g., by a particle density n) is effectively also characterised by T_L which unlike T_H plays a major role in local physics.

The wavelength of the mode has to be much smaller than the distance to the black hole. Moreover, it must be much smaller than a detector size l_D. This implies $m_e^2/eT_H \gg |\mathbf{k}|$ and $|\mathbf{k}| \gg 1/l_D \gg 10^2\omega_{\rm p}$, where $\omega_{\rm p} \equiv \frac{1}{3}eT_L$ is a plasma-like frequency (Emelyanov 2016a) and the upper threshold is to suppress higher loop corrections (Emelyanov 2016c & 2016a). In this case, plasmons are not propagating modes (Le Bellac 1996), whereas photons acquire an effective mass

$$m_\gamma \approx \frac{1}{\sqrt{6}}eT_L \approx 2.5\times10^{-9}\left(\frac{1\,\rm m}{R}\right)\rm eV\,. \quad (2.5)$$

In the physical plasma, plasmons are quasi-particles, i.e. collective excitations mediated by plasma particles. In our case, the vacuum is full of *virtual* particles, rather than real ones. Thus, the result of having no plasmons seems to be physically sensible.

A static electric field $\mathbf{E}$ sourced by a point-like charge q is screened in the physical plasma - a phenomenon known as the Debye screening. This effect is due to a collective response of the plasma particles on putting the charge q in the system. The linear response theory in our case yields a similar result, namely $\mathbf{E} = -\nabla\varphi$ with the electromagnetic potential $\varphi = (q/r)\,e^{-r/r_L}$, where

$$r_L = \left(\sqrt{2}m_\gamma\right)^{-1} \approx 3.5\times10^2 R\,. \quad (2.6)$$

Thus, the Debye-like length r_L is much bigger than R. It means the screening is a tiny effect, because $l_D \ll R$ should hold for our approximation be reliable (at distances $|r - r'| \gtrsim R$ the propagator (2.3) is invalid). With the technology of Williams *et al.* (1971) this can in principle be tested if such small black holes do exist in nature and one of them is within $R \lesssim 250\,\mathrm{km}$ from the detector. For a practical purpose, however, the probability of having such a black hole per cubic centimeter should be estimated to decide whether it is realistic to probe this effect during a reasonable amount of time.

If the black-hole evaporation lasts till its complete disappearance, then we should set a lower bound on the black-hole mass M to guarantee the quasi-equilibrium approximation we have exploited above. It would be of order of 10^{10} g corresponding to the black-hole lifetime of one day. The *initial* mass of this primordial hole (i.e. 10^{10} years ago) should be in the range $3\times10^{14}\,\mathrm{g} \lesssim M_0 \ll 10^{16}\,\mathrm{g}$, which corresponds to $10^{10}\,\mathrm{g} \lesssim M \ll 10^{16}\,\mathrm{g}$ today.

3. Discussions

We have considered three examples when the quantum vacuum can be envisaged as possessing medium-like properties. The *virtual* particles were understood as constituents of the vacuum like the *real* particles of ordinary media - the point of view which seems to be widely-accepted by the researchers.

As with any analogy, this viewpoint cannot be taken literally. Indeed, the real particles can transfer energy, whereas the virtual particles seem not be capable of doing that, because these are just fluctuations, although quantum in nature.

To clarify this aspect of the quantum fluctuations, one may put an extra conducting plate between the Casimir plates. It may be taken to be freely moving and at the distance $l_C/2$ from each plates at $t < 0$. Its motion is however fixed by $v(t) = v\,\theta(t)\theta(t - \tau) \ll 1$. It follows from energy conservation that a part of the vacuum energy in both cavities will by absorbed by the middle plate during its motion. In other words, the vacuum energy is partially dissipated in the middle plate and partially redistributed between the cavities. The dissipative part could result in heating of the middle plate with an increase in temperature of $\Delta T \propto V(v\tau)^2/l_C^6$ in case $v\tau \ll l_C$, where V is its volume. This *gedankenexperiment* might imply the virtual particles are capable of transferring energy.

Acknowledgements

I am grateful to the organisers of IAU Symposium 324 for the financial support. It is a pleasure to thank the anonymous referee for his physically relevant comments/questions.

References

Bittencourt, J. A., 2005, *Fundamentals of Plasma Physics*, Springer-Verlag

Dittrich, W. & Gies, H., 2000, *Probing the quantum vacuum. Perturbative effective action approach in Quantum Electrodynamics and its application*, Springer-Verlag

Drummond, I. T. & Hathrell, S. J., 1980, *Phys. Rev. D*, 22, 343

Emelyanov, S., 2016a, arXiv:1603.01148

Emelyanov, S., 2016b, arXiv:1608.05318

Emelyanov, S., 2016c, *Nucl. Phys. B*, 913, 318

Haag, R., 1996, *Local Quantum Physics. Fields, Particles, Algebras*, Springer-Verlag

Le Bellac, M., 1996, *Thermal field theory*, Cambridge University Press

Melrose, D. B. & Stoneham, R. J., 1976, *Nuovo Cim. A*, 32, 435

Mukhanov, V. F. & Chibisov, G. V., 1981, *JETP Lett.*, 33, 532

Peskin, M. E. & Schroeder, D. V., 1995, *An Introduction to Quantum Field Theory*, Addison-Wesley Publishing

Weldon, H. A., 1982, *Phys. Rev. D*, 26, 1394

Williams, E. R., Faller, J. E., & Hill, H. A., 1971, *Phys. Rev. Lett.* 26, 721

New Frontiers in Black Hole Astrophysics
Proceedings IAU Symposium No. 324, 2016
Andreja Gomboc, ed.

doi:10.1017/S1743921316012515

Spectroscopy of candidate electromagnetic counterparts to gravitational wave sources

Iain A. Steele, Chris M. Copperwheat and Andrzej S. Piascik

Astrophysics Research Institute, Liverpool John Moores University, L3 5RF, UK

Abstract. A programme of worldwide, multi-wavelength electromagnetic follow-up of sources detected by gravitational wave detectors is in place. Following the discovery of GW150914 and GW151226, wide field imaging of their sky localisations identified a number of candidate optical counterparts which were then spectrally classified. The majority of candidates were found to be supernovae at redshift ranges similar to the GW events and were thereby ruled out as a genuine counterpart. Other candidates ruled out include AGN and Solar System objects. Given the GW sources were black hole binary mergers, the lack of an identified electromagnetic counterpart is not surprising. However the observations show that it is possible to organise and execute a campaign that can eliminate the majority of potential counterparts. Finally we note the existence of a "classification gap" with a significant fraction of candidates going unclassified.

Keywords. gravitational waves

1. Introduction

As part of the effort to discover and characterise astronomical gravitational wave (GW) sources, a worldwide programme of electromagnetic (EM) follow-up has been established (Abbot *et al.* 2016a). The programme is organised under the auspices of a series of Memoranda of Understanding (MOU) with over 70 groups who have access to observing resources that can participate in the follow-up. The MOU maintains confidentiality until the discoveries are announced. In the first aLIGO/Virgo observing run (O1) a number of candidate optical counterparts were identified by various wide field optical imaging facilities. In this paper we mainly discuss the work carried out by the 2.0 metre robotic Liverpool Telescope (LT) to spectroscopically follow up a number of those candidates. We also draw some broad conclusions about a potential spectroscopic "classification gap". More details of the work presented here can be found in Copperwheat *et al.* (2016).

2. Expected Optical Signatures

It is a reasonable expectation that binary mergers involving one or more neutron stars (NS) should show a transient EM signature due to energetic outflows. An EM signature is less likely for the merger of two black holes (BH). For NS+NS and NS+BH mergers we can anticipate a number of scenarios based on the assumption that the energetic outflow will have a jet configuration:

• If the observer is within the jet opening angle, a "prompt" spectral signature similar to a short GRB (e.g. Berger 2014) might be expected.

• If the observer is outside the jet opening angle, then (infra-)red kilo-nova emission from radioactive decay of heavy elements synthesised in the merger ejecta is predicted (e.g. Li & Paczyński 1998). Such emission is likely to be delayed with respect to the GW signal.

Table 1. Classifications of transients candidates followed by the LT in response to event GW151226. Where we obtain a secure supernova classification we provide a redshift, the time t since peak, and the percentage of matching templates in the SNID database which are consistent with the spectrum.

Candidate ID	Comments
iPTF-15fed	No transient detected to limiting magnitude of $R{\sim}19.1$
iPTF-15fel	Supernova Type Ia, $z = 0.038$, $t = +40$ d, 97.7 per cent template fit
iPTF-15fev	Supernova Type Ia, $z = 0.023$, $t = +50$ d, 94.7 per cent template fit
iPTF-15ffh	Possible supernova Type Ia, $z = 0.061$, $t = +15$d
iPTF-15ffi	Supernova Type Ia, $z = 0.085$, $t = +3$ d, 89.1 per cent template fit
iPTF-15ffk	Supernova Type Ia, $z = 0.102$, $t = +5$ d
iPTF-15ffm	Supernova Type Ia, $z = 0.094$, $t = +36$ d
iPTF-15ffz	Emission lines consistent with AGN at $z{\sim}0.07$
iPTF-15fgy	Supernova Type Ia, $z = 0.076$, $t = +20$ d, 84.7 per cent template fit
iPTF-15fhd	Possible supernova Type Ia, $z = 0.091$, $t = +11$ d
iPTF-15fhl	Possible supernova Type Ib, $z = 0.043$, $t = +18$ d
iPTF-15fhp	Possible supernova Type Ic, $z = 0.129$, $t = +1$ d
iPTF-15fhq	Narrow emission lines, consistent with AGN at $z = 0.043$
iPTF-15fib	Slow moving asteroid
LSQ15bvw	No transient detected to limiting magnitude R$\sim$19.5
MASTER OTJ020906	No transient detected to limiting magnitude R$\sim$20
UGC 1410 transient	No transient detected. ID'd as minor planet 2 606 Odessa

• If the jet Lorentz factor is low, we may find a "failed GRB" orphan afterglow which lacks high energy emission but still has an X-ray/optical/radio signature (Lamb & Kobayashi 2016).

3. Methodology

The aLIGO/Virgo consortium have put in place a number of mechanisms for communication of information about GW events and the subsequent follow-up activity between the MOU partners. These include GCN Notices (machine readable information packets that can be rapidly distributed), GCN Circulars (human readable information that is distributed by email), and a machine and human read/write database system (GraceDB) that distributes information on burst times, localisations (sky maps) and false alarm rates. All MOU partners are able to use the GCN and GraceDB systems to provide information regarding their follow-up activities (e.g. search footprints and candidate counterparts) to other MOU partners.

The natural follow-up sequence for the search is to use wide field facilities to identify potential EM counterparts in the area covered by the GW localisation sky map followed by more detailed observations with narrower field instruments. Examples of some of the optical wide field instruments used during O1 include iPTF, PanSTARRS, SkyMapper, VISTA, MASTER, TOROS, TAROR, VST, DECam and Pi of the Sky. At higher energies facilities included Fermi, INTEGRAL, SWIFT. In the radio regime MWA, ASKAP and LOFAR were used. Follow-up of counterparts was carried out by facilities including Keck, PESSTO, UH2.2 and LT (optical spectroscopy) and VLA (radio).

4. Results

During O1 three GW triggers were distributed to the MOU partners:

• GW150914 (Abbott *et al.* 2016b). This comprised a $36^{+5}_{-4} + 29^{+4}_{-4}\mathrm{M}_{\odot}$ black hole

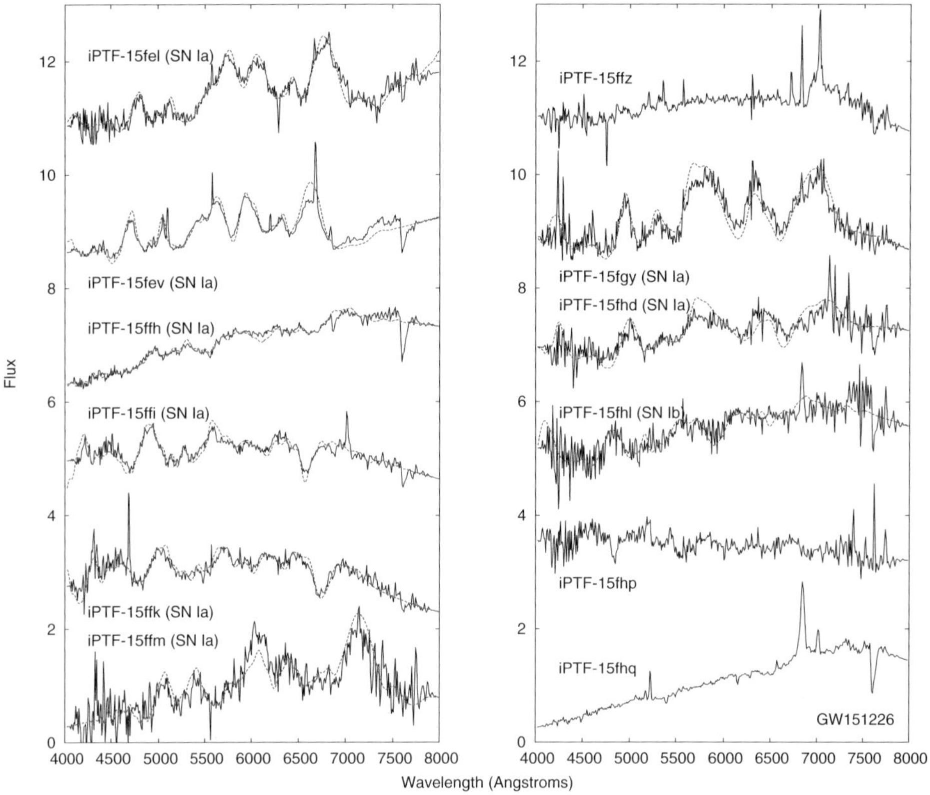

Figure 1. LT spectra obtained during the follow-up of GW151226. We omit candidates for which our observations showed no evidence of a transient. For objects where a supernova identification is obtained we overplot the best template fit.

merger at redshift $z \sim 0.09$. It was detected on 2015 Sept 14 (just before the official start of O1) and an alert issued to the MOU partners on 2015 Sept 16.

• G194575 (LIGO Scientific Collaboration and Virgo 2015). This alert was issued on 2015 Oct 22 but retracted on 2015 Nov 11 when the false alarm rate was recomputed and the event was determined not likely to be a real event.

• GW151226 (Abbott *et al.* 2016c). This comprised a $14.2^{+8.3}_{-3.7} + 7.5^{+2.3}_{-2.3} M_{\odot}$ black hole merger also at redshift $z \sim 0.09$. It was detected on 2015 Dec 26 and an alert issued on 2015 Dec 27.

For all three of these triggers extensive follow-up was carried out. A summary of the GW150914 follow-up campaign is presented in Abbott *et al.* (2016d). Figure 1 shows a selection of the spectra obtained by the LT using the low resolution ($R \sim 350$) SPRAT spectrograph (Piascik *et al.* 2014) in the follow-up of GW151226. A summary of all of the LT classifications for that event is presented in Table 1. Examining the table shows the majority of candidate counterparts detected are supernovae in the redshift range $z \sim 0.02 - 0.13$. Some Solar System sources and AGN are also identified, as well as a number of objects where the transient source had faded back into the galaxy background before a spectrum could be obtained. None of the candidates could be associated with GW151226.

5. Conclusions

Given that the O1 GW sources were both BH+BH systems (where we do not expect an EM signature) the overall results of the follow-up and classification programme are encouraging. We have shown that a 2-metre class telescope with a high-throughput, low resolution spectrograph can eliminate many candidate counterparts at redshifts similar to the GW sources. In future aLIGO/Virgo observing runs it is proposed that a distance estimate ("3d sky localisation" - Singer *et al.* 2016) and "EM-BRIGHT" flag (indicating the possible presence of a neutron star) will be distributed with the alerts. Localisations will also improve as more GW stations come on-line. All of this will help with targeted follow-up of candidates associated with potential host galaxies in the correct redshift range.

For GW151226 a total of 77 candidate optical counterparts were announced via GCN. We note that (over all of the facilities involved):

- 37 of these received a firm spectral classification,
- a further 18 had a more tentative classification based on photometric light curves,
- there were 3 cases where the transient had faded into the host galaxy before spectroscopy could be attempted.

It follows that 19 candidates were not followed up. It is therefore clear that there is a significant danger of a "classification gap" opening up, where potential counterparts will be discovered at a significantly faster rate than can be spectroscopically followed up. While localisation error boxes are anticipated to reduce in size (from the current hundreds of square degrees to tens of square degrees) as more GW stations come on line, it is also anticipated that event numbers will increase by an order of magnitude over the next few years as detector sensitivities improve. The gap is therefore likely to remain a problem. In addition we note that new sources of transients such as LSST (predicted to discover $\sim 10^6$ transients/night – LSST Science collaboration *et al.* 2009) will also be in competition for spectroscopic follow-up time.

To reduce the classification gap we propose the community must (a) start moving to more automated and efficient methods of triggering spectroscopic follow-up of candidates using technologies such as RTML (Hessman 2006) and VOEvent (Williams & Seaman 2006) and (b) consider the construction of optimized spectroscopic follow-up facilities with large apertures and fast slew speeds (e.g. Liverpool Telescope 2 - Copperwheat *et al.* 2015).

References

Abbott, B. P., Abbott, R., Abbott, T. D., *et al.* 2016a, *Living Reviews in Relativity*, 19, 1
Abbott, B. P., Abbott, R., Abbott, T. D., *et al.* 2016b, *Phys. Rev. Lett.*, 116, 061102
Abbott, B. P., Abbott, R., Abbott, T. D., *et al.* 2016c, *Phys. Rev. Lett.*, 997
Abbott, B. P., Abbott, R., Abbott, T. D., *et al.* 2016d, *ApJ*, 826, L13
Berger, E. 2014, *ARAA*, 52, 43
Copperwheat, C. M., Steele, I. A., Piascik, A. S., *et al.* 2016, *MNRAS*, 462, 3528
Copperwheat, C. M., Steele, I. A., Barnsley, R. M., *et al.* 2015, *Experimental Astron.*, 39, 119
Hessman, F. V., 2006, *AN*, 327, 751
Lamb, G. P & Kobayashi, S. 2016, *ApJ*, 829, 112
Li, L.-X. & Paczyński, B. 1998, *ApJ*, 507, L59
LIGO Scientific Collaboration and Virgo 2015a, *GRB Coordinates Network*, 18442
LSST Science Collaboration, *et al.*, 2009, arXiv:0912.0201
Piascik, A. S., Steele, I. A., Bates, S. D., *et al.*, 2014, *Proc. SPIE*, 9147, 91478H
Singer, L. P., Chen, H.-Y., Holz, D. E., *et al.* 2016, *ApJ*, 829, L15
Williams, R. D. & Seaman, R., 2006, *ASPC*, 351, 637

New Frontiers in Black Hole Astrophysics
Proceedings IAU Symposium No. 324, 2016
Andreja Gomboc, ed.

doi:10.1017/S1743921317001430

MAGIC electromagnetic follow-up of gravitational wave alerts

Barbara De Lotto[1],
Stefano Ansoldi[1,6], Angelo Antonelli[2,4], Alessio Berti[3],
Alessandro Carosi[2,4], Francesco Longo[3], Antonio Stamerra[5]
on behalf of the MAGIC Collaboration[7]

[1]University of Udine & INFN Trieste, Italy
[2]INAF National Institute for Astrophysics, Rome, Italy
[3]University & INFN Trieste, Italy
[4]ASI Science Data Center, I-00133 Rome, Italy
[5]INAF Osservatorio Astrofisico di Torino & Scuola Normale Superiore, Pisa, Italy
[6]Department of Physics, Kyoto University, Kyoto, 606-8502 Japan
[7]http://wwwmagic.mppmu.mpg.de

Abstract. The year 2015 witnessed the first direct observations of a transient gravitational-wave (GW) signal from binary black hole mergers by the Advanced Laser Interferometer Gravitational-wave Observatory (aLIGO) Collaboration with the Virgo Collaboration. The MAGIC two 17m diameter Cherenkov telescopes system joined since 2014 the vast collaboration of electromagnetic facilities for follow-up of gravitational wave alerts. During the 2015 LIGO-Virgo science run we set up the procedure for GW alerts follow-up and took data following the last GW alert. MAGIC results on the data analysis and prospects for the forthcoming run are presented.

Keywords. electromagnetic follow-up, gravitational wave alert, cherenkov telescope

1. Introduction

From September 2015 to January 2016 the first science run (O1) of the recently upgraded Advanced Laser Interferometer Gravitational-wave Observatory (Aasi *et al.*2015) and Advanced Virgo (Acernese *et al.*2015) experiments took place: the era of GW astronomy was opened with the first ever direct detections of gravitational waves (Abbot *et al.*2016a), (Abbot *et al.*2016b).

One of the most likely sources of gravitational waves detectable by the second generation of GW detectors is the coalescence of a compact binary, i.e. one containing neutron stars (NS) or stellar-mass black holes (BH) (Abadie *et al.*2012).

Electromagnetic signatures (EM) are expected, due to energetic matter outflows at different timescales and wavelengths. Together with the gravitational wave signal, the detection of an EM counterpart allows to build a complete picture of the physics of the event. While GWs are ideal probes for the innermost region of the newly formed black hole (Nakano *et al.*2016), the detection of EM radiation would allow to sharpen our understanding of the progenitor and post-merger environments.
An EM follow-up program has been set up between the LIGO and Virgo scientifc Collaboration (LVC) and a broad astronomy community with access to ground and space-based facilities. As one of these facilities, the Major Atmospheric Gamma-ray Imaging Cherenkov telescopes (MAGIC) system (Aleksić *et al.*2016a) participates in a shared bulletin board to announce, coordinate, and visualize the footprints and wavelength coverage of follow-up observations (GW-EM).

2. Motivations

Observing the EM counterparts will play a key role in localizing the GW source, in establishing the properties of the environment of the binary system, and in ascertaining the association to cosmic phenomena, such as gamma-ray bursts (GRBs). The EM and GW signals are essentially complementary in understanding the progenitor.
Even though BH mergers, such as the ones observed by LIGO, are not expected to have EM counterparts, it is still important to promptly react with follow-up observations. The Fermi GBM detector identified a weak gamma-ray transient 0.4 s after the GW signal with consistent sky localization (Connaughton *et al.*2016). Whether the association between GW150914 and the Fermi GBM candidate is real or not (Savchenko *et al.*2016), it opened a debate about the possibility that BH-BH mergers may also produce EM counterparts (Loeb 2016, Perna *et al.* 2016, Yamazaki *et al.* 2016, Zhang2016, see also Kamble & Kaplan 2013).
Of special interest are binaries that consist of at least one neutron star: NS-NS or NS-BH are ideal systems for a rich transient EM emission (Bartos *et al.* 2013, Berger 2014, Metzger & Berger 2012, Fernandez & Metzger 2015, Cowperthwaite & Berger 2016) as well as for other messengers, such as cosmic rays or neutrinos (Murase *et al.* 2013). The detection of an EM counterpart could confirm NS-NS merger as the progenitor of a short GRB (Veres & Meszaros 2014, Takami *et al.* 2014).

3. MAGIC observations

3.1. *The MAGIC telescopes system*

MAGIC is an array of two imaging atmospheric Cherenkov telescopes (IACTs) designed for the detection of gamma-rays in the energy band between few tens of GeV and few tens of TeV. It is located in the Canary island of La Palma (Spain) at 2250 m above the sea level. The two telescopes have a 17 m diameter reflective surface and fine pixelized cameras with a 3.5^o field of view (Aleksić *et al.*2016a). Currently the array has an energy threshold as low as 50 GeV for low zenith angle observations, and an integral sensitivity above 300 GeV of 0.6% of the Crab nebula flux in 50 hours of observation. The energy resolution is $15 - 17$ % at 1 TeV (Aleksić *et al.* 2016b).

3.2. *MAGIC follow-up of GW alerts*

The MAGIC response to external triggers relies on a dedicated GRB/transient alert monitor that manages the communication within the Gamma-ray Coordinates Network (GCN) and the MAGIC central control. Before and during the O1 run, the alert system has been adapted to receive/validate GW alerts.

During the first observing run O1 the Advanced LIGO observatory announced 3 GW triggers to the EM follow-up partners. The first two did not trigger MAGIC observations. The first one (GW150914), detected at the end of the Advanced LIGO engineering run immediately prior to O1, is the first ever direct detection of GW (Abbot *et al.*2016a): it was mostly outside the MAGIC field of view. The second one (GW151012), later on determined by the LIGO-Virgo offline analysis not to be a real GW event, could not be observed by MAGIC because of bad weather conditions.

On 2015 December 26 at 03:38:53.648 UT the Advanced Laser Interferometer Gravitational wave Observatory detected a high significance candidate GW event designated GW151226 (Abbot *et al.*2016b); one day later provided spatial location information in the form of probability sky maps via a private GCN circular (GCN18728). The candidate was identified by an expanded low-latency pipeline configuration that is sensitive to

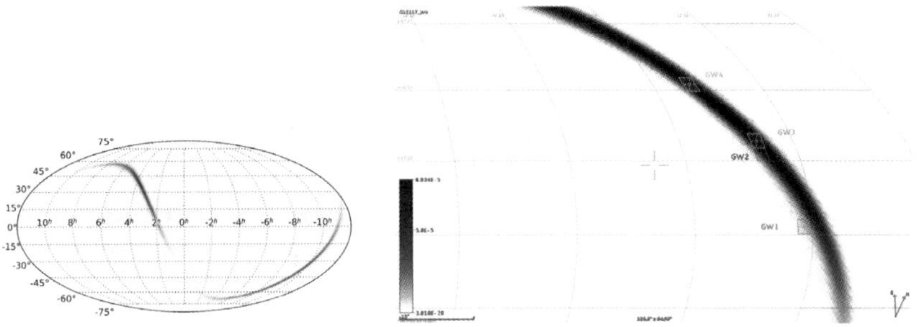

Figure 1. Left: The LIGO localization probability skymap of GW151226. Right: Zoom on the region with the four MAGIC pointing positions of GW151226.

Table 1. The four pointing positions followed up by MAGIC. Observations on target 3 and 4 have been performed under moderate-to-strong moon illuminations. This made necessary a dedicated analysis to account for the higher level of the night sky background. (a) V. Lipunov *et al.*, GCN Circular 18804 (2016). (b) E. Brocato *et al.*, GCN Circular 18734 (2016).

#	Target	RA	Dec (J2000)	Duration	Zenith angle
1	$PGC1200980(MASTER\ OT)(a)$	02 h 09m 05.800 s	$+1^o$ $38'$ 03.00"	48 min	$[27^o, 30^o]$
2	*strip from GW map*	02 h 38m 38.930 s	$+16^o$ $36'$ 59.27"	59 min	$[13^o, 24^o]$
3	*Field VST* (b)	02 h 38m 02.210 s	$+19^o$ $13'$ 12.00"	30 min	$[22^o, 30^o]$
4	*Field VST* (b)	03 h 18m 23.712 s	$+31^o$ $13'$ 12.00"	30 min	$[19^o, 27^o]$

stellar-mass NSNS, NSBH, and BBH mergers. Its false alarm rate (FAR), as determined by the online analysis, passed the stated alert threshold of $\sim$ 1/month, later refined to a FAR lower then one per hundred years (Abbot *et al.* 2016). The event probability sky map (Singer & Price 2016) was concentrated in two long, thin arcs on opposite directions in the sky (Fig. 1, left). The 50% credible region spans about 430 deg^2 and the 90% region about 1400 deg^2.

Four sky pointing positions (Fig. 1, right) for MAGIC observations were manually selected in the region showing maximum probability according to the visibility, observations of EM-partners, and overlap with existing catalogs. Observations started on 2015 December 28 at around 21 UT, covering the four $\sim$ 2.5 x 2.5 deg^2 target regions reported in Table 1. The observations were performed in the so-called wobble mode, where the pointing position is offset by 0.4^o from the camera center.

4. Results

The data analysis was performed using the standard MAGIC Analysis and Reconstruction Software package (MARS) adapted to stereoscopic observations (Aleksić *et al.* 2016b). From neither of the 4 pointed regions we detected significant emission above the instrument energy threshold. Furthermore, no VHE gamma-ray counterpart emission was detected within the MAGIC Field of View (FoV). Integral upper limits (ULs, 95% CL) above the energy threshold in the observed FoV are being estimated. The standard MARS routine calculates the flux UL from the pointing position (nominal position) of the telescopes. However the case of interest requires a different approach to determine meaningful flux ULs, since the source position in the MAGIC FoV is not known. This is a consequence of the fact that there is not a precise localization of the event and only a probability skymap is provided to search for an EM emission.

5. Discussion and outlook

The MAGIC telescopes system, with its $\sim$ 50 GeV energy threshold, is filling the gap between the space and ground-based gamma-ray instruments. Thanks to its fast re-positioning capabilities, it is well suited for fast transient follow-up observations, like Gamma Ray Bursts. MAGIC performed a follow-up observation of the third GW alarm released to the EM partners by the aLIGO team during the O1 science run. No very high energy gamma-ray counterpart was found. Integral upper limits for four selected regions of the GW probability map will be estimated.
The MAGIC field of view is $\sim 3.5^o$. For follow-up observations of GW alerts, a more precise localization of the source position would be advantageous. The joining of Advanced Virgo for the forthcoming science run O2, expected to take place in Spring 2017, and of Kagra (Somiya *et al.* in the future, will greatly improve the localization capabilities of the interferometers (Aasi *et al.* 2016). Information provided by other high-energy space based wide-field instruments, such as Fermi-GBM & LAT, would also improve the follow-up capabilities; a more strict collaboration between the EM follow-up community is then desirable.

Future EM follow-ups of GW sources will shed light on the presence or absence of firm EM counterparts and astrophysical processes that may trigger EM emission from these systems. MAGIC is ready to contribute to the quest.

References

Aasi, J., *et al.* 2015, *Classical and Quantum Gravity*, 32, 074001
Aasi, J., *et al.* , 2016, *Living Rev.Rel.*, 19, 1
Abadie, J., *et al.* 2012, *Phys. Rev. D*, 85, 082002
Abbot, B. P., *et al.* 2016a, *Phys. Rev. Lett.*, 116, 061102
Abbot, B. P. et al. 2016b, *Phys. Rev. Lett.*, 116, 241103
Abbot B. P. et al. 2016 *Phys. Rev. Lett.*, 116, 131103
Acernese, F., *et al.* 2015, *Classical and Quantum Gravity*, 32, 024001
Aleksić J. et al. 2016a, *Astroparticle Physics* 72, 61-75
Aleksić J. et al. 2016a, *Astroparticle Physics* 72, 76-94
Bartos I. et al., 2013, *Classical and Quantum Gravity*, 32, 123001
Berger E. , 2014, *A&A*, 52, 43
Connaughton, V., *et al.* 2016, *The Astrophysical Journal Letters*, 826, 1
Cowperthwaite, P. S. & Berger S. , 2016, *The Astrophysical Journal Letters*, 814, L25
Fernandez R. & Metzger B.D. , 2015 *preprint arXiv:1512.05435*
LIGO Scientific Collaboration and Virgo, 2015, *Gamma Ray Coordinate Network Circular* 18728
GW-EM: See program description at *http://www.ligo.org/scientists/GWEMalerts.php*
Kamble A. & Kaplan D. L. A. , 2013, *Int. Jour. of Mod. Phys. D*, 22, 1341011
Loeb A. 2016, *The Astrophysical Journal Letters*, 819, L21
Metzger B. D. , Berger E. , 2012, *The Astrophysical Journal*, 746, 48
Murase K. et al., 2013, *Phys. Rev. Lett.*, 111, 131102
Nakano, H., Nakamura T. & Tanaka T. 2016, *Progr. Theor. Exp. Phys.*, 031E02
Perna R. , Lazzati D. & Giacomazzo B. , 2016, *The Astrophysical Journal Letters*, 821, L18
& Rolke W. A. , Lopez, A. M. , Conrad J. , 2005, *Nucl. Instrum. Meth. A*, 551, 493
Savchenko V. et al., 2016, *The Astrophysival Journal Letters*, 820, L36
Singer L. & Price L. R. , 2016, *Phys. Rev. D*, 93, 024013
Somiya K. [KAGRA Collaboration], 2012, *Class. Quant. Grav.*, 29, 124007
Takami K. et al., 2014, *Phys. Rev. D*, 89, 063006
Yamazaki R. , Asano K. & Ohira Y. , 2016, *preprint arXiv:1602.05050*
Veres P. & Meszaros P. , 2014, *The Astrophysical Journal*, 787, 168
Zhang B. , 2016, *preprint arXiv:1602.04542*

New Frontiers in Black Hole Astrophysics
Proceedings IAU Symposium No. 324, 2016
Andreja Gomboc, ed.

doi:10.1017/S174392131700223X

On the gamma-ray burst – gravitational wave association in GW150914

Agnieszka Janiuk[1], **Szymon Charzynski**[2] **and Michal Bejger**[3]

[1]Center for Theoretical Physics, Polish Academy of Sciences,
Al. Lotnikow 32/46, 02-668 Warsaw, Poland
email: agnes@cft.edu.pl

[2]Chair of Mathematical Methods in Physics, University of Warsaw
ul. Pasteura 5, 02-093 Warszawa, Poland
email: szycha@fuw.edu.pl

[3]Copernicus Astronomical Center, Polish Academy of Sciences,
Bartycka 18, 00-716 Warsaw, Poland
email: bajger@camk.edu.pl

Abstract. Hyperaccreting disks around black holes are the engines that drive outflows and jets in gamma ray bursts (GRBs). The torus formed after the core collapse or a compact binary merger is composed of free nucleons, Helium, electron-positron pairs, and is cooled by neutrinos rather than photon emission. Hyperaccretion powers the ultra-relativistic jets, where the GRB prompt emission originates. The neutrons produced in the disk and also in the outflowing material are necessary for the production of heavier nuclei. We discuss here the observable consequences of nucleosynthesis and we also apply the scenario of hyperaccretion to the gravitational wave source, GW150914. Temporal coincidence reported by the Fermi satellite suggested that the black hole merger might be accompanied with a GRB. We propose that a collapsing massive star and a black hole in a close binary could lead to such event. Gravitational wave emission due to the merger of collapsed core and the companion black hole might then coincide with a weak GRB.

Keywords. Black hole physics, accretion disks, gamma ray bursts, gravitational waves, neutrinos

---

1. Introduction

Gamma Ray Bursts (GRBs), are extremely energetic, transient events, visible from the most distant parts of the Universe. When a newly born black hole forms, either via the core collapse supernova, or after the merger of two compact stars, a large amount of matter is accreted onto it within the timescale between tens of milliseconds, to thousands of seconds, with highly hyper-Eddington rates. In such conditions, the nuclear reactions inside the plasma lead to creation of neutrinos, mainly via the electron-positron capture on nucleons. The annihilation of neutrino-antineutrino pairs is a source of power to the relativistic jets launched along the axis of the disk-black hole system. This can be comparable to the power extracted at the cost of the black hole spin, with the disk mediated by the magnetic fields. Both neutrinos and magnetic fields in the accreting matter can drive the wind outflow from the disk surface, and act as a collimation mechanism for the relativistic jets (e.g., Popham *et al.* 1999, Di Matteo *et al.* 2002, Janiuk *et al.* 2007), Janiuk & Yuan 2010). More recently, accretion flows in GRBs were described by fully relativistic, magneto-hydrodynamic (MHD) computations (e.g., Barkov & Komissarov 2010, Barkov & Baushev 2011, Janiuk *et al.* 2013a).

In 2013, before the gravitational waves were detected, we have speculated, that in a certain type of progenitors, such as the high -mass X-ray binaries, the core collapse of a massive star might coincide with the companion black hole entering the star's envelope (Janiuk *et al.* 2013b). In this scenario, the collapsing core is spun up by the transfer of angular momentum from the companion, and the merger of a spinning black hole just after the collapse, with this companion black hole, produces the gravitational wave signal. The event can be accompanied by the gamma ray burst, if one, or two jets are launched at the cost of accretion onto a rotating black hole. The GRB may be observed prior, and/or after the black hole merger, depending on the black hole spin, hence the jet orientation with respect to the line of sight.

2. Black hole meger and the hyper-accretion flow for GW 150914

The first discovered source of gravitational waves, GW150914 event, was interpreted to be a merger of two black holes (BHs) of the masses of $36^{+5}_{-4}\ M_{\odot}$ and 29^{+4}_{-4}. The final BH mass was $62^{+4}_{-4}\ M_{\odot}$ and its spin was $0.67^{+0.05}_{-0.07}$ (Abbott *et al.* 2016). Probabilities that the angles between spins and the normal to the orbital plane are between 45° and 135° are about 0.8 for each component BH, while the spin magnitudes are smaller than 0.7 and 0.8 at 90% probability. Assumption of a strict co-alignment of spins with the orbital angular momentum results in an upper limit of 0.2 and 0.3 for the spins. The Fermi satellite detected a weak gamma ray signal coincident with the GW event (Connaughton *et al.* 2016).

Binary black hole merger simulations and constraints for the engine. We made several runs for BH mass and range of spins, constrained from the Advanced LIGO data. These simulations were performed in vacuum approximation, and the numerical package we used is `Einstein Toolkit`† It is a family of codes for use in relativistic astrophysics, based on finite difference computation on a gridded mesh (Loeffler *et al.* 2012), supported by a distributed model, combining core support of software, tools, and documentation. We use the source code which we configured, compiled and run in the computer cluster in the Warsaw University (ICM). The code is developed on the basis of the theoretical achievements of the last decade, which made possible to establish the field of numerical relativity (see also, e.g., Pretorius 2005, Campanelli *et al.* 2006).

In Figure 1, on the left, we plot the example result from the run with BH mass ratio of 0.82, and aligned spins of 0.2 and 0.3. The waveform is plotted using the Weyl scalar ψ_4. The final BH spin obtained in this simulation was 0.68. On the right, we show the 3-D visualisation of the torus accreting onto a single black hole, which presumably formed after the merger, and was composed from the remnants of the star's envelope.

GRB central engine. To model the accretion and the GRB engine, we use the code *HARM-2D* (High Accuracy Magnetohydrodynamics, see Gammie *et al.* (2003)), with the numerical EOS taken from Janiuk *et al.* (2007), and neutrino cooling implemented into MHD evolution. The black hole parameters were constrained by the post-merger conditions, and the torus parameters matched by the weak GRB properties. From our simulation, we estimate the neutrino and Blandford-Znajek luminosities available to power the GRB jet in this source (see Janiuk *et al.* 2017, and Janiuk 2016 for details).

The HARM code works in the stationary metric around black hole and integrates the total energy equation and updates the set of 'conserved' variables, i.e. comoving density, energy-momentum, and magnetic field, assuming a force-free approximation. Here, we first incorporate this detailed microphysics into the heart of the general relativistic

† `http://www.einsteintoolkit.org`

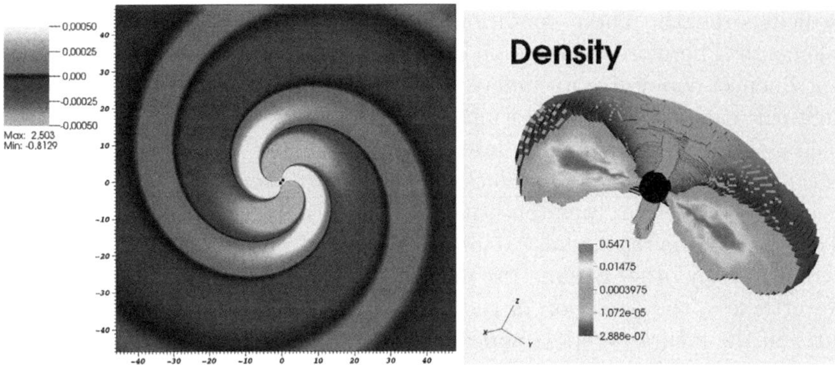

Figure 1. The spacetime curvature, depicted in the orbital plane, just before the black hole merger (left). The plotted quantity is the Weyl scalar ψ_4, which shows the outgoing gravitational radiation for the asymptotically flat spacetime. Distribution of density in the torus accreting onto a spinning black hole (right). The BH in this scenario is a product of the binary merger.

MHD scheme. We allow for the partial degeneracy of relativistic nucleons, electrons and positrons in the accreting plasma, and we compute the neutrino cooling rate due to the weak interaction processes, electron-positron pair annihilation, as well as bremsstrahlung and plasmon decay.

We found that for the BH mass of $M = 62M_{\odot}$, the spin $a = 0.7$, and disk mass $M_{\rm d} = 15M_{\odot}$, and also assuming the low efficiencies of conversion between neutrino annihilation, and conversion between the jet kinetic and radiative power, this scenario is able to meet quantitatively the limits put by the Fermi data.

Neutrino emission from GRB engine and efficiency of jet power supply. The maximum of the neutrino luminosity obtained in our simulations is reached about 0.4 seconds after an equilibrium torus, prescribed by our initial conditions, had formed. This may tentatively give the lower limit for the timescale when the jet appears after the BH merger. The Blandford-Znajek luminosity is of the order of $10^{50} - 10^{51}$ ergs/s, and only occasionally non-zero for low spins (the jet is powered by magnetic field only episodically). We find therefore, that the GRB luminosity inferred from our simulations can be reconciled with the observational upper limits, for moderate spins of the final BH ($a = 0.6 - 0.8$).

3. Heavy elements produced in the GRB engine

In the outer parts of the accretion flow in the GRB central engine, the heavy elements can be formed, and then the isotopes are prone to the radioactive decay. We compute the abundance pattern of these elements with the thermonuclear reaction network code, based on the profiles of density, temperature, and electron fraction in the accretion torus, obtained in our HARM simulations. We determined the abundances of heavy elements (above Helium, up to the mass number of ~ 130), which are synthesized. The decay of some of the unstable isotopes may in the future bring observable effects on the measured hard X-ray spectra of GRBs (Janiuk 2014) as well as the additional peaks in the blue and infrared bands a few days after the prompt GRB emission. The observational signatures of radioactive decay of some of the unstable isotopes can be, for instance, the emission lines seen in the X-rays (in principle, in the *NuSTAR* range), or in the faint emission in lower energy band continuum (i.e. the 'macronova' scenario). In addition, the gamma ray bursts contribute in this way to the galactic chemical evolution.

The physical conditions in the engine depend on the global parameters of the flow, such as the amount of matter and accretion rate, black hole mass, its spin, and the magnetic field strength. These conditions in turn will affect the total abundances of heavier elements. Therefore, a detection of such gamma ray burst signatures, apart from its prompt fluence, could give an independent constraint on the black hole parameters. These, in turn, can now be also estimated, if the black hole was born via a merger detectable by the gravitational wave interferometer.

Constraints from and for the gravitational wave astronomy. So far, for GW150914, the electromagnetic counterpart was tentatively reported only by the Fermi satellite, which detected a weak gamma ray signal. No infrared emission was reported, and also the signal was not confirmed by other gamma-ray missions for the case of this burst (Greiner *et al.* 2016). Nevertheless, the detection of the electromagnetic counterpart of a gravitational wave source in the future events is appealing.

In the case of the relatively small rate of the black hole rotation, such as estimated for GW150914, the hyperaccretion results in a rather small electromagnetic flux dragged through the black hole horizon and hence small power of the Blandford-Znajek process. The gravitational waves discovery may reveal a new population of weak gamma ray bursts originating from the low Lorentz factor jets. In this case, the prompt electromagnetic emission may be difficult to discover, but the observation of a radioactive decay of elements produced in the engine may be confronted with our simulation predictions. The black hole parameters, its spin and mass, determine the shape of the gravitational wave signal.

Acknowledgements

This work was supported in part by the grants no. DEC-2012/05/E/ST9/03914 and UMO-2014/14/M/ST9/00707 from the Polish National Science Center. We also acknowledge support from the Interdisciplinary Center for Mathematical Modeling of the Warsaw University, through the computational grant G53-5.

References

Abbott, B. P., *et al.*, 2016, *Phys. Rev. Lett.*, 116, 061102
Barkov, M. V. & Baushev, A. N., 2011, *New Astr.*, 16, 46
Barkov, M. V. & Komissarov, S. S., 2010, *MNRAS*, 401, 1644
Blandford, R. D., & Znajek, R. L., 1977, *MNRAS*, 179, 433
Campanelli, M., Lousto, C. O., Marronetti, P., & Zlochower, Y., 2006, *Phys. Rev. Lett.*, 96, 11
Connaughton, V., *et al.*, 2016, *ApJ Lett.*, 826, 6
Di Matteo, T., Perna, R., & Narayan, R., 2002, *ApJ*, 579, 706
Fishbone, L. G., & Moncrief, V., 1976, *ApJ*, 207, 962
Gammie, C. F., McKinney, J. C., & Tth, G., 2003, *ApJ*, 589, 444
Greiner, J, Burgess, J. M., Savchenko, V., & Yu, H.-F., 2016, *ApJ Lett.*, 827, 38
Janiuk, A., Bejger, M., Charzynski, S., & Sukova, P., 2017, *New Astr*, 51, 7
Janiuk, A., 2014, *A&A*, 568, 105
Janiuk, A., Mioduszewski, P., & Moscibrodzka, M., 2013, *ApJ*, 776, 105
Janiuk, A., Charzynski, S., & Bejger, M., 2013b, *A&A*, 560, 25
Janiuk, A., & Yuan, Y. F., 2010, *A&A*, 509, 55
Janiuk, A., Yuan, Y. F., Perna, R., & Di Matteo, T., 2007, *ApJ*, 664, 1011
Janiuk, A., 2016, submitted, arXiv:1609.09361
Löffler, F., *et al.* 2012, *Class. Q. Grav.*, 29, 11
Li, L. X., & Paczynski, B. P., 1998, *ApJ*, 507, 59
Mc Kinney, J. C., & Gammie, C. F. , 2004, *ApJ*, 611, 977
Noble, S. C., Gammie, C. F., McKinney, J. C., & Del Zanna, L., 2006, *ApJ*, 641, 626
Pretorius F., 2005, *Class. Q. Grav.*, 22, 425
Popham, R., Woosley, S. E. & Fryer, C. 1999, *ApJ*, 518, 356

New Frontiers in Black Hole Astrophysics
Proceedings IAU Symposium No. 324, 2016
Andreja Gomboc, ed.

doi:10.1017/S1743921317001454

Search for UHE neutrinos in coincidence with LIGO GW150914 event with the Pierre Auger Observatory

Lili Yang[a], **for the Pierre Auger Collaboration**[b]

[a]Laboratory for Astroparticle Physics, University of Nova Gorica, Slovenia
[b]Observatorio Pierre Auger, Av. San Martin Norte 304, 5613 Malargue, Argentina
(Full author list: http://www.auger.org/archive/authors_2016_06.html)
email: Lili.Yang@ung.si

Abstract. The first gravitational wave transient GW150914 was observed by Advanced LIGO on September 14th, 2015 at 09:50:45 Universal Time. In addition to follow-up electromagnetic observations, the detection of neutrinos will probe deeply and more on the nature of astrophysical sources, especially in the ultra-high energy regime. Neutrinos in the EeV energy range were searched in data collected at the surface detector of the Pierre Auger Observatory within ± 500 s and 1 day after the GW150914 event. No neutrino candidates were found. Based on this non-observation, we derive the first and only neutrino fluence upper limit at EeV energies for this event at 90% CL, and report constraints on existence of accretion disk around mergers.

Keywords. gravitational waves, neutrinos, black hole physics

1. Introduction

The LIGO collaboration recently announced the first detection of gravitational wave event GW150914, occurring on September 14th, 2015 (Abbott *et al.* 2016). This event was produced by the inspiral and merger of a pair of black holes lying at a luminosity distance $D = 410^{+160}_{-180}$ Mpc, according to the waveform analysis. The merger can result in a black hole, relic magnetic fields and debris remaining from the formation of the black holes, providing the environment to accelerate cosmic rays to ultrahigh energies (Kotera & Silk 2016). As byproducts, ultrahigh energy gamma rays and neutrinos can also be generated by the hadronic interactions of cosmic rays and ambient surroundings. Even though such a scenario is unlikely with current consensus, the *Fermi* Gamma-ray Burst Monitor (GBM) revealed a weak transient with the location consistent with GW150914, and 0.4 s after (Connaughton *et al.* 2016). However, no other gamma-ray or X-ray detector found any coincident signal. The debates have been raised about whether this GBM detection is associated with GW150914 (Xiong 2016) and whether the merger of two black holes will produce any electromagnetic emission (Liu & Zhang 2009).

The discovery of GW150914 and the follow-up detections of GW151226 and LVT 151012 (Abbott *et al.* 2016) opened a new era for multi-messenger astronomy and motivated neutrino follow-up search at all energies. At PeV energies, it has been performed by ANTARES and IceCube neutrino observatories, and no neutrino candidate was found in both temporal and spatial coincidence during the event period set by (Adrian-Martinez *et al.* 2016). In the EeV energy ranges, with the surface detector (SD) array, the Pierre Auger Observatory is able to detect and identify ultrahigh energy (UHE) neutrinos at high zenith angles (Abreu *et al.* 2013). In this work, we present results of an UHE neutrino follow-up search for GW150914 with the SD and put stringent constraints on the source of GW150914 event.

2. UHE neutrinos coincidence search

The placed array covers an area of 3000 km^2, consisting of 1660 water-Cherenkov stations in a triangular grid with 1.5 km spacing (Allekotte *et al.* 2008). With the SD, neutrino induced air showers can be distinguished from the background of air showers initiated by hadronic cosmic rays at energies above 100 PeV. The idea for identification is quite simple, the neutrinos can generate showers deeply in the atmosphere, while nucleons and photons interact shortly after they enter the atmosphere. Therefore, at large zenith angles ($\theta > 60°$), neutrino showers keep a considerable amount of electromagnetic component at the ground ("young" showers), with signals spread in time over hundreds of nano-seconds; on the other hand, at large zenith angles, the atmosphere is thick enough so that electromagnetic component of the nucleonic cosmic ray showers gets absorbed, leaving the muonic components dominated at ground level ("old showers"), with narrower signals spreading over tens of ns. Based on this characteristic of young and old shower front traces, with the 25 ns time resolution of Flash Analog-Digital converters (FADC), neutrino shower signals can be distinguished from background. In this work, two types of neutrino-induced showers are searched for above 100 PeV:
(1). Earth-skimming (ES) showers at zenith angles $90° \leqslant \theta \leqslant 95°$. They are induced by tau neutrinos ν_τ, which travel in a slightly upward direction with respect to ground, and can skim the Earth's crust and interact near the surface, inducing a tau lepton which escapes the Earth and decays in flight in the atmosphere, close to the SD. Typically, only ν_τ-induced showers with zenith angles $90° \leqslant \theta \leqslant 95°$ may be identified.
(2). Downward-going high (DGH) zenith angle showers of any flavor moving down at $75° \leqslant \theta \leqslant 90°$ with respect to the vertical. They are initiated deeply in the atmosphere close to the SD array through charged-current or neutral-current interactions.

The SD array is located at the latitude of 35.2° (Aab *et al.* 2015), sensitive to specific portion of the sky at each instant time. During the searching periods of ±500 s around GW150914 and one day after, the SD was monitored every minute and was in stable operation, with $\sim$ 97.5% of stations active. The search window of ±500 s is an estimated upper bound based on the observation of time duration of GRB prompt emission (Baret *et al.* 2011), when UHE neutrinos are expected from the interactions of accelerated protons with the observed gamma-rays (Moharan *et al.* 2016). On the other hand the choice of one day after the event is based on the conservative time bound accounting for the GRB afterglow emissions.

In Fig. 1, the reconstructed 90% confidence level (CL) probability density contour of GW150914 is shown in equatorial coordinates, and derived from the full parameter estimation of the signal, covering 590 deg^2. The position is not well constrained (Adrian-Martinez *et al.* 2016). In the same figure, the field of view (FOV) of ES and DGH channels of Auger are also plotted. Here, we assume that within a 1000 s time interval, the FOVs do not change with time. Within ±500 s around GW150914, only a small portion of the event position is overlapping with the FOV of the DGH channel and not overlapping with the ES channel.

Later, we calculated the fractions of one sidereal day of the source at declination δ visible in the ES and DGH angular ranges. In Fig. 2, the sky map of the GW150914 event at 90% CL is shown and the color scales indicate the fractions changing with source direction. For ES and DGH channels, up to $\sim$ 17% and $\sim$ 35% of this source can be seen with Auger in one day, respectively. These numbers of visible fractions strongly depend on the duration and sky position of the sources, and can be significantly different for various sources. As we can see, the declination of GW150914 visible to Auger is $-1.0°$ to $-14.5°$ and $-38.5°$ to $-78.0°$. We applied the neutrino selection criteria which

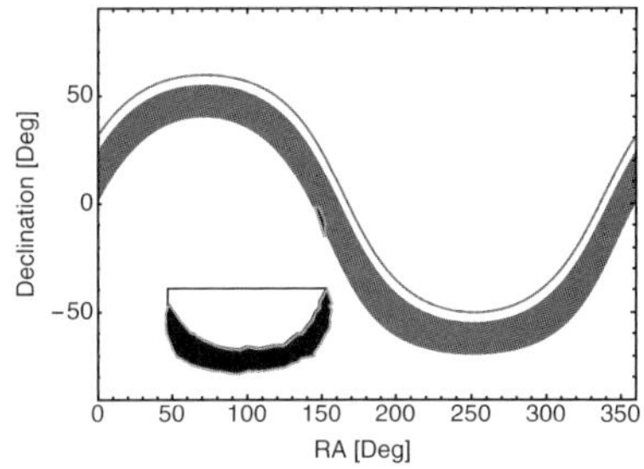

Figure 1: The upper blank (lower shaded) sinusoid bands represent the Auger instant field of view of ES (DGH) channels at the moment of GW150914 transient in equatorial coordinates. The two irregular spots in the lower left corner show the reconstructed probability density contour of GW150914 at 90% CL.

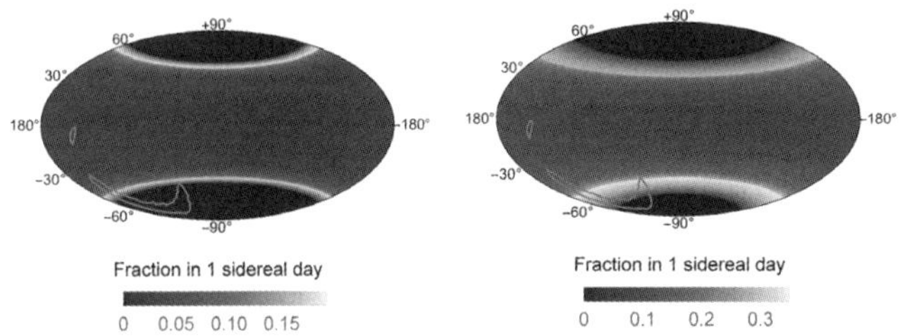

Figure 2: The sky map of the GW150914 event at 90% CL in solid curves. The color scales indicate the fractions of one sidereal day the source is visible at each declination changing with source direction in ES (left panel) and DGH (right panel) channels

were established as in reference Abreu *et al.* (2013) and searched for UHE neutrinos for both time windows. No neutrino candidates were found in temporal coincidence with GW150914.

3. Source constraints

Since there was no coincident neutrino candidate found, we were allowed to calculate the upper limits to the UHE neutrino spectral fluence as a function of source declination δ. We assumed a standard $\mathrm{d}N/\mathrm{d}E = kE^{-2}$ source model, hence a 90% CL upper limit on normalization k can be obtained as $k(\delta) = \frac{2.39}{\int E_\nu^{-2}\epsilon(E_\nu,\delta)\mathrm{d}E_\nu}$, where $\epsilon(E_\nu, \delta)$ is the exposure to UHE neutrinos, which was computed following the reference (Abraham *et al.* 2008) for ES and DGH channels separately. With this equation, we obtain upper limits to the normalization k_{ES} and k_{DGH} for both channels and a combined upper limits to k with the formula $k^{-1}=(k_{\mathrm{ES}})^{-1}+(k_{\mathrm{DGH}})^{-1}$. Next, the upper limits on the UHE neutrino spectral fluence $E^{-2}\mathrm{d}N/\mathrm{d}E \times T = k(\delta) \times T$, where $T=$ 1 day+500 s is the search window, are obtained and shown in the left plot in Fig. 3. The shaded bands indicate that the limits are only valid in these declination range at 90% CL. Fluences above the solid line are excluded at 90% CL from the non-observation of UHE neutrino events in Auger.

With the assumed neutrino spectrum $\mathrm{d}N/\mathrm{d}E = kE^{-2}$, we integrate the emission flux over energy from 100 PeV to 25 EeV and obtain the constraint $F(\delta)$. Finally, the total emitted energy in neutrinos expressed as $E_{tot} = F(\delta) \times 4\pi D^2$ is evaluated at three luminosity distances and shown in the right plot of Fig. 3. Energies above each line with respect to the corresponding luminosity distance to the source are excluded at the 90% CL from the non-observation of UHE neutrinos in Auger.

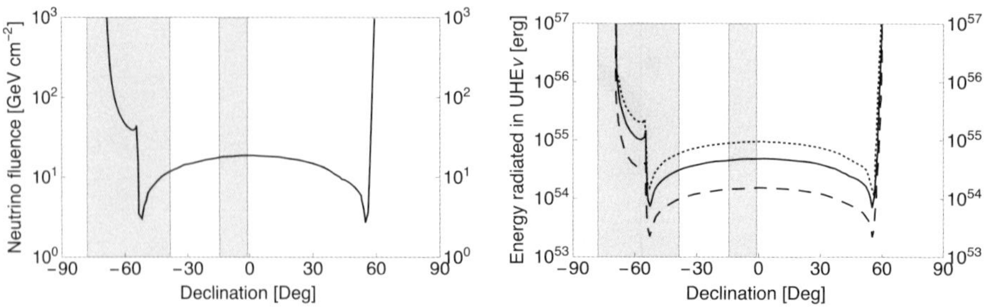

Figure 3: (a) Upper limit on the UHE neutrino spectral fluence per flavor from GW150914 as a function of equatorial declination. The shaded regions indicate the declination ranges which are visible in Auger at 90% CL. (b) Upper limits on the total energy radiated in UHE neutrinos per flavor from GW150914 as a function of equatorial declination. The dotted line represents the constraints when the source is farther away at $D = 410 + 160$ Mpc, and the dashed line shows when the source is closer to Earth at $D = 410 - 180$ Mpc, corresponding to the 90% CL interval of possible distances to the source.

4. Conclusion

In this work, we present the most stringent upper limit on the neutrino emissions from an identified GW transient in the energies above 100 PeV. Considering the energy radiated in gravitational wave from GW150914 is $\sim 5.4\times10^{54}$ erg (Abbott *et al.* 2016), and our most restrictive upper limit on the total energy emitted per flavor in UHE neutrinos achieved at declination $-53°$ is $<7.7\times10^{53}$ erg, we obtain the upper limit on the fraction of energy radiated in UHE neutrinos relative to the energy emitted in GW150914 is $\sim$ 14.3%. We have also searched for signals at ±500 s around and 1 day after GW151226 and LVT151012 and did not find any neutrino candidate, more details given in the paper Aab *et al.* (2016).

The first detection of GW event has opened a new era for multi-messenger studies involving the joint observations of neutrinos, gamma rays and GWs. This joint analysis will provide strong constraints on the properties of these astrophysical phenomena and provide more accurate directional information. The inferred black hole merger rate is in the range of 9 to 240 $\text{Gpc}^{-3}\text{yr}^{-1}$, and more gravitational wave detections are expected in the near future (Abbott *et al.* 2016).

References

B. P. Abbott *et al.*, *Phys. Rev. Lett.* 116, 061102 (2016)
K. Kotera & J. Silk, *Astrophys. J.* 823, L29 (2016)
V. Connaughton *et al.*, *Astrophys. J.* 826, L6 (2016)
S. Xiong, arXiv:1605.05447 [astro-ph.HE]
Y. Liu & S. N. Zhang, *Phys. Lett. B* 679, 88 (2009)
B. P. Abbott *et al.* Phys. Rev. Lett. 116, 241103 (2016)
S. Adrian-Martinez *et al.*, Phys. Rev. D 93, 122010 (2016)
P. Abreu *et al.*, Adv. High Energy Phys. 2013, 708680 (2013)
I. Allekotte *et al.*, Nucl. Instrum. Meth. A 586, 409 (2008)
A. Aab *et al.*, Nucl. Instrum. Meth. A 798, 172 (2015)
B. Baret *et al.*, Astropart. Phys. 35, 1 (2011)
R. Moharana, S. Razzaque, N. Gupta & P. Meszaros, Phys. Rev. D 93, 123011 (2016)
J. Abraham *et al.*, *Phys. Rev. Lett.* 100, 211101 (2008)
B. P. Abbott *et al.*, arXiv:1606.04856 [gr-qc].
A. Aab *et al.*, [arXiv:1608.07378 [astro-ph.HE]]

New Frontiers in Black Hole Astrophysics
Proceedings IAU Symposium No. 324, 2016
Andreja Gomboc, ed.

doi:10.1017/S1743921317001752

What was the Initial Mass of Merging Black Holes in GW150914 ?

Hiromichi Tagawa[1] **and Masayuki Umemura**[2]

[1]University of Tokyo,
7-3-1 Hongo Bunkyo, Tokyo 113-0033, Japan
email: tagawahr@nao.ac.jp

[2]Center for Computational Sciences, University of Tsukuba,
Tsukuba, Ibaraki 305-8577, Japan
email: umemura@ccs.tsukuba.ac.jp

Abstract. Recently, the Laser Interferometer Gravitational-Wave Observatory (LIGO) has detected the gravitational wave (GW) event, GW150914, as a result of the merger of a $\sim 30\ M_\odot$ black hole (BH) binary. So far, the merger of stellar-mass BHs has been thought to result from the evolution of binary stars. Here, we propose a novel path of the merger stemming from non-binary isolated stars. In our previous studies, we have found that multiple non-binary stellar-mass BHs whose separations are larger than 1000 AU can merge with each other under the gas-rich environments through the gas dynamical friction and three-body interaction. In this case, a considerable amount of gas can accrete onto BHs before the merger, that is, the initial mass of BHs can be lower than 30 $M_\odot$. To explore this possibility, we perform post-Newtonian N-body simulations on mergers of accreting stellar-mass BHs. Based on our simulations, we find that the BH merger in GW150914 from smaller seed BHs is likely to occur in galactic nuclear regions or dense interstellar cloud cores. Furthermore, we roughly estimate event rates to be $\sim 0.4\ \mathrm{yr}^{-1}$ in galactic nuclear regions and $\sim 8\ \mathrm{yr}^{-1}$ in dense interstellar cloud cores.

Keywords. stars: black holes – gravitational waves – methods: numerical

1. Introduction

Recently, the advanced LIGO (aLIGO) observatory reported the first direct detection of GWs (Abbott *et al.* 2016), GW 150914. The observed signal indicates that the masses of merged BHs are $m_1 = 36^{+5}_{-4}\ M_\odot$ and $m_2 = 29^{+4}_{-4}\ M_\odot$. The merger of stellar-mass BHs has been hitherto thought to originate from inherent binary stars. However, it is unclear how such a heavy close BH binary as implied in the GW150914 event can be formed. In this paper, we newly propose a merger scenario from non-binary isolated BHs.

The multiple non-binary stellar-mass BHs are expected to form in first generation objects. Furthermore, according to the hierarchical merger history of galaxies, multiple non-binary stellar-mass BHs are likely to be incorporated after galaxy mergers as depicted in Figure 1. From our previous study (Tagawa *et al.* 2015), the multiple non-binary stellar-mass BHs whose typical extension is larger than a thousand AU can merge with each other under the gas-rich environments through the gas dynamical friction and the three-body interaction. In galaxies, we can expect several gas rich environments (Figure 1) such as galactic center regions of $\lesssim 1$ pc ($n_{\rm gas} \gtrsim 10^7\ \mathrm{cm}^{-3}$), dense interstellar cloud cores ($n_{\rm gas} = 10^{5-7}\ \mathrm{cm}^{-3}$), and interstellar clouds ($n_{\rm gas} < 10^5\ \mathrm{cm}^{-3}$). In these situations, a considerable amount of gas can accrete onto BHs before the merger (Tagawa *et al.* 2016). Hence, the initial mass of BHs can be lower than 30 $M_\odot$, and the initial separation of BHs can be much larger than several AU. Therefore, mergers of non-binary isolated BHs in gas rich environments may explain the GW150914 event. In our study, to investigate

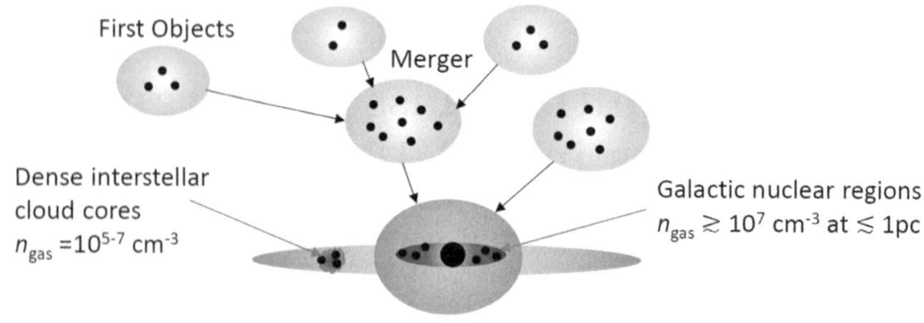

Figure 1. The formation processes of non-binary stellar-mass BHs under gas rich environments in a galaxy.

the conditions and environments under which BHs can merge from smaller mass BHs and large separation, we perform post-Newtonian N-body simulations on mergers of accreting stellar-mass BHs.

2. Post-Newtonian N-body simulations

Tagawa *et al.* (2016) gives a detailed description of the simulations used here. The equations of motion are integrated using the fourth-order Hermite scheme. The simulations incorporate the effects of gas dynamical friction, gas accretion, and the general relativity that is given by the post-Newtonian prescription up to 2.5PN term. 1PN and 2PN terms correspond to the pericentre shift, and 2.5PN term to the GW emission.

The key parameters in our simulations are the initial typical extension of BH distribution ($r_{\rm typ}$), the gas number density ($n_{\rm gas}$), and the initial BH mass (m_0). To parameterize the accretion rate, we set the gas accretion rate ($\dot{m}_i$) to be the accretion efficiency ($\epsilon \leqslant 1$) times the Hoyle-Lyttleton accretion rate ($\dot{m}_{\rm HL}$). The effect of the radiation pressure on Hoyle-Lyttleton accretion is incorporated as Tagawa *et al.* (2016).

We set the finite radius of a gas sphere whose mass is 10^5 $M_\odot$ and temperature is 1000 K. We assume that two BHs merge, when their separation is less than 100 times the sum of their Schwarzschild radii. The evolution is pursued for 10 Gyr, since we consider the merger within the cosmic time. Also, we terminate the simulation, when the first BH merger occurs. We give the initial positions of BHs randomly in the $x-y$ plane within $r_{\rm typ}$. Also, the velocity of each BH is given as the sum of a circular component and a random component. The circular velocity is given to balance against the gravity of gas in the $x-y$ plane. Besides, we give the random velocity in the xyz space according to a Gaussian distribution with the same dispersion as the circular velocity.

3. Confrontation to GW150914

We show the results of simulations for the mergers of non-binary BHs in Figure 2. Left panel of Figure 2 shows the masses of binary BHs just before the merger in our simulations. We find several parameters in which BHs merge in the mass range inferred from GW150914. For these matched parameters in hatched regions, the initial BH mass is higher than 20 $M_\odot$. Right panel of Figure 2 shows the accreted mass of binary BHs. Mostly, the accreted mass is smaller than a half of the initial BH mass in this range. Therefore, if the initial BH mass is lower than 20 $M_\odot$, then the merger rate at the masses of GW150914 is reduced. Also, the accretion rate is expected to be around 0.01 $\dot{m}_{\rm HL}$, and the initial extension of BH distributions ($r_{\rm typ}$) is to be mostly less than 1 pc.

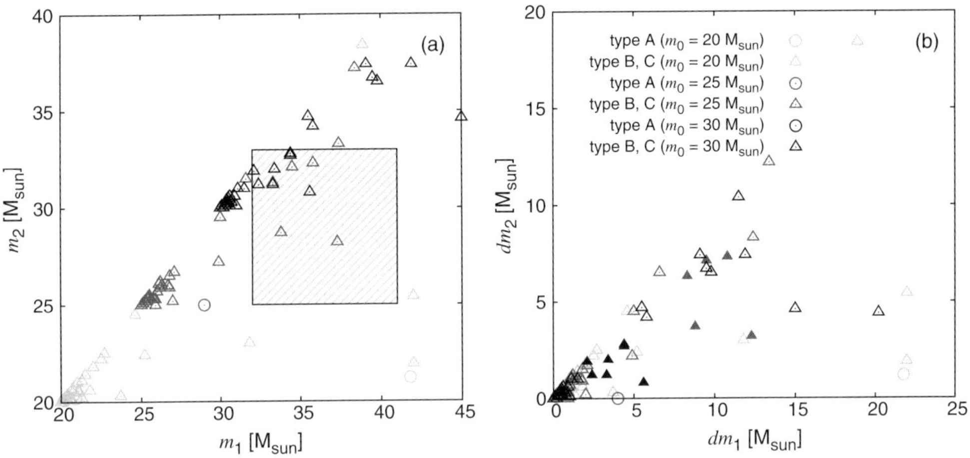

Figure 2. (a) The BH masses ($m_1 > m_2$) and (b) accreted masses in a binary just before the first merger in each run. Light gray, deep gray, black plots represent the initial masses (m_0) of 20, 25 and 30 $M_\odot$, respectively. Circles and triangles represent the gas drag-driven mergers (type A), and interplay-driven mergers (type B) or three body-driven mergers (type C), respectively. The masses of GW150914 with their uncertainties are indicated in panel (a) by the hatched region. The filled symbols in panel (b) are the parameters where the masses of merged BHs match those of GW150914.

In Tagawa *et al.* (2016), the merger mechanisms from non-binary BHs are categorized into four types: a gas drag-driven merger (type A), an interplay-driven merger (type B), a three body-driven merger (type C), and an accretion-driven merger (type D). In type A, the dynamical friction by gas effectively decays the orbit before the GW radiation drives the merger. In both types B and C, the strong disturbance of the orbit is induced by the three-body interaction during the first merger. In type D, significant gas accretion occurs before the first merger. Here, the discrimination between types B and C is determined by subsequent mergers. However, in this study, since simulations are terminated at the first merger, types B and C are not discriminated. From Figure 2, the merger mechanisms in the GW150914 event is likely to be type B or C, that is, the interplay of gas drag-driven and three-body driven merger or purely three body-driven merger.

Next, we discuss preferable environments for our scenario. Because the environments could vary within 10^8 yr due to the active galactic nucleus (AGN) duty cycle and/or the galactic rotation timescale, the merger time should be shorter than 10^8 yr. Besides, the merger time is almost inversely proportional to the gas density (Tagawa *et al.* 2015). From these conditions, the gas number density should be higher than 10^5 cm^{-3}. Therefore, galactic nuclear regions and dense interstellar cloud cores are favorable sites for the merger of non-binary isolated BHs.

4. Event rates

We roughly estimate the event rates based on our scenario for the first observing run of LIGO advanced detectors. First, we estimate the event rates for the merger in dense interstellar cloud cores. Roughly, there are 10^8 stellar-mass BHs in each Milky Way (MW)-sized galaxy (Remillard & McClintock 2006). Here, the volume of the MW Galaxy is about 10 kpc $\times$ 10 kpc $\times$ 1 kpc $=$ 100 kpc^3, and the total volume of all giant molecular clouds (GMCs), whose number is supposed to be about 1000 in the MW Galaxy, is about 10 pc $\times$ 10 pc $\times$ 10 pc $\times$ 1000 $=$ 10^{-3} kpc^3. Thus, the expected number of BHs that exist in GMCs is roughly $10^8 \times (10^{-3}\ \mathrm{kpc}^3/100\ \mathrm{kpc}^3) = 1000$. If we consider the Poisson distributions, about 100 GMCs are likely to possess multiple BHs. Since the merger time in GMCs of $n_{\rm gas} \sim 10^6 \mathrm{cm}^{-3}$ is about 20 Myr (Tagawa *et al.* 2015), the merger rate in a MW-sized galaxy is $100/20$ Myr $= 5\ \mathrm{Myr}^{-1}$. For the binary with total mass of 65 $M_\odot$, we adopt a horizon distance of $D_h \approx 3$ Gpc ($z \approx 0.3$) for the first advanced LIGO observation run (Belczynski *et al.* 2016), corresponding to a comoving volume $V_c \approx 50\ \mathrm{Gpc}^3$. By assuming the number density of the MW-sized galaxies as about $2 \times 10^6\ \mathrm{Gpc}^{-3}$, the number of MW-sized galaxies involved in this volume is $50\ \mathrm{Gpc}^3 \times 2 \times 10^6\ \mathrm{Gpc}^{-3} = 1 \times 10^8$. We also assume that ratio of massive BHs to all BHs is 0.1, and the fraction of massive BH pairs to all BH pairs is supposed to be 0.01. From these assumptions, the event rates in GMCs is estimated to be $5\ \mathrm{Myr}^{-1} \times 0.01 \times 1 \times 10^8 = 5\ \mathrm{yr}^{-1}$.

Next, we estimate the event rates for the merger in galactic nuclear regions (GNRs). In the Milky Way Galaxy, about 3×10^6 stars exist in 1 pc from the supermassive BH (Alexander 2005). If we assume that about one BH exists per 1000 stars, 3×10^3 BHs may exist in GNRs. Based on our simulations, we envisage that all stellar-mass BHs in GNRs can merge with each other in the AGN phase ($\sim$ 100Myr) because of the high density of GNRs in the AGN phase. From these conditions, we estimate that the event rates in GNRs are $\sim 0.3\ \mathrm{yr}^{-1}$ by following the estimate of the event rates in GMCs.

Thus far, we estimate the event rates of mergers for the Pop I/II remnant BHs. Besides, we also estimate that the event rates of mergers for Pop III remnant BHs whose number is assumed to be $\sim 10^6$ in MW-sized galaxies are $\sim 3\ \mathrm{yr}^{-1}$ in GMCs and $\sim 0.1\ \mathrm{yr}^{-1}$ in GNRs. As a result, the event rates are about $\sim 8\ \mathrm{yr}^{-1}$ for the mergers in GMCs and $\sim 0.4\ \mathrm{yr}^{-1}$ for the mergers in GNRs. Here, since our estimates contain large uncertainties such as the uncertainty of the number and distribution of BHs, the estimated event rates are possibly accompanied by the uncertainties of one or two order of magnitude. Furthermore, our scenario prefers disk galaxies as host galaxies of GW events, because gas rich environments are required. To confirm our scenario, additional GW observations are desired.

Acknowledgements

This research was supported by a grant from the Hayakawa Satio Fund awarded by the Astronomical Society of Japan. Numerical computations and analyses were carried out on Cray XC30 and computers at Center for Computational Astrophysics, National Astronomical Observatory of Japan, respectively. This research was also supported in part by Interdisciplinary Computational Science Program in Center for Computational Sciences, University of Tsukuba, and Grant-in-Aid for Scientific Research (B) by JSPS (15H03638).

References

Abbott, B., *et al.*, 2016, *Phys. Rev. Lett.*, 116, 061102
Alexander, T., 2005, *Physics Reports*, 419, 65
Belczynski, K., Holz, D. E., Bulik, T., & O'Shaughnessy Richard., 2016, *Nature*, 534, 512
Remillard, R. A. & McClintock, J. E., 2006, *ARA&A*, 44, 49
Tagawa, H., Umemura, M., Gouda, N., Yano, T., & Yamai, Y., 2015, *MNRAS*, 451, 2174
Tagawa, H., Umemura, M., & Gouda, N., 2016, *MNRAS*, 462, 3812

New Frontiers in Black Hole Astrophysics
Proceedings IAU Symposium No. 324, 2016
Andreja Gomboc, ed.

doi:10.1017/S1743921316012904

Black holes formed by direct collapse: observational evidences

I. F. Mirabel[1,2]

[1]Institute of Astronomy and Space Physics. CONICET - Universidad de Buenos Aires, Ciudad Universitaria, Av. Cantilo S/N , 1428 Buenos Aires - Argentina
email: mirabel@iafe.uba.ar

[2]Laboratoire AIM-Paris-Saclay, CEA/DSM/Irfu-CNRS, CEA-Saclay, pt courrier 131, 91191 Gif-sur-Yvette, France
email: felix.mirabel@cea.fr

Abstract. Binary black holes as the recently detected sources of gravitational waves can be formed from massive stellar binaries in the field or by dynamical interactions in clusters of high stellar density, if the black holes are the remnants of massive stars that collapsed without natal kicks that would disrupt the binary system or eject the black holes from the cluster before binary black hole formation. Here are summarized and discussed the kinematics in three dimensions of space of five Galactic black hole X-ray binaries. For Cygnus X-1 and GRS 1915+105 it is found that the black holes of $\sim$15 $M_{\odot}$ and $\sim$10 $M_{\odot}$ in these sources were formed in situ, without energetic kicks. These observations suggest that binary black holes with components of $\sim$10 $M_{\odot}$ may have been prolifically produced in the universe.

Keywords. black hole physics, gravitational waves, X-rays: binaries, supernovae: general

1. Introduction

It is assumed that a binary black hole (BBH) with components of 30-40 $M_{\odot}$ as the source of gravitational waves GW150914 (Abbott *et al.* 2016a), can be formed from a relatively isolated binary of massive stars if both black holes (BHs) are formed by implosion, namely, by complete or almost complete collapse of massive stars with no energetic supernovae (SNe) accompanied by a sudden large mass loss that would significantly reduce the mass of the compact objects, and in most cases unbind the binary system (Belczynski *et al.* 2016). BBHs can also be formed by dynamical interactions in the high stellar density environments (Rodriguez *et al.* 2016) of typical globular clusters of $\leqslant 10^7$ $M_{\odot}$, if the BHs are formed with no energetic kicks that will eject the BHs out from the cluster before BBH formation. Theoretical models set progenitor masses for BH formation by implosion, but observational evidences have been elusive.

The question on how stellar black holes are formed is of topical interest for the incipient Gravitational-Wave Astrophysics. Whether BHs are formed through energetic natal supernova kicks or by implosion will determine the final evolutionary stage of massive stars, the numbers of BBHs that would be formed, and therefore the merger rates of BBHs that will be detected by gravitational-wave (GW) observations with LIGO, VIRGO and other GW collaborations. In fact, from population synthesis models it is inferred that the disruption rate of massive stellar binaries increases by $\sim$20 varying from the assumption of BH formation with no kicks to that of a kick distribution typical of neutron stars (Dominik *et al.* 2012). Besides, the scape velocity from a typical globular cluster of $\leqslant 10^7$ $M_{\odot}$ is few tens of km s^{-1}, and if formed with a kick distribution typical of neutron stars most BHs would be kicked out from typical globular clusters.

The kinematics of BH X-ray binaries can provide clues on the formation of BHs. If a compact object is accompanied by a mass-donor star in an X-ray binary, it is possible to determine the distance, proper motion, and radial velocity of the center of mass of the system, from which can be derived the velocity in three dimensions of space, and in some cases, infer the site of birth of

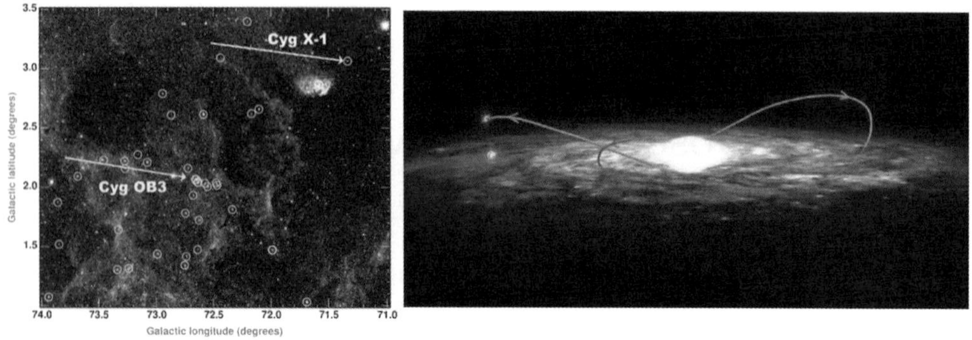

Figure 1. Left: Optical image of the sky around the BH X-ray binary Cygnus X-1 and the association of massive stars Cygnus OB3. The arrows show the magnitudes and directions of the motion in the plane of the sky of the radio counterpart of Cygnus X-1 and the average Hipparcos motion of the massive stars (circled) of Cygnus OB3 for the past 0.5 Myrs. After the formation of the BH, Cygnus X-1 remained anchored in the parent association of massive stars Cygnus OB3. From Mirabel & Rodrigues (2003). **Right**: Schematic Galactic orbit of the runway BH X-ray binary XTE J1118+480 during the last orbital period of the Sun around the Galactic Center (240 Myr). The source left the plane towards the northern Galactic hemisphere with a peculiar space velocity of 217 ± 18 km s^{-1} relative to the Galactic disk frame, and a component perpendicular to the plane of 126 ± 18 km s^{-1}. The galactic orbit of XTE J1118+480 has an eccentricity of 0.54. At the present epoch XTE J1118+480 is at a distance from the Sun of only 1.9 ± 0.4 kpc flying through the Galactic local neighborhood with a velocity of 145 km s^{-1}. From Mirabel *et al.* (2001)

the BH. Here are summarized the recently improved determination, mostly by VLBI at radio wavelengths, of the kinematics of five Galactic BH X-ray binaries.

2. Black holes formed by implosion

Cygnus X-1 is an X-ray binary at a distance of 1.86 ± 0.1 kpc composed by a BH of 14.8 ± 1.0 $M_\odot$ and a 09.7 lab donor star of 19.2 ± 1.9 $M_\odot$ with an orbital period of 5.6 days and eccentricity of 0.018 ± 0.003. Cygnus X-1 appears to be at comparable distance and moving together with the association of massive stars Cygnus OB3 (Mirabel & Rodrigues 2003). Therefore, it has been proposed that the BH in Cygnus X-1 was formed in situ and did not receive an energetic trigger from a natal or nearby supernova. The recent VLBI measurements of the parallax and proper motion of Cygnus X-1 (Reid *et al.* 2011), and the new reduction of the Hipparcos data to infer the mean distance and proper motion of Cygnus OB3, show that within the errors, the distance and proper motion, -as well as the radial velocity of the black hole X-ray binary barycenter-, are consistent with those of the association of massive stars, which reaffirmed the conjecture that Cygnus OB3 is the parent association of Cygnus X-1. The motions on the plane of the sky of Cygnus X-1 and Cygnus OB3 are shown on the left side of Figure 1.

The upper limit of the velocity in three dimensions of Cygnus X-1 relative to the mean velocity of Cygnus OB3 is 9 ± 2 km s^{-1}, which is typical of the random velocities of stars in expanding stellar associations. From the equations for spherical mass ejection in BH formation (Nelemans *et al.* 1999) it is estimated that the maximum mass that could have been suddenly ejected to accelerate the binary without disruption to a velocity of 9 ± 2 km s^{-1} is less than 1 ± 0.3 $M_\odot$. Indeed, there are no observational evidences for a SN remnant in the radio continuum, X-rays, and atomic hydrogen surveys of the region where Cygnus X-1 was most likely formed. It was estimated that the initial mass of the progenitor is more than 40 ± 5 $M_\odot$ and that it may have lost $\sim$6 $M_\odot$ by stellar winds during a Wolf-Rayet stage (Mirabel & Rodrigues (2003). The lower mass limit of $\sim$40 $M_\odot$ for the progenitor is the same as the mass theoretically predicted for the transition from fall-back to complete collapse for a BH progenitor of solar metallicity (Fryer *et al.* 2012).

GRS 1915+105 is a low-mass X-ray binary containing a BH of 10.1 ± 0.6 $M_\odot$ and a donor star of spectral type K-M III of 0.5 ± 0.3 $M_\odot$ with a 34 day circular orbital period. The companion overflows its Roche lobe and the system exhibits episodic superluminal radio jets. Using a decade of astrometry with the NRAO Very Long Baseline Array it was measured a parallax distance of 8.6 ± 1.8 kpc and a proper motion for GRS 1915+105, that together with the published radial velocity of the binary barycenter, it is inferred a modest peculiar velocity of 22 ± 24 km s^{-1} (Reid *et al.* 2014), which is consistent with the earlier proposition that the BH in GRS 1915+105 was formed without a strong natal kick (Mirabel *et al.* 2001). The modest peculiar speed of 22 ± 24 km s^{-1} and a donor star in the giant branch suggest that GRS 1915+105 is an old system that has orbited the Galaxy many times, acquiring a peculiar velocity component on the galactic disk of 20-30 km s^{-1}, consistent with the velocity dispersions of ∼20 km s^{-1} of old stellar systems in the thin disk, due to galactic diffusion by random gravitational perturbations from encounters with spiral arms and giant molecular clouds.

The kinematics in three dimensions of Cygnus X-1 relative to the parent association of massive stars Cygnus OB3, and the kinematics of GRS 1915+105 relative to its Galactic environment, suggest, irrespective of their origin in isolated massive stellar binaries or in typical dense clusters of less than 10^7 $M_\odot$, that the black holes in the x-ray binaries Cygnus X-1 and in GRS 1915+105, were formed in situ by complete or almost complete collapse of massive stars, with no energetic kicks from natal supernovae. These observational results are consistent with theoretical models, and in particular, with one of the most recent models, where the black holes of ∼15 $M_\odot$ in Cygnus X-1, and of ∼10 $M_\odot$ in GRS 1915+105, may be formed by complete or almost complete collapse of stellar helium cores (Sukhbold *et al.* 2016).

3. Runaway black hole x-ray binaries

GRO J1655-40 is an X-ray binary with a BH of 5.3 ± 0.7 $M_\odot$ and a F6-F7 IV donor star with a runaway velocity of 112 ± 18 km s^{-1} moving in a highly eccentric (e=0.34 ± 0.05) Galactic orbit (Mirabel *et al.* 2002). The runaway linear momentum of the X-ray binary is similar to that of a solitary neutron star with a runaway velocity of ∼420 km s^{-1}.

XTE J1118+480 is a high-galactic-latitude (l=157°.78, b=+62°.38) x-ray binary with a BH of 7.6 ± 0.7 $M_\odot$ and a 0.18 $M_\odot$ donor star of spectral type K7 V to M1 V, moving in a highly eccentric orbit around the Galactic center region, as some ancient stars and globular clusters in the halo of the Galaxy. On the right panel of Figure 1 is shown the orbital path of this X-ray binary relative to the Galactic disk (Mirabel *et al.* 2001).

V404 Cyg (GS 2023+338) is a low mass X-ray binary system composed of a BH of 9.0 ± 0.6 $M_\odot$ and a 0.75 $M_\odot$ donor of spectral type K0 IV. From astrometric VLBI observations, it was measured for this system a parallax that corresponds to a distance of 2.39 ± 0.14 kpc and proper motion (Miller-Jones *et al.* 2009a), from which it is derived a peculiar velocity of 39.9 ± 5.5 km s^{-1}, with a component on the Galactic plane of 39.6 km s^{-1} (Miller-Jones *et al.* 2009b), that is ∼2 times larger than the expected velocity dispersion in the Galactic plane.

Discussion on the origin of the runaway kinematics of black hole X-ray binaries: GRO J1655-40 and V404 Cyg are in the Galactic disk (b $\leqslant$ 3.2^o; z $\leqslant$ 0.15 kpc) where likely were formed, except XTE J118+480 which is in the Galactic halo (b = 62.3^o; z = 1.5 kpc) and could either have been propelled to its present position from the Galactic disk by a supernova explosion, or have been formed in a globular cluster from which it could have escaped with a mild velocity of tens km s^{-1}. If these X-ray binaries were formed from binary stars in a field of relative low density, one would conclude that the three BHs with $\leqslant$10 $M_\odot$ were formed with significant natal kicks, whereas the BHs with $\geqslant$ 10 $M_\odot$ were formed with no energetic supernovae. But if the runaway BH X-ray binaries were formed in dense stellar clusters, the anomalous velocities of the X-ray binaries barycenter could have been caused by dynamical interactions in the stellar cluster, or by the explosion of a nearby massive star, rather than by the explosion of the progenitor of the runaway BH. In this context, the runaway velocities and supernova nucleosynthetic products in the atmospheres of the donor stars in GRO J1655-40, XTE 1118+480 and V404 Cyg could be due to the explosion of a nearby star, rather than to the explosion of the BH progenitor. In fact, the three BH runaway X-ray binaries have low mass donors, and their linear momenta are comparable to those of runaway massive stars ejected from multiple stellar systems by the Blaauw mechanism. Therefore, without knowing the origin of a runaway X-ray binary, it is not possible to constrain from its peculiar velocity alone, the strength of a putative natal supernova kick to the compact object in the x-ray binary. It is expected that future parallax distances and proper motions of BHs and their environment, determined at radio wavelengths by VLBI and at optical wavelengths with GAIA, will provide further observational constraints on the physical mechanisms of stellar BH formation.

4. Conclusion

Stars of solar metallicity and $\sim$40 $M_\odot$ may collapse directly to form black holes without energetic supernova explosions. The kinematics of Cygnus X-1 shows that the black hole of $\sim$15 solar masses did not receive a strong natal kick, and was formed by implosion from a progenitor of $\sim$40 $M_\odot$, that by exchange of mass with the binary companion and stellar winds in a Wolf-Rayet phase lost $\sim$25 $M_\odot$. The kinematics of GRS 1915+105 shows that its black hole of $\sim$10 $M_\odot$ was also formed in situ with no strong natal kick. Recent core-collapse models of massive stars based on neutrino-powered explosions are consistent with these observations.

The formation by implosion of black holes of $\sim$10 $M_\odot$ in X-ray binaries is consistent with the observation of GW151226, produced by fusion of black holes of $\sim$10 $M_\odot$ (Abbott *et al.* 2016b).

The interpretation of the runaway velocity of an X-ray binary as a kick velocity from a natal supernova imparted to the compact object in the runaway binary is uncertain without knowledge of the binary origin. X-ray binaries may be originated in different environments and their runaway velocities be caused by a variety of physical mechanisms.

Parallax distances and proper motions of black hole X-ray binaries determined by VLBI observations at radio wavelengths, and with GAIA at optical wavelengths, will allow to determine accurate velocities in three dimensions for larger samples of sources, and therefore, will allow to better constraint models on stellar black hole formation.

A short review on the current optical and infrared observations that may provide insights on the nature of the stellar progenitors of black holes can be found in Mirabel (2017).

References

Abbott, B. P., *et al.* 2016a, *PRL*, 116, 061102
Abbott, B. P., *et al.* 2016b, *PRL*, 116, 241103
Belczynski, K, Holz, D. E., Bulik, T., *et al.* 2016, *Nature*, 534, 512
Dominik, M. & Belczynski, K, Fryer, C. *et al.* 2012, *ApJ*, 759, 52
Fryer, C. L., Belczynski, K., Wiktorowicz, G., *et al.* 2012, *ApJ*, 749, 91
Miller-Jones, J. C. A.., Jonker, P. G., Dhawan, V., *et al.* 2009a, *ApJ*, 706, L230
Miller-Jones, J. C. A.., Jonker, P. G., Nelemans, G., *et al.* 2009b, *MNRAS*, 394, 1440
Mirabel, I. F. 2017, *New Frontiers in Black Hole Astrophysics. Proceedings IAU Symposium 324. Andreja Gomboc, & Carole Mundell, eds.*
Mirabel, I. F., Dhawan, V., Mignani, R. P., *et al.* et al. 2001, *Nature*, 413, 139
Mirabel, I. F., Mignani, R., Rodrigues, I., *et al.* 2002, *A&A*, 395, 595
Mirabel, I. F. & Rodrigues, I. 2003, *Science*, 300, 1119
Nelemans, G., Tauris, T. M., & van den Heuvel, E. P. J. 1999, *A&A*, 352, L87
Reid, M. J., McClintock, J. E., Narayan, R. *et al.* 2011, *ApJ*, 742, 83
Reid, M. J., McClintock, J. E., Steiner, J. F. *et al.* 2014, *ApJ*, 796, 2
Rodriguez, C. L., Haster, C-J, Chatterjee, S., *et al.* 2016, *ApJL*, 824, L8
Sukhbold, T., Ertl, T. Woosley, S. E., *et al.* 2016, *ApJ*, 821, 38

Topic 6: Technology Drivers and Future Capabilities

New Frontiers in Black Hole Astrophysics
Proceedings IAU Symposium No. 324, 2016
Andreja Gomboc, ed.

doi:10.1017/S1743921317002289

Black hole astrophysics with HAWC, the High Altitude Water Cherenkov γ-ray observatory

Alberto Carramiñana[1] ***for the HAWC Collaboration***

[1]Instituto Nacional de Astrofísica, Óptica y Electrónica
Luis Enrique Erro 1, Tonantzintla, Puebla, México
email: alberto@inaoep.mx

Abstract. The HAWC gamma-ray observatory is a wide field of view and high duty cycle γ-ray detector investigating the 0.1 - 100 TeV energy range. It has detected supermassive black holes in the near Universe, and is seeking to detect black hole related objects like gamma-ray bursts, Galactic binary systems, primordial black holes and gravitational wave mergers. Daily light curves of the BL Lac objects Mrk 421 and Mrk 501 are presented here, together with a compilation of studies of black hole related objects.

1. Introduction: investigating black holes with HAWC

Black holes, an early prediction of general relativity, are objects of extreme astrophysical interest (Schwarzschild 1916; Oppenheimer & Snyder 1939). They can exist in radically different mass regimes, having originated in very different processes: primordial black holes ranging from the mass of a mountain, formed in the first instants of the Universe; stellar mass black holes, found in binary systems, are products of the stellar evolution of massive stars; and supermassive black holes, believed to be the power engines of active galactic nuclei, are thought to have built up in the early stages of cosmological evolution. The existence of black holes was debated for decades, until the discovery of the extragalactic nature and rapid variability of active galactic nuclei (Salpeter 1964; Ozernoi & Chertoprud 1966), the determination of the mass of the binary stellar system Cygnus X-1 (Webster & Murdin 1972), and the kinematic evidence of a 4×10^6 $M_\odot$ in the center of the Galaxy (Genzel *et al.* 1996). Recently, the Laser Interferometer Gravitational-Wave Observatory (LIGO) provided unmistakable evidence of gravitational waves emitted from unexpected $\sim 10-30\,M_\odot$ binary system mergers (Abbott *et al.* 2016a, 2016b).

For half a century, the strong gravity inherent to black holes prompted their potential as high energy sources (Salpeter 1964). The recent availability of the High Altitude Water Cherenkov (HAWC) observatory, a wide-field of view and high duty cycle γ-ray TeV detector of unprecedented sensitivity, now provides unique opportunities for black holes astrophysics. HAWC has been monitoring two thirds of the sky every sidereal day since it started early operations in August 2013, with upgraded sensitivity to detect sources with the TeV flux of the Crab in a single transit once it reached full operations in March 2015. HAWC has studied blazars and very high energy γ-ray sources in the Galaxy for over a year. Its science case includes investigating TeV emission from gamma-ray bursts, primordial black holes and X-ray binary systems. In addition, HAWC is an ideal instrument for multi-messenger follow-ups of neutrino sources or for the search for electromagnetic counterparts of gravitational wave events.

2. Gamma-ray detectors

Gamma rays, the most extreme form of electromagnetic radiation, are produced in violent astrophysical phenomena. They have been observed up to at least 100 TeV, with searches conducted up to 10^{18} eV (Aab *et al.* 2014). The atmosphere of the Earth is opaque to γ rays, prompting their observation with space borne observatories in the MeV and GeV ranges. Individual very high energy γ rays, as well as cosmic rays with $E \gtrsim 100$ GeV, have a perceptible effect in the atmosphere, triggering secondary particle cascades that can be detected with ground based instruments. In consequence, the γ-ray region of the electromagnetic spectrum is nowadays investigated with three types of instruments of complementary capabilities.

2.1. *Space borne telescopes*

Gamma-ray space telescopes detect photons through either the Compton effect, at low MeV energies, or pair production, at higher energies, in the GeV band. These instruments have wide fields of view which allow for complete sky surveys. Up to now, there has been only one sky survey in low energy γ rays, specifically in the 0.7–30 MeV range by *CGRO*-COMPTEL, in operations between 1991 and 2000 (Schönfelder *et al.* 2000). In contrast, several pair production telescopes have scanned the sky with increasing depth, with the Large Area Telescope (LAT) on-board the *Fermi γ-ray Space Telescope* detecting over 3000 sources, in the deepest sky survey from 100 MeV–300 GeV summarized in the 3FGL catalog (Acero *et al.* 2015); furthemore, *Fermi*-LAT has extended its performance to the 50 GeV–2 TeV in the 2FHL catalog (Ackermann *et al.* 2016). *Fermi*-LAT has confirmed blazars as the dominant type of celestial γ-ray source, as previously proved by *CGRO*-EGRET (Hartman *et al.* 1999).

2.2. *Air Cherenkov telescopes*

Air Cherenkov telescopes (ACTs) study the very high energy (VHE) range, from about 30 GeV to 30 TeV, using the atmosphere as part of the detector system. Most secondary charged particles produced in atmospheric cascades travel faster than the speed of light in the air, $v > c/n$, emitting Cherenkov radiation confined to a narrow $\sim 2^\circ$ cone. This radiation is detectable from the ground. Working on dark night-time background during cloudless nights, ACTs can perform deep observations of TeV point sources down to the flux of the Crab in less than a minute (Aharonian *et al.* 2006). ACTs have narrow fields of view and low duty cycle, as their operation is restricted to nights with clear and moonless conditions. Their high sensitivity, in particular for point sources, make them optimal for detailed studies of targeted objects, producing high resolution spectra and sub-arcminute source localizations. They can partially compensate their narrow field-of-view through their high sensitivity in order to perform dedicated surveys of areas of interest, covering $\lesssim 1000$ deg^2 over several years (Aharonian *et al.* 2013). Observations from different ACT observatories are summarized in the TeVCat, a heterogeneous but useful catalog of TeV γ-ray sources found with different TeV γ-ray observatories (Wakely & Horan 2008).

2.3. *Extensive Air Shower arrays*

Extensive air shower arrays (EAS) detect secondary particles reaching the detector array. They were developed for cosmic-ray studies in the 1950s, with water Cherenkov detectors in use already in the mid-1960s (Galbraith & Jelley 1953; Wilson *et al.* 1963). Current EAS like HAWC and Tibet AS-γ, have the ability to separate γ-ray primaries from hadronic cosmic rays, previously developed by the MILAGRO Collaboration (Atkins *et al.* 2003). These instruments have large fields of view ($\lesssim 2$ sr) and duty cycles close to 100%, which permit surveys of $> 50\%$ of the sky transiting through their zenith-pointed instantaneous fields of view. Ideally, they operate with very little interruption, performing very shallow daily unbiased surveys that build in depth as their data accumulate. In particular, HAWC reaches down to a 1σ sensitivity level of ~ 0.2 Crab $(t/\text{day})^{-1/2}$. EAS are best suited for transient source monitoring, the study of extended diffuse emission regions and reaching high photon energies, up to 100 TeV.

3. The HAWC γ-ray observatory and its data

3.1. *The HAWC γ-ray observatory*

The HAWC observatory was installed and is currently operated by a collaboration of about thirty institutions of Mexico and United States, including over a hundred scientists, with the recent incorporation of the Max Planck Institute at Heidelberg, the Cracow University and the associate membership of the University of Costa Rica. HAWC is a second generation water Cherenkov observatory (WCO), built from the experience gained during the construction and operation of the MILAGRO observatory that operated in New Mexico, between 1999 and 2008. HAWC is located at a higher altitude (4100 m compared to 2600 m) and benefits of an improved design. It has a sampling area of $22000\,\text{m}^2$, four times larger than MILAGRO; a ten times larger muon detection area, compared to MILAGRO; and includes optical isolation of its detector elements. The result is a γ-ray detector 15 times more sensitive than its predecessor.

The HAWC site is located in the Northern slope of Volcán Sierra Negra, also named Tliltepetl, inside the Parque Nacional Pico de Orizaba, in the Mexican state of Puebla. The geographical coordinates of the site are latitude of $+18.99°$N and longitude $97.31°$W. The more equatorial location of HAWC, relative to MILAGRO, allows for a coverage of two-thirds of the sky, reaching down to the declination of the Galactic Center. HAWC shares the basic infrastructure developed between 1998 and 2006 for the Gran Telescopio Milimétrico Alfonso Serrano (GTM; also known as LMT), located on the 4600 m top of Sierra Negra. The GTM/LMT access road, electrical power line and optical fiber were extended to the HAWC site in 2010 and the array platform was prepared in 2011. With project funding available, the installation of the water Cherenkov array started in 2012, finishing in December 2014. The complete HAWC γ-ray observatory was inaugurated in March 2015, and has been in full operation since.

HAWC consists of 300 individual water Cherenkov detectors (WCD). Each WCD is made of a cylinder of corrugated steel of 7.2 m diameter and 5 m height, with an opaque bladder inside holding 180,000 liters of water and isolating it from environmental light, during day and night. Four photo-multiplier tubes (PMT) are located on the interior base of every WCD: three 8" PMTs that were previously used in MILAGRO, and a central fourth 10" PMT added to the design through the support of LANL-DoE. The PMT system provides fast response to Cherenkov light with high quantum efficiency in the blue to ultraviolet. Every single PMT is connected to the central counting house, where the experiment is controlled, data are acquired and stored, and the calibration system is run. A network of 180 km of underground cables connects each WCD to the counting house.

3.2. *HAWC data*

HAWC is continuously sampling the ensemble of almost 1200 PMT channels, seeking for multiple coincident signals to declare "an event". Data are reconstructed event per event, considering events with more than 30 channels hit, nHit > 30. Relative timing allows to locate the arrival direction of primary cosmic and γ rays in the sky. Shower fronts

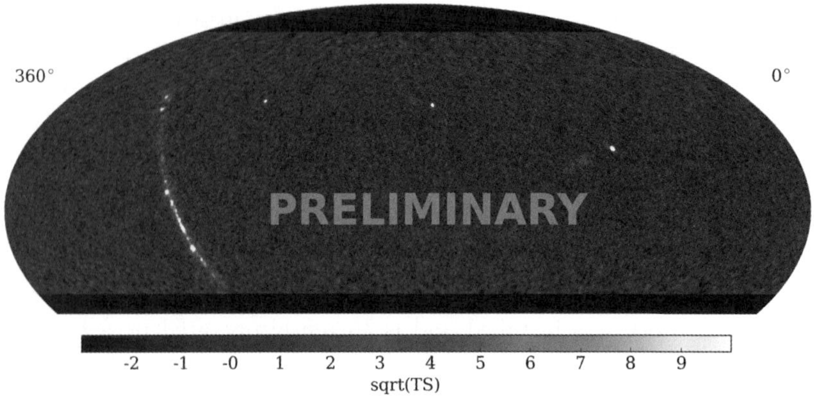

Figure 1. Map of the sky seen in TeV γ rays obtained with HAWC. The map is in celestial coordinates, covering declinations between $+70^\circ$ and -30°. The tilted emission band clearly visible on the left corresponds to the first Galactic Quadrant. The three outstanding point sources from left to right, Mrk 501, Mrk 421 and the Crab Nebula. Next to the Crab, with lower significance, is the extended emission coincident with Geminga.

are fitted with sub-nanosecond residuals, leading to localizations of $\sim 0.25^\circ$ in the sky validated with observations of the Crab Nebula. A datum includes the charge deposits on each PMT channel, used for energy estimations and for γ/hadron discrimination through algorithms based on core location and the topology of the deposits in the array - including muon identification. HAWC registers over 20,000 cosmic rays per second, generating 2 TB of data per day - every day. The data are physically transported to the HAWC-MX datacenter, at the Instituto de Ciencias Nucleares of UNAM, then transferred through the Internet to the HAWC-US datacenter at the University of Maryland.

The HAWC Collaboration has analyzed the first 17 months of data. Sources have been searched for assuming point sources with spectral index $k = -2.7$. Three distinct features can be identified in the HAWC skymap, shown in Figure 1: (1) the Galactic anti-center, highlighted by the Crab Nebula which is detected with a cumulative signal to noise of $\sim 100\sigma$ and is used to validate the performance of the observatory relative to Montecarlo simulations; (2) two distinct active galaxies, Mrk 421 and Mrk 501, detected with sufficient strength as to perform daily monitoring of their TeV γ-ray fluxes; (3) the Galactic Plane, specifically its 1st Quadrant, *i.e.* longitudes $0 \lesssim \ell \lesssim 90^\circ$, where HAWC has detected over forty sources.

4. Black hole astrophysics with HAWC

4.1. *HAWC observations of supermassive black holes*

Supermassive black holes are currently believed to exist in the center of all galaxies. When accreting significant amounts of matter they power the nuclei of active galaxies, which in turn can produce cosmic and γ rays in powerful, relativistic, jets. *CGRO*-EGRET and *Fermi*-LAT observations established BL Lacs and Flat Spectrum Radio Quasars (FSRQ) as the dominant types of GeV sources, several of them confirmed to extend their emission in the TeV range by ACTs. HAWC has made clear detections of Mrk 421 and Mrk 501, the two closest BL Lac objects known, in the 17-month data taken between November 2014 and February 2016. These are two of only four sources detected in the 0.6-2.0 TeV band by *Fermi*-LAT out of the Galactic plane, $|b| > 5^\circ$, and at low redshift,

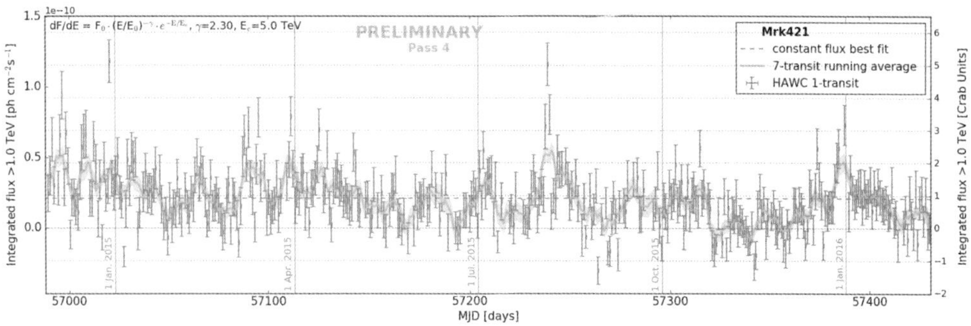

Figure 2. HAWC light curve of the BL Lac object Mrk 421 spanning from November 2014 to February 2016. Five periods of three months are indicated by the vertical grey lines. High states, in which the TeV flux of Mrk 421 was stronger than that of the Crab, are clearly noticeable.

$z < 0.3$ (Ackermann *et al.* 2016). HAWC detected their emission with significances of 33σ and 23σ, respectively, allowing for their daily monitoring.

Mrk 421 is the nearest BL Lac object known, at $d_L = 134\ Mpc$, as estimated from its redshift, $z = 0.031$. Mrk 421 was originally detected at GeV energies in 1991-1992, during the early observations of *CGRO*-EGRET (Fichtel *et al.* 1994). In follow-up observations by the Whipple collaboration, it became the first AGN found at TeV energies (Punch *et al.* 1992), and in fact the *only* of the EGRET detected at TeV energies at the time, providing the first indications of the attenuation of TeV γ rays by pair absorption with the extragalactic background light. Mrk 421 was detected by MILAGRO, with a 7.1σ significance on 2.5 years of data taken between September 2005 and March 2008 (Abdo *et al.* 2014). The HAWC light curve of Mrk421, including 387 selected transits, represents the first daily monitoring of a TeV blazar with the sensitivity to actually detect flaring on a single transit. The data are inconsistent with a constant flux with $p-\text{value} \lesssim 10^{-10}$. Frequent high states of fluxes larger than 1 or 2 Crab have been observed. Recently, HAWC, FACT and *Swift*-XRT reported the joint observation of a four day activity of Mrk 421, which reached 2 Crab units above 1 TeV (Biland *et al.* 2016).

At a distance $d_L = 143$ Mpc, deduced from its redshift $z = 0.033$, Markarian 501 is the second nearest known BL Lac object. With a GeV luminosity lower than Mrk 421, it remained undetected all through the *CGRO* mission (Hartman *et al.* 1999), despite that ACTs discovered its TeV emission (Quinn *et al.* 1996). Its GeV detection was achieved by *Fermi*-LAT, that has now measured it up to the 0.585-2.0 TeV band, as reported in the 2FHL catalog (Ackermann *et al.* 2016). Mrk 501 was clearly detected with HAWC in the 17-month dataset, allowing to generate the corresponding daily light curve of 390 transits. The light curve, shown in Figure3, is inconsistent with a constant flux with $p \lesssim 10^{-10}$. More recently, on the 6th of April 2016, the online monitor system of HAWC measured a two-day flare from this object, which reached a flux 2.2 times that of the Crab (Sandoval, Lauer, Wood 2016).

HAWC has proven its capability to provide timely alerts to the astronomical community. While these two Markarian galaxies have been observed by HAWC with high significance, even daily detections, the current sensitivity of the observatory below 1 TeV has impeded the detection of more distant and less luminous sources, for which the $\gamma\gamma$ absorption of TeV γ rays by the extragalactic background light (EBL) becomes relevant. Mrk 421 and 501 are detected up to at least 10 TeV, providing fine probes to study the EBL; in addition, HAWC keeps searching for the long term averaged emission and for flaring episodes of TeVCat and 2FHL sources with $z < 0.3$, in order to set limits or detect their luminosities and flaring activity at TeV energies.

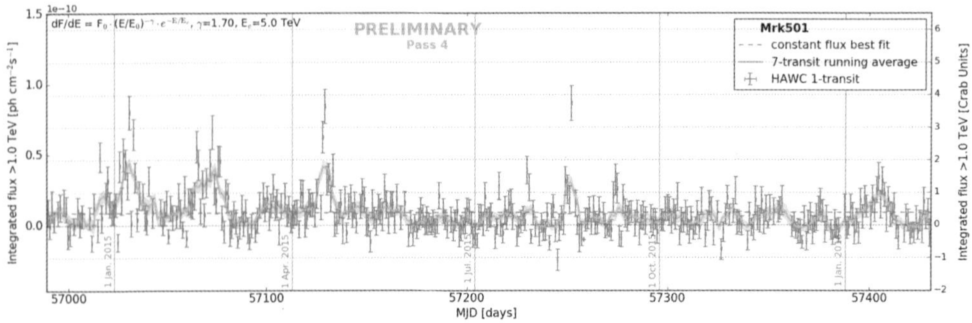

Figure 3. Daily HAWC light curve of Mrk 501 between November 26, 2014 and February 12, 2016. The fluxes are calculated assuming a spectral index of 1.7 and an exponential cut-off at 5 TeV. Episodes of flaring activity, with fluxes up to 3× the Crab are clearly seen.

4.2. *The HAWC search for gamma-ray bursts*

Gamma-ray bursts (GRBs) are believed to be caused by stellar collapse, in the case of long-duration events, or by the merger of two compact objects, like two neutron stars, in the case of short GRBs. These processes can produce stellar mass black holes. While GRBs are particularly strong at MeV energies (Hanlon *et al.* 1994), their exploration of GeV band had to wait for the advent of *Fermi*-LAT, which has reported at least 7 GRBs with $E > 30$ GeV photons in 8 years; the highest energy GRB photon recorded so far stands at 95 GeV, from GRB 130427A (Ackermann *et al.* 2014).

HAWC has the capability to detect TeV emission from GRBs (Abeysekara *et al.* 2012). This can be done through the main data acquisition system or with the scaler systems. HAWC is currently performing both triggered and untriggered searches. Of the seven GRBs reported by *Fermi*-LAT with photons above 30 GeV, only occurred after November 2014, when HAWC started taking data with 250 WCDs; however, with a declination $\delta = +76°$, GRB 160509A was out of the reach of HAWC (Lennarz 2016). Prior to full operations, HAWC provided limits for GRB 130427A (Abeysekara *et al.* 2015).

4.3. *Galactic black holes*

Black holes in X-ray binary systems are known Galactic VHE γ-ray emitters (Wakely & Horan 2008); consistent with being young systems, most of them are located along the Galactic Plane, where HAWC has detected about 40 sources, of which $\sim 25\%$ were previously undetected. A large fraction of the HAWC sources appear to have extended emission and can be positionally associated with objects like SNR (Hui 2016). Four TeV binary systems are in the declination range of HAWC: LSI +61 303, HESS J0632+057, HESS J1832–093 and LS 5039. While LSI+61 303 and HESS J0632+057 are clearly undetected, both HESS J1832–093 and LS 5039, are located too close to strong HAWC Galactic Plane sources, and their search is affected by source confusion.

The Galactic Center hosts a 4×10^6 $M_\odot$ black hole, which has been put forward by HESS observations as the preferred source of PeV cosmic rays in our Galaxy, produced through hadronic processes (Aharonian *et al.* 2016). Unfortunately, the HAWC access to the Galactic Center, which culminates at 46° from the zenith at Sierra Negra, limits its sensitivity for this relevant source. As the response of HAWC at large zenith angles is investigated, the data currently collected may makes it feasible to have its eventual detection at multi-TeV energies.

4.4. *Primordial black holes*

Primordial black holes (PBHs) are potential remnants of the very early Universe, created as results of topological defects, domain wall / cosmic string collisions (Carr 2005). According to Hawking (1974), PBHs should radiate thermally with a temperature $T_{bh} \sim 10^{-7}\,\mathrm{K}\ M^{-1}$, in a timescale $\tau \sim 10^{64}\,\mathrm{yr}\ M^3$, with M measured in Solar units. Black holes of mass $\sim 10^{14}$ g would evaporate in a timescale equivalent to the current age of the Universe, with $T \sim 10^{12}$ K. As the Hawking radiation rapidly becomes an important mass-loss mechanism, the black hole is set to evaporate in a catastrophic event. HAWC can set the most stringent limits for PBHs evaporating on a wide range of timescales, from 1 ms to 100s, going down to evaporation rates $\lesssim 10^4\,\mathrm{pc^{-3}yr^{-1}}$, an order of magnitude below the current MILAGRO limits (Abdo *et al.* 2015).

4.5. *Follow-up on gravitational events from black hole mergers*

The Laser Interferometer Gravitational-Wave Observatory (LIGO) has opened a fresh new field of astrophysical research with the spectacular discovery of the black-hole merger event of the 14th of September 2015, GW150914, very likely located in the Southern hemisphere (Abbott *et al.* 2016a). On December 2016, LIGO detected another clear gravitational-wave signal, whose localization contours intersected the field of view of HAWC at the time of the merger observation (Abbott *et al.* 2016b). Even though the localization uncertainty of the GW151226 event is very large, the precise time information allowed for a search of the event on a very specific time window ± 10 s within the HAWC data, with null results (Wood 2016).

Since 2014 the HAWC has worked in coordination with the LIGO/VIRGO observatories to search for high energy electromagnetic counterparts of gravitational wave events. The large field of view and high duty cycle make HAWC suitable for multi-frequency and messenger follow-ups. The next LIGO run (O2) is due to start at the time of this writing and last about six months. The run will feature not only improvements in the performance of LIGO, but also the eagerly expected addition of VIRGO that will allow for localization errors of $\lesssim 20\,\mathrm{deg}^2$, fitting well inside the HAWC field-of-view (Abbott *et al.* 2013).

5. Summary and next steps

The HAWC γ-ray observatory is in full operation, surveying on daily basis the most energetic phenomena in the sky. Its observations are relevant for our knowledge of black holes on different mass regimes, ranging from the supermassive black holes powering Mrk 421 and Mrk 501, to the 10^{14} g primordial black holes that may have formed in the very early Universe. As the data accumulate, HAWC will be providing a deeper and deeper view of the sky, probing the extreme physical processes powered by astrophysical black holes.

Acknowledgements

The High Altitude Water Cherenkov γ-ray observatory is a collaboration of over thirty institutions in Mexico and the United States, with recent additional participation of Germany, Poland and Costa Rica. HAWC has been possible thanks to the generous support of the National Science Foundation, the US Department of Energy and the Consejo Nacional de Ciencia y Tecnología in Mexico.

This research has made use of the TeVCat online source catalog (tevcat.uchicago.edu);

of the SIMBAD database, operated at CDS, Strasburg, France (Wenger *et al.* 2000); and of NASA's Astrophysics Data System (www.adsabs.harvard.edu).

References

The Pierre Auger and Telescope Array Collaborations: Aab, A. *et al.* 2014, *ApJ* 794, 172.
LIGO Scientific Collaboration and Virgo Collaboration, Abbott, B. P., *et al.* 2013, *ArXiV* 1304.0670v3.
LIGO Scientific Collaboration and Virgo Collaboration, Abbot, B. P., *et al.* 2016, *Phys. Rev. Lett* 116, 061102.
LIGO Scientific Collaboration and Virgo Collaboration, Abbot, B. P., *et al.* 2016, *Phys. Rev. Lett* 116, 241103.
Abdo, A. A., *et al.* 2014, *ApJ* 782, 110.
Abdo, A. A., *et al.* 2015, *Astrop. Phys.* 64, 4.
Abeysekara, A. U., *et al.* 2012, *Astrop. Phys.* 35, 641.
Abeysekara, A. U., *et al.* 2015, *ApJ* 800, 78.
Acero, F., *et al.* 2015, *ApJS* 218, 23.
Ackermann, M., *et al.* 2014, *Science* 343, 42.
Ackerman, M., *et al.* 2016, *ApJS* 222, 5.
Aharonian, F., *et al.* 2006, *A&A* 457, 899.
Aharonian, F., *et al.* 2013, *ArXiV* 1307.4690 from ICRC 2013.
Aharonian, F., *et al.* 2016, *Nature* 531, 476.
Atkins, R., *et al.* 2003, *ApJ* 595, 803.
Biland, A., *et al.* 2016, *The Astronomer's Telegram* 9137.
Carr, B. J., 2005, *Proc. of "Inflating horizon of particle astrophysics and cosmology"*, astro-ph/0511743.
Fichtel, C. E., *et al.* 1994, *ApJS* 94, 551.
Galbraith, W. & Jelley, J. V. 1953, *Nature* 171, 349.
Genzel, R., *et al.* 1996, *ApJ* 472, 153.
Hanlon, L., *et al.* 1994, *A&A* 285, 16.
Hartman, R. C., *et al.* 1999, *ApJS* 123, 79.
Hawking, S. W. 1974, *Nature* 248, 30.
Hui, M. 2016, *APS April Meeting*, U4.001.
Lennarz, D. & for the HAWC Collaboration, 2016, *GCN Circular* 19423.
Oppenheimer, J. R. & Snyder, H. 1939, *Phys. Rev.* 56, 455.
Ozernoi, L. M. & Chertoprud, V. E. 1966, *Sov. Astron.* 10, 15.
Punch, M. *et al.* 1992, *Nature* 358, 477.
Quinn, J. *et al.* 1996, *ApJL* 456, 83
Salpeter, E. 1964, *ApJ* 140, 796.
Sandoval, A., Lauer, R., Wood., & J, for the HAWC Collaboration, 2016, *The Astronomer's Telegram* 8922.
Schönfelder, V., *et al.* 2000, A&A
Schwarzschild, K. 1916, *Sitzungberichte der K. Preussischen Akademie der Wissenschaften zu Berlin* 1, 189.
Wakely, S. P. & Horan, D., 2008, ICRC 3, 1341.
Webster, B. L. & Murdin, P. 1972, *Nature* 235, 37.
Wenger, M., *et al.* 2000, *A& AS* 143, 9.
Wilson, J. G., *et al.* 1963, *ICRC* 4, 27.
Wood, J. 2016, *GCN Circular* 19156.

New Frontiers in Black Hole Astrophysics
Proceedings IAU Symposium No. 324, 2016
Andreja Gomboc, ed.

doi:10.1017/S1743921317002332

Acceleration of particles up to PeV energies at the galactic centre

Stefano Gabici[1], Felix A. Aharonian[2,3], Emmanuel Moulin[4] and Aion Viana[2]

[1]APC, Univ Paris Diderot, CNRS/IN2P3, CEA/Irfu, Obs de Paris, Sorbonne Paris Cité, France – email: gabici@apc.in2p3.fr

[2]Dublin Institute for Advanced Studies, 31 Fitzwilliam Place, Dublin 2, Ireland

[3]Max-Planck-Institut für Kernphysik, Postfach 103980, D-69029 Heidelberg, Germany

[4]DRF/Irfu, Service de Physique des Particules, CEA Saclay, F-91191 Gif-Sur-Yvette Cedex, France

Abstract. Recent very high energy observations of the galactic centre region performed by H.E.S.S. revealed the presence of a powerful PeVatron. This is the first of such objects detected, and its most plausible counterpart seems to be associated to Sgr A*, the supermassive black hole in the centre of our galaxy. The implications of this discovery will be discussed, in particular in the context of the problem of the origin of galactic cosmic rays.

Keywords. Gamma rays: observations, Gamma rays: theory, Acceleration of particles

1. Introduction

The observed spectrum of Cosmic Ray (CR) protons extends up to the PeV energy domain, and then it steepens originating a feature called knee (Antoni *et al.* 2015, Bartoli *et al.* 2015). CRs of energy below that of the knee are believed to have a galactic origin, and this implies that our Galaxy must contain CR PeVatrons. The acceleration of PeV particles can be revealed indirectly by means of observations in the very high energy gamma-ray domain: protons of energy 1 $E_{\rm PeV}$ PeV interact with the ambient gas and produce neutral pions that in turn decay into gamma rays of typical energy $\sim 100\ E_{\rm PeV}$ TeV (Aharonian 2004). Remarkably, the detection of a gamma-ray spectrum extending without significant attenuation up to the multi-TeV domain would constitute a proof of proton acceleration up to PeV energies, because competing leptonic mechanisms (most notably, inverse Compton scattering) are inefficient at such large energies (Aharonian 2004, Gabici & Aharonian 2007).

According to a widely accepted (but not proven) picture, CRs are accelerated at supernova remnant shocks via first order Fermi mechanism (Drury 2012). However, the acceleration of CR protons up to PeV energies remains a challenge for theoretical models of supernova remnants (Bell *et al.* 2013). In addition to that, it is very difficult to determine observationally whether a gamma-ray source is a PeVatron or not, due to the meagre photon statistics in the multi-TeV domain. For this reason, the issue of PeV particle acceleration in the Galaxy is still open.

A breakthrough came recently, when the H.E.S.S. Collaboration (2016) anounced the discovery of the first galactic PeVatron. It is located in the very central region of the Galaxy (within the inner ~ 10 pc) and is most plausibly associated with Sgr A*, the supermassive black hole located at the galactic centre. This discovery has a strong impact on the discussion about the origin of CRs, since it demonstrates that PeVatrons other than (the hypothetical) supernova remnants do exist.

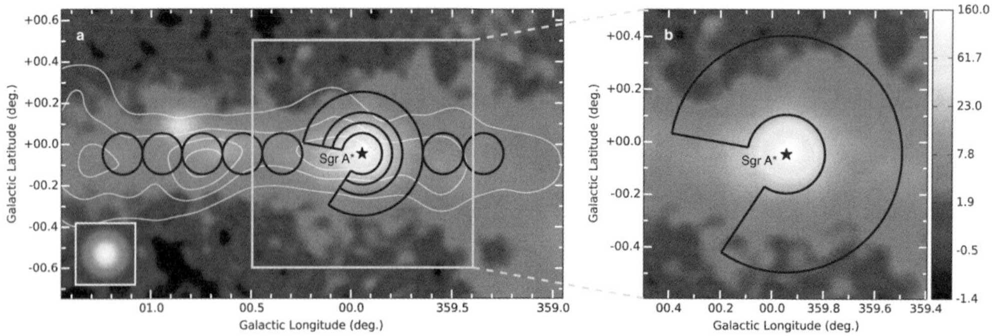

Figure 1. Very-high-energy γ-ray image of the Galactic Centre region as seen by H.E.S.S. The colour scale indicates counts per $0.02^\circ \times 0.02^\circ$ pixel. **Left panel:** The black lines outline the regions used to calculate the CR energy density throughout the central molecular zone. White contour lines indicate the density distribution of molecular gas, as traced by its CS (carbon monosulfide) line emission (Tsuboi *et al.* 1999). The inset shows the simulation of a point-like source. **Right panel:** Zoomed view of the inner $\sim$ 70 pc and the contour of the region used to extract the spectrum of the diffuse emission shown in Figure 3.

2. H.E.S.S. observations and their interpretation

Figure 1 shows the very high-energy gamma-ray image of the Galactic Centre (GC) region, resulting from a 10 years long observational campaign performed by H.E.S.S. (Aharonian *et al.* 2006, Aharonian *et al.* 2009, Aharonian, F. *et al.* 2004, H.E.S.S. Collaboration 2016). A diffuse very-high-energy gamma-ray emission is visible in correspondence with the Central Molecular Zone (CMZ), which contains predominantly molecular gas and extends (in projection) out to $r \sim 250$ pc at positive galactic longitudes and $r \sim 150$ pc at negative longitudes. The strong spatial correlation (although not linear, see discussion below) between gamma rays and gas points towards a hadronic origin of the diffuse emission.

The energy density of multi-TeV CRs in the CMZ is found to be roughly an order of magnitude larger than that of the sea of CRs that universally fills the Galaxy, while the energy density of low energy (GeV) cosmic rays in this region has a level comparable to it (Yang *et al.* 2015). This requires the presence of one or more accelerators of multi-TeV particles operating in the CMZ. From the angular distribution of gamma rays and the spatial 3D distribution of gas in the CMZ it is possible to derive the spatial distribution of multi-TeV CRs (Figure 2), which is consistent with the solution of the transport equation of CRs injected continuously at a constant rate Q at the GC ($R = 0$) and characterised by an isotropic and homogeneous particle diffusion coefficient D:

$$\frac{\partial n_{CR}}{\partial t} = D\nabla^2 n_{CR} + Q\delta(R) \tag{2.1}$$

which reads:

$$n_{CR}(E, R) = \frac{Q(E)}{4\pi D(E) R} \tag{2.2}$$

where n_{CR} represents the differential energy distribution of CRs per unit volume. Equation 2.2 is valid under the condition that the source located at the GC injected CRs for a time Δt long enough to fill with them the entire CMZ (otherwise we would observe a deviation from the $1/r$ profile derived in Figure 2). A lower limit for such time scale can be estimated by equating the CR diffusion length to the size of the CMZ ($r_{CMZ} \sim 200$

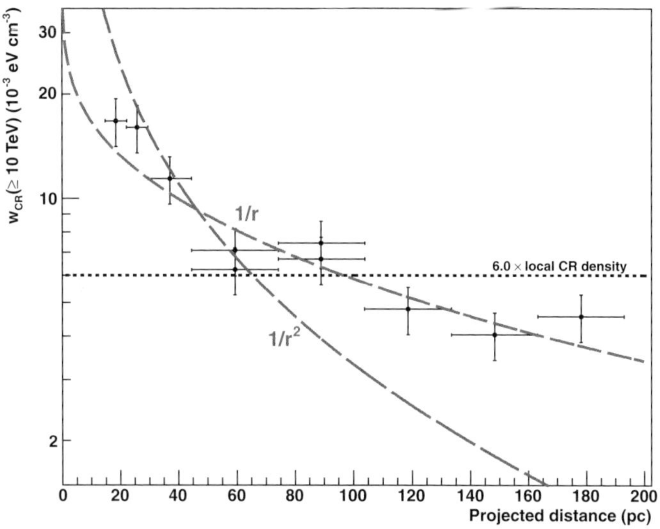

Figure 2. Spatial distribution of the CR density versus projected distance from Sgr A*. The vertical and horizontal error bars show the 1σ statistical plus systematical errors and the bin size, respectively. A fit to the data of a $1/r$, $1/r^2$ and an homogeneous (horizontal dotted line) CR density radial profiles are shown. The best fit of a $1/r^\alpha$ profile to the data is found for $\alpha = 1.10 \pm 0.12$. Thus, the $1/r$ radial profile is clearly preferred by the H.E.S.S. data.

pc):

$$r_{CMZ} \lesssim \sqrt{6\ D(E)\ \Delta t} \tag{2.3}$$

which gives:

$$\Delta t \gtrsim 2 \times 10^3 \left(\frac{D}{10^{30}\ \mathrm{cm^2 s^{-1}}} \right)^{-1} \mathrm{yr} \tag{2.4}$$

where D is normalised to the characteristic value of the diffusion coefficient of multi-TeV CRs in the Galactic Disk (Strong *et al.* 2007).

By fitting the data shown in Figure 2 with Equation 2.2 one can obtain Q or, equivalently, the total energy injection rate of CRs at the GC:

$$W(> 10\ \mathrm{TeV}) \approx 4 \times 10^{37} \left(\frac{D}{10^{30} \mathrm{cm^2/s}} \right) \mathrm{erg/s}\ . \tag{2.5}$$

The particle diffusion coefficient is very unknown in most astrophysical contexts, and this introduces a significant uncertainty in the estimate of the injection power. However, an upper limit for $W(> 10$ TeV) can be obtained after recalling that the $1/r$ profile of the CR spatial distribution is a characteristic feature of diffusive propagation. In order to be in the diffusive regime the following condition has to be satisfied:

$$\frac{R^2}{6D} \gg \frac{R}{c} \tag{2.6}$$

where c is the speed of light. When the condition above is not satisfied CRs propagate ballistically rather than diffusively. Combining Equation 2.5 and 2.6 one finally gets:

$$W(> 10\ \mathrm{TeV}) \ll 1.2 \times 10^{38}\ \mathrm{erg/s}\ . \tag{2.7}$$

Figure 3 shows the energy spectrum of the diffuse gamma-ray emission, which can be fitted by a power law $E^{-\alpha}$ extending with a photon index $\alpha \sim 2.3$ up to energies of tens of TeV. No indication of a cutoff or a break has been found in the spectrum, indicating

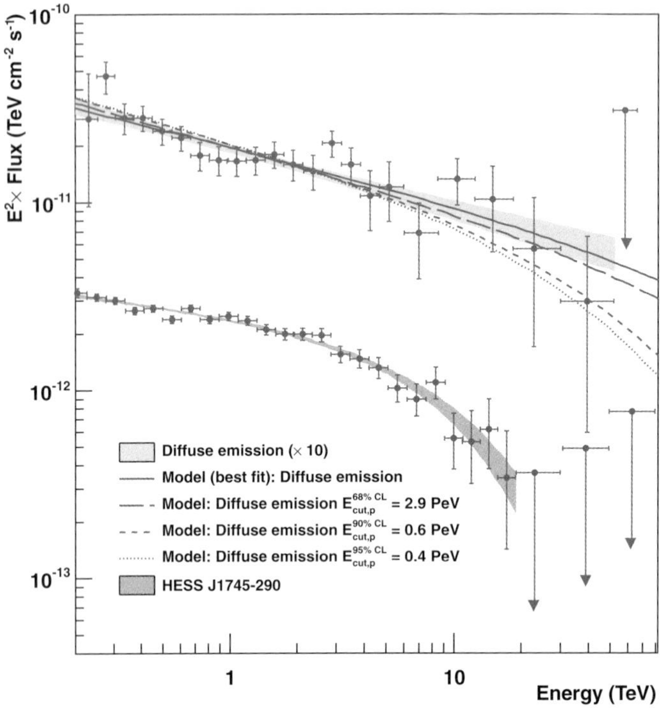

Figure 3. Very-high-energy gamma-ray spectra of the diffuse emission and of HESS J1745-290 (the source at the GC) in units of TeV cm^{-2} s^{-1}. The 1σ confidence bands of the best-fit spectra of the diffuse and HESS J1745-290 are shown in the upper and lower shaded areas, respectively. The lines show the numerical computations assuming that gamma rays result from the decay of neutral pions produced by proton-proton interactions. For the sake of clarity, the fluxes of the diffuse emission spectrum and models are multiplied by 10.

that the emission is hadronic (neutral pion decay) and that particle acceleration proceeds up to PeV energies. This is the first time that such a γ-ray spectrum is detected. The best fit to data is found for a proton spectrum following a pure power-law with index $\alpha \sim 2.4$. If the H.E.S.S. data are fitted with a power law spectrum of protons plus an exponential cutoff at an energy E_{cut}, then the corresponding secondary γ-ray spectrum deviates from data at 68%, 90% and 95% confidence levels for E_{cut} = 2.9 PeV, 0.6 PeV and 0.4 PeV, respectively. We can conclude, then, that a PeVatron is present within $\sim$ 10 pc (the H.E.S.S. angular resolution at a distance of 8.5 kpc) from the GC. The H.E.S.S. Collaboration (2016) discussed several scenarios in order to identify the most plausible site for the acceleration of PeV particles, and concluded that most likely the production of PeV particles is connected, in a way or another, to the supermassive black hole (SMBH) in the GC (Sgr A*). The H.E.S.S. Collaboration (2016) also discussed possible links between the diffuse gamma-ray emission and the GC point-like gamma-ray source (whose spectrum is shown in Figure 3). The reader is referred to that paper for more details.

If the central SMBH is indeed the source of PeV particles, then the required acceleration rate of about $10^{37} - 10^{38}$ erg/s derived in Equation 2.5 and 2.7 would exceed by two to three orders of magnitude the current bolometric luminosity of Sgr A* (Genzel *et al.* 2010). Equivalently, such luminosity can be expressed as about 1 % of the current accretion power of the SMBH. These are quite tight figures, but one has to recall that the accretion rate in the central SMBH currently is relatively modest, and that at certain epochs in the past the central SMBH, of mass 4×10^6 Solar masses, might have operated at much higher accretion rate. Thus, one can speculate that in the past the CR production rate was higher than the present one (see e.g. Istomin 2014 and references therein). In this regard it is interesting to note that an average acceleration rate of 10^{39} erg/s of $E > 10$ TeV protons over the last several millions years would be sufficient to explain the flux of CRs around the knee at 1 PeV. If so, this could be a solution to the problem of the origin of PeV CRs, and provide an alternative to the widely accepted paradigm of the supernova remnant origin of Galactic CRs (Cristofari *et al.* 2013, Parizot 2014). Remarkably, such an acceleration rate would also suffice to explain the flux of very-high-energy neutrinos recently detected by IceCube (Taylor *et al.* 2014).

References

Aharonian, F. A. *Very high energy cosmic gamma radiation : a crucial window on the extreme Universe*, River Edge, NJ: World Scientific Publishing (2004)
Aharonian, F., *et al.* [H. E. S. S. Collaboration] 2009, *A&A*, 503, 817
Aharonian, F., *et al.* [H. E. S. S. Collaboration] 2006, *Nature* 439, 695
Aharonian, F., *et al.* [HESS Collaboration] 2004, *A&A Lett.*, 425, L13
Antoni, T., *et al.* 2005, *Astropart. Phys.*, 24, 1
Bartoli, B., *et al.* 2015, *Phys. Rev. D*, 92, 092005
Bell, A. R., Schure, K. M., Reville, B., & Giacinti, G. 2013, *MNRAS*, 431, 415
Cristofari, P., Gabici, S., Casanova, S., Terrier, R., & Parizot, E. 2013, *MNRAS*, 434, 2748
Drury, L. O'C. 2012, *Astropart. Phys.*, 39, 52
Gabici, S. & Aharonian, F. A. 2007, *ApJ Lett.*, 665, L131
Genzel, R., Eisenhauer, F., & Gillessen, S. 2010, *Rev. Mod. Phys.*, 82, 3121
H. E. S. S. Collaboration 2016, *Nature*, 531, 476
Istomin, Y. N. 2014, *New Astronomy*, 27, 13
Parizot, E. 2014, Nucl. Phys. Proc. Suppl., 256, 197
Strong, A. W., Moskalenko, I. V. & Ptuskin, V. S. 2007 *ARNPS*, 57, 285
Taylor, A. M., Gabici, S., & Aharonian, F. A. 2014, *Phys. Rev. D*, 89, 103003
Tsuboi, M., Handa, T., & Ukita, N. 1999, *ApJ Supp.* 120, 1-39
Yang, R., Jones, D. I., & Aharonian, F. 2015, *A&A*, 580, A90

New Frontiers in Black Hole Astrophysics
Proceedings IAU Symposium No. 324, 2016
Andreja Gomboc, ed.

doi:10.1017/S1743921317002307

Neutrino Astronomy with IceCube

Kevin J. Meagher[1] **on behalf of the IceCube Collaboration**[2]
[1]Université Libre de Bruxelles, Science Faculty CP230,
B1050 Bruxelles, Belgium
email: kmeagher@ulb.ac.be
[2]website: http://icecube.wisc.edu

Abstract. The IceCube Neutrino Observatory is a cubic kilometer neutrino telescope located at the Geographic South Pole. Cherenkov radiation emitted by charged secondary particles from neutrino interactions is observed by IceCube using an array of 5160 photomultiplier tubes embedded between a depth of 1.5 km to 2.5 km in the Antarctic glacial ice. The detection of astrophysical neutrinos is a primary goal of IceCube and has now been realized with the discovery of a diffuse, high-energy flux consisting of neutrino events from tens of TeV up to several PeV. Many analyses have been performed to identify the source of these neutrinos: correlations with active galactic nuclei, gamma-ray bursts, and the galactic plane. IceCube also conducts multi-messenger campaigns to alert other observatories of possible neutrino transients in real-time. However, the source of these neutrinos remains elusive as no corresponding electromagnetic counterparts have been identified. This proceeding will give an overview of the detection principles of IceCube, the properties of the observed astrophysical neutrinos, the search for corresponding sources (including real-time searches), and plans for a next-generation neutrino detector, IceCube–Gen2.

Keywords. acceleration of particles, neutrinos, galaxies: active, gamma rays: bursts

---

1. Neutrino Astronomy

From radio waves to gamma-rays, electromagnetic radiation has been the source of a wealth of information about the universe. Unfortunately, photons with energies above 1 TeV are absorbed on the extra-galactic background light making it impossible to see extra-galactic sources at these energies. In order to study the universe above this cut-off we need to find an alternative to photons. Cosmic rays tell us that charged particles are accelerated by astrophysical objects up to at least 10^{20} eV, but since charged particles are deflected by magnetic fields, the origin of these particles still remains unclear. Since neutrinos interact solely via the weak force, they can traverse the universe completely unimpeded and therefore hold the potential to open a new window on astronomy.

2. The IceCube Neutrino Observatory

Neutrinos' small cross-section, the same property that allows them to arrive at Earth unimpeded, also makes them difficult to detect. Observing neutrinos requires a large target mass to make up for the small cross-section. In addition, the medium must be transparent in order to observe the light from the secondary particles. The IceCube Neutrino Observatory was built in the Antarctic glacier under the South Pole Station.

The fundamental unit of IceCube is the digital optical module (DOM). Each DOM contains a 25 cm photomultiplier tube, high voltage power supply, and digitization and communication electronics. The 60 DOMs are aligned on vertical structures called strings, with a vertical spacing of 17 m between a depth of 1450 m and 2450 m. There are 86 strings for a total of 5160 DOMs. The strings form a triangular grid with a spacing of 125 m, except for 8 strings arrayed in the center to form a denser formation referred to as DeepCore.

There are three main detector channels used for neutrino astronomy: muon tracks, cascades, and HESE. Muon tracks are produced from charged current ν_μ interactions. At these energies, muons can travel tens of kilometers from the interaction vertex. The observed 1 km tracks allow for muons to be reconstructed with good angular resolution, $0.7°$ for energies above 10 TeV. Including muons produced far from the detector greatly increases the effective area of this channel. The energy that the muon deposits in the detector can be measured, but this may be significantly less than the energy of the original neutrino. However, it is difficult to distinguish neutrinos from muon bundles produced by cosmic ray air showers, so the track channel has sensitivity primarily in the Northern Hemisphere where air shower bundles are screened out by the Earth.

The cascade channel uses events from either neutral current interactions or charged current interactions from electron and tau neutrinos. These interactions create secondary particles which lose energy on length scales much smaller than the spacing of the strings and thus appear to be near point-like emissions of light from the interaction vertex. Since all the energy is deposited near the vertex, these events have much better energy resolution than tracks, but at the cost of relatively poor angular resolution.

The high energy starting-event (HESE) channel observes events which start in the detector volume, by only selecting events where the initial light occurs on DOMs within the interior of the detector, and vetoing events which start near the edge. Although the events in this channel may be considered either tracks or cascades, from an analysis point of view it is often easier to consider HESE as a separate channel.

IceCube experiences two main kinds of background: muon bundles from cosmic ray air showers referred to simply as muons and neutrinos from the same air showers which are referred to as atmospheric neutrinos. Muons are only a background in the Southern Hemisphere as the Earth screens out muons from the Northern Hemisphere. But atmospheric neutrinos represent an irreducible background which can only be distinguished from astrophysical neutrinos by their energy spectrum. The atmospheric neutrino spectrum is very soft, being described with a power-law of $E^{-3.7}$ or steeper while acceleration mechanisms for astrophysical objects are expected to produce a spectral index of 2–2.5.

3. Observation of High-Energy Neutrinos

Using the HESE channel, an analysis performed on 4 years of data found 54 neutrino candidate events with a statistical significance of 6.5σ (Aartsen *et al.* 2015a). In order to describe the data, a maximum likelihood, forward-folding fit of all components (atmospheric muons, atmospheric neutrinos from π/K decay, atmospheric neutrinos from charm decay and an astrophysical flux assuming a 1:1:1 flavor ratio) was performed on the energy spectrum. The result of the fit, shown in Figure 1 (left), is $\mathrm{d}N/\mathrm{d}E = (2.2 \pm 0.7) \times 10^{-18} \cdot (E/100\,\mathrm{TeV})^{-2.58\pm0.25}\mathrm{GeV}^{-1}\mathrm{cm}^{-2}\mathrm{s}^{-1}\mathrm{sr}^{-1}$. A maximum likelihood clustering method was used to look for any neutrino point sources in this sample. This test, shown in Figure 1 (right), did not yield significant evidence of clustering with p-values of 44% and 58% for the shower-only and the all-events tests, respectively. A test for galactic plane clustering was also performed. Assuming a galactic width of $2.5°$ around the plane resulted in a p-value of 7% and a variable galactic width scan resulted in a p-value 2.5%.

A separate diffuse spectral analysis was performed using six years of data with the muon track channel (Aartsen *et al.* 2016a). At energies between 191 TeV and 8.3 PeV a

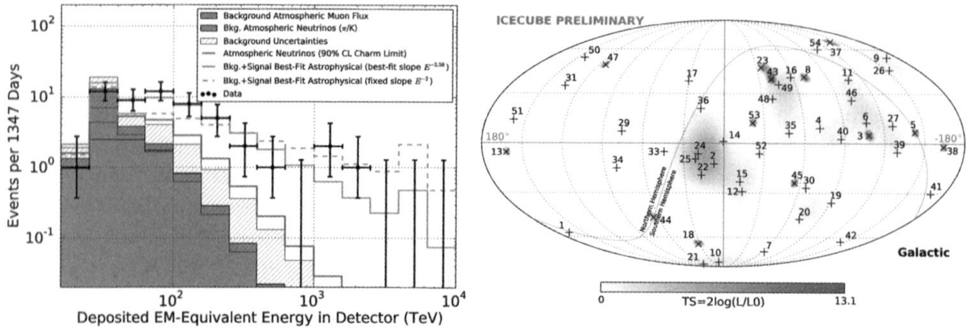

Figure 1. **Left:** Deposited energy spectrum of high energy starting-event (HESE) channel (Aartsen *et al.* 2015a). The hashed region shows uncertainties on the sum of all backgrounds. Muons are computed from simulation to overcome statistical limitations in our background measurement and scaled to match the total measured background rate. Atmospheric neutrinos and uncertainties thereon are derived from previous measurements of both the π/K and charm components of the atmospheric ν_μ spectrum. **Right:** Arrival directions of the same sample in galactic coordinates. Shower-like events are marked with "+" and those containing tracks with "×". Colors show the test statistics for the point-source clustering test at each location. No significant clustering was found.

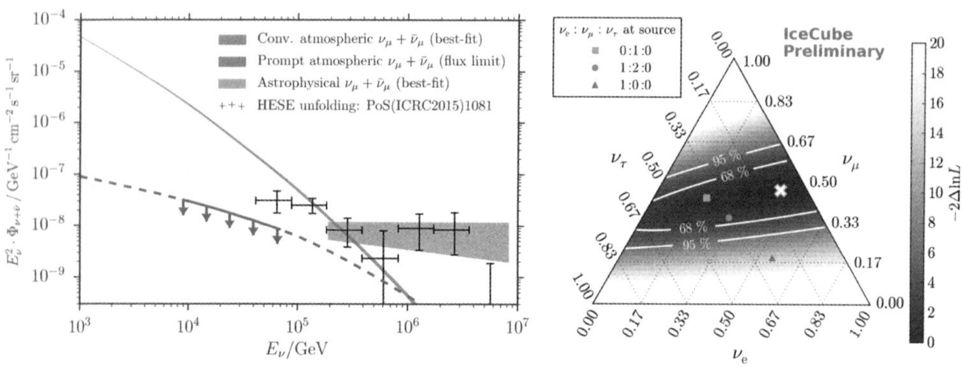

Figure 2. **Left:** Best-fit neutrino spectra for the unbroken power-law model using muon neutrino interactions. The conventional and astrophysical neutrino fluxes are represented by shaded regions indicating one sigma error on the measured spectrum where as the solid line represents the upper limit on the prompt neutrino model in Enberg *et al.*(2008). The horizontal width of the astrophysical shaded region denotes the range of neutrino energies which contribute 90% to the total likelihood ratio between the best-fit and the conventional atmospheric-only hypothesis. The crosses show the unfolded spectrum of the high-energy sample discussed above. **Right:** The results of the profile likelihood scan of the flavor composition at Earth. Each point in the triangle corresponds to a ratio $\nu_e : \nu_\mu : \nu_\tau$ as measured on Earth, the individual contributions are read off the three sides of the triangle. The best-fit composition is marked with "×", 68% and 95% confidence regions are indicated. The ratios corresponding to three flavor composition scenarios at the sources of the neutrinos are marked by the square for pion-decay (0:1:0), circle for muon-damped (1:2:0), and triangle for neutron-beam (1:0:0) sources respectively.

significant astrophysical contribution was observed with a significance of 5.6σ. As shown in Figure 2 (left), the data was well-described by a power-law: $\mathrm{d}N/\mathrm{d}E = (0.90^{+0.30}_{-0.27}) \times 10^{-18} \cdot (E/100\,\mathrm{TeV})^{-2.13\pm0.13}\,\mathrm{GeV}^{-1}\mathrm{cm}^{-2}\mathrm{s}^{-1}\mathrm{sr}^{-1}$. Compared to HESE, this result has a harder spectrum. However, this analysis has a lower energy threshold, possibly indicating a break in the astrophysical neutrino spectrum. This analysis included the detection of a track with a reconstructed muon energy of 4.5 ± 1.2 PeV which implies a probability of less than 0.005% for this event to be of atmospheric origin. In addition, this analysis

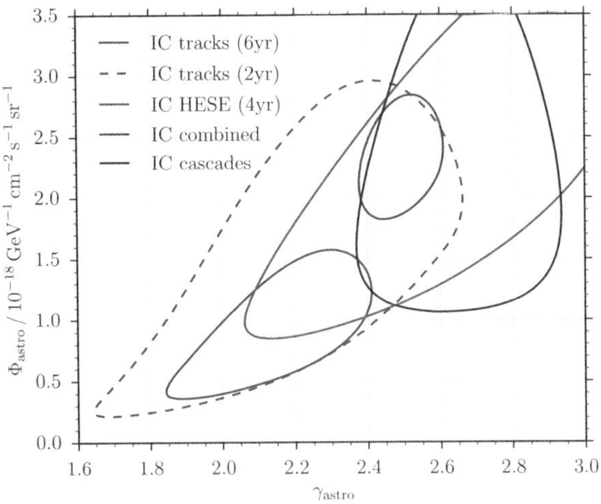

Figure 3. Results of different IceCube analyses measuring the astrophysical flux parameters Φ_{astro} and γ_{astro}. The contour lines show the 90% CL. The result of the muon track diffuse analysis (Aartsen *et al.* 2016a) is shown by the bottom-left solid contour line. The contour obtained by the previous measurement using the same method but using 2 years of data is the bottom-left dashed line. The results of the HESE analysis (Aartsen *et al.* 2015a), cascade channel (Aartsen *et al.* 2015b), and the combined analysis (Aartsen *et al.* 2016a) are also shown.

was able to constrain the contribution from a prompt atmospheric muon neutrino flux originating from charmed meson decays below 1.06 of the flux normalization from Enberg *et al.* (2008).

The ratio of different neutrino flavors can give important clues to acceleration mechanisms of the source. In Aartsen *et al.*(2015b) we also performed a measurement of the flavor composition of the astrophysical neutrino flux, in which the normalizations of all three flavors were allowed to vary independently. The results, shown in Figure 2 (right), are consistent with pion-decay sources and muon-damped sources but disfavor neutron-beam sources with a significance of 3.6σ.

In the cascade channel, in an analysis of the first two years of data a total of 172 events were observed with energies between 10 TeV and 1 PeV (Aartsen *et al.* 2015b). The astrophysical component is also well described by a power-law: $\mathrm{d}N/\mathrm{d}E = (2.3^{+0.7}_{-0.6}) \times 10^{-18} \cdot (E/100\,\mathrm{TeV})^{-2.67\pm0.13}\,\mathrm{GeV}^{-1}\mathrm{cm}^{-2}\mathrm{s}^{-1}\mathrm{sr}^{-1}$. The background-only hypothesis is rejected with a significance of 4.7σ.

The results of these analyses along with the results of 3 other diffuse analyses were combined into a global spectral analysis (Aartsen *et al.* 2016a). Assuming the astrophysical neutrino flux to be isotropic and to consist of equal flavors at Earth, the all-flavor spectrum with neutrino energies between 25 TeV and 2.8 PeV is well described by an unbroken power law with a best-fit spectral index 2.50 ± 0.09 and a flux at 100 TeV of $6.7^{+1.1}_{-1.2} \cdot 10^{-18}\,\mathrm{GeV}^{-1}\,\mathrm{s}^{-1}\,\mathrm{sr}^{-1}\,\mathrm{cm}^{-2}$. Under the same assumptions, an unbroken power law with index 2 is disfavored with a significance of 3.8σ (p=0.0066%) with respect to the best fit. This significance is reduced to 2.1σ (p=1.7%) if instead we compare the best fit to a spectrum with index 2 that has an exponential cut-off at high energies. Allowing the electron neutrino flux to deviate from the other two flavors, we find a ν_e fraction of 0.18 ± 0.11 at Earth. The sole production of anti-electron neutrinos, which would be characteristic of neutron-decay dominated sources, is rejected with a significance of 3.6σ (p=0.014%). The results of the combined sample spectral fit along with the previously mentioned analyses are shown in Figure 3. Significant tension is seen between the

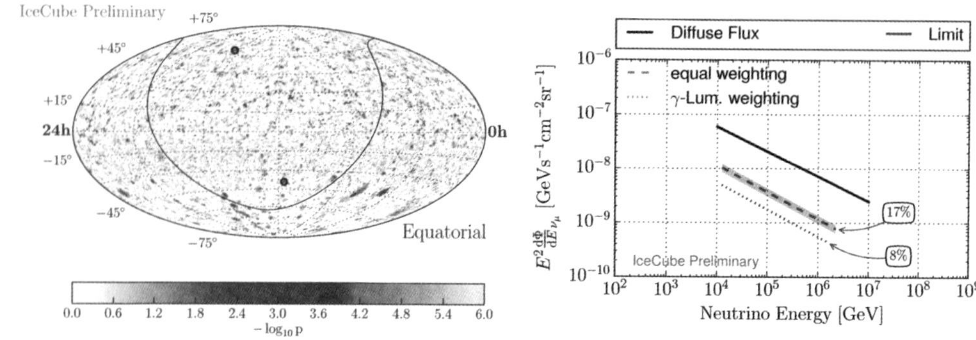

Figure 4. Left: Pre-trial significance skymap of the all-sky point source scan (Aartsen *et al.* 2015d). The color indicates the negative logarithm of the pre-trial p-value assuming no clustering as null-hypothesis. Shown in Equatorial (J2000) coordinates, a line indicates the galactic plane. The most significant fluctuation in each hemisphere is marked. **Right:** Results of the stacked blazar analysis. Neutrino flux upper limits for an $E^{-2.5}$ spectrum from Glüsenkamp (2016) compared to diffuse bestfit (solid line) from Aartsen *et al.* (2016a). Two seperate signal weighting schemes are shown: equal weighting (dashed line) where blazars are considered to contribute equally to the neutrino flux and weighitng by blazars' observed gamma-ray luminosity (dotted line).

different analyses. Since each analysis has a different sensitivity as a function of energy and zenith angle, the difference may be due to the energy spectrum changing shape at higher energy or possibly due to a population of harder sources in the Southern Hemisphere.

4. The Search for Astrophysical Sources

To identify the source of the neutrino populations described in the previous section, many analyses have been performed. To date none of them have identified any association with known or unknown astrophysical sources. In seven years of data, from 2008–2015, using an unbinned maximum-likelihood search for local clustering in the muon channel, no significant clustering of neutrinos above background expectation was observed (Aartsen *et al.* 2015d). The map generated by this analysis is shown in Figure 4 (left). The negative result of this analysis excludes point sources with a flux of $E^2 \mathrm{d}\Phi/\mathrm{d}E = 10^{-12}$ TeV cm^{-2} s^{-1}.

Blazars have been proposed as a possible source of high-energy neutrinos. To investigate this a stacked analysis was performed with blazars from the 2nd Fermi-LAT AGN catalogue (2LAC) (Glüsenkamp 2016). No significant excess is observed, constraining the total population of 2LAC blazars to contributing 27% or less of the observed astrophysical neutrino flux, assuming equipartition of neutrino flavors at Earth and the currently favoured power-law spectral index for the neutrino flux of 2.5. As shown in Figure 4 (right), the 2LAC blazars (and sub-populations) are excluded as being the dominant sources of the observed neutrinos up to a spectral index as hard as 2.2.

Another astrohphysical source considered to be a likely source of neutrinos are gamma-ray bursts (GRBs). An analysis incorporating 4 years of muon track events and 506 observed Northern Hemisphere bursts found no correlation more significant than expected from background (Aartsen *et al.* 2015c). In addition, a 3 year search using the cascade channel correlated with 807 GRBs from the whole sky found similar results (Aartsen *et al.* 2016b). The limits on the neutrino flux set by these analyses (see Figure 5,left) disfavor much of the parameter space for the theories on neutrino emission from

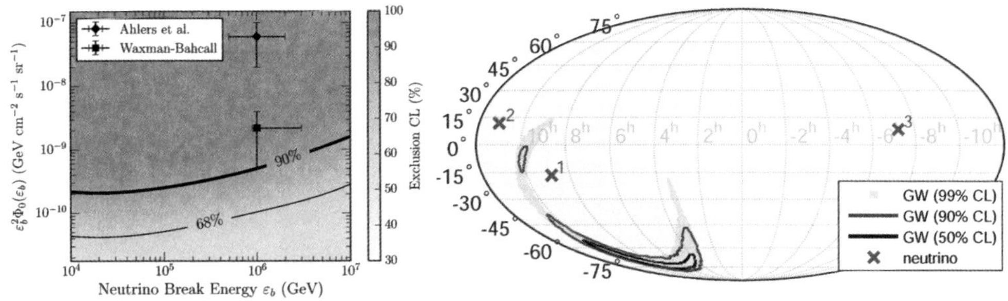

Figure 5. Left: Constraint on generic doubly broken power law neutrino flux models as a function of first break energy ε_b and normalization Φ_0. The model by Ahlers *et al.* (2011) assumes that only neutrons escape from the GRB fireball to contribute to the UHECR flux. The Waxman & Bahcall model (1997), which allows all protons to escape the fireball, has been updated to account for more recent measurements of the UHECR flux (Katz *et al.* 2009) and typical gamma break energy (Goldstein *et al.* 2012). Exclusion contours, calculated from the combined 4 year all-sky analysis (Aartsen *et al.* 2016b) with the 4 year Northern Hemisphere analysis (Aartsen *et al.* 2015c), are shown. **Right:** Gravitational wave skymap in equatorial coordinates, showing the reconstructed probability density contours of the GW event at 50%, 90% and 99% CL, and the reconstructed directions of high-energy neutrino candidates detected by IceCube (crosses) during a ± 500 s time window around gravitational wave event GW150914. The neutrino directional uncertainties are $< 1^\circ$ and are not shown. Gravitational wave shading indicates the reconstructed probability density of the gravitaitonal wave event, darker regions corresponding to higher probability.

GRBs. Combined, these analyses find that no more than 1% of the observed astrophysical neutrino flux consists of prompt emission from GRBs that are observable by existing satellites.

Another considered source was the first gravitational wave transient GW150914 observed by the Advanced LIGO detectors on Sept. 14th, 2015. The analysis was perfomed by looking for neutrino candidates within 500 s of the gravitational wave event. As shown in Figure 5 (right) and consistent with background, 3 events were observed within this time window, none of them within the region triangulated by LIGO (Adrián-Martánez *et al.* 2016).

In order to alert other astronomers about possible neutrino transient events, the IceCube collaboration has developed several real-time alert programs. The neutrino data is processed in real-time at the South Pole station and the most interesting neutrino events are selected to trigger observations with optical and X-ray telescopes aiming for the detection of an electro-magnetic counterpart such as a GRB afterglow or a rising SN light curve. The program is capable of triggering follow-up observations in less than a minute. The optical follow-up program (Abbasi *et al.* 2012) has been sending such alerts to optical telescopes since 2008 and to X-ray telescopes since 2009. The gamma-ray follow-up program has been running since 2012 (Aartsen *et al.* 2016c), sending triggers to the MAGIC and VERITAS gamma-ray telescopes. This program focuses on blazar flares by monitoring a predefined list of known blazars and looks for excesses of neutrino events on timescales of up to three weeks.

5. Future Instrumentaion

Although IceCube has positivetly identified neutrinos of astrophysical origin, the ability of IceCube to be an efficient tool for neutrino astronomy over the next decade is limited by the modest numbers of cosmic neutrinos measured, even with a cubic

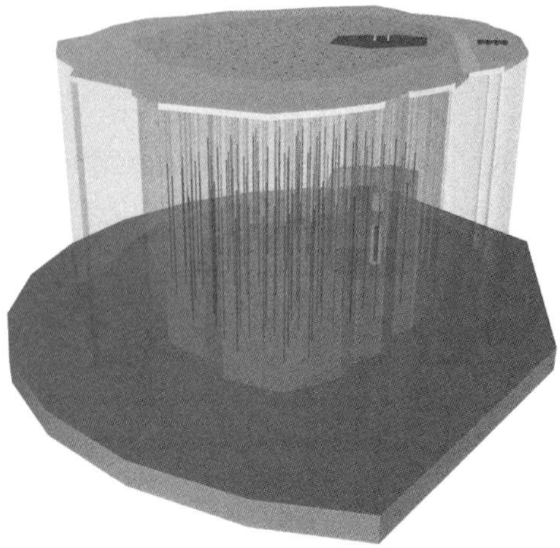

Figure 6. Conceptual schematic of the proposed high-energy extension to the IceCube detector. The current IceCube is shown in red with DeepCore in green. 120 additional string are added to increase the instrumented volume to $\sim 10\,\text{km}^3$. A veto detector comprised of scintillator detectors is also envisioned at the surface.

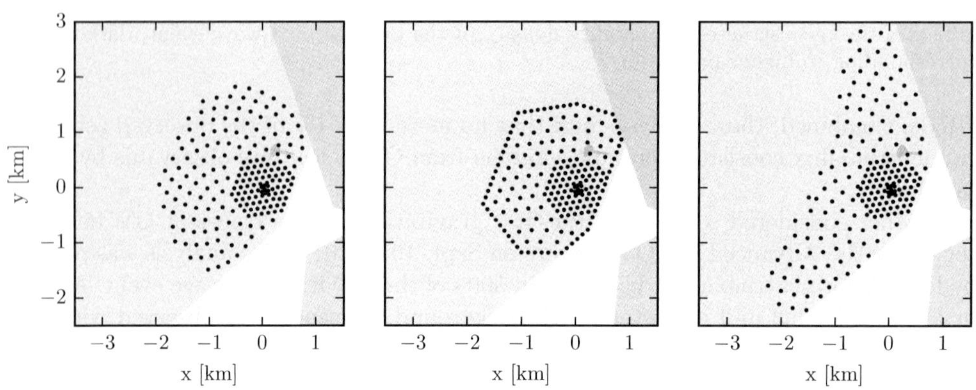

Figure 7. Example benchmark detector string layouts under study for the high energy extension to IceCube. Each expands about IceCube by adding 120 strings constrained to the South Pole Dark Sector (shaded in light green). For the left panel, uniform string spacing of 240 m is shown. The central panel represents a string layout with a denser edge weighting for improved veto efficiency. The right panel shows the so-called banana geometry which seeks to create a very long detector for certian muon tracks.

kilometer array. Design studies to increase IceCube's sensitivity with additional strings outside the current volume are currently underway (Aartsen *et al.* 2014a). This section will describe this effort, referred to as the IceCube–Gen2 High-Energy Array. The design, shown in Figure 6, seeks to increase the instrumented volume to $\sim 10\,\text{km}^3$. The high-energy array is proposed to complement the high-density, low-energy sub-array known as PINGU (Aartsen *et al.* 2014b). PINGU targets precision measurements of the atmospheric oscillation parameters and the determination of the neutrino mass hierarchy.

The light properties of deep Antarctic ice allow string spacing to be increased to 300 m for energies exceeding 10 TeV. Since angular resolution for muon tracks is porportional to the length of the lever arm, by increasing the size of the detector, the angular resolution will also be improved, further improving point-source sensitivity. Studies to find the optimum geometry and string spacing are currently underway, some of the geometries can be seen in Figure 7. All of the designs add 120 strings to the detector within the region of the South Pole station designated as the dark sector. Uniform string spacings of 200 m, 240 m, and 300 m, which instrument volumes of 6.0 km^3, 8.0 km^3, and 11.9 km^3 respectively, have been studied. Alternative array designs are also under study. The edge weighted geometry was conceived to improve the efficiency at which the outer layers aid in identifying interactions which occur inside the detector volume. It would have edge strings spaced at 125 m, while interior strings would be spaced at 240 m, with an instrumented volume of 6.2 km^3. The so-called banana geometry would create one direction in which muon tracks would traverse a very long instrumented distance which will improve angular resolution along that axis. Tracks which traverse perpendicular to this direction would not benefit from the increased angular resolution but would see an increased detector size. In addition, IceCube–Gen2's reach may further be enhanced by exploiting the air-shower detection and vetoing capabilities of an extended surface array, or by including an extended 100 km^2 radio array to achieve improved sensitivity to neutrinos in the 10^{16}–10^{20} eV energy range, including GZK neutrinos.

While the design details remain to be finalized, IceCube–Gen2 will reveal an unobstructed view of the universe at PeV energies where most of the universe is opaque to high-energy photons. It will operate simultaneously with next generation of electromagnetic and gravitational wave detectors, allowing for more multimessenger analyses. With its unprecedented sensitivity and improved angular resolution, this instrument will enable detailed spectral studies, significant point source detections and new discoveries.

References

Aartsen, M. G. *et al.* 2014a, `arXiv:1412.5106`.

Aartsen, M. G. *et al.* 2014b, `arXiv:1401.2046v1`.

Aartsen, M. G. *et al.* 2015a, in: *Proceedings of the 34th International Cosmic Ray Conference*. PoS(ICRC2015)1081. `arXiv:1510.05223v2`.

Aartsen, M. G. *et al.* 2015b, in: *Proceedings of the 34th International Cosmic Ray Conference*. PoS(ICRC2015)1109. `arXiv:1510.05223v2`.

Aartsen, M. G. *et al.* 2015c, *ApJ* 805 L5. `arXiv:1412.6510`

Aartsen, M. G. *et al.* 2015d, in: *Proceedings of the 34th International Cosmic Ray Conference*. PoS(ICRC2015) 1047. `arXiv:1510.05222v3`.

Aartsen, M. G. *et al.* 2016a, accepted for publication in *ApJ*. `arXiv:1607.08006`.

Aartsen, M. G. *et al.* 2016b, *ApJ* 824 115. `arXiv:1601.06484`.

Aartsen, M. G. *et al.* 2016c, submitted to *J. Inst.*

Abbasi, R. *et al.* 2012, *A&A* 539 A60. `arXiv:1111.7030`.

Adrián-Martánez, S. *et al.* 2016, *Phys. Rev. D* 93, 122010. `arXiv:1602.05411`.

Ahlers, M., Gonzalez-Garcia, M., & Halzen, F. 2011, *APh* 35, 87. `arXiv:1103.3421`

Enberg, R., Reno, M. H., & Sarcevic, I. 2008, *Phys.Rev.D* 78, 043005. `arXiv:0806.0418`.

Glüsenkamp, T. 2016, in: Roma International Conference on Astroparticle Physics 2014 (RICAP-14), EPJ Web of Conferences Volume 121. `arXiv:1502.03104`.

Goldstein, A. *et al.* 2012, *ApJS* 199, 19. `arXiv:1201.2981`.

Katz, B., Budnik, R., & Waxman, E. 2009, *JCAP* 2009, 020. `arXiv:0811.3759`

Waxman, E. & Bahcall, J. 1997, *PhRvL* 78, 2292. `astro-ph/9701231`

New Frontiers in Black Hole Astrophysics
Proceedings IAU Symposium No. 324, 2016
Andreja Gomboc, ed.

doi:10.1017/S1743921316012424

LSST survey: millions and millions of quasars

Željko Ivezić

Department of Astronomy, University of Washington,
Box 351580, Seattle, WA 98195-1580, USA
email: ivezic@astro.washington.edu

Abstract. The Large Synoptic Survey Telescope (LSST), the next-generation optical imaging survey sited at Cerro Pachon in Chile, will provide hundreds of detections for a sample of more than ten million quasars with redshifts up to about seven. The LSST design, with an 8.4m (6.7m effective) primary mirror, a 9.6 sq. deg. field of view, and a 3.2 Gigapixel camera, will allow about 10,000 sq. deg. of sky to be covered twice per night, every three to four nights on average, with typical 5-sigma depth for point sources of r=24.5 (AB). With about 1000 observations in *ugrizy* bands over a 10-year period, these data will enable a deep stack reaching r=27.5 (about 5 magnitudes deeper than SDSS) and faint time-domain astronomy. The measured properties of newly discovered and known astrometric and photometric transients will be publicly reported within 60 sec after closing the shutter. In addition to a brief introduction to LSST, I review optical quasar selection techniques, with emphasis on methods based on colors, variability properties, and astrometric behavior.

Keywords. surveys, galaxies: active, quasars: general, stars: variables, stars: statistics

1. Introduction

The selection of large samples of active galactic nuclei, including quasars as their high-luminosity tail, is required in many astrophysical areas, such as galaxy evolution, black hole growth, and the large-scale structure of the universe. The available quasar samples have increased by over two orders of magnitude in less than two decades and this rapid progress is expected to continue (for example, simulations predict that LSST will deliver a sample of about 10 million quasars; see Section 3.1). Here I briefly review optical quasar selection methods based on SDSS data and discuss how they will be improved with the aid of more precise time-resolved photometry expected from LSST. My main sources, where more detailed information about LSST and AGN/quasar science with LSST data can be found, include the so-called LSST overview paper (Ivezić *et al.* 2008) and the LSST Science Book (LSST Science Collaboration 2009).

2. Finding quasars with SDSS

The similarity of unresolved quasars to stars in optical imaging surveys poses a difficulty in their identification. Quasars can be reliably identified by their spectra, and the SDSS provided a spectroscopically complete survey to $i < 19$ with ~106,000 quasars (Schneider *et al.* 2010). This is the largest homogeneous quasar sample with high-quality optical spectra assembled to date (Pâris *et al.* 2012 recently reported an additional ~74,000 objects selected with selection criteria modified to enable the Lyman-α forest analysis). These objects were selected as quasar candidates using colors measured from the *ugriz* SDSS imaging data, and also using radio continuum 20 cm data from the FIRST survey (Becker *et al.* 1995). After SDSS collected enough multi-epoch imaging data, it was demonstrated that photometric variability-based selection is even more

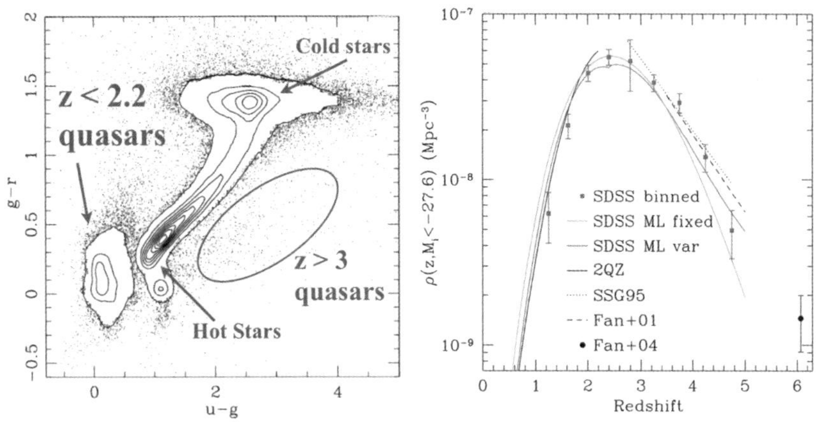

Figure 1. The left panel illustrates the color-based selection method for quasar candidates. Essentially, the algorithm selects all sources whose colors place them outside the main stellar locus, seen in the middle of the panel. Quasars with redshifts below 2.2 have distinctively blue $u-g$ colors. The right panel shows the variation of volume number density of luminous quasars. The local maximum at about $z = 2.4$ is clearly visible. Adapted from Richards *et al.* (2006).

efficient than color-based selection. These two selection methods are briefly summarized here.

2.1. *Color selection of quasar candidates*

A small fraction of SDSS sources unresolved in imaging data were automatically targeted as quasar candidates for spectroscopic followup (Richards *et al.* 2002). The quasar targeting algorithm selects all point sources with $15 < i < 19.1$ (i is the apparent magnitude in the SDSS i band) whose colors place them outside the main stellar locus (it is a bit more complicated than this – for details please see figure 1 in Richards *et al.* 2002). For quasars at redshifts below about two, the most discriminatory color is the $u-g$ color: compared to stars of the same visual color (e.g., the $g-r$ color), quasars show an excess of ultraviolet flux in the u band and thus have bluer $u-g$ colors (see the left panel in Figure 1); this method is often called the UV-excess selection. The colors of quasars at higher redshifts are typically significantly different from stellar colors. A thorough analysis of the colors of quasars in the SDSS photometric system was presented in Richards *et al.* (2001). In addition, all unresolved sources from the same magnitude range that are within 2 arcsec of a FIRST radio detection are also targeted (for analysis of these radio quasars, see Kimball *et al.* 2011 and references therein). Some quasars were also targeted fortuitously via the algorithms for selecting galaxies (Strauss *et al.* 2002). The completeness of the resulting quasar sample is above 90% (the confirmed fraction of all quasars within the adopted flux limits and within the surveyed area) and the efficiency of color selection is about 65% (that is, about 35% of selected quasar candidates turned out not to be quasars). This homogeneously selected sample spans a large redshift range (there are 56 quasars at redshifts beyond 5 in the Schneider *et al.* sample) and has enabled numerous quasar studies. For example, the peak in the comoving volume number density of luminous quasars (dominated by type I objects) is now reliably and precisely determined (see the right panel in Figure 1), and the luminosity functions of quasars and AGN galaxies (selected using emission line strengths) appear mutually consistent despite grossly different selection procedures (Hao *et al.* 2005). We note that several hundred candidate type II quasars (high-luminosity analogs of type 2 Seyfert galaxies) were found in the SDSS spectroscopic survey of galaxies (Zakamska *et al.* 2003, 2004).

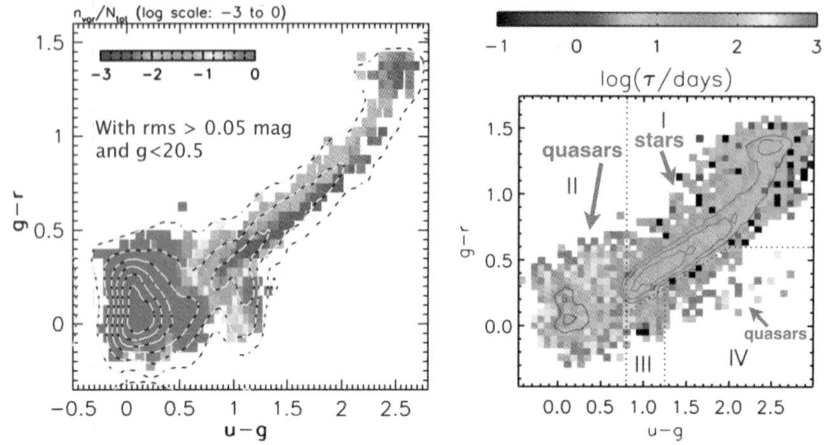

Figure 2. The left panel shows the fraction of point sources with $g < 20.5$ from SDSS Stripe 82 dataset that are variable (the root-mean-square variation of at least 0.05 mag; adapted from Sesar *et al.* 2007). The two regions where essentially all point sources are variable correspond to low-redshift quasars ($u - g \sim 0$) and RR Lyrae stars ($u - g \sim 1$). The right panel shows the median characteristic timescale τ for all variable point sources in SDSS Stripe 82 dataset that are well described by the damped random walk model, as a function of $u - g$ and $g - r$, following the legend at the top (adapted from MacLeod *et al.* 2011).

The relatively simplistic color-based selection algorithm employed by the original SDSS spectroscopic target selection pipeline has been significantly improved using modern data mining and machine learning methods. For example, Richards *et al.* (2009) introduced a kernel density estimator and a non-parametric Bayesian classification method, and Bovy *et al.* (2012) introduced a Gaussian mixture model to recognize quasar candidates. These methods have yielded large samples of candidates (of order a million) and with improved completeness and efficiency tradeoff; Richards *et al.* (2009) reported efficiency of up to 97% while maintaining fairly high completeness levels above 70%. More details about the performance of these modified selection algorithms are available in Pâris *et al.* (2012) and Ross *et al.* (2012).

2.2. *Variability selection of quasars*

Quasars are variable sources with optical amplitudes of several tenths of a magnitude for time scales longer than a few months, and this behavior can be used for their selection (Hawkins & Veron 1995; Ivezić *et al.* 2003). Sesar *et al.* (2007) showed using SDSS Stripe 82 data (a $\sim$300 deg^2 equatorial region imaged about 60 times) that practically all quasars spectroscopically confirmed by SDSS are also variable in SDSS imaging data (see left panel in Figure 2), and Kozłowski *et al.* (2010) demonstrated that variability can be used to separate quasars from most variable stars even in the dense stellar environments of the Magellanic Clouds.

A number of studies used SDSS light curves for close to 10,000 quasars from Stripe 82 to quantify the structure function for quasar variability (Ivezić *et al.* 2004; MacLeod *et al.* 2010; Schmidt *et al.* 2010; Butler & Bloom 2011; Palanque-Delabrouille *et al.* 2011). The structure function as a function of time lag Δt, SF(Δt), as defined in recent quasar studies, is equal to the standard deviation of the distribution of the magnitude difference $m(t_2) - m(t_1)$ evaluated at many different times t_1 and t_2, such that time lag $\Delta t = t_2 - t_1$ (and divided by $\sqrt{2}$ because of differencing). The structure function is directly related to the autocorrelation function, which makes a Fourier pair with the power spectral density function (PSD). When the structure function SF $\propto (\Delta t)^\alpha$, then PSD $\propto 1/f^{(1+2\alpha)}$.

The power-law index α is a good discriminator between variable stars and quasars† (Schmidt *et al.* 2010; Butler & Bloom 2011). The key insight is that for time lags below a year or so quasars have a much steeper structure function ($\alpha \sim 0.4 - 0.5$) than most variable stars. In other words, compared to their variability at a time lag of, say, one year, quasars vary by 1-2 orders of magnitude less at time lags below a month or so – this is not true for the overwhelming majority of variable stars. For UV-excess selected objects, variability-based methods that utilize α select quasars with a completeness of 90% and a purity of 95% (Schmidt *et al.* 2010). Furthermore, this performance level is maintained in the redshift range $2.5 < z < 3$ where color selection is known to be problematic.

The power-law dependence of the structure function for quasar variability cannot be extrapolated beyond a few years, as shown using SDSS and POSS data by Ivezić *et al.* (2003) and later using also Stripe 82 data by MacLeod *et al.* (2012). It has been demonstrated that quasar variability can be statistically described using a stochastic model called damped random walk (Kelly, Bechtold & Siemiginowska 2009; Kozłowski *et al.* 2010, MacLeod *et al.* 2010, 2011, 2012; Zu *et al.* 2012). Also known in the physics literature as the Ornstein–Uhlenbeck process, and as the continuous autoregressive process in the statistics literature, this model includes three parameters: the mean value (of the quasar magnitude), the characteristic time scale τ, and the asymptotic (at time scales longer than τ) root-mean-square variability, SF_∞. Alternatively, the model (as well as its modifications) can be described via a covariance matrix (see MacLeod *et al.* 2010). The predicted structure function for the damped random walk process is $SF(\Delta t) = SF_\infty\, [1 - \exp(-\Delta t/\tau)]^{1/2}$. At small time lags, $SF(\Delta t) \propto \Delta t^{1/2}$, and thus a damped random walk is equivalent to an ordinary random walk for $\Delta t < \tau$ (for a random walk, $PSD \propto 1/f^2$; the "damped" aspect manifests itself as a flat PSD for $\Delta t > \tau$).

The time span of SDSS data from Stripe 82 is sufficiently long to constrain τ for the majority of the ~10,000 quasars with light curves (see left panel in Figure 2). MacLeod *et al.* (2011) showed that this additional time scale information can be used to improve significantly the selection based only on the slope of the structure function for small time lags. For example, it is possible to construct samples with only 15% stellar contamination even when using *only* variability-based constraints, and 2% stellar contamination when a UV-excess constraint is added (see Figure 12 in MacLeod *et al.* 2011), while maintaining a completeness of 90%.

3. Finding quasars with LSST

The last decade has seen fascinating observational progress in optical imaging surveys. The SDSS dataset is currently being greatly extended by the ongoing surveys such as Pan-STARRS (Kaiser *et al.* 2010) and the Dark Energy Survey (Flaugher 2008). The Large Synoptic Survey Telescope (LSST; for a brief overview see Ivezić *et al.* 2008) is the most ambitious survey currently planned in the visible band. The unparalleled LSST survey power is due to its large étendue (see Figure 3).

LSST will extend the faint limit of SDSS by about 5 magnitudes and will have unique survey capability in the faint time domain. In particular, LSST will revolutionize our understanding of the growth of supermassive black holes with cosmic time, AGN fueling mechanisms, the detailed physics of accretion disks, the contribution of AGN feedback to galaxy evolution, the cosmic dark ages, and gravitational lensing (for a detailed discussion, see Chapter 10 in the LSST Science Book, Abell *et al.* 2009).

† This parameter is also a good discriminator of various models for the origin of quasar variability, see Kawaguchi *et al.* (1998).

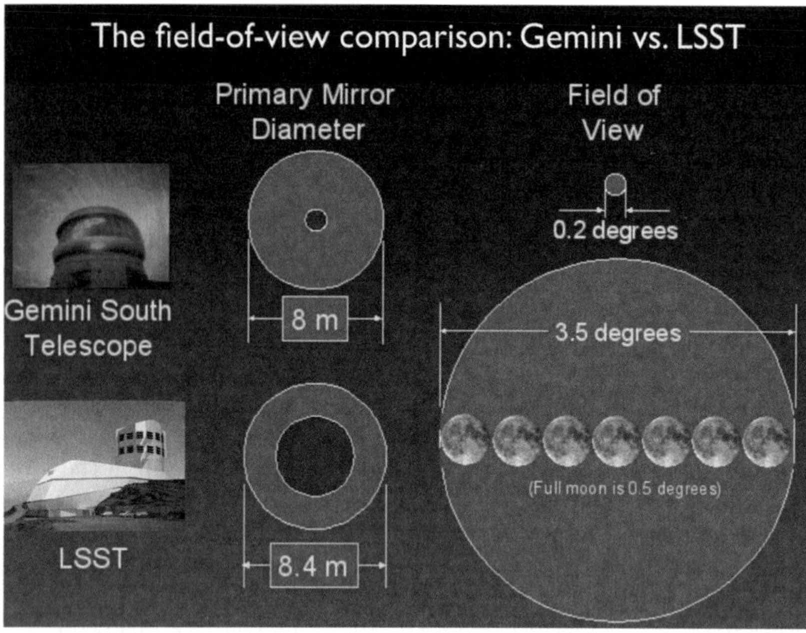

Figure 3. A comparison of the primary mirror size and the field-of-view size for LSST and Gemini South telescopes. The product of the primary mirror size and the field-of-view size, the so-called étendue (or grasp), a characteristic that determines the speed at which a system can survey a given sky area to a given flux limit, is much larger for LSST. Figure courtesy of Chuck Claver.

While no massive spectroscopic followup of quasar candidates will be attempted as part of the LSST project, the time-resolved aspect of LSST photometric and astrometric data will enable significant improvements in the completeness and efficiency of resulting quasar samples compared to single-epoch measurements. We first briefly describe anticipated LSST surveys and then discuss how these data will be used to construct quasar samples with up to about 10 million objects.

3.1. *Brief overview of anticipated LSST surveys*

The LSST design is driven by four main science themes: probing dark energy and dark matter, taking an inventory of the Solar System, exploring the transient optical sky, and mapping the Milky Way. LSST will be a large, wide-field ground-based system designed to obtain multiple images covering the sky that is visible from Cerro Pachón in Northern Chile. The project is scheduled to have first light around 2019 and the beginning of survey operations by 2021.

The system design, with an 8.4m (6.5m effective) primary mirror, a 9.6 deg^2 field of view, and a 3.2 Gigapixel camera, will allow about 10,000 deg^2 of sky to be covered using pairs of 15-second exposures in two photometric bands every three nights on average, with typical 5σ depth for point sources of $r \sim 24.5$. The system is designed to yield high image quality as well as superb astrometric and photometric accuracy. The survey area will include 30,000 deg^2 with $\delta < +34.5°$, and will be imaged multiple times in six bands, *ugrizy*, covering the wavelength range 320–1050 nm. About 90% of the observing time will be devoted to a deep-wide-fast survey mode which will observe an 18,000 deg^2 region over 800 times (summed over all six bands) during the anticipated 10 years of operations, and yield a coadded map to $r \sim 27.5$. These data will result in databases including about

20 billion galaxies and a similar number of stars, and will serve the majority of science programs. The remaining 10% of the observing time will be allocated to special programs such as a Very Deep and Fast time-domain survey. More details about various science programs that will be enabled by LSST data can be found in the LSST Science Book (Abell *et al.* 2009).

3.2. *Color and variability selection of quasars*

The existing selection algorithms based on colors and photometric variability developed with the aid of SDSS data will be applicable to LSST data, too. Although LSST will not obtain simultaneous multi-band photometry like SDSS did, the averaging of many observations (ranging from about 50-60 in the u band to 180-190 in the r and i bands) will result in sufficiently precise color measurements to recognize easily color offsets from the main stellar locus. The addition of variability information will boost the sample efficiency to levels comparable to those obtained for spectroscopic samples. Two additional selection methods, enabled by the multi-epoch LSST imaging and described below, will further improve the resulting samples.

Detailed simulations of the quasar luminosity function and light curves, the LSST observing cadence, and the LSST photometric error distribution (Abell *et al.* 2009; MacLeod *et al.* 2011; Palanque-Delabrouille et al. 2013) show that the LSST quasar sample will include close to 10 million objects, and will be complete for $M < -23$ objects (a formal absolute magnitude definition cutoff of quasars) to redshifts beyond 3. Notably, LSST will discover about 1000 quasars with redshifts beyond 7 (using the z-band drop-out technique) which will represent a valuable sample for studying the epoch of reionization.

3.3. *Additional constraints: proper motion and differential chromatic refraction*

In addition to photometry, astrometric measurements can help to distinguish quasars from stars. First, measurable proper motion will reject about 2/3 of all stars, even before any color or photometric variability criteria are applied. LSST proper motion errors will be 0.5 mas/yr for sources with $r = 23$ and 1.0 mas/yr for $r = 24$ (Ivezić *et al.* 2008). Simulations of the proper motion distribution for Milky Way stars (for model description, see Ivezić, Beers & Jurić 2012) indicate that over ~80% of stars with $r < 23$ and over ~70% of stars with $r < 24$ will have proper motions three times larger than expected measurement errors (with very little dependence on Galactic coordinates, see Figure 4). These high fractions of rejecteable stars are also expected for various stellar subpopulations, such as brown dwarfs (contaminants for very high-redshift quasar candidates) and white dwarfs (contaminants of $z < 2.2$ candidate samples).

An additional astrometric effect that is sensitive to detailed differences in the spectral energy distributions between quasars and stars is the differential chromatic refraction (DCR). Due to the wavelength dependence of atmospheric refraction, in images astrometrically calibrated using stars, quasars show astrometric offsets in a given bandpass as a function of airmass, and also show astrometric offsets between different bandpasses even at a fixed airmass (if larger than one, of course). As shown in Figure 5 from Kaczmarczik *et al.* (2009), the astrometric offsets of quasars from Stripe 82 between the u and g bands, and between the g and r bands, can be up to about 20 mas (depending on redshift), even for moderate airmass (~1.1-1.2). The measurement errors for these offsets anticipated for the LSST main survey will be below 3 mas for objects brighter than $r = 22$, about 5 mas for objects with $r \sim 23$, and 10-15 mas for objects with $r \sim 24$. Therefore, the DCR offsets for quasars will represent an additional quantity, indpendent of color, variability and proper motion, to help distinguish quasars from stars.

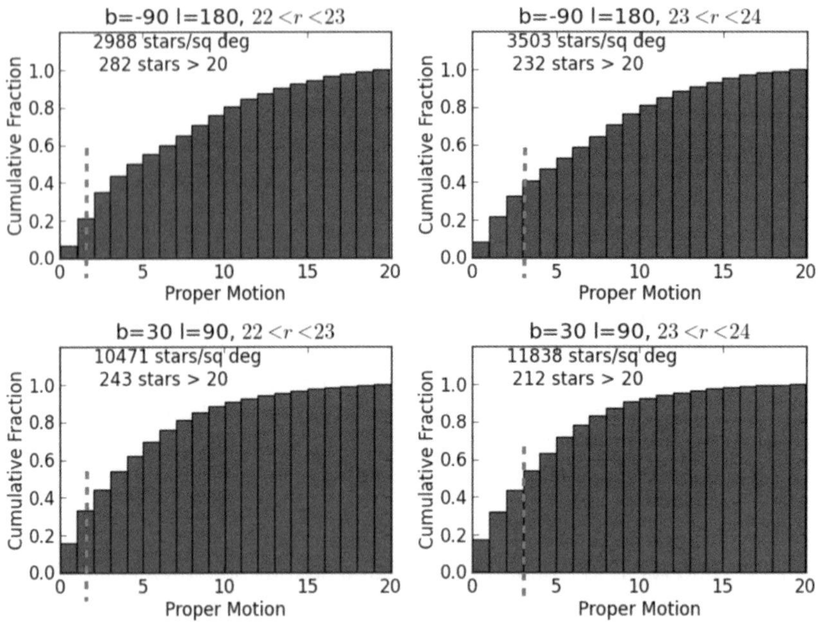

Figure 4. The distributions of proper motions for stars selected from two magnitude bins (left: $22 < r < 23$; right: $23 < r < 24$) and two sky locations (top: south celestial pole; bottom: Galactic longitude $l = 90°$, and latitude $b = 30°$). The vertical dashed lines mark the 3σ cutoffs, where σ is the expected proper motion measurement error for the selected magnitude bins for the main 10-year long LSST survey. The legends in each panel give stellar counts, as well as the counts of stars with anticipated Gaia measurements that provide excellent astrometric calibration. Figure courtesy of Peter Yoachim.

Last but not least, both color and DCR measurements can be used to estimate quasar redshifts with a precision comparable to that expected for LSST galaxies. Using SDSS data, Kaczmarczik *et al.* (2009) obtained a measurement precision for this "photo-astro-z" of 0.03, with the fraction of quasars with redshifts correct to within 0.1 of order 90%.

Acknowledgements

This material is based upon work supported in part by the National Science Foundation through Cooperative Agreement 1258333 managed by the Association of Universities for Research in Astronomy (AURA), and the Department of Energy under Contract No. DE-AC02-76SF00515 with the SLAC National Accelerator Laboratory. Additional LSST funding comes from private donations, grants to universities, and in-kind support from LSSTC Institutional Members.

References

Becker, R. H., White, R. L. & Helfand, D. J. 1995, *ApJ*, 450, 559
Bovy, J., Myers, A. D., Hennawi, J., *et al.* 2012, *ApJ*, 749, 41
Butler, N. R. & Bloom, J. S. 2011 *AJ*, 141, 93
Flaugher, B. 2008, In *A Decade of Dark Energy: Spring Symposium, Proceedings of the conferences held May 5-8, 2008 in Baltimore, Maryland. (USA). Ed. by N. Pirzkal & H. Ferguson.*
Hao, L., Strauss, M. A., Fan, X., *et al.* 2005, *AJ*, 129, 1795
Hawkins, M.R.S. & Veron, P. 1995, *MNRAS*, 275, 1102
Ivezić, Ž., Lupton, R. H., Johnston, D. E., *et al.* 2003, ArXiv:*astro-ph/0310566*

Ivezić, Ž., Lupton, R. H., Jurić, M., *et al.* 2004, ArXiv:*astro-ph/0404487*
Ivezić, Ž., Tyson, J. A., Acosta, E., *et al.* 2008, ArXiv:*0805.2366*
Ivezić, Ž., Beers, T. C. & Jurić, M. 2012, *ARAA*, 50, 251
Kaczmarczik, M. C., Richards, G. T., Mehta, S. S. & Schlegel, D. J. 2009, *AJ*, 138, 19
Kaiser, N., Burgett, W., Chambers, K., *et al.* 2010, *Proc. SPIE 7733, Ground-based and Airborne Telescopes III*, vol. 7733, 77330E
Kawaguchi, T., Mineshige, S., Umemura, M., & Turner, E. L. 1998, *ApJ*, 504, 671
Kelly, B. C., Bechtold, J. & Siemiginowska, A. 2009, *ApJ*, 698, 895
Kimball, A., Ivezić, Ž., Wiita, P. J. & Schneider, D. P. 2011, *AJ*, 141, 182
Kozłowski, S., Kochanek, C. S., Udalski, A., *et al.* 2010, *ApJ*, 708, 927
LSST Science Collaboration 2009, LSST Science Book, http://www.lsst.org/lsst/SciBook, arXiv:0912.0201
MacLeod, C. L., Ivezić, Ž., Kochanek, C. S., *et al.* 2010, *ApJ*, 721, 1014
MacLeod, C. L., Brooks, K., Ivezić, Ž., *et al.* 2011, *ApJ*, 728, 26
MacLeod, C. L., Ivezić, Ž., Sesar, B., *et al.* 2012, *ApJ*, 753, 106
Palanque-Delabrouille, N., Yéche, Ch., Myers, A. D., *et al.* 2011, *A&A*, 530, 122
Palanque-Delabrouille, N., Magneville, Ch., Yéche, Ch., *et al.* 2013, *A&A*, 551, 29
Pâris, I., Petitjean, P., Aubourg, É., *et al.* 2012, *A&A*, 548, 66
Richards, G. T., Fan, X., Schneider, D. P., *et al.* 2001, *AJ*, 121, 2308 (see also arXiv:1311.4870)
Richards, G. T., Fan, X., Newberg, H. J., *et al.* 2002, *AJ*, 123, 2945
Richards, G. T., Strauss, M. A., Fan, X., *et al.* 2006, *AJ*, 131, 2766
Richards, G. T., Myers, A. D., Gray, A. G., *et al.* 2009, *ApJS*, 180, 67
Ross, N. P., Myers, A. D., Sheldon, E. S., *et al.* 2012 *ApJS*, 199, 3
Sesar, B., Ivezić, Ž., Lupton, R. H., *et al.* 2007, *AJ*, 134, 2236
Schmidt, K. B., Marshall, P. J., Rix, H.-W., *et al.* 2010, *ApJ*, 714, 1194
Schneider, D. P., Richards, G. T., Hall, P. B., *et al.* 2010, *AJ*, 139, 2360
Strauss, M. A., Weinberg, D. H., Lupton, R. H., *et al.* 2002, *AJ*, 124, 1810
Zakamska, N. L., Strauss, M. A., Krolik, J. H., *et al.* 2003, *AJ*, 126, 2125
Zakamska, N. L., Strauss, M. A., Heckman, T. M., *et al.* 2004, *AJ*, 128, 1002
Zu, Y., Kochanek, C. S., Kozłowski, S. & Udalski, A. 2013, *ApJ*, 765, 106

New Frontiers in Black Hole Astrophysics
Proceedings IAU Symposium No. 324, 2016
Andreja Gomboc, ed.

doi:10.1017/S1743921317003416

X-ray polarimetry – A new Window on Black Hole Systems

René W. Goosmann on behalf of the XIPE consortium
Université de Strasbourg, CNRS, Observatoire astronomique de Strasbourg,
UMR 7550, F-67000 Strasbourg, France
email: rene.goosmann@astro.unistra.fr

Abstract. Three dedicated X-ray polarimetry mission projects are currently under phase A study at NASA and ESA. The need for this new observational window is more apparent than ever. On behalf of the consortium behind the X-ray Imaging Polarimetry Explorer (XIPE) we present here some prospects of X-ray polarimetry for our understanding of supermassive and stellar mass black hole systems. X-ray polarimetry is going to discriminate between leptonic and hadronic jet models in radio-loud active galactic nuclei. For leptonic jets it also puts important constraints on the origin of the seed photons that constitute the high energy emission via Comptonization. Another important application of X-ray polarimetry allows us to clarify the accretion history of the supermassive black hole at the Galactic Center. In a few Black Hole X-ray binary systems, X-ray polarimetry allows us to estimate in a new, independent way the angular momentum of the black hole.

Keywords. polarization, relativity, space vehicles, techniques: polarimetric, galaxies: active, X-rays: binaries

---

1. Introduction

Electromagnetic radiation remains our most important messenger from the Universe. From the radio band to very high photon energies light contains multiple layers of information that can be analyzed by the means of imaging, spectroscopy or timing. Since the early 1970s X-ray data is being collected from space-borne observatories and has greatly improved our knowledge about the extreme Universe. The physics of black hole and neutron star systems, of extra-galactic jets, or of violent explosions in space would still be in the dark without X-ray astronomy.

Nonetheless, X-ray observatories do not yet explore the complete set of information that is comprised in X-rays from space. In addition to the intensity flux, I, the two Stokes parameters Q and U can now be measured systematically. This became possible by the development of a new generation of X-ray polarimeters that are based on Gas Pixel Detectors (GPDs, Bellazzini *et al.* 2003). Such detectors are much more efficient than the Bragg reflection polarimeter that was mounted on the only preceding X-ray polarimeter aboard the OSO-8 satellite more than four decades ago (Weisskopf *et al.* 1976). This pioneering experiment gave us a solid measurement of the soft X-ray polarization emitted by the Crab Nebula and upper limits for less than a handful X-ray binary systems.

Currently, a total of three dedicated X-ray polarimetry missions are considered for a phase A study by NASA and ESA. Based on the GPD or related technologies we are now at the threshold to enter a new era of X-ray astronomy that includes polarization resolved in energy, time and position in the sky. This note describes some key prospects of X-ray polarimetry for black hole systems. The examples and simulations presented here are based on the study of the X-ray Imaging Polarimetry Explorer (XIPE) being a mission candidate in ESA's M4 competition.

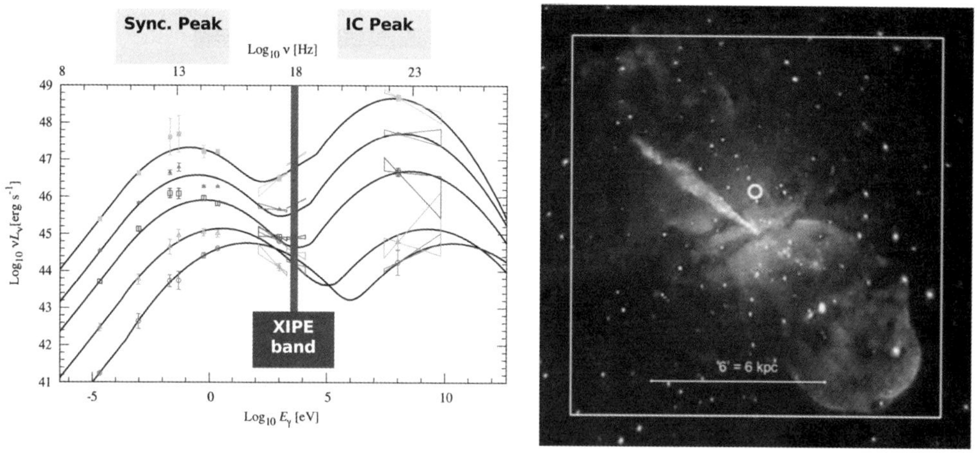

Figure 1. Left: the blazar sequence with respect to the spectral band of XIPE. From top to bottom, the broad band blazar spectra evolve from low synchrotron peak to high synchrotron peak sources. The data points are collected from various radio, optical, X-ray and gamma ray telescopes. Depending on where a given blazar is located in the sequence, XIPE is going to measure the polarization of the low energy or the high energy peak. In the former case it can constrain the level of turbulence of the magnetic field. In the latter case XIPE is going to discriminate between hadronic and leptonic jets. Right: multicolor image of the radio-loud AGN Centaurus A. The collimated jet component is colored in white and the XIPE resolution limit is denoted by the white circle. Multiple imaging with XIPE will allow us to probe the magnetic field of the jet at various distances from the nucleus.

2. Astrophysical jets launched in Active Galactic Nuclei

Radio-loud Active Galactic Nuclei (AGN) show kpc-scale jets often reaching out far beyond the host galaxy. Depending on the observer's viewing angle with respect to the jet, radio-loud AGN are divided into blazars, with a line of sight close to the jet axis, and radio-loud quasars / radio galaxies seen at larger viewing angle. The broad band spectrum of the jet emission shows two bumps. Strong polarization tells us that the low-energy bump centered on the radio-to-sub-mm range is due to synchrotron emission. The origin of the high-energy bump covering the X-ray and γ-ray bands depends on the underlying jet scenario. If the jet is leptonic, the high energy bump is interpreted as Comptonized radiation. In hadronic jets, the high energy bump is due to efficient particle acceleration and subsequent synchrotron radiation by protons or secondary pions.

There are several open questions for AGN jets and also for astrophysical jets in related objects: are jets hadronic or leptonic? What can we say about the level of turbulence in the magnetic field along the jet? What is the dominating emission mechanism in leptonic jets, synchrotron self-Compton (SSC) or inverse Compton radiation with external seed photons (e.g. coming from the accretion disk)?

X-ray polarimetry can give unbiased answers to these questions, in particular when combined with polarimetry at optical and radio wavelengths. In high synchrotron peaked blazars, XIPE can probe the synchrotron polarization and thereby the level of turbulence of the magnetic field at the location of the hottest electrons, i.e. close to the injection sites. Comparison with optical polarization then allows us to draw conclusions on the size of the emission sites at different wavelengths. The X-ray-to-optical polarization degree of low synchrotron peaked blazars can discriminate hadronic from leptonic scenarios. In the latter case, the X-ray polarization can also disentangle an SSC origin of the seed photons from external Comptonization scenarios.

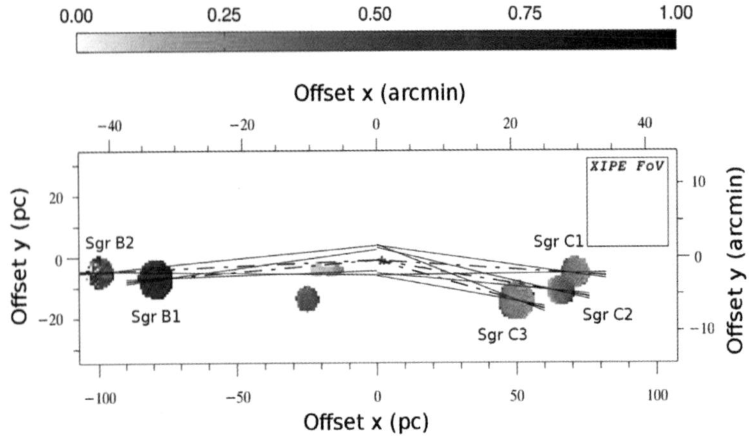

Figure 2. Simulated XIPE observation of the X-ray reflection nebulae around the supermassive black hole at the Galactic Center (Marin *et al.* 2015). The color scale represents polarization fraction. The imaging capability is going to allow us to pin down the position angle of the scattering plane for each cloud. If the primary emission comes indeed from Sgr A*, these planes must coincide at the location of the black hole (coordinates 0, 0). The field of view (FoV) of XIPE would enable us to observe the cloud complexes Sgr B, Sgr C and the Bridge region (located to the lower left of Sgr A*) by only three separate observations.

3. Past activity of the supermassive black hole at the Galactic Center

The supermassive black hole at the Galactic Center, known as Sgr A*, is surrounded by a number of X-ray reflection nebulae (XRN) sitting inside a radius of about 100 pc. X-rays of unknown origin emerge from these prominent molecular clouds. No corresponding source or emission mechanism is known inside or nearby the clouds and therefore the hypothesis was made that the XRN are irradiated by the remnant X-ray emission associated to an active phase of Sgr A* that took place hundreds of years ago (Sunyaev *et al.* 1993, Koyama *et al.* 1996).

Nonetheless, some alternative mechanisms were suggested such as a population of X-ray transients situated in the vicinity of the clouds (Sunyaev *et al.* 1993) or interaction of low energy cosmic ray electrons with the molecular gas (Valinia *et al.* 2000). Both alternative scenarios seem somewhat fine-tuned but still may contribute to some extent to the X-ray emission emerging from the XRN.

The imaging capabilities of XIPE would allow us to measure the polarization position angle of individual cloud complexes in the vicinity of Sgr A*, i.e. Sgr B, Sgr C and the so-called Bridge region. The scattering scenario with a past active state of Sgr A* then implies that the observed flux from a molecular cloud is polarized perpendicularly to the direction towards the supermassive black hole (Churazov *et al.* 2002). XIPE has sufficient sensitivity to test this scenario and to eventually identify Sgr A* as the unambiguous primary X-ray source (Marin *et al.* 2015).

4. Measuring black hole spin in X-ray binaries

A census of the supposedly primordial angular momentum (spin) of stellar mass black holes in X-ray binary systems has strong implications for our understanding of black hole formation. For X-ray binary systems, such measurements have been carried out in the past and meanwhile three different methods to measure black hole spin are well established. The classical spectroscopic method using skewed iron fluorescence lines (Fabian

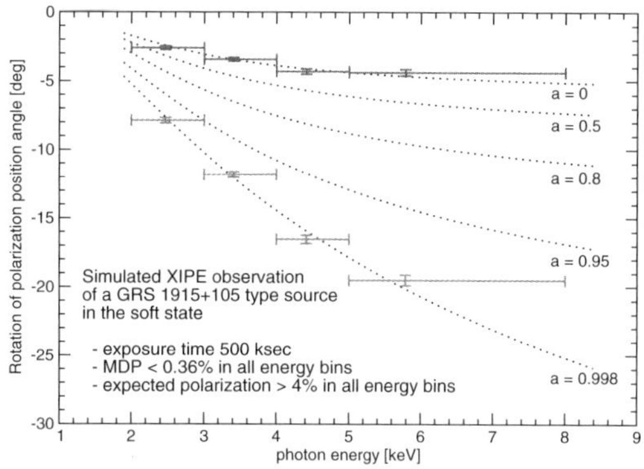

Figure 3. Simulated XIPE observations based on a relativistic ray tracing model (dashed lines). The two sets of simulated data points trace the rotation of the X-ray polarization angle with photon energy for the case of a Schwarzschlid ($a = 0$, upper set) and a maximally rotating Kerr black hole ($a = 0.998$, lower set).

et al. 1989), spectral fits to the continuum emission of the accretion disk (Li *et al.* 2005), and estimates based on quasi-periodic oscillations (Abramowicz & Kluźniak 2001). Unfortunately, the three methods tend do disagree in many cases. With X-ray polarimetry it is possible to put independent constraints on the black hole spin that are based on a rotation of the polarization position angle with energy.

In this independent approach we exploit the effect that the polarization direction rotates while it is parallel-transported along a null geodesics connecting the vicinity of a black hole with the distant observer (Connors *et al.* 1980). The rotation angle depends on the starting location of the geodesics and since the temperature of the accretion disk also varies with the distance from the black hole, the net polarization angle integrated across the inner accretion disk rotates with energy.

The radius of the so-called innermost stable circular orbit, which is assumed to coincide with the inner edge of the accretion disk in a thermal state, depends on the angular momentum of the black hole. For higher black hole spin, the accretion disk extends down to smaller distances from the black hole and therefore produces higher temperatures. As a consequence, the rotation of the polarization position angle with photon energy depends on black hole spin (see Fig. 3). Therefore, X-ray polarimetry gives us a handle on constraining the angular momentum (Dovčiak *et al.* 2008, Schnittman & Krolik 2009).

References

Abramowicz, M. A. & Kluźniak, W. 2001, *A&A*, 374, L19
Bellazzini, R., Baldini, L., Brez, A., *et al.* 2003, *Nuclear Instruments and Methods in Physics Research A*, 510, 176
Churazov, E., Sunyaev, R., & Sazonov, S., 2002, *MNRAS*, 330, 817
Connors, P. A., Stark, R. F., & Piran, T. 1980, *ApJ*, 235, 224
Dovčiak, M., Muleri, F., Goosmann, R. W., Karas, V., & Matt, G. 2008, *MNRAS*, 391, 32
Fabian, A. C., Rees, M. J., Stella, L., & White, N. E. 1989, *MNRAS*, 238, 729
Koyama, K., Maeda, Y., Sonobe, T., *et al.* 1996, *PASJ*, 48, 249
Li, L.-X., Zimmerman, E. R., Narayan, R., & McClintock, J. E. 2005, *ApJS*, 157, 335
Marin, F., Muleri, F., Soffitta, P., Karas, V., & Kunneriath, D. 2015, *A&A*, 576, A19
Schnittman, J. D. & Krolik, J. H. 2009, *ApJ*, 701, 1175
Sunyaev, R. A., Markevitch, M., & Pavlinsky, M. 1993, *ApJ*, 407, 606
Valinia, A., Tatischeff, V., Arnaud, K., Ebisawa, K., & Ramaty, R. 2000, *ApJ*, 543, 733
Weisskopf, M. C., Cohen, G. G., Kestenbaum, H. L., *et al.* 1976, *ApJL*, 208, L125

New Frontiers in Black Hole Astrophysics
Proceedings IAU Symposium No. 324, 2016
Andreja Gomboc, ed.

doi:10.1017/S1743921316013004

KETJU: Post-Newtonian-Accurate Supermassive Black Hole Dynamics in GADGET-3

Antti Rantala, Pauli Pihajoki and Peter H. Johansson

Department of Physics, University of Helsinki
Gustaf Hällstrmin katu 2A,00560, Helsinki,Finland
email: `antti.rantala@helsinki.fi`

Abstract. We present KETJU, a new regularized tree code based on algorithmic chain regularization and implemented into Gadget-3. This new code is able to follow simultaneously galactic-scale dynamical and astrophysical processes and the small-scale supermassive black hole binary dynamics. We present here the general idea of this new code and show a test simulation of black hole binary dynamics in a galaxy merger of two massive elliptical galaxies. The separation of the black holes at the time of the merger is several orders of magnitude smaller in KETJU than when compared to ordinary Gadget-3 simulations. The merger timescale is also longer by 100 − 200 Myr.

Keywords. black hole physics, galaxies: kinematics and dynamics, galaxies: nuclei, methods: n-body simulations

---

1. Introduction

All massive early-type galaxies and stellar bulges host supermassive black holes (SMBH) in the mass range of $10^6 M_\odot \lesssim M_\bullet \lesssim 10^{10} M_\odot$ in their centres (e.g. Kormendy & Richstone 1995). In addition, galaxies do not live in isolation but frequently interact and merge with each other shaping their evolution (White & Rees 1978, Johansson *et al.* 2009). Galaxy mergers also bring multiple SMBHs into the same stellar system. As very close quasar pairs appear to be observationally rare there should be an efficient physical mechanism which drives the two SMBHs rapidly into coalescence. Indeed, dynamical friction from the stars and gas in the host galaxy causes the SMBHs to sink to the center of the merger remnant. Forming a wide ($a \sim 10$ pc) binary SMBH, the black holes now interact with the central stars slingshotting them away in three-body encounters. In this process the binary loses orbital energy at the expense of the ejected stars and becomes 'harder'. At sub-parsec separations, the energy loss by gravitational wave (GW) emission drives the binary into coalescence (e.g. Merritt & Milosavljević 2005).

Future space-borne GW detection experiments, such as LISA (Amaro-Seoane *et al.* 2012), have generated interest in estimating the SMBH merger rates using numerical simulations. Naively, the SMBH merger rate depends on the frequency of galaxy mergers containing multiple SMBHs and the SMBH merging timescale after the binary has formed. However, there are considerable uncertainties in the estimates, especially in the duration of the slingshot-hardening phase. In the hardening phase, the inverse semi-major axis of the binary is expected to evolve as $d(1/a)/dt \propto \rho_\star(t)/\sigma_\star(t)$, where $\rho_\star$ and $\sigma_\star$ are the stellar density and velocity dispersion at the center of the galaxy (Quinlan 1996). Thus, a simulation predicting the SMBH merger timescale should accurately describe both the global kiloparsec-scale evolution of the host galaxy of the SMBH binary and also model the SMBH binary evolution at sub-parsec scales. In this proceedings contribution, we present a new regularized tree code KETJU, based on algorithmic chain regularization and implemented into the widely used galaxy simulation code Gadget-3 (Springel 2005).

2. Regularization

2.1. *Background*

In the case of point-like simulation particles, the Newtonian gravitational force is manifestly divergent when two particles collide: the gravitational force $|\bar{F}_{ij}| \rightarrow \infty$ as their separation $|\bar{r}_i - \bar{r}_j| \rightarrow 0$. The most straightforward manner to circumvent this collision problem is to introduce a small softening length ϵ in order to render $|\bar{F}_{ij}|$ finite at particle collisions and close encounters. However, this procedure has the unfortunate consequence that the softening scale ϵ, typically 1 pc $< \epsilon <$ 100 pc in modern galactic-scale simulations, becomes the natural spatial resolution limit of the simulation. This is highly undesirable as we want to simulate SMBH binary evolution at sub-parsec scales until the GW-driven coalescence. Another possibility is the regularization of the close encounters between simulation particles to avoid the Newtonian force divergences. The Keplerian two-body problem was regularized a century ago by the pioneering work of Levi-Civita. A practical three-dimensional KS-regularization method including external perturbing forces was developed by Kustaanheimo & Stiefel (1965). Subsequent development in the field yielded several new regularization methods, such as the Aartseth-Zare method and Heggie's global regularization scheme (see e.g. Aarseth 1999 and references therein). An important chain-based formulation of the KS regularization was found by Mikkola & Aarseth (1993). Finally, the work of Mikkola & Tanikawa (1999) and Preto & Tremaine (1999) on time-transformed regularization methods resulted in the formulation of the Algorithmic Regularization Chain (ARCHAIN) algorithm, which is also our choice of regularization method for KETJU ('ketju'='chain' in Finnish).

2.2. *The ARCHAIN algorithm*

The inner workings of the ARCHAIN algorithm consist of three aspects. First, the equations of motion are time-transformed. As opposed to regularization methods reviewed in the previous section, no spatial coordinate transformation is performed: the singularities remain in the equations of motion but are avoided algorithmically by using a leapfrog integrator, hence the name of the algorithm. The new time coordinate, which is an independent variable, is defined as

$$ds = [\alpha U + \beta \Omega + \gamma]\, dt, \tag{2.1}$$

where U is the negative of the gravitational potential energy and the function Ω is generally used to regularize the encounters between particles with very different masses, if an extrapolation method is not used. The triplet (α, β, γ) determines the exact regularization procedure. Note that constants α and β are dimensional so that the expression inside the brackets is dimensionless. See Mikkola & Merritt (2006) and Mikkola & Merritt (2008) for a detailed description of the time transformation and the different parameter choices. The equations of motion are now transformed using the new time coordinate ds. The algorithm also allows for Post-Newtonian corrections as velocity-dependent perturbations. The corrections are added up to the order PN3.5. The second aspect of ARCHAIN is the use of chained inter-particle vectors for each chain subsystem instead of a global reference frame. This proceduce greatly reduces the numerical round-off error

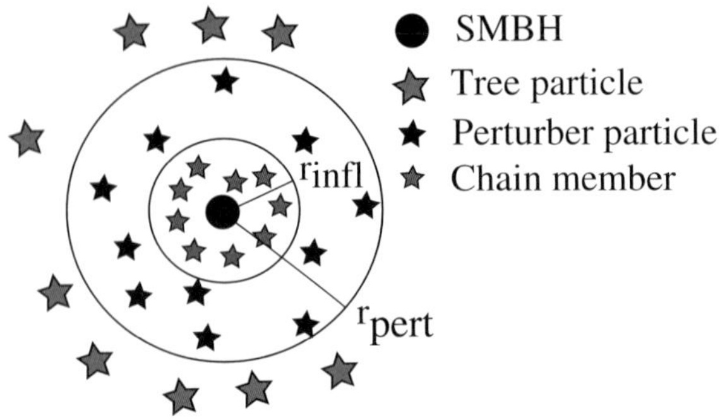

Figure 1. A schematic illustration of a chain subsystem surrounding a SMBH in the KETJU code.

even without the time transformation. Finally, an extrapolation algorithm is applied in order to push the error in the dynamical variables to a user-defined level. We utilize the Gragg-Bulirsch-Stoer (GBS) extrapolation method. The central idea of the GBS extrapolation here is to compute time evolution of the dynamical variables during a time interval H with a different number of subdivision counts n and finally extrapolate the variable values to $n \to \infty$. This procedure, although somewhat time-consuming, ensures that the numerical errors are kept at a very low level which is essential in computing the temporal evolution of chain systems containing SMBHs and surrounding stars.

3. Implementation

3.1. *Chain subsystems and perturber particles*

In developing KETJU (Rantala *et al.* 2016, in prep.) we follow and extend the ideas behind the rVINE code of Karl *et al.* (2015). In Gadget-3, the Newtonian gravitational force between the simulation particles is computed using a TreePM algorithm, including gravitational softening. We regularize the gravitational dynamics around the SMBHs in the following way. First, we give every SMBH a parametrized influence radius which is proportional to the mass of the SMBH: $r_{\rm infl} \propto \alpha M_\bullet$. Every stellar particle inside the influence radius of a SMBH belongs to the chain subsystem of the particular SMBH. If the volumes of several subsystems overlap, the subsystems are merged. The dynamics of the particles in the chain (chain particles hereafter) is computed using the algorithmic chain integrator as opposed to the ordinary Gadget-3 particles (tree particles hereafter). During the chain integration, the chain particles feel the tidal perturbation of the tree particles which lie within the perturber radius $r_{\rm pert} = \gamma r_{\rm infl} (m/M_\bullet)^{1/3}$, where m is the mass of the perturber particle and γ is an adjustable free parameter. A simplified illustration of the situation is presented in Figure 1. The chain subsystem itself is treated as a single 'macro' particle in the tree code. Both the macro particle and the neighbouring perturber particles receive a correction to their gravitational accelerations due to the internal structure of the chain subsystem. The macro particles are set to be always at the lowest active timestep level in the tree code and the tree/chain memberships of the particles are updated every tree timestep.

3.2. *A new SMBH merging criterion*

Traditionally, two SMBHs are merged together in softened tree codes, such as Gadget-3, if their smoothing lengths overlap and their relative velocity is less than a fraction times the local sound speed. For typical resolutions employed in galaxy-scale simulations this

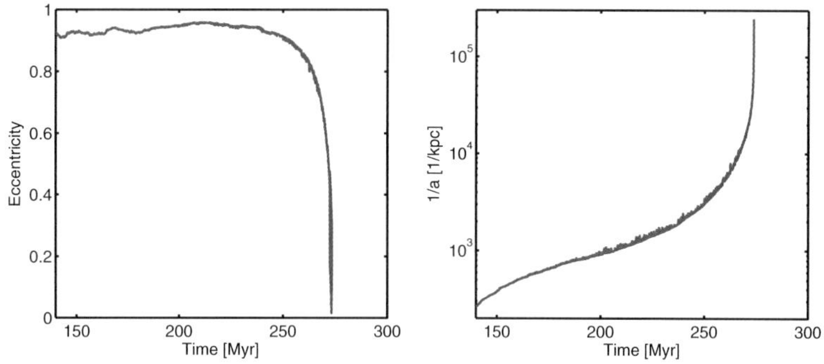

Figure 2. The evolution of eccentricity (left panel) an the inverse semi-major axis (right panel) of the SMBH binary until the GW-driven coalescence. The binary circularizes and hardens as the gravitational waves carry energy and angular momentum away from the system. The separation at the time of the SMBH merger is $\sim$ 1000 AU in this particular test run.

amounts usually to $\sim$ 10 pc. With KETJU, we can follow the SMBH binary to much smaller separations and thus a new SMBH merging criterion is required. The new criterion is based on the analytical GW coalescence timescale by Peters & Mathews (1963) formula valid at PN2.5. The estimate of the merging timescale is $t_c = a/4\dot{a}$, where $\dot{a}$ is obtained from the Peter's formula:

$$\dot{a} = -\frac{64}{5}\frac{G^3 M_{\bullet,1} M_{\bullet,2}(M_{\bullet,1} + M_{\bullet,2})}{c^5 a^3}\frac{1 + \frac{73}{24}e^2 + \frac{37}{96}e^4}{(1-e^2)^{7/2}}, \tag{3.1}$$

where $M_{\bullet,1}$ and $M_{\bullet,2}$ are the masses of the two SMBHs and e is the constant eccentricity. The timescale estimate is accurate enough for our purposes as $e_{\text{final}} \sim 0.01 << 1$. See e.g. Maggiore (2007) for a more rigorous timescale estimate. The SMBHs coalesce if t_c is smaller than the next global Gadget-3 timestep. The Post-Newtonian evolution of a SMBH binary at the late stages of a gasless ('dry') merger of two massive elliptical galaxies ($M_\star = 10^{11} M_\odot$, $M_{\text{DM}} = 10^{13} M_\odot$, Hernquist scale radius $a = 1.5$ kpc) is shown in Figure 2. Due to the high eccentricity of the binary inherited from the merger geometry of the merger progenitor galaxies the GW emisson drives the SMBHs to coalescence $\sim$ 150 Myr after the formation of the binary. The separation at the time of the SMBH merger was $r \sim 0.005$ pc $\sim$ 1000 AU in this particular simulation.

4. Conclusions

We have developed a new regularized tree code KETJU using algorithmic chain regularization and implemented into Gadget-3. Using KETJU, we are able to follow the evolution of a SMBH binary in a galactic-scale simulation until the GW-driven coalescence and study simultaneously the structural evolution of the host galaxy affected by the evolution of the SMBH binary.

References

Aarseth, S. J. 1999, *PASP*, 111, 1333
Amaro-Seoane, P., Aoudia, S., Babak, S., *et al.* 2012, *Classical and Quantum Gravity*, 29, 124016
Johansson, P. H., Naab, T., & Burkert, A. 2009, *APJ*, 690, 802
Karl, S. J., Aarseth, S. J., Naab, T., Haehnelt, M. G., & Spurzem, R. 2015, *MNRAS*, 452, 2337
Kormendy, J. & Richstone, D. 1995, *ARAA*, 33, 581
Kustaanheimo, P. & Stiefel, E. 1965, *J. Reine Angew. Math*, 218, 204
Maggiore, M. 2007, Gravitational Waves: Volume 1: Theory and Experiments
Merritt, D. & Milosavljević, M. 2005, *Living Reviews in Relativity*, 8,

Mikkola, S. & Valtonen, M. J. 1992, *MNRAS*, 259, 115
Mikkola, S. & Aarseth, S. J. 1993, *Celestial Mechanics and Dynamical Astronomy*, 57, 439
Mikkola, S. & Tanikawa, K. 1999, *MNRAS*, 310, 745
Mikkola, S. & Merritt, D. 2006, *MNRAS*, 372, 219
Mikkola, S. & Merritt, D. 2008, *AJ*, 135, 2398
Preto, M. & Tremaine, S. 1999, *AJ*, 118, 2532
Peters, P. C. & Mathews, J. 1963, *Physical Review*, 131, 435
Quinlan, G. D. 1996, *New Astronomy*, 1, 35
Springel, V. 2005, *MNRAS*, 364, 1105
White, S. D. M. & Rees, M. J. 1978, *MNRAS*, 183, 341

New Frontiers in Black Hole Astrophysics
Proceedings IAU Symposium No. 324, 2016
Andreja Gomboc, ed.

doi:10.1017/S1743921316012990

Ray-tracing and polarized radiative transfer in General Relativity

Pauli Pihajoki, Antti Rantala and Peter H. Johansson

University of Helsinki, Department of Physics,
Gustaf Hällströmin katu 2a, 00560 Helsinki, Finland
email: pauli.pihajoki@iki.fi

Abstract. We discuss the problem of polarized radiative transfer in general relativity. We present a set of equations suitable for solving the problem numerically for the case of an arbitrary space-time metric, and show numerical solutions to example problems. The solutions are computed with a new ray-tracing code, ARCMANCER, developed by the authors.

Keywords. relativity, radiative transfer, polarization, methods: numerical

1. Introduction

In curved space-times, light is observed to propagate along curved paths. As such, computing mock observations of scenes affected by strong gravity requires taking general relativity fully into account. A conceptually simple approach in the limit of geometric optics is to compute the paths of light, given by null geodesics. This can be done numerically or analytically, in the case of highly symmetric space-times such as the Kerr space-time. This approach has been often used in the literature since the late 1960's (e.g. Cunningham & Bardeen 1973; Luminet 1979; Dexter 2016). A complete solution of the polarized radiative transfer problem in curved space requires not only solving the geodesic, but solving the general relativistic polarized radiative transfer equation along this geodesic as well.

Currently, there are several codes capable of computing general relativistic polarized radiative transfer: GRTRANS (Dexter 2016), ASTRORAY (Shcherbakov & McKinney 2013) and the code described by Broderick & Blandford (2003a,b). However, none of codes above support arbitrary space-time metrics, and are instead restricted to the Kerr metric. To remedy this, we have developed ARCMANCER (Pihajoki et al. 2017, *in prep*): a C++ library for computing geodesics and polarized radiative transfer in space-times with arbitrary user defined metrics. In the following, we will briefly describe the general relativistic polarized radiation transfer problem and show some promising initial results.

2. Polarized radiative transfer in curved space-time

To compute radiative transfer along the path of light, given by the null geodesic γ, we need to first solve it numerically. A geodesic can be given as a curve $\gamma : \mathbb{R} \supseteq I \to M$, where M is the space-time manifold, which satisfies the equations of motion. The equations of motion can be given in two complementary forms. For initial conditions $(x, k^a) \in TM$, where $x \in M$ is the initial position and $k^a \in TM_x$ is the initial tangent vector of the geodesic, given on the tangent bundle, we have in a coordinate frame

$$\dot{x}^\alpha = k^\alpha \tag{2.1}$$

$$\dot{k}^\alpha = -g^{\alpha\beta}\Gamma_{\alpha\beta\delta}k^\beta k^\delta \tag{2.2}$$

$$\Gamma_{\alpha\beta\delta} = \frac{1}{2}\left(\partial_\beta g_{\alpha\delta} + \partial_\delta g_{\alpha\beta} - \partial_\alpha g_{\beta\delta}\right) \tag{2.3}$$

where we define $\dot{x} = \mathrm{d}x/\mathrm{d}\lambda$, and where λ is the affine parameter of the geodesic, g_{ab} is the metric and $\Gamma_{\alpha\beta\delta}$ are the Christoffel symbols of the first kind. For initial conditions $(x, p_a) \in T^*M$, where $p_a \in T^*M_x$ is the initial four-momentum of the geodesic, given on the cotangent bundle, we can define a Hamiltonian H and the Hamiltonian equations of motion a coordinate frame as

$$H(x, p_a) = \frac{1}{2}g(x)^{bc}p_b p_c \quad (= 0, \text{ for null geodesics}) \tag{2.4}$$

$$\dot{x}^\alpha = \frac{\partial H}{\partial p_\alpha} = g(x)^{\alpha\beta}p_\beta \tag{2.5}$$

$$\dot{p}_\alpha = -\frac{\partial H}{\partial x^\alpha} = -\frac{1}{2}\partial_\alpha g(x)^{\beta\delta}p_\beta p_\delta. \tag{2.6}$$

The two different parametrizations are connected by the natural bijection $p_a = g_{ab}k^b$. However, it should be noted that when plasma effects cannot be ignored, the path of light is not given by a geodesic following the dispersion relation (2.4), but a dispersion relation depending on the polarization and local properties of the plasma (Broderick & Blandford 2003b). These effects are not important for the applications presented in this paper, so we can safely ignore them here.

To compute polarized radiative transfer along a geodesic, we first need to fix a polarization frame along each point of the geodesic, and connect these frames to the polarization frame at the point of observation. Conceptually the simplest method is to define a frame of two four-vectors $\{h^a, l^a\}$, representing local horizontal and vertical directions, in the spatial subspace of the observer, who is assumed to move with a four-velocity u^a. If h^a and l^a are orthogonal with respect to each other as well as with respect to to the tangent vector of the geodesic, k^a, they define a proper polarization frame. This frame can then be parallel propagated along the geodesic to obtain a well defined polarization frame at each point of the geodesic. Next, the redshift $\mathcal{G}$ between a point on the geodesic and the observation point is given by

$$\mathcal{G} = \frac{\nu_0}{\nu} = \frac{{u_0}^a k(x_0)_a}{u^a k(x)_a}, \tag{2.7}$$

where k^a is the tangent vector of the geodesic, ν_0 and ν are the observed and emitted frequencies, u_0^a and u^a are the observer and emitter four-velocities, and x_0 and x are the observation and emission points, respectively. Finally, we need a relativistic generalization of the flat space radiative transfer equation

$$\frac{\mathrm{d}\boldsymbol{I}_\nu}{\mathrm{d}s} = \boldsymbol{J}_\nu - \mathbf{M}\boldsymbol{I}_\nu, \tag{2.8}$$

where s is the physical distance through the medium, $\boldsymbol{I} = (I, Q, U, V)$ is the Stokes vector, $\boldsymbol{J} = (j_I, j_Q, j_U, j_V)$ is the vector of corresponding emissivities and

$$\mathbf{M} = \begin{pmatrix} \alpha_I & \alpha_Q & \alpha_U & \alpha_V \\ \alpha_Q & \alpha_I & r_V & -r_U \\ \alpha_U & -r_V & \alpha_I & r_Q \\ \alpha_V & r_U & -r_Q & \alpha_I \end{pmatrix} \tag{2.9}$$

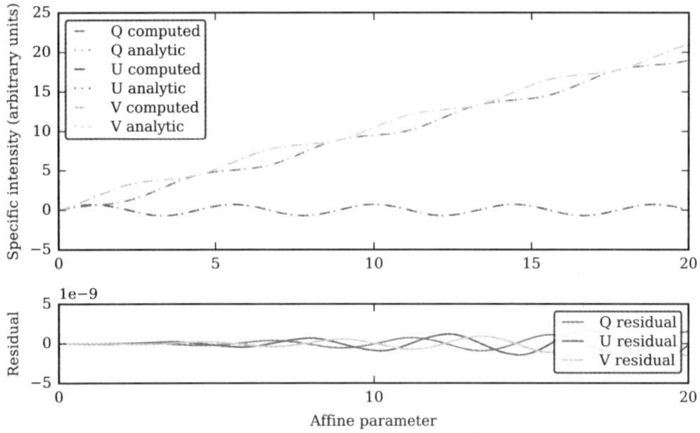

Figure 1. Comparison of a numerical (points) and an analytic (dashed lines) solution to polarized radiative transfer with constant coefficients in Minkowskian metric. Upper panel show the solutions in arbitrary units and lower panel show the residuals between numerical and analytic solutions. The panels show the solution to a problem with constant emission $\mathcal{J} = (0, 1, 1, 1)$ and Faraday rotation and conversion factors $r_V = r_Q = 1$, with rest of the $\mathcal{M}$ entries identically zero, starting with initial values $\mathcal{I} = (0, 0, 0, 0)$. I is zero to within numerical precision and is not shown. For the analytic solutions, see Dexter (2016).

is the Müller matrix containing the absorption coefficients α_i and the Faraday conversion and mixing coefficients r_i for the various Stokes parameters. When the four-velocity of the local medium (i.e. astrophysical plasma) and a polarization reference direction b^a (given e.g. by the local direction of the magnetic field B) is specified, we can use the parallel propagated polarization frame of the observer to generalize equation (2.8), yielding

$$\frac{d\mathcal{I}_{\nu_0}}{d\lambda} = \mathcal{J}_{\nu_0} - \mathcal{M}_{\nu_0}\mathcal{I}_{\nu_0}, \text{ where} \tag{2.10}$$

$$\mathcal{I}_{\nu_0} = \frac{\mathcal{G}^3}{\nu_0^3}\boldsymbol{I}_{\nu_0/\mathcal{G}}, \qquad \mathcal{J}_{\nu_0} = \frac{\mathcal{G}^2}{\nu_0^2}R(\chi)\boldsymbol{J}_{\nu_0/\mathcal{G}}, \tag{2.11}$$

$$\mathcal{M}_{\nu_0} = \frac{\mathcal{G}}{\nu_0}R(\chi)\mathbf{M}_{\nu_0/\mathcal{G}}R(-\chi) \quad \text{and} \quad R(\chi) = \begin{pmatrix} 0 & 0 & 0 & 0 \\ 0 & \cos(2\chi) & -\sin(2\chi) & 0 \\ 0 & \sin(2\chi) & \cos(2\chi) & 0 \\ 0 & 0 & 0 & 0 \end{pmatrix}, \tag{2.12}$$

which are in general functions of frequency ν, the angle θ between the four-velocity k^a of the geodesic and the local reference direction b^a, and the angle χ between the parallel propagated polarization frame and the local reference polarization frame, projected orthogonal to k^a. See Shcherbakov & Huang (2011) for details. Given initial conditions for the intensity $\mathcal{I}$, the emissivity $\boldsymbol{J}$, the response tensor $\mathbf{M}$, the local four-velocity u^a and the local polarization reference direction b^a on all points of a geodesic, equation (2.10) can be numerically integrated along the geodesic, yielding the solution to the general relativistic polarized radiative transfer problem.

3. Example problems and future plans

The equations described above have been implemented in the Arcmancer library. Using the library we can compute results to some example problems. Figure 1 shows the comparison between numerical and analytic results to two examples of polarized radiative transfer problems in Minkowskian flat space. We see that the analytic results are well matched within the given numerical tolerance. Figure 2 shows numerically computed images of a Novikov–Thorne accretion disc (Novikov & Thorne 1973) around a rotating

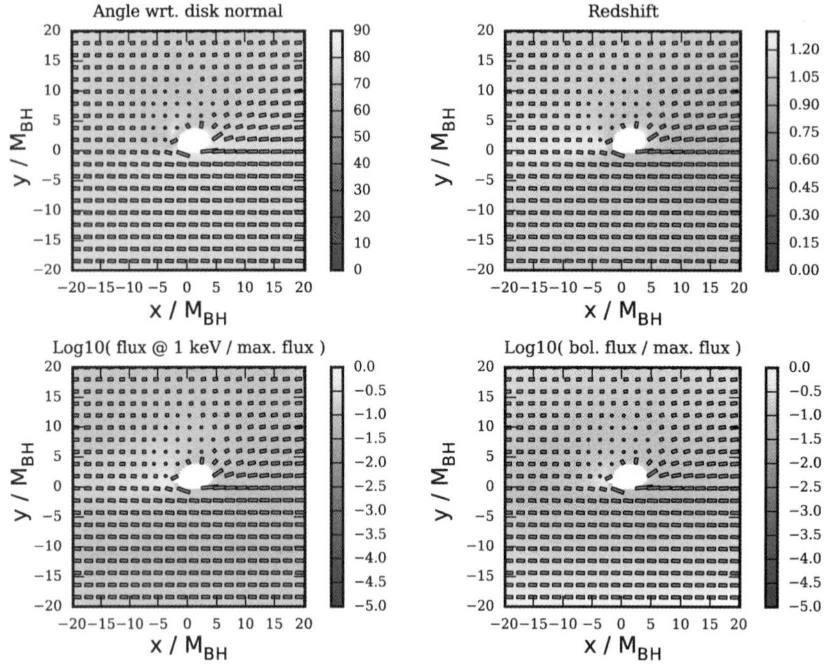

Figure 2. Ray-traced images of a Kerr black hole with a mass of $M = 10\mathrm{M}_\odot$, accretion rate of $\dot{M} = 0.1 M_{\mathrm{Edd}}$ and a dimensionless spin of $a/M = 0.998$. The surrounding accretion disc is optically thick and geometrically thin, modelled with the Novikov–Thorne equations (Novikov & Thorne 1973). The emission is modelled with a semi-infinite electron scattering atmosphere (Chandrasekhar 1960). The direction of linear polarization is shown by the red bars, and the degree of polarization is indicated by the bar lengths, with the degree of polarization at the bottom edges corresponding to $\sim 4\%$. *Top left*: The angle between the geodesic and the disc normal as measured by an observer co-moving with the disc matter. *Top right*: The redshift factor, $\mathcal{G}$, between the disc matter and the observation point *Bottom left:* Observed flux at 1 keV, integrated over a 10 keV window, relative to the maximum flux, in logarithmic units. *Bottom right:* Observed bolometric flux, relative to the maximum flux, in logarithmic units.

Kerr black hole. The relativistic effects on the direction of the linear polarization and on the degree of polarization are clearly visible.

Our future plans are focused on testing the ARCMANCER code with a larger and more demanding set of applications, such as optically thin black hole accretion flows and black hole jet synchrotron plasmas. Eventually, the full code together with the Python interface will be made freely available for the general public.

References

Broderick, A., & Blandford, R. 2003a, *MNRAS*, 342, 1280
Broderick, A., & Blandford, R. 2003b, *Ap&SS*, 288, 161
Chandrasekhar, S. 1960, Radiative transfer (Dover Publications)
Cunningham, C. T. & Bardeen, J. M. 1973, *ApJ*, 183, 237
Dexter, J. 2016, *MNRAS*, 462, 115
Luminet, J.-P. 1979, *A&A*, 75, 228
Novikov, I. D. & Thorne, K. S. 1973, in: C. DeWitt & B. S. DeWitt (eds.) *Astrophysics of Black Holes*, Black Holes (New York: Gordon and Breach), p. 343
Shcherbakov, R. V. & Huang, L. 2011, *MNRAS*, 410, 1052
Shcherbakov, R. V. & McKinney, J. C. 2013, *ApJ (Letters)*, 774, L22

New Frontiers in Black Hole Astrophysics
Proceedings IAU Symposium No. 324, 2016
Andreja Gomboc, ed.

doi:10.1017/S1743921317002368

Observing shadow of the Schwarzschild black hole in presence of a plasma

Farruh Atamurotov

National University of Uzbekistan, Tashkent 100174, Uzbekistan
Ulugh Beg Astronomical Institute, Astronomicheskaya 33, Tashkent 100052, Uzbekistan
Inha University in Tashkent, Tashkent 100170, Uzbekistan
email: atamurotov@yahoo.com

Abstract. We have investigated particle motion around Schwarzschild black holes in the presence of a plasma with radial power-law density profile has been shown that the photon sphere around a spherically symmetric black hole is unchanged under the influence of the plasma; however, the Schwarzschild black hole shadow size is reduced due to the refraction of electromagnetic radiation in the plasma environment of the black hole.

1. Introduction

The study of astrophysical processes in the plasma medium surrounding a black hole becomes very interesting and important due to the evidence for the presence of black holes at the centres of the galaxies. For example, the gravitational lensing in inhomogeneous and homogeneous plasma around black holes has been recently studied in Bisnovatyi-Kogan and Tsupko (2010) as an extension of vacuum studies.

The shape of the black hole is determined through the boundary of the shadow which can be studied by application of the null geodesic equations in Atamurotov *et al.* (2016). The presence of a plasma in the vicinity of black holes changes the equations of motion of photons which may lead to the modification of the black hole shadow. In this paper our goal is to consider the silhouette of the shadow of an axially symmetric black hole using the equations of motion for photons in a plasma with radial power-law density profile. We would like to underline that recently, the influence of a non-magnetized cold plasma with the radially dependent density to black hole shadow has been studied by Perlick *et al.* (2015) using a different approach.

2. Photon motion around kerr black hole

The rotating black hole is described by the spacetime metric, which in the standard Boyer-Lindquist coordinates can be written in the form

$$\begin{aligned} ds^2 &= -\left(1-\frac{2Mr}{\Sigma^2}\right)dt^2+\frac{\Sigma^2}{\Delta}dr^2+ \\ &\quad \Sigma^2 d\theta^2-\frac{4aMr\sin^2\theta}{\Sigma^2}dtd\phi+ \\ &\quad \sin^2\theta\left[\Sigma^2+\frac{a^2(\Sigma^2+2Mr)\sin^2\theta}{\Sigma^2}\right]d\phi^2, \end{aligned} \tag{2.1}$$

where

$$\begin{aligned} \Sigma^2 &= r^2+a^2\cos^2\theta, \\ \Delta &= r^2-2Mr+a^2. \end{aligned}$$

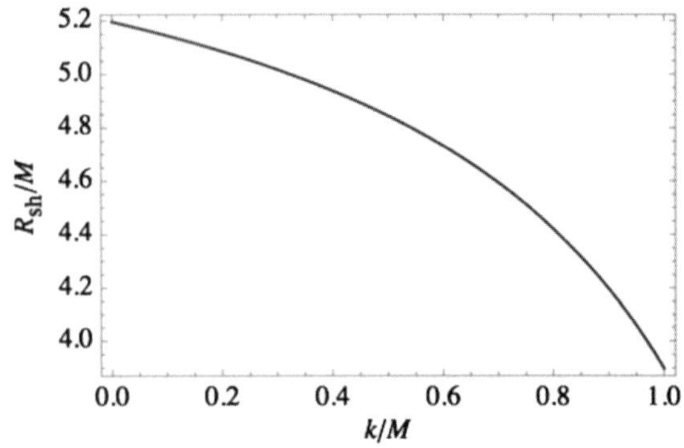

Figure 1. The dependence of the Schwarzschild black hole shadow on the plasma frequency parameter in Atamurotov *et al.* (2015)

Using the celestial coordinates one can easily describe the shadow in Vázquez and Esteban (2004), Atamurotov (2016):

$$\alpha = \lim_{r_0 \to \infty} \left(-r_0^2 \sin\theta_0 \frac{d\phi}{dr} \right) , \tag{2.2}$$

and

$$\beta = \lim_{r_0 \to \infty} r_0^2 \frac{d\theta}{dr} , \tag{2.3}$$

In Fig. 1, shows that the radius of the shadow of black hole surrounded by an inhomogeneous plasma decreases in Atamurotov *et al.* (2015).

3. Conclusion

In order to extract the pure effect of the influence of a plasma on the black hole image, the particular case of the Schwarzschild black hole has also been investigated. It was shown that under the influence of a plasma, the observed size of the shadow of the spherically symmetric black hole becomes smaller than that in the vacuum case due to the refraction of the electromagnetic radiation in the plasma.

References

Bisnovatyi-Kogan, G. S. & Tsupko, O. Yu. (2010), *Mon. Not. R. Astron. Soc.*, **404**, 1790.
Perlick, V., Tsupko, O. Yu. & Bisnovatyi-Kogan, G. S. (2015) *Phys. Rev. D*, **404**, 1790.
Atamurotov, F., Ahmedov, B., & Ghosh, S. G. (2016) *Eur. Phys. J. C*, **76**, 273.
Atamurotov, F. (2016) *Proceedings of the International Astronomical Union, IAU Symposium*, **312**, pp. 135-136.
Atamurotov, F., Ahmedov, B., & Abdujabbarov, A. (2015) *Phys. Rev. D*, **92**, 084005.
Vázquez S. & Esteban, E. (2004) Nuovo Cim. **119B**, 489.

New Frontiers in Black Hole Astrophysics
Proceedings IAU Symposium No. 324, 2016
Andreja Gomboc, ed.

doi:10.1017/S1743921317001557

Circular motion and Polish Doughnuts in NUT spacetime

Paul I. Jefremov

ZARM – Centre for Applied Space Technology and Microgravity, University of Bremen,
Am Fallturm, 28359 Bremen, Germany
email: paul.jefremow@zarm.uni-bremen.de

Abstract. The astrophysical relevance of the NUT spacetime(s) is a matter of debate due to pathological properties exhibited by this solution. However, if it is realised in nature, then we should look for the characteristic imprints of it on possible observations. One of the major sources of data on black hole astrophysics is the accretion process. Using a simple but fully analytical "Polish Doughnuts" model of accretion disk one gets both qualitative and quantitative differences from the Kerr spacetime produced by the presence of the gravitomagnetic charge. The present paper is based on our work Jefremov & Perlick (2016).

Keywords. Accretion disks, hydrodynamics, relativity, black hole physics

1. Introduction

Taub-NUT metric is a solution of Einstein field equations [see e. g. Griffiths & Podolský (2009)] describing a source possessing in addition to the "gravitoelectric charge" (usual mass, M) also the "gravitomagnetic monopole" charge, n. In spherical coordinates (t, r, θ, ϕ) the NUT metric reads

$$g_{\mu\nu} dx^\mu dx^\nu = \frac{\Delta}{\Sigma}\Big(\mathrm{d}t - 2n(\cos\theta + C)\mathrm{d}\phi\Big)^2 - \frac{\Sigma}{\Delta}\mathrm{d}r^2 - \Sigma(\mathrm{d}\theta^2 + \sin^2\theta \mathrm{d}\phi^2) \tag{1.1}$$

where $\Delta = r^2 - 2Mr - n^2$ and $\Sigma = r^2 + n^2$. In this expression C is the Manko-Ruiz constant introduced by Manko & Ruiz (2005), the physical identification of which is not clear. Geodesics in the NUT space-time were studied by Kagramanova *et al.* (2010). Using their results we derived parameters of circular geodesics and in particular the surface where they are situated (see Jefremov & Perlick (2016) and the dashed black line on Fig. 1). The location of the circular geodesic orbits is important for the Polish Doughnuts accretion model considered in the next section, since the centre of the configuration moves on a geodesic.

2. Polish Doughnuts in NUT Spacetime

The Polish Doughnuts model of matter with constant specific angular momentum rotating around a black hole is remarkable for its fully analytical expression for the density distribution in the resulting toroidal configuration if one assumes the polytropic relation ($p = K\rho^\Gamma$) for the flow [see e. g. Rezzolla & Zanotti (2013)]:

$$\rho(r, \theta) = \left[\frac{\Gamma - 1}{K\Gamma}[\exp(W(r_{in}, \theta_{in}) - W(r, \theta)) - 1]\right]^{1/(\Gamma-1)}, \tag{2.1}$$

where the effective potential $W(r, \theta)$ is introduced as $W(r, \theta) := \ln\sqrt{\frac{g_{t\phi}^2 - g_{tt}g_{\phi\phi}}{-g_{\phi\phi} - 2g_{t\phi}\ell(r_c) - g_{tt}\ell(r_c)^2}}$, ℓ is the specific angular momentum, K is the polytropic constant, Γ is the polytropic

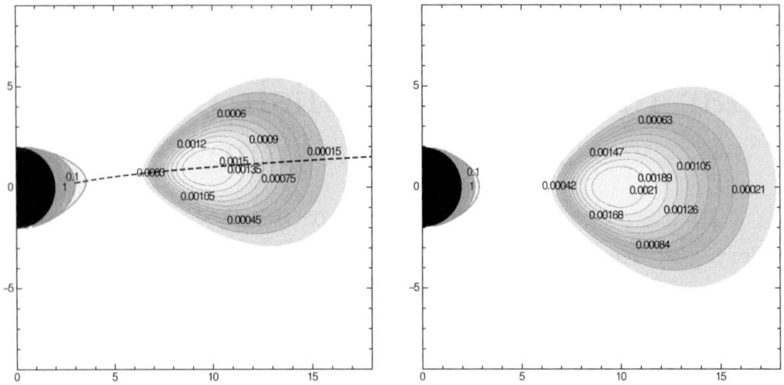

Figure 1. Polish Doughnut with $r_{\rm cen} = 9.7\,M$, $r_{\rm in} = 6.5\,M$ and $\Gamma = 5/3$ and $K = 0.2$ in NUT space-time with $n = 0.2\,M$ (left) and in Kerr space-time with $a = 0.2\,M$ (right). M is the unit on the axes. Different colours correspond to different values of the matter density ρ in units $[1/M^2]$. The parameters $r_{\rm cen}, r_{\rm in}, \Gamma, K$ are chosen to satisfy the absence of self-gravity and special relativistic effects conditions (Jefremov & Perlick (2016)). The dashed line on the left picture denotes the surface of circular orbits for the given value of n.

$\frac{n}{M}$	$\frac{a}{M}$	$m/(10^{-2}M)$	$\rho_{\rm cen}/(10^{-4}M^{-2})$	$T_{\rm cen}/(10^9 K)$
0	0	5.71	2.05	7.46
0.2	0	5.28	1.99	7.01
0.4	0	7.74	1.81	7.60
0	0.2	7.91	2.52	8.74

Table 1. Torus mass m, density in the centre $\rho_{\rm cen}$ and temperature in the centre $T_{\rm cen}$ for different values of NUT and Kerr parameters with the values of model parameters as in Fig. 1. The values for n, a (Kerr parameter), m, ρ_{cen} are given in terms of the black-hole mass M, the temperature is given in Kelvins.

exponent and r_{in} (inner edge) and r_c (centre) are parameters of the configuration. Choosing values of these parameters one obtains the density distribution in the torus (Fig. 1) and assuming further the ideal gas equation of state, the temperature and other characteristics of it. One clearly sees the qualitative difference in the shape of the torus which has its centre outside of the equatorial plane, where it lies in Kerr spacetime, and quantitative dependence of the physical properties of the torus on n (see Table 1).

3. Conclusions

NUT spacetime exhibits qualitatively different properties compared to the Kerr spacetime in the considered model of the accretion disc. The presence of the NUT parameter makes the configuration less massive and less hot in the centre than in the Kerr or Schwarzschild case.

References

Jefremov, P. I. & Perlick, V. 2016, *arXiv:1608.06218*, submitted to *Class. Quant. Grav.*
Griffiths, J. B. & Podolský, J., *Exact Space-Times in Einstein's General Relativity*, CUP (2009).
Manko, V. S. & Ruiz, E. 2005, *Class. Quantum Grav.*, 22, 3555.
Kagramanova, V., Kunz, J., Hackmann, E., & Lämmerzahl, C. 2010, *Phys. Rev. D*, 81, 124044.
Rezzolla, L. & Zanotti, O., *Relativistic Hydrodynamics*, OUP, 2013.

New Frontiers in Black Hole Astrophysics
Proceedings IAU Symposium No. 324, 2016
Andreja Gomboc, ed.

doi:10.1017/S1743921316012837

A solution to unconstrained Einstein's equations for a relativistic radiation sphere

Luboš Neslušan

Astronomical Institute, Slovak Academy of Sciences,
05960 Tatranská Lomnica, Slovakia; email: ne@ta3.sk

Abstract. We describe an application of a solution to unconstrained Einstein's field equations (Ni, 2011) as a qualitative model for a relativistic radiation sphere, an object that may resemble a quasar with an extended galactic-scale 'corona' that resembles behaviors explained by dark matter.

Keywords. Relativity, (galaxies:) quasars: general, (cosmology:) dark matter

In 2011, Ni published the solution of the Einstein (1915; 1916) field equations (EFEs), solved in course to model a stable neutron core with no upper mass limit. Ni could achieve this result because he did not establish some additional, post-Einstein postulates, which constrained ('newtonized') the general relativity (GR) in its application to model the relativistic compact objects. With the help of the Ni's solution, we can show that the upper mass limit of neutron stars is not any consequence of the GR, as generally believed, but this limit and, consequently, the black holes are the consequence of the constraint of GR. The upper mass limit does not occur in the continuum of the Ni's solution, which is the super-class to the Oppenheimer and Volkoff solutions. (For a further reasoning, see Neslušan 2015; sabotin.ung.si/~agomboc/IAUS324_presentations/Posters/Poster28_Neslusan.pdf.)

Here, we outline the concept of stable super-massive object consisting of radiation - relativistic radiation sphere (RRS). Using the EFEs for the spherical symmetry (the same as Oppenheimer and Volkoff used) with the equation of state for the radiation, $E = 3P$ (E – energy density, P – pressure), we construct an example of such object. The behaviors of its E, u, and g_{tt} are shown in Fig. 1. Quantity u is other form of g_{rr}, whereby $u = r(1 + 1/g_{rr})/2$. g_{rr} and g_{tt} are the components of metric tensor. Note: Equation $E = 3P$ is also valid for a gas consisting of fermions, which have the rest energy negligible in comparison to their total energy.

In the E-behavior (Fig. 1a), we see a central condensation (CC) having the radius (5.55 AU) slighly larger than the radius of the Jupiter's orbit. The total energy inside the CC equals $W = 5.55 \times 10^{59}$ J ($W/c^2 = 3.10 \times 10^{12}\ M_{\odot}$). It is such a large energy that the CC photosphere emitting the radiaton with the luminosity of bright quasar (e.g. 3C 273 having luminosity $\sim 3.0 \times 10^{40}\ \mathrm{J\,s^{-1}}$) during the age of the universe (13.799 Gyr) would spend only 2.3% of this energy.

At the border of the CC, E suddenly decreases about 6 orders of magnitude. Beyond the border, it is non-zero, but negligible, in a certain interval of distance, in comparison to that inside the CC, therefore the metrics outside can be well approximated with the outer Schwarzschild metrics (yellow curves in Fig. 1b and 1c). The trajectory of an object in a vicinity of the CC would practically be a Keplerian cone section. In our example, the orbit would be like that in the Newtonian potential generated by the central body of mass $2.8 \times 10^8\ M_{\odot}$ (mass of 3C 273).

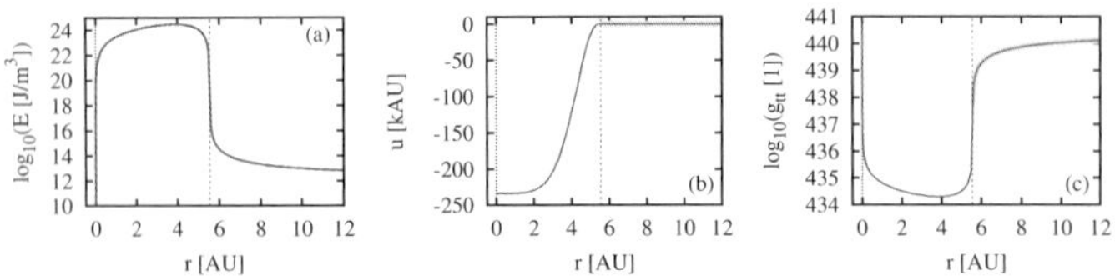

Figure 1. The behaviors of energy density, E (plot a), auxiliary function u related to g_{rr}-component of metric tensor (b), and g_{tt}-component of metric tensor (c) in an example of the RRS. The behaviors are plotted by the solid blue curves. The fit of the behavior by the corresponding outer Schwarzschild metrics in plots b and c is shown with the thick yellow curve. The dashed red vertical line indicates the border of the CC of object.

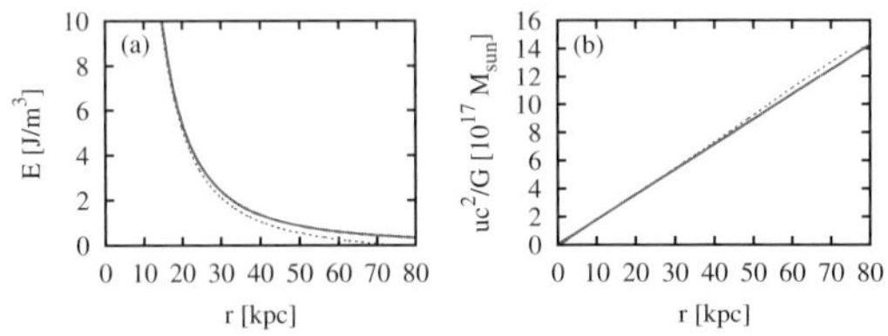

Figure 2. The behaviors of energy density, E, and quantity uc^2/G in the scale of tens of kiloparsecs. The behavior for the zero value of cosmological constant Λ is plotted with the solid blue line and that for this constant equal to 8×10^{-44} m^{-2} with the dashed red curve.

The energy of the RRS is spread to a large distance. In the distance scale of several kiloparsecs, the net energy of the RRS in form of a 'corona' exceeds the energy concentrated in the CC and increases linearly with the radial distance (Fig. 2a). The quantities uc^2/G (c is the speed of light and G is the gravitational constant) and g_{tt} also lineraly increase with this distance in the RRS corona when the cosmological constant Λ is zero (the behavior of uc^2/G is shown in Fig. 2b). The trajectories of some objects, e.g. stars, moving around the RRS's center in the kiloparsec-scale distance are considerably different from the Keplerian orbits. The gravitational effect by the corona resembles that we assign to the dark matter in a galaxy.

The extremely concentrated energy in the CC implies an extremely high luminosity of the CC, like that observed at quasars. In the concept of quasar as the CC of RRS, the assumption of physically problematic super-Eddington accretion, which is accepted for some quasars, would not be necessary.

Here, we presented the RRS as an hypothetical object. One can nevertheless ask if some real counterparts could form in the universe. We can speculate that the RRSs could form in the cosmological era of radiation, when the universe was filled in with a fluid similar to that inside the CC of RRS. Our modeling demonstrates that the most massive RRSs could contain enough energy, in their CCs, to survive until the present and the CCs could permanently emit the radiation with a quasar luminosity.

Acknowledgements

This work was supported by VEGA - the Slovak Grant Agency for Science, grant No. 02/0031/14.

References

Einstein, A. 1915, *Sitzungsberichte der Preussischen Akad. der Wissenschaften zu Berlin*, p. 844
Einstein, A. 1916, *Annalen der Physik*, 354, 769
Neslušan, L. 2015, *J. Modern Phys.*, 6, 2164
Ni, J. 2011, *Sci. China: Phys. Mechan. Astron.*, 54, 1304
Oppenheimer, J. R. & Volkoff, G. M. 1939, *Phys. Rev.*, 55, 374

New Frontiers in Black Hole Astrophysics
Proceedings IAU Symposium No. 324, 2016
Andreja Gomboc, ed.

doi:10.1017/S1743921317002253

MOPTOP: Multi-colour Optimised Optical Polarimeter

Helen Jermak[1,2], **Iain A. Steele**[1] **and Robert J. Smith**[1]

[1]Astrophysics Research Institute, Liverpool John Moores University,
146 Brownlow Hill, Liverpool, L3 5RF, UK.
email: `i.a.steele@ljmu.ac.uk`, `r.j.smith@ljmu.ac.uk`

[2]Dept. of Physics, Lancaster University, Bailrigg Campus,
Lancaster, LA1 4YW, UK.
email: `h.e.jermak@ljmu.ac.uk`

Abstract. Polarimetric measurements are essential for the study of jetted sources associated with black holes, such as γ-ray bursts and blazars. The relativistic jets launched from regions close to the black hole are threaded with magnetic fields, which produce synchrotron emission, and can be studied with polarimetric measurements. The multi-colour, optimised, optical polarimeter (MOPTOP) is a multi-band imaging instrument designed for use on the Liverpool Telescope. By replacing the rotating polaroid with a half wave plate and beam splitter, the instrument utilises twice as much of the incoming beam of light from the telescope compared to its predecessor, Ringo3. MOPTOP also builds on the successful introduction of dichroic mirrors to perform simultaneous multi-waveband polarimetric and photometric analysis in Ringo3, and enhances the sensitivity of the instrument with sCMOS cameras to use all photons as efficiently as possible.

Keywords. polarization, instrumentation: polarimeters, techniques: polarimetric, etc.

1. Introduction

With the increase in demand for polarimetric observations using the Ringo3 polarimeter (Arnold *et al.* 2012), a new polarimeter has been designed to meet a list of requirements for an optimised instrument on the 2 metre, fully robotic and autonomous, Liverpool Telescope (Steele *et al.* 2004). These specifications include: maximum sensitivity, very low systematic errors (<0.1%), rapid observations of variable or fading sources, measuring all sources in the widest possible field of view without prior identification or prior knowledge of source brightness, multi-wavelength polarimetric observations, readiness for response to transient triggers and best possible spatial resolution.

2. Overview

The fundamental difference between the optical design of Ringo3 and MOPTOP is the replacement of the rotating polaroid with a half-wave plate and wire-grid beam-splitter. This means that instead of ~50% light being discarded by the polaroid, the full beam is utilised, splitting the two polarisation states and directing them to separate detectors.

The detectors of choice are sCMOS cameras; they maintain the rapid readout (>100 frames per second) and ultra low read noise (<0.9 electrons) advantages of electron-multiplying charge-coupled devices (EMCCDs) but add greater sensitivity (>70% vs ~40% for EMCCDs; because EMCCDs have a charge multiplication noise component that effectively reduces the quantum efficiency by a factor of 2). In addition sCMOS detectors are available in much larger formats (~4 Mpix vs ~1 Mpix from EMCCDs) allowing the full 8 arcmin telescope field of view to be sampled at optimal (Nyquist

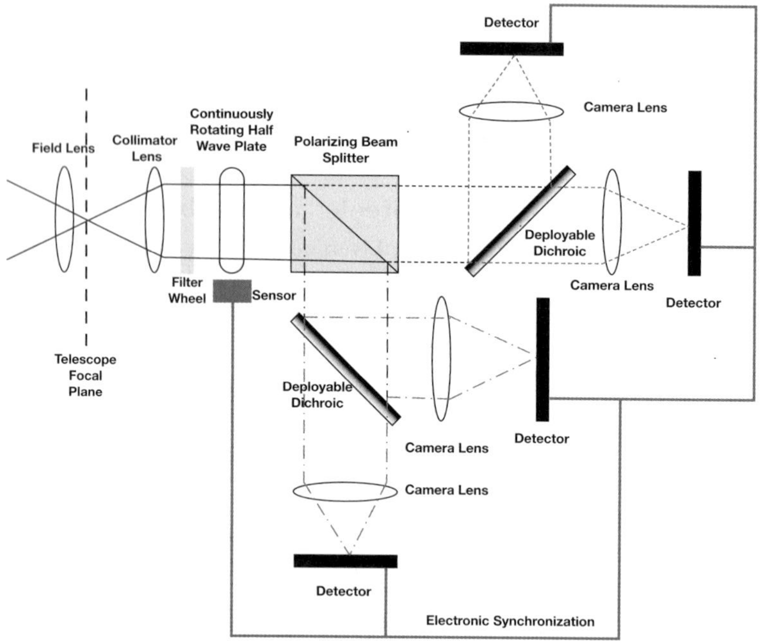

Figure 1. The two colour design of MOPTOP. The light from the telescope enters the instrument from the left where the unvignetted instrument field of view is increased by the field lens. The light enters a lens to collimate the light before passing through a filter wheel (for use in calibrating the instrument to standard sources). The polarisation of the beam is then modified by the continuously rotating half wave plate and is split to its ordinary and extraordinary beams by the beam splitter. The light is then directed through a deployable dichroic mirror and split to two electronically synchronous sCMOS detectors. If the dichroic mirrors are not deployed the instrument can record a more sensitive result in a broader wavelength range.

sampled) resolution. MOPTOP will allow a throughput improvement from the Ringo3 value of ~20% to ~60% and a field of view improvement by a factor of 4. The improved optical image quality from the use of better lenses, combined with the other improvements will give an increase in sensitivity of ~1.5 magnitudes (from V=17.0 to V=18.5 for 1% polarisation accuracy), which is equivalent to quadrupling the telescope size. This improvement reduces the exposure times required for polarimetric accuracy, and also vastly improves the number of sources available to the instrument; particularly optically detected γ-ray burst afterglows for which the Ringo series of instruments were designed (see Mundell *et al.* (2013) for example).

The optical design is shown in Figure 1. The instrument will be built for use on the Liverpool Telescope, however, the design will be made fully public domain, with the aim of encouraging copies to be placed on other (smaller) robotic facilities around the world such as the all sky network operated by LCOGT (`http://www.lcogt.net`).

References

Arnold, D. M., Steele, I. A., Bates, S. D., Mottram, C. J., & Smith, R. J. 2012, in *Society of Photo-Optical Instrumentation Engineers (SPIE) Conference Series*, Vol. 8446, Society of Photo-Optical Instrumentation Engineers (SPIE) Conference Series

Mundell, C. G., Kopač, D., Arnold, D. M., *et al.* 2013, *Nature*, 504, 119

Steele, I. A., Smith, R. J., Rees, P. C., *et al.* 2004, in *Society of Photo-Optical Instrumentation Engineers (SPIE) Conference Series*, Vol. 5489, Ground-based Telescopes, ed. J. M. Oschmann, Jr., 679–692

Author index